New Advances in Fluid Structure Interaction

New Advances in Fluid Structure Interaction

Editors

Wenli Chen
Zifeng Yang
Gang Hu
Haiquan Jing
Junlei Wang

MDPI • Basel • Beijing • Wuhan • Barcelona • Belgrade • Manchester • Tokyo • Cluj • Tianjin

Editors

Wenli Chen
Harbin Institute of Technology
China

Zifeng Yang
Wright State University
USA

Gang Hu
Harbin Institute of Technology
China

Haiquan Jing
Central South University
China

Junlei Wang
Zhengzhou University
China

Editorial Office
MDPI
St. Alban-Anlage 66
4052 Basel, Switzerland

This is a reprint of articles from the Special Issue published online in the open access journal *Applied Sciences* (ISSN 2076-3417) (available at: https://www.mdpi.com/journal/applsci/special_issues/FStructure_Interaction).

For citation purposes, cite each article independently as indicated on the article page online and as indicated below:

LastName, A.A.; LastName, B.B.; LastName, C.C. Article Title. *Journal Name* **Year**, *Volume Number*, Page Range.

ISBN 978-3-0365-4639-1 (Hbk)
ISBN 978-3-0365-4640-7 (PDF)

Contents

About the Editors

Wenli Chen

Prof. Wenli Chen received his PhD in civil engineering from the Harbin Institute of Technology (HIT) in 2009 and spent two years visiting the Department of Aeronautics at Iowa State University in the US, from 2011 to 2013. He was appointed as a professor in 2015 and was awarded the National Natural Science Foundation of China—Outstanding Youth Fund and the Key Project in 2017 and 2021, respectively. His research interests include wind engineering and engineering applications of bridges, focusing on wind effects and the wind vibration control of large-span bridge structures. He has achieved systematic research results and engineering applications for complex wind effects such as wind vibration and the multi-modal vortex vibration of large span bridges on cable-stayed bridges and has developed corresponding active/passive flow control methods. He has presided over six national projects, including the "National Natural Science Foundation of China—Key Projects and Excellent Youth", and has been selected for the "Young Scientist Workshop" and "Young Talent Program" of the HIT. He has published more than 100 papers, including more than 90 SCI-indexed papers; received 18 national patents and soft publications; co-edited 2 monographs and textbooks; co-edited 3 specifications; and has delivered 7 conference and invited presentations. He was awarded the first prize of the Science and Technology Award of the Chinese Highway Society (top the list), the first prize of the Natural Science of Heilongjiang Province (second the list), the "Youth Science and Technology Award" of the Chinese Vibration Engineering Society, and the first prize of the Meteorological Science and Technology Progress Achievement Award (fourth) and has been listed in the top 2% of scientists in the world for two consecutive years, from 2020 to 2021. He is a member of the Board of Directors of the Bridge and Structural Engineering Branch of the Chinese Highway Society, a member of the Youth Editorial Board of the Chinese Journal of Highways, a member of the Youth Editorial Board of the Journal of Vibration Engineering, and a member of the Topics Editorial Board of *Applied Sciences*.

Zifeng Yang

Dr. Zifeng Yang has worked as an associate professor in the Department of Mechanical and Materials Engineering at Wright State University since 2017. Dr. Yang was appointed as an assistant professor at the same institute. He graduated from Iowa State University (ISU) with a PhD degree in Aerospace Engineering in 2009. Then, he was appointed as a postdoc research associate at ISU until 2011. His research was mainly focused on the development of advanced flow diagnostic techniques, experimental fluid dynamics, bioflows, and wind energies. His research was funded by NIH, RMD Inc, Ronald Houck II and IHE, LLC, and Premier Health Boonshoft School Endowment Funding. He has published 35 peer-reviewed journal articles, 22 peer-reviewed conference papers, and 31 conference oral presentations and invited talks. He has been an APS member since 2008. He was the technical co-chair for the 33rd Applied Aerodynamics Conference at Aviation 2015 and served as the session chairs for multiple APS, ASME, and AIAA conferences. He was the vice chair for the 47th AIAA Dayton-Cincinnati Aerospace Sciences Symposium (DCASS) organizing committee and chair for the 48th DCASS. Dr. Yang was a guest-editor for *Applied Sciences* from 2021 to 2022. Dr. Yang has been serving as the Vice Chair for the American Physical Society East Great Lakes Section (EGLS) since 2022.

Gang Hu

Dr. Gang Hu is a Professor at the School of Civil and Environmental Engineering at the Harbin Institute of Technology, Shenzhen. He obtained his bachelor, MPhil, and PhD degrees from Central South University (2009), the Harbin Institute of Technology, Shenzhen (2011), and Hong Kong University of Science and Technology (2015), respectively. In September 2015, he joined the CLP Power Wind/Wave Tunnel Facility at HKUST as a research associate and served as a postdoc fellow at Department of Civil and Environmental Engineering at HKUST from September 2016 to November 2017. From December 2017 to December 2019, he was employed as a postdoc research associate at the School of Civil Engineering, the University of Sydney. He joined the Harbin Institute of Technology, Shenzhen, in December 2019. He is an executive committee member and secretary of the Hong Kong Wind Engineering Society and a member of the Australian Wind Engineering Society. His research interests include structural wind engineering, wind energy, bluff body aerodynamics by using wind tunnel test, CFD, and AI techniques. To date, he has published more than 70 papers in international reputable journals.

Haiquan Jing

Dr. Haiquan Jing is an Associate Professor in the School of Civil Engineering at Central South University. He received his B.E degree and Master's degree from the School of Transportation Science and Engineering at the Harbin Institute of Technology in 2010 and 2012, respectively, and his Ph.D from the Department of Civil and Environmental Engineering at the Hongkong Polytechnic University in 2016. He joined Central South University in March 2017. His main research interests include the rain–wind induced vibration of stay cables, vibration suppression of stay cables, and aerodynamic characteristics of PV modules. Dr. Jing is the two-time recipient National Natural Science Foundation of China (NSFC) Grants No: 51708559 and 52078502. He also received the science and technology innovation Program of Hunan Province award (2021RC3017). He has published more than 50 journal papers, including 28 SCI papers. He won second prize in the national science and technology progress award in 2019, first prize of the science and technology progress of China Railway Society in 2018, and first prize of science and technology progress of China Highway Society in 2021. He also won the Mao Yisheng Medal of Technology Progress in Railway in 2020.

Junlei Wang

Dr. Junlei Wang is a Professor in the School of Mechanical and Power Engineering at Zhengzhou University. He received his B.E degree and Ph.D from the School of Power Engineering at Chongqing University in 2009 and 2014, respectively. He was also a visiting professor of Mechanical Engineering at the University of Auckland, New Zealand, from 2019 to 2020. He joined Zhengzhou University in December 2014. His main research interests include flow-induced vibrations suppression, flow energy harvesting, and triboelectric, piezoelectric, and hybrid energy harvesting technology. Dr. Wang is a two-time recipient of the National Natural Science Foundation of China (NSFC) Grant No: 51606071 and 51977196. He also received the Natural Science Foundation of Excellent Youth of Henan Province award. He has published more than 80 SCI papers in international journals (more than 50 during the past three years) and had more than 10 ESI High-Cited Papers or Hot papers. His H-index is 27, and he has received more than 2500 Google citations. He has won the most influential 100 papers in China (2019), the most influential researchers of Springer (2020), and a featured article award of Applied Physics letters (2021). He is also the Lead guest editor of *ASCE JEE* (2022–present).

applied sciences

Editorial

New Advances in Fluid–Structure Interaction

Wenli Chen [1,*], Zifeng Yang [2], Gang Hu [3], Haiquan Jing [4] and Junlei Wang [5]

1 School of Civil Engineering, Harbin Institute of Technology, Harbin 150090, China
2 Department of Mechanical and Materials Engineering, Wright State University, Dayton, OH 45435, USA; zifeng.yang@wright.edu
3 School of Civil and Environmental Engineering, Harbin Institute of Technology, Shenzhen 518055, China; hugang@hit.edu.cn
4 School of Civil Engineering, Central South University, Changsha 410075, China; hq.jing@csu.edu.cn
5 School of Mechanical and Power Engineering, Zhengzhou University, Zhengzhou 450001, China; jlwang@zzu.edu.cn
* Correspondence: cwl_80@hit.edu.cn

Fluid–structure interactions (FSI) play a crucial role in the design, construction, service and maintenance of many engineering applications, e.g., aircraft, towers, pipes, offshore platforms and long-span bridges. The old Tacoma Narrows Bridge (1940) is probably one of the most infamous examples of serious accidents due to the action of FSI. Aircraft wings and wind-turbine blades can break because of FSI-induced oscillations. To alleviate or eliminate these unfavorable effects, FSI must be dealt with in ocean, coastal, offshore and marine engineering to design safe and sustainable engineering structures. In addition, the act of wind on plants and its resultant wind-induced motions are an example of FSI in nature.

To meet the objectives of progress and innovation in FSI in various scenarios of engineering applications and control schemes, this book includes 15 research studies and collects the most recent and cutting-edge developments on these relevant issues. The topics cover different areas associated with FSI, including wind loads [1–3], flow control [4–9], energy harvesting [10], buffeting and flutter [11,12], complex flow characteristics [13], train–bridge interactions [14] and the application of neural networks in related fields [15]. In summary, these complementary contributions in this publication provide a volume of recent knowledge in the growing field of FSI.

Citation: Chen, W.; Yang, Z.; Hu, G.; Jing, H.; Wang, J. New Advances in Fluid–Structure Interaction. *Appl. Sci.* **2022**, *12*, 5366. https://doi.org/10.3390/app12115366

Received: 20 May 2022
Accepted: 25 May 2022
Published: 26 May 2022

Author Contributions: All authors contributed equally to the preparation of this manuscript. All authors have read and agreed to the published version of the manuscript.

Funding: The financial support from the National Natural Science Foundation of China (52008140 and 51978222) is gratefully acknowledged.

Conflicts of Interest: The authors declare no conflict of interest.

References

1. Yang, J.; Zhang, J.; Li, C. Research on Equivalent Static Load of High-Rise/Towering Structures Based on Wind-Induced Responses. *Appl. Sci.* **2022**, *12*, 3729. [CrossRef]
2. Jing, H.; Ji, X.; He, X.; Zhang, S.; Zhou, J.; Zhang, H. Dynamic Characteristics of Unsteady Aerodynamic Pressure on an Enclosed Housing for Sound Emission Alleviation Caused by a Passing High-Speed Train. *Appl. Sci.* **2022**, *12*, 1545. [CrossRef]
3. Wu, B.; Xue, G.; Feng, J.; Laima, S. The Effects of Aerodynamic Interference on the Aerodynamic Characteristics of a Twin-Box Girder. *Appl. Sci.* **2021**, *11*, 9517. [CrossRef]
4. Huang, Y.; Chen, W. An Experimental Investigation of Passive Jet Control Method on Bridge Tower Wake. *Appl. Sci.* **2022**, *12*, 4691. [CrossRef]
5. Wang, Z.; Hu, G.; Zhang, D.; Kim, B.; Xu, F.; Xiao, Y. Aerodynamic Characteristics of a Square Cylinder with Vertical-Axis Wind Turbines at Corners. *Appl. Sci.* **2022**, *12*, 3515. [CrossRef]
6. Yu, H.; Yang, Z. Effect of the Extended Rigid Flapping Trailing Edge Fringe on an S833 Airfoil. *Appl. Sci.* **2022**, *12*, 444. [CrossRef]
7. Liu, X.; Bai, W.; Xu, F. Study on Traveling Wave Wall Control Method for Suppressing Wake of Flow around a Circular Cylinder at Moderate Reynolds Number. *Appl. Sci.* **2022**, *12*, 3433. [CrossRef]
8. Song, T.; Liu, X.; Xu, F. Moving Surface Boundary-Layer Control on the Wake of Flow around a Square Cylinder. *Appl. Sci.* **2022**, *12*, 1632. [CrossRef]
9. Chen, G.; Chen, W. Experimental Investigation and Validation on Suppressing the Unsteady Aerodynamic Force and Flow Structure of Single Box Girder by Trailing Edge Jets. *Appl. Sci.* **2022**, *12*, 967. [CrossRef]
10. Shi, T.; Hu, G.; Zou, L. Aerodynamic Shape Optimization of an Arc-Plate-Shaped Bluff Body via Surrogate Modeling for Wind Energy Harvesting. *Appl. Sci.* **2022**, *12*, 3965. [CrossRef]
11. Su, Y.; Di, J.; Li, S.; Jian, B.; Liu, J. Buffeting Response Prediction of Long-Span Bridges Based on Different Wind Tunnel Test Techniques. *Appl. Sci.* **2022**, *12*, 3171. [CrossRef]
12. Feng, J.; Wu, B.; Laima, S. Effects of the Configuration of Trailing Edge on the Flutter of an Elongated Bluff Body. *Appl. Sci.* **2021**, *11*, 10818. [CrossRef]
13. Wang, Z.; Zou, Y.; Yue, P.; He, X.; Liu, L.; Luo, X. Effect of Topography Truncation on Experimental Simulation of Flow over Complex Terrain. *Appl. Sci.* **2022**, *12*, 2477. [CrossRef]
14. Wang, H.; Li, H.; He, X. Aerodynamics of a Train and Flat Closed-Box Bridge System with Train Model Mounted on the Upstream Track. *Appl. Sci.* **2022**, *12*, 276. [CrossRef]
15. Li, K.; Li, H.; Li, S.; Chen, Z. Fully Convolutional Neural Network Prediction Method for Aerostatic Performance of Bluff Bodies Based on Consistent Shape Description. *Appl. Sci.* **2022**, *12*, 3147. [CrossRef]

applied sciences

MDPI

Article

Research on Equivalent Static Load of High-Rise/Towering Structures Based on Wind-Induced Responses

Junhui Yang [1], Junfeng Zhang [1] and Chao Li [2,*]

[1] School of Civil Engineering, Zhengzhou University, Zhengzhou 450001, China; 18530099573@163.com (J.Y.); brilliantshine@163.com (J.Z.)

[2] School of Civil and Environmental Engineering, Harbin Institute of Technology Shenzhen, Shenzhen 518055, China

* Correspondence: lichaosz@hit.edu.cn; Tel.: +86-13510655957

Abstract: A method of assessing equivalent static wind loads that can represent all the real ultimate states of a high-rise building and towering structure has still not been fully determined in wind engineering. Based on random vibration theory, the wind-induced response and equivalent static wind loading of high-rise buildings and towering structures are investigated using the vibration decomposition method. Firstly, the structural wind-induced mean response, background response, resonant response and background and resonant coupled response are studied in the time and frequency domains. Secondly, a new gust load factor (*GLF*) assessment method suitable for wind-induced displacement, bending moment and shear force response at any height of the structure is proposed, and a typical high-rise building is used as an example for comparison with the previous research results, in order to verify the effectiveness of the method in this paper. The results show the following: for high-rise buildings and towering structures, the percentage of the coupled components in the total pulsation response is less than 2%, and the influence can be ignored; the *GLF* based on bending moment (*MGLF*) and the *GLF* based on shear force (*QGLF*) increase significantly with height, and the traditional *GLF* methods underestimate the maximum wind effects.

Keywords: time domain method; frequency domain method; background and resonance coupled components; wind induced dynamic responses; equivalent static wind load

Citation: Yang, J.; Zhang, J.; Li, C. Research on Equivalent Static Load of High-Rise/Towering Structures Based on Wind-Induced Responses. *Appl. Sci.* **2022**, *12*, 3729. https://doi.org/10.3390/app12083729

Academic Editor: José A. F. O. Correia

Received: 2 March 2022
Accepted: 30 March 2022
Published: 7 April 2022

1. Introduction

Wind load is one of the loads that must be considered in engineering design. For highly flexible, low-damping and light-mass structures, such as high-rise buildings [1,2], bridges [3,4], circular cylinder structures [5,6], wind turbines [7–9], railway catenary [10], cables [11,12] and transmission towers [13], the wind vibration responses are very sensitive and highly susceptible to wind vibration disasters, and determining their wind vibration response and equivalent wind load is one of the core problems of structural wind engineering. Furthermore, the structural wind-induced response is the basis for the study of equivalent static wind loads. In the past, the wind loads of structures were mainly determined directly by anemometers. In the 1960s, Davenport AG [14,15] first introduced random vibration theory into the analysis of the along-wind vibration response of high-rise buildings, divided the structure's total pulsation response into background and resonance components, and proposed the gust load factor method (GLF) based on the first-order displacement response, in order to calculate the equivalent static wind load of the structure. Zhang Xiangting [16] argued that the equivalent pulsating wind load on the structure could instead be assessed by the inertial force of the first-order pulsating displacement response, and then proposed the inertial wind load method (IWL), which had been adopted into the Chinese code. Kasperski et al. [17,18] proposed an accurate assessment method for the structural background response of low-rise buildings with high stiffness, named the load–response correlation (LRC) method. Zhou Yin et al. [19,20] proposed the basal moment array wind load factor (MGLF) method, which is characteristic of the advantages of LRC and IWL, and has been adopted by the American code. Subsequently, equivalent static wind load assessment methods were proposed by Holmes, Gu Ming, Xie Zhuangning, Lin Yongjun, Ke Shitang and others [21–29]. The above-mentioned methods all take the responses at key locations, such as peak displacement and peak internal force, as the equivalent value, without considering other locations. However, with the development of building shape and height, considerations of only the first-order vibration mode cannot guarantee the security of the structure. On the other hand, with progress in the theory of the wind vibration response calculation, not only the higher-order mode of the wind vibration but also the background and resonant coupling components can be taken into consideration, so there is still a need to investigate the equivalent static wind loads.

The main research methods of structural wind-induced response studies include theoretical analysis, field measurements, wind tunnel experiments and numerical simulations [1,2,5,7,30], and the theoretical analysis can be classified into two types: the time domain method and the frequency domain method. The advantage of the time domain method is that it has a wide range of applications, given its ability to obtain the dynamic response of the displacement, bending moment and shear force, and it can take into account the influence of material's nonlinearity, but the disadvantage is its time-consuming method of calculation. The advantage of the frequency domain method is that the concept is clear and the computational cost is small; the disadvantage is that it cannot consider nonlinear problems. In fact, with the development of computer hardware and frequency-domain computational theory, the disadvantages of the time domain method are gradually being overcome, and the results of the frequency domain method are becoming more and more accurate. Gu Ming et al. [31] proposed an along-wind vibration response analysis method that takes the higher-order modes and inter-modal coupling terms into consideration, and it is based on non-constant load tests performed in the frequency domain. Li Shouke et al. [32] derived a refined assessment method for wind-induced response, considering background, resonance, and background–resonance coupling components, which is based on the combination of the modal acceleration method and the principle of stochastic dynamics two-input single-output system. Zou Lianghao et al. [33] proposed a simplified assessment method for the wind vibration response of high-rise buildings, considering the second-order mode in conjunction with the Chinese code. Zhang Junfeng et al. [34] analyzed the division of the background and resonant wind responses in the time domain, and established a method to calculate the background and resonant coupling components in the time domain.

This paper presents a method for calculating the background component, the resonance component, and the background and resonance coupled component of the wind-driven vibration response of a structure in the time and frequency domains, and presents an equivalent static wind load assessment method for any location on the structure. The method combines the advantages of the traditional MGLF method and the IWL method. The equivalent static wind load at each position of the structure is obtained directly from the wind-induced displacement, the bending moment, and the shear force response at that position, instead of considering the response at a specific position, so that the equivalent static wind load at any position of the structure can be obtained accurately and the result can be more reasonable. The accuracy of the method is also verified using a high-rise building as an example. The work of this paper provides a reference for the study of equivalent static wind load.

2. Refinement of Structural Wind-Induced Response Assessment Method

For a multi-degree-of-freedom structure, the basic equation of motion under wind load is

$$[M]\ddot{Y}(z,t) + [C]\dot{Y}(z,t) + [K]Y(z,t) = P(z,t) \tag{1}$$

where [M], [C] and [K] are the mass matrix, damping matrix and stiffness matrix of the structure, respectively, $P(z, t)$ is the pulsating wind load time course at node z, and $\ddot{Y}(z,t)$, $\dot{Y}(z,t)$ and $Y(z,t)$ are the acceleration time course, velocity time course and total pulsating displacement time course, respectively.

Decoupling the above structural dynamic equilibrium equations yields N mutually independent equations expressed by generalized coordinates. The normalization of the vibration mode in relation to mass gives the following equations:

$$\ddot{q}_j(t) + 2\zeta_j\omega_j\dot{q}_j(t) + \omega_j^2 q_j(t) = F_j(t) \tag{2}$$

$$Y(z,t) = \sum_{j=1}^{N} \phi_j(z) q_j(t) \tag{3}$$

$$F_j(t) = \sum_{i=1}^{N} P(z_i,t)\phi_{ji} = \sum_{i=1}^{N} \rho C_D(z) A_i \overline{U}(z_i) u(z,t)\phi_{ji} \tag{4}$$

where $q_j(t)$ is the jth modal generalized coordinate time course, ζ_j is the jth modal damping ratio, ω_j is the jth modal natural vibration angular frequency, $F_j(t)$ is the jth modal generalized load, ϕ_{ji} is the vibration displacement component of the ith node of the jth vibration mode, ρ is the air density, generally taken as 1.25 kg/m^3, C_D is the quasi-constant wind pressure coefficient, A_i is the windward area of the ith node, $\overline{U}(z_i)$ is the average wind speed of incoming flow at height z_i, and $u(z_i, t)$ is the time course of incoming pulsating wind speed at height z_i.

2.1. The Full Three-Component Expressions in the Frequency Domain

The average response of a high-rise building under wind loads is

$$\overline{r}(z) = \int_0^H \overline{P}(z')i(z,z')dz' \tag{5}$$

where $\overline{P}(z')$ is the average wind load at height z', and $i(z, z')$ is the influence function (including the influence of displacement, bending moment and shear force), which represents the response generated at position z when a unit load is applied at position z' [23].

The frequency domain method is mostly used for the analysis of the structural wind vibration response, given its advantages of clear physical concepts and speed. When the wind speed at one spatial point reaches a maximum, it usually does not reach a maximum at another point in a certain range at the same time. Therefore there is a certain spatial correlation between the wind speeds at two points in the space [35]. From Equation (4), the structure's generalized wind load mutual spectrum $S_{FiFj}(f)$ can be expressed as:

$$S_{FiFj}(f) = \sum_{k=1}^{N}\sum_{l=1}^{N} \phi_{ik}\phi_{jl}\rho^2 {C_D}^2 A_k \overline{U}_k A_l \overline{U}_l R_X R_Z \sqrt{S_{uk}(f) S_{ul}(f)} \tag{6}$$

where R_X, R_Z are the horizontal and vertical coherence function, $\overline{U}_i$ is the average wind speed at the ith node, and $S_{ui}(f)$ is the pulsating wind speed spectrum.

According to Equation (2), if the generalized wind load mutual spectrum $S_{FiFj}(f)$ is known, then the total pulsation response spectrum $S_{qj}(f)$ of the modal generalized coordinates q_j can be expressed as Equation (7), and then using the modal superposition principle (Equation (3)), the response spectrum $S_{yi}(f)$ at node i of the structure can be obtained, as in Equation (8).

$$S_{qj}(f) = \sum_{i=1}^{N} S_{FiFj}(f) H_i^*(f) H_j(f) \tag{7}$$

$$S_{yi}(f) = \sum_{j=1}^{N}\sum_{k=1}^{N} \phi_{ji}\phi_{ki} H_j^*(f) H_k(f) S_{FjFk}(f) \tag{8}$$

$$H_k(f) = \frac{1}{K_k^*} \cdot \frac{1}{1-(f/f_k)^2 + \mathrm{i}2\zeta_k(f/f_k)} \tag{9}$$

where $H_k(f)$ is the kth modal complex frequency response function, and $H_j^*(f)$ is the conjugate of $H_j(f)$. The response spectra of the bending moment and shear force can be obtained by replacing the vibration displacement ϕ in Equation (8) with bending moment or shear force.

$$\sigma_{Ti} = \left(\int_0^\infty |S_{yi}(f)|df\right)^{1/2} \tag{10}$$

The total pulsation response σ_{Ti} can be calculated according to Equation (10), and it contains three components: the background component σ_B, the resonance component σ_R, and the background–resonance coupling component σ_{BR}. Because of the complicated nature of the calculation for σ_T, researchers often choose to determine each component, and then calculate σ_T according to Equation (11). The expressions for each component are as follows.

$$\sigma_{Ti} = (\sigma_{Bi}^2 + \sigma_{Ri}^2 + \sigma_{BRi}^2)^{1/2} \tag{11}$$

$$\hat{\sigma}_{\mathrm{B}i} = g_u \left(\int_0^\infty \int_0^H \int_0^H \int_0^W \int_0^W \overline{p(x_1,z_1,f)p(x_2,z_2,f)} i(z_i,z_1) i(z_i,z_2) dx_1 dx_2 dz_1 dz_2 df \right)^{1/2} \tag{12}$$

$$\hat{\sigma}_{Ri} = g_R \left(\int_0^\infty \sum_{j=1}^{N}\sum_{k=1}^{N} \phi_{ji}\phi_{ki} S_{FjFk}(f)[H_j^*(f) - \frac{1}{k_j^*}][H_k^*(f) - \frac{1}{k_k^*}]df \right)^{1/2} \tag{13}$$

$$\hat{\sigma}_{BRi} = 2\rho_{BR} \left(\int_0^\infty \sum_{j=1}^{N}\sum_{k=1}^{N} \phi_{ji}\phi_{ki} S_{FjFk}(f)[\frac{1}{k_j^*}H_k^*(f) + \frac{1}{k_k^*}H_j^*(f) - \frac{2}{k_j^* k_k^*}]df \right)^{1/2} \tag{14}$$

$$\overline{p(x_1,z_1,f)p(x_2,z_2,f)} = (\rho C_{\mathrm{D}} \overline{U}_H)^2 (\frac{z_1}{H})^\alpha (\frac{z_2}{H})^\alpha R_Z R_X S_u(f) \tag{15}$$

$$g_R = \sqrt{2\ln(f_1 T)} + 0.5772/\sqrt{2\ln(f_1 T)} \tag{16}$$

$$\rho_{BR} = \sigma_{BR}^2/(\sigma_B \sigma_R) \tag{17}$$

where $\overline{p(x_1,z_1,f)p(x_2,z_2,f)}$ is the pulsating wind pressure mutual spectrum at the structure's surface points (x_1, z_1) and (x_2, z_2), g_u is the peak wind speed factor, generally taken as 3.5, g_{R} is the peak factor of the resonant response, and ρ_{BR} is the background resonant coupled response mutual relationship number [32].

The above equation is the full three-component expression of the structural wind response in the frequency domain, and it can be seen that the expressions for σ_R and σ_{BR} are quite complicated, and are not convenient for engineering applications. As high and flexible structures such as high-rise buildings and towering structures have a sparse natural frequency, the wind vibration response of such structures is mainly related to the first few modes, so the simplification of the calculation of such buildings can be carried out by ignoring the contribution of the coupling term between each structural vibration mode and the coupling terms of background and resonance. At this point, the total pulsation response of the structure can be expressed via the following equation.

$$\sigma_T = \left(\sigma_B^2 + \sum_j^n \sigma_{Rj}^2 \right)^{1/2} \tag{18}$$

$$\sigma_{Rj} = g_R\left(S_{Fj}(f_j)\int_0^\infty |H_j(f)|_2 df\right)^{1/2} = g_R\frac{1}{K_j^*}\left(S_{Fj}(f_j)\frac{\pi f_j}{4\zeta_j}\right)^{1/2}\varphi_j \quad (19)$$

2.2. The Full Three-Component Expressions in Time Domain

$Y(z,t)$ can be obtained by the direct solution of the above dynamic equilibrium equation. From the concept and development of the background and resonant responses, it is known that the background and resonant responses can be calculated by the following equation:

$$Y_B(z,t) = P(z,t)/[K] \quad (20)$$

$$Y_R(z,t) = Y(z,t) - Y_B(z,t) \quad (21)$$

After obtaining the $Y(z,t)$, $Y_B(z,t)$ and $Y_R(z,t)$, the variances of the respective responses σ_Y^2,σ_B^2 and σ_R^2 can be calculated. It is important to note that there must also be coupling effects in the vicinity of the structural natural frequency, since both $Y_B(z,t)$ and $Y_R(z,t)$ contain components of this frequency [34].

$$\sigma_{BR} = \sigma_T^2 - \sigma_B^2 - \sigma_R^2 \quad (22)$$

So far, the assessment methods for the structural background component, resonance component and background–resonance coupling component in the time domain wind response have been elaborated. The time domain calculation can directly give a dynamic response, such as displacement, bending moment and shear force.

2.3. Comparison of Time–Frequency Domain Results for the Full Three Components

In this section, the wind-induced vibration response of a single-pole lightning rod structure will be studied to compare the frequency domain method and time domain method. The study will include displacement, bending moment and shear force. A lightning rod is a typical towering structure with a small natural frequency and significant wind-induced vibration effect, which often causes engineering accidents [36]. The physical parameters of the lightning rod investigated in this section are: total height 60 m, tower tip height 2 m, tower height 58 m. It is composed of seven sections of variable-thickness thin-walled steel pipe, with a diameter from 1.5 m tapered to 0.15 m, a wall thickness from 16 mm tapered to 5 mm; the material is Q345 steel, and the steel pipes are connected by plug-in adder flange. The steel pipe size is shown in Figure 1b. The simulation was carried out in the ANSYS finite element software, using Beam188 unit, and each of the lower six sections of steel pipe were divided into five units; each upper section was equivalent to 1 unit, and a total of 31 units were used in the model. In order to simplify the calculation, it does not include the tip of the tower, and does not consider the overlap areas between the steel pipes; the steel cylinder is only the interface connection and rigid connection. The five orders of vibration pattern for structure X-direction are shown in Figure 1c, with frequencies of 0.75, 2.09, 4.31, 7.46 and 11.66 Hz. Only the along-wind wind-induced response is analyzed in the simulation.

The lightning rod operates in a Class B site with a basic wind speed U_{10} = 35.8 m/s, wind profile power index α = 0.15, turbulence I_{10} = 0.14, and air density ρ = 1.25 kg/m^3. The section drag coefficient C_D takes the value of 0.6 according to the load code [37], the Davenport wind spectrum was considered in this case, and the Davenport coherence function exponential decay coefficient is C_Z = 7, C_X = 8. The harmonic superposition method is used to simulate the pulsating wind load time course, and the other parameters are selected according to the specifications. Rayleigh damping is used for the simulation and the modal damping ratio is taken as ζ = 0.01.

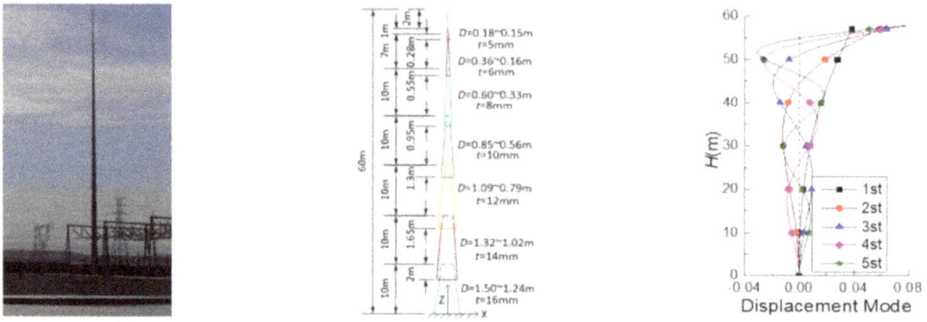

(**a**) Picture of the prototype; (**b**) dimensional illustration; (**c**) first 5 orders of vibration.

Figure 1. Geometry parameters and the FE model of the lightning rod.

Figure 2 shows the curve of the displacements, bending moments and shear forces (σ_T, σ_B and σ_R, respectively), obtained by the frequency domain method and the time domain method, as well as the percentage of the coupled components in the total pulsation response obtained by the time domain method. As shown in the picture, the σ_T values obtained from the frequency domain and time domain methods are in good agreement, and the deviations in the displacement of the tower tip, the base bending moment and the base shear force are all within 1.5%. As regards the displacement and bending moment responses, the background response obtained by the frequency domain method is smaller, and the resonance response is larger, than those obtained by the time domain method. The percentage of σ_{BR} in the total pulsation response for each variable is less than 2%, so the effect of neglecting σ_{BR} on the total pulsation response is small.

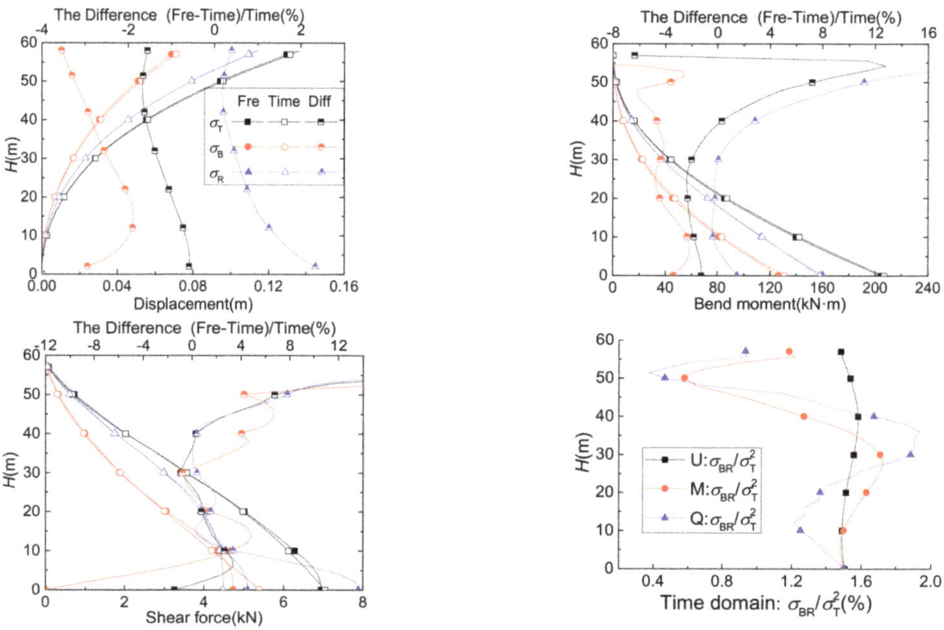

Figure 2. Comparison of time/frequency domain results and the percentage of coupling components.

3. The Assessment Method for the Equivalent Static Wind Load of Arbitrary Structural Response

Theoretical Analysis

As in previous studies, for high-rise buildings and towering structures, this paper still assumes that the equivalent static wind load on the structure is equal to the product of the mean wind load and the gust load factor, which can be expressed as follows:

$$\hat{P}(z) = GLF(z) \cdot \overline{P}(z) \tag{23}$$

$$GLF(z) = 1 + \frac{\sigma_T(z)}{\overline{r}(z)} = 1 + \sqrt{\frac{\sigma_B(z)^2}{\overline{r}(z)^2} + \sum_i^n \frac{\sigma_{Ri}(z)^2}{\overline{r}(z)^2}} = 1 + \sqrt{G_B(z)^2 + \sum_i^n G_{Ri}(z)^2} \tag{24}$$

$$\sigma_B(z) = g_u\left(\int_0^\infty \int_0^H \int_0^H \int_0^W \int_0^W \overline{p(x_1, z_1, f)p(x_2, z_2, f)} i(z, z_1) i(z, z_2) dx_1 dx_2 dz_1 dz_2 df\right)^{1/2} \tag{25}$$

$$\sigma_{Rj}(z) = \int_0^H \hat{P}_{Rj}(z') i(z, z') dz' \tag{26}$$

$$\begin{aligned} \hat{P}_{Rj}(z) &= m(z)(2\pi f_j)^2 \sigma_{YRj} \varphi_j(z) \\ &= g_R\left(S_{Fj}(f_j)\frac{\pi f_j}{4\zeta_j}\right)^{1/2} \varphi_j(z) \\ &= g_R\left(\sum_{k=1}^N \sum_{l=1}^N \phi_{ik}\phi_{jl}\rho^2 C_D{}^2 A_k \overline{U}_k A_l \overline{U}_l R_X R_Z \sqrt{S_u(f_j)S_u(f_j)}\frac{\pi f_j}{4\zeta_j}\right)^{1/2} \varphi_j(z) \end{aligned} \tag{27}$$

where $\hat{P}_{Rj}(z)$ is the jth order modal resonant equivalent wind load at height z of the structure, and its value is equal to the inertia force caused by the jth order modal resonant displacement [38].

The background response $\sigma_B(z)$ and resonant response $\sigma_R(z)$ can be obtained according to Equations (25) and (26), respectively, and then bringing them into the *GLF* expression (Equation (24)) can give the expressions of the $G_B(z)$ and $G_R(z)$ at any position on the structure.

The background gust load factor $G_B(z)$ of the structure's response when under wind load at any location is

$$\begin{aligned} G_B(z) &= \frac{\sigma_B(z)}{\overline{r}(z)} \\ &= \frac{g_B\sqrt{\int_0^\infty \int_0^H \int_0^H \int_0^W \int_0^W \overline{p(x_1, z_1, f)p(x_2, z_2, f)} i(z, z_1) i(z, z_2) dx_1 dx_2 dz_1 dz_2 df}}{\int_0^H \overline{P}(z') i(z, z') dz'} \\ &= \frac{2g_B I_H H^\alpha}{\int_0^h z'^{2\alpha} i(z, z') dz' + \int_h^H z'^{2\alpha} i(z, z') dz'} \sqrt{\int_0^\infty S_u^*(f) |K_Z(\alpha, z, f)|^2 |J_X(f)|^2 df} \end{aligned} \tag{28}$$

$$\begin{aligned} |K_Z(\alpha, z, f)|^2 = &\int_0^z \int_0^z z_1^\alpha z_2^\alpha R_Z(z_1, z_2, f) i(z, z_1) i(z, z_2) dz_1 dz_2 \\ &+ \int_0^z \int_z^H z_1^\alpha z_2^\alpha R_Z(z_1, z_2, f) i(z, z_1) i(z, z_2) dz_1 dz_2 \\ &+ \int_z^H \int_0^z z_1^\alpha z_2^\alpha R_Z(z_1, z_2, f) i(z, z_1) i(z, z_2) dz_1 dz_2 \\ &+ \int_z^H \int_z^H z_1^\alpha z_2^\alpha R_Z(z_1, z_2, f) i(z, z_1) i(z, z_2) dz_1 dz_2 \end{aligned} \tag{29}$$

Following the method of *GLF* based on displacement (DGLF), Davenport [15] defined the integral term in the structural force spectrum with respect to height z as a vertical joint receiver function. To make the above expression formally identical to the DGLF method, this paper also draws on this idea, and defines the integral term with respect to height z as a new function (Equation (29)), but the difference is that we use four integral terms.

The jth order modal resonant gust wind load factor $G_{Rj}(z)$ of the structure, in terms of all responses at any location under wind load, is

$$G_{Rj}(z) = \frac{\sigma_{Rj}(z)}{\overline{r}(z)} = \frac{\int_0^H \hat{P}_{Rj}(z')i(z,z')dz'}{\int_0^H \overline{P}(z')i(z,z')dz'} = \frac{g_R \int_0^H \left(\sum_{k=1}^{N} \sum_{l=1}^{N} \phi_{ik}\phi_{jl}\rho^2 C_D{}^2 A_k \overline{U}_k A_l \overline{U}_l R_X R_Z \sqrt{S_u(f_j) S_u(f_j) \frac{\pi f_j}{4\zeta_j}} \right)^{1/2} \varphi_j(z) i(z,z')dz'}{\frac{1}{2}\rho \overline{U}_H^2 C_D W \int_0^H \left(\frac{z'}{H}\right)^{2\alpha} i(z,z')dz'} \quad (30)$$

According to Equations (28) and (30), $G_B(z)$ and $G_R(z)$ can be determined by the height z and the influence function $i(z, z')$.

4. Example and Analysis

Yin Zhou [20] used a high-rise building as an example to illustrate the correctness of their proposed method of basal bending moment gust wind load factor (MGLF). In the following, the *GLF*, G_B and G_{Rj} of the same structure are calculated according to the method given in this paper. The physical parameters of the structure are: dimensions $H \times W \times D$ = 200 × 50 × 40 m, natural frequency f_1 = 0.22 Hz; damping ratio ξ = 0.01; first order vibration mode $\varphi_1(z) = (z/H)^\beta$; mass distribution $m(z) = m_0(1 - \lambda(z/H))$; $m_0 = 5.5 \times 10^5$ kg/m; section drag coefficient C_D = 1.3. The wind environment parameters are: fundamental wind speed $\overline{U}_{10} = 30$ m/s; wind profile power index $\alpha = 0.15$; and turbulence $I_{10} = 0.2$, while the wind spectrum type is Davenport wind spectrum, and the coherence function parameter $C_X = C_Z = 11.5$. The following four operating conditions are obtained by adjusting the values of the vibration index β and the mass discount factor λ, respectively (Table 1).

Table 1. The values of the parameters for the four working conditions [20].

Condition	Vibration Index β	Mass Discount Factor λ
1	1.0	0.0
2	1.6	0.0
3	1.0	0.2
4	1.6	0.2

Zhou Yin (2001) calculated the values of *GLF*, G_B and G_{R1} via the DGLF method proposed by Davenport, as well as the QGLF method, and the MGLF method, which was proposed by him. The DGLF method directly takes the first-order vibration mode as linear, i.e., the vibration index $\beta = 1$ (of course, the DGLF method can also obtain results when $\beta \neq 1$), but the DGLF method itself cannot consider the mass discount of the structure along with the height. Therefore, Zhou could only derive the result of one of the above four conditions using the DGLF method. The MGLF method also divides the structure σ_T into two parts: the σ_B is calculated directly based on the wind load and influence function, and the σ_R is calculated by the indirect method, which distinguishes the influence of β and λ, so the results of the four conditions are different. The results are presented in reference [20].

According to Equations (24), (28) and (29), the *GLF*, G_B and G_{Rj} values at arbitrary positions are calculated, and when the height is taken as 0 or H, the *GLF*, G_B and G_{R1} values at these positions can also be obtained—the results are shown in Table 2. The values in parentheses are the results relative to the results of Zhou Yin [20]. It can be seen that the *GLF* values of the base bending moments obtained by the method proposed in this paper are identical to those obtained by Zhou Yin [20]. The values of base shear force *GLF* and top displacement *GLF* deviate slightly, but the maximum deviation is only 0.6%, meaning they are consistent with each other. The expressions of the basal bending moment *GLF* and top displacement *GLF* obtained according to the method of this paper are identical to those of Zhou Yin [20]. The deviations in the result may be caused by numerical integration. Due to the different values derived for the respective integration step and the upper integration limit when numerical integration is performed, the integration results will be somewhat deviant. The specific values of these parameters are not given by Zhou Yin [20]. The first-order frequency of this high-rise building is f_1 = 0.22 Hz; the trapezoidal integration formula is used in this paper for the numerical integration calculation, and the frequency integration step is df = 1/1000 Hz, while the upper integration limit is f_{MAX} = 6 Hz. Other values are also taken (f_{MAX} = 1 Hz, 3 Hz and 9 Hz; df = 1/500 Hz and 1/2000 Hz) for verification to ensure the calculation's accuracy. The results show that the resulting G_B decreases slightly when df = 1/500 Hz compared to df = 1/1000 Hz, but it remains unchanged when df = 1/2000 Hz, and the resulting G_B decreases slightly when f_{MAX} = 1 Hz and 3 Hz, compared to f_{MAX} = 6 Hz, while it remains unchanged when f_{MAX} = 9 Hz, which indicates that the df and f_{MAX} used in the numerical integration are reasonable.

Table 2. *GLF* of feature locations obtained in this paper.

	Top Displacement			Basal Bend Moment			Basal Shear Force		
Condition	G_{YB}	G_{YR}	*DGLF*	G_{MB}	G_{MR}	*MGLF*	G_{QB}	G_{QR}	*QGLF*
1	0.6520 (1.000)	0.9761 (0.998)	2.1738 (0.999)	0.6520 (1.000)	0.9761 (1.000)	2.1738 (1.000)	0.6560 (0.994)	0.8275 (1.002)	2.0560 (1.000)
2	0.6591	1.0302	2.2230	0.6520 (1.000)	0.9532 (1.000)	2.1549 (1.000)	0.6560 (0.994)	0.7460 (1.003)	1.9934 (0.999)
3	0.6520	0.9761	2.1738	0.6520 (1.000)	0.9761 (1.000)	2.1738 (1.000)	0.6560 (0.994)	0.8438 (1.001)	2.0688 (0.999)
4	0.6591	1.0302	2.2230	0.6520 (1.000)	0.9589 (1.000)	2.1596 (1.000)	0.6560 (0.994)	0.7612 (1.002)	2.0049 (1.000)

The values of the *GLF* of the tip displacement and its components G_{YB} and G_{YR}, obtained under the four working conditions, are the results for condition 1. Since the effects of β and λ can all be considered when calculating the displacement response using the method in this paper, different results can be obtained under each of the four conditions, and because of that, the results under conditions 2, 3 and 4 will not be compared.

From Table 2, it can be seen that under each working condition, the top displacement *GLF* is not less than the base bending moment *GLF*, with a maximum deviation of about 3.1%, and the base bending moment *GLF* is greater than the base shear force *GLF*, with a maximum deviation of 7.5%. The σ_B values of the base bending moment and shear force are obtained directly from the load and influence function (Equation (25)), which does not contain parameters β and λ; therefore, the bending moment and shear force G_B are not affected by β and λ, and G_B is constant under all four conditions. Further, the base bending moment G_B is slightly smaller than the base shear force G_B, since the displacement influence function is obtained according to the assumption of the first-order vibration mode, and G_{YB} increases with the increase of β. Since the σ_R values of the base bending moment and shear force are obtained according to the resonant displacement inertia force $\hat{P}_R(z')$ and the influence function (Equation (26)), and $\hat{P}_R(z')$ contains the effects of β and λ, the σ_R response of each condition is not same, and both G_R values decrease with the increase in β. For the displacement of σ_R, although both $\hat{P}_R(z')$ and the influence function contain the parameters β and λ, the parameter λ can be removed, so G_{YR} is not affected by parameter λ, and only increases with the increase in β. The G_{MR} decreases with the increase in β, but remains unchanged or increases when λ increases from 0 to 0.2, which is mainly because λ can be approximately removed when β = 1. When only the first-order vibration mode is considered, the G_{QR} of base shear force decreases with the increase in β, and increases with the increase in λ.

Taking the above high-rise building as an example, the *GLF* values of each response at any position of the structure under different working conditions are obtained, and the different MGLF values and QGLF values relative to DGLF for each working condition are given in terms of DGLF values. The results are shown in Figure 3.

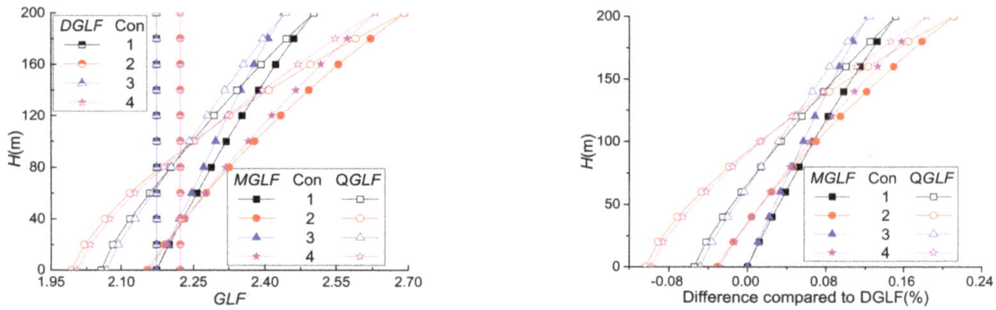

Figure 3. *GLF* values obtained by the method in this paper and their difference rates.

From Figure 3, it can be seen that with the increase in building height, the *MGLF* and *QGLF* corresponding to the four working conditions increase significantly, and the *DGLF* remains unchanged along the height. This is mainly because this paper uses the influence function based on the first-order vibration mode to calculate the *DGLF*, so the total pulsation response and the average response are in the form of the first-order vibration mode, and for *MGLF* and *QGLF*, the increase rate increases with the increase in β and decreases with the increase in λ. Under the same working conditions, the *MGLF* is larger than both the *QGLF* and the *DGLF*, and only the value of *DGLF* at the basal level is larger (slightly) than the *MGLF*, but the deviation is within 3%. Therefore, if *DGLF* and *QGLF* are used as the equivalent wind load in the structural design, the moment response of the structure may be underestimated; however, the use of the *MGLF* value for the equivalent wind load calculation is more important, because the design of the structure is biased towards safety, which is corroborated by the popular idea of using the bending moment as the control parameter for towering structures. Under the same working conditions, the basal *QGLF* value is the smallest, but as the building's height increases, the *QGLF* value gradually increases, and when the height is above 100 m, the *QGLF* under each working condition is greater than the DGLF. As the building's height increases, the difference between the *MGLF* and *QGLF* gradually decreases, until the tops tend to be equal, and the MGLF is greater than the *DGLF*. Until the top is reached, the difference between the *MGLF* value and the *DGLF* value gradually increases, and the maximum deviation between the *MGLF* and *QGLF* is about 22%, compared with *DGLF*. The position of maximum deviation is the top of the structure, so the wind effect of the structure will be seriously underestimated when the wind load design is based only on the base moment *GLF*.

5. Conclusions

This paper presents an assessment method for the equivalent static wind load based on random vibration theory and quasi-steady aerodynamic theory, and investigated a method for analyzing the background response, the resonant response, and the coupling components between these two responses, to wind vibration in the time domain and the frequency domain. The main conclusions are as follows:

(1) The displacement and bending moment responses obtained by the time domain and frequency domain methods are in good agreement, and the deviation in the base shear force response of the structure is slightly increased. However, the deviation in the total pulsation response is within 1.5%, and the percentage of σ_{BR} in the total pulsation response of each response is within 2%, so the influence of neglecting σBR on the total pulsation response is small;

(2) The *MGLF* and *QGLF* obtained by the new assessment method in this paper increase significantly with height, while the traditional GLF methods underestimate the maximum wind effects. The *DGLF* is constant as the height increases, and the *MGLF* at the top of the structure is about 22% larger than the *DGLF*. At this time, if only the *DGLF* or the base location *MGLF* values are considered, the wind effect of the structure will be seriously underestimated. The *DGLF*, *MGLF* and *QGLF* values increase as the β increases, and the *MGLF* and *QGLF* values decrease as the λ increases;

(3) The differences between the *GLF* obtained in this paper and the traditional gust load factor method are compared through the case of a high-rise building, and the validity of this paper's method is verified. The GLF assessment method of this paper considers the higher-order vibration pattern of the structure and the whole wind-induced response, including displacement, bending moments and shear forces. It has a wide range of applications.

Author Contributions: J.Z. and C.L. conceived the idea and designed the framework; J.Z. set up the numerical model; J.Y. analyzed the simulation data; and J.Y. and C.L. writing—review and editing the paper. All authors have read and agreed to the published version of the manuscript.

Funding: Shenzhen Basic Research Program (JCYJ20190806145216643) and National Natural Science Foundation of China (51778200).

Institutional Review Board Statement: Not applicable.

Informed Consent Statement: Informed consent was obtained from all subjects involved in the study.

Acknowledgments: This research has received support from the Shenzhen Basic Research Program (JCYJ20190806145216643) and National Natural Science Foundation of China (51778200), all of which are gratefully acknowledged.

Conflicts of Interest: The authors declare no conflict of interest.

References

1. Longarini, N.; Cabras, L.; Zucca, M.; Chapain, S.; Aly, A.M. Structural Improvements for Tall Buildings under Wind Loads: Comparative Study. *Shock Vib.* **2017**, *2017*, 2031248. [CrossRef]
2. Chapain, S.; Aly, A.M. Vibration attenuation in high-rise buildings to achieve system-level performance under multiple hazards. *Eng. Struct.* **2019**, *197*, 109352. [CrossRef]
3. Scanlan, R. The action of flexible bridges under wind, II: Buffeting theory. *J. Sound Vib.* **1978**, *60*, 201–211. [CrossRef]
4. Su, Y.; Di, J.; Li, S.; Jian, B.; Liu, J. Buffeting Response Prediction of Long-Span Bridges Based on Different Wind Tunnel Test Techniques. *Appl. Sci.* **2022**, *12*, 3171. [CrossRef]
5. Brownjohn, J.; Carden, E.; Goddard, C.; Oudin, G. Real-time performance monitoring of tuned mass damper system for a 183m reinforced concrete chimney. *J. Wind Eng. Ind. Aerodyn.* **2010**, *98*, 169–179. [CrossRef]
6. Gao, D.; Chen, G.; Chen, W.; Huang, Y.; Li, H. Active control of circular cylinder flow with windward suction and leeward blowing. *Exp. Fluids* **2019**, *60*, 26. [CrossRef]
7. Bernuzzi, C.; Crespi, P.; Montuori, R.; Nastri, E.; Simoncelli, M.; Stochino, F.; Zucca, M. Resonance of steel wind turbines: Problems and solutions. *Structures* **2021**, *32*, 65–75. [CrossRef]
8. Rahman, M.; Ong, Z.C.; Chong, W.T.; Julai, S.; Khoo, S.Y. Performance enhancement of wind turbine systems with vibration control: A review. *Renew. Sustain. Energy Rev.* **2015**, *51*, 43–54. [CrossRef]
9. Sari, D.P.; Cho, K.-P. Performance Comparison of Different Building Shapes Using a Wind Tunnel and a Computational Model. *Buildings* **2022**, *12*, 144. [CrossRef]
10. Song, Y.; Zhang, M.; Øiseth, O.; Rønnquist, A. Wind deflection analysis of railway catenary under crosswind based on nonlinear finite element model and wind tunnel test. *Mech. Mach. Theory* **2021**, *168*, 104608. [CrossRef]
11. Gao, D.; Chen, W.; Zhang, R.; Huang, Y.; Li, H. Multi-modal vortex- and rain–wind- induced vibrations of an inclined flexible cable. *Mech. Syst. Signal Process.* **2019**, *11*, 8245–8258. [CrossRef]
12. Chen, W.-L.; Zhang, Q.-Q.; Li, H.; Hui, L. An experimental investigation on vortex induced vibration of a flexible inclined cable under a shear flow. *J. Fluids Struct.* **2015**, *54*, 297–311. [CrossRef]
13. Zhong, Y.; Li, S.; Jin, W.; Yan, Z.; Liu, X.; Li, Y. Frequency Domain Analysis of Alongwind Response and Study of Wind Loads for Transmission Tower Subjected to Downbursts. *Buildings* **2022**, *12*, 148. [CrossRef]
14. Davenport, A.G. The Application of Statistical Concepts to the Wind Loading of Structures. *Proc. Inst. Civ. Eng.* **1961**, *19*, 449–472. [CrossRef]
15. Davenport, A.G. Gust Loading Factors. *J. Struct. Div.* **1967**, *93*, 11–34. [CrossRef]

16. Zhang, X. *Wind Load Theory and Wind-Resistance Design Manual for Engineering Structures*; Tongji University Press: Shanghai, China, 1990; pp. 92–98.
17. Kasperski, M.; Niemann, H.-J. The L.R.C. (load-response-correlation)-method a general method of estimating unfavourable wind load distributions for linear and non-linear structural behaviour. *J. Wind Eng. Ind. Aerodyn.* **1992**, *43*, 1753–1763. [CrossRef]
18. Kasperski, M. Design wind loads for low-rise buildings: A critical review of wind load specifications for industrial buildings. *J. Wind Eng. Ind. Aerodyn.* **1996**, *61*, 169–179. [CrossRef]
19. Zhou, Y. *Theoretical and Experimental Study of High-Rise Building Static Equivalent Wind Loads and Response*; Tongji University: Shanghai, China, 2001.
20. Zhou, Y.; Kareen, A. Gust loading factor: A new model. *ASCE J. Struct. Eng.* **2001**, *127*, 168–175. [CrossRef]
21. Holmes, J. Effective static load distributions in wind engineering. *J. Wind Eng. Ind. Aerodyn.* **2002**, *90*, 91–109. [CrossRef]
22. Holmes, J.D. *Wind Loading of Structure*; Spon Press: London, UK, 2001.
23. Gu, M.; Ye, F. Characteristics of wind induced responses and equivalent static wind loads of tall building. *Eng. Mech.* **2006**, *23*, 93–98. [CrossRef]
24. Ye, F.; Gu, M. Along-wind background response and background equivalent wind loads of tall buildings. *J. Build. Struct.* **2002**, *23*, 58–60. [CrossRef]
25. Xie, Z.N.; Fang, X.D.; Ni, Z.H. Equivalent static wind liads on tall building the extended load response correlation (ELRC) approach. *J. Vib. Eng.* **2008**, *21*, 398–403. [CrossRef]
26. Lin, Y.; Lin, C.; Lin, X.; Zhang, J.; Song, J. Research on equivalent static load of high- rise buildings based on wind-induced responses. *J. Southwest Jiaotong Univ.* **2019**, *54*, 137–144.
27. Ke, S.; Ge, Y. Refined research on wind-induced response of roof structure of a large museum based on consistent coupled method. *J. Build. Struct.* **2012**, *33*, 111–117.
28. Cao, S.; Ke, S.; Zhang, W.; Zhao, L.; Ge, Y.; Cheng, X. Load–response correlation-based equivalent static wind loads for large cooling towers. *Adv. Struct. Eng.* **2019**, *22*, 2464–2475. [CrossRef]
29. Zhang, J.-F.; Ge, Y.-J.; Zhao, L.; Zhu, B. Wind induced dynamic responses on hyperbolic cooling tower shells and the equivalent static wind load. *J. Wind Eng. Ind. Aerodyn.* **2017**, *169*, 280–289. [CrossRef]
30. Solari, G. Wind Loading of Structures: Framework, Phenomena, Tools and Codification. *Structures* **2017**, *12*, 265–285. [CrossRef]
31. Gu, M.; Zhou, X.; Huang, P. A study on wind-induced buffeting responses of large-span roof structures. *China Civ. Eng. J.* **2006**, *39*, 37–42. [CrossRef]
32. Li, S.K.; Li, S.Y.; Fang, X.L.; Sun, H.X.; Li, H.L. Refined study on wind-induced response of structure by stochastic wind load. *Eng. Mech.* **2015**, *32*, 111–115. [CrossRef]
33. Zou, L.; Shi, T.; Liang, S.; Song, J.; Li, F. Simplified calculation of along-wind dynamic responses of high-rise buildings considering second mode. *J. Build. Struct.* **2018**, *39*, 18–25. [CrossRef]
34. Zhang, J.; Liu, Q.; Yang, J.; Tu, B. A method of separating the background and resonant components of wind dynamic effects in time domain. *J. Disaster Prev. Mitig. Eng.* **2020**, *40*, 365–371.
35. Zeng, J.; Li, M.; Li, S. Spatial correlation of downwind pulsating wind loads in rectangular high-rise buildings. *J. Harbin Inst. Technol.* **2017**, *49*, 150–155.
36. Yu, P. *Research on Wind Vibration Response of Substation Lightning Rod Structure Downwind*; Zhengzhou University: Zhengzhou, China, 2018.
37. Ministry of Housing and Urban-Rural Development of the People's Republic of China. *GB 50009-2012. Code for Structural Loading of Buildings*; China Construction Industry Press: Beijing, China, 2012.
38. Gu, M.; Ye, F. Along-wind equivalent wind loads and responses on tall buildings. *J. Wind. Eng.* **2001**, *89*, 609–612.

applied sciences

MDPI

Article

Dynamic Characteristics of Unsteady Aerodynamic Pressure on an Enclosed Housing for Sound Emission Alleviation Caused by a Passing High-Speed Train

Haiquan Jing [1,2], Xiaoyu Ji [1,2], Xuhui He [1,2,*], Shifeng Zhang [3], Jichao Zhou [3] and Haiyu Zhang [3]

1 School of Civil Engineering, Central South University, Changsha 410075, China; hq.jing@csu.edu.cn (H.J.); xiaoyuj@csu.edu.cn (X.J.)
2 National Engineering Laboratory for High-Speed Railway Construction, Changsha 410075, China
3 China Railway Design Corporation, Tianjin 300308, China; zhangshifeng@crdc.com (S.Z.); zhoujichao@crdc.com (J.Z.); zhanghaiyu@crdc.com (H.Z.)
* Correspondence: xuhuihe@csu.edu.cn

Abstract: Train speed is increasing due to the development of high-speed railway technology. However, high-speed trains generate more noise and discomfort for residents, enclosed housing for sound emission alleviation is needed to further reduce noise. Because these enclosed housings for sound emission alleviation restrain the air flow, strong and complicated aerodynamic pressures are generated inside the housing for sound emission alleviation when a train passes through at a high speed. This train-induced aerodynamic pressure, particularly its dynamic characteristics, is a key parameter in structural design. In the present study, the train-induced unsteady aerodynamic pressure in an enclosed housing for sound emission alleviation is simulated using the dynamic mesh method, and the dynamic characteristics of the aerodynamic pressure are investigated. The simulation results show that when the train is running in the enclosed housing for sound emission alleviation, the unsteady aerodynamic pressure is complicated and aperiodic, and after the train leaves the housing for sound emission alleviation, the aerodynamic pressure reverts to periodic decay curves. Two new terms, the duration of the extreme aerodynamic pressure and the pressure change rate, are proposed to evaluate the dynamic characteristics when the train passes through the barrier. The dominant frequency and decay rate are adopted to express the dynamic characteristics after the train exits. When the train runs in the enclosed housing for sound emission alleviation, the longest durations of the positive and negative extreme aerodynamic pressures are in the middle section, and the maximum change rate of aerodynamic pressure occurs at the entrance area. After the train exits the housing for sound emission alleviation, the pressure amplitude at the central region is always higher than those close to the entrance/exit. The dominant frequency of the aerodynamic pressure is identified and explained using wave propagation theory, the decay rate of the aerodynamic pressure at all sections is close.

Keywords: high-speed train; enclosed housing for sound emission alleviation; pressure wave; unsteady aerodynamic pressure; load patterns

Citation: Jing, H.; Ji, X.; He, X.; Zhang, S.; Zhou, J.; Zhang, H. Dynamic Characteristics of Unsteady Aerodynamic Pressure on an Enclosed Housing for Sound Emission Alleviation Caused by a Passing High-Speed Train. *Appl. Sci.* **2022**, *12*, 1545. https://doi.org/10.3390/app12031545

Academic Editor: Luís L. Ferrás

Received: 1 December 2021
Accepted: 19 January 2022
Published: 31 January 2022

1. Introduction

In recent decades, the vigorous development of high-speed railway technology has promoted an increase in train speed. When the train runs at high speed, it generates more vibrational and aerodynamic noise and discomforts the residents living close to the railway lines. To improve residential comfort, various housings for sound emission alleviation have been invented and installed on high-speed railway lines.

Traditional housings for sound emission alleviation are mainly vertical or curved open-style barriers. They are simple, easy to install and cost-effective structures. However, when high-speed trains pass through residential areas or villages, the noise reduction effectiveness of these traditional housings for sound emission alleviation is insufficient. In recent years, enclosed housings for sound emission alleviation have been proposed and are becoming increasingly popular because of their good noise reduction performance.

However, because the housings for sound emission alleviation restrain the air flow, when a train passes through at a high speed, strong and complicated aerodynamic pressures are generated in the housing for sound emission alleviation. The aerodynamic pressure can cause structural failure. For instance, housings for sound emission alleviation installed on the Cologne-Frankfurt line in Germany were damaged due to train-induced transient pressures in 2003 [1]. Therefore, to ensure the operational safety of railway lines, the transient pressure on housing for sound emission alleviation caused by a running train has been a key factor for the structural design of housing for sound emission alleviation.

Many researchers have investigated the transient pressure on traditional housing for sound emission alleviation caused by running trains. Baker et al. [2] conducted an experiment on a moving model train rig using a 1/25 scale moving model device and measured the transient wind loads on the housing for sound emission alleviation, bridges, station canopies and trestle platforms caused by three different shapes of train models with different nose types. Lü et al. [3] and Xiong et al. [4] carried out field measurements and investigated the influence of the train speed, distances between the train and housing for sound emission alleviation, train types, and train marshaling length on the aerodynamic characteristics of the train-induced aerodynamic pressure on the housing for sound emission alleviation in the Datong-Xi'an and Beijing-Shanghai high-speed railway lines. They found that higher train speed and shorter distance increase the wind load acting on the housing for sound emission alleviation, the wind load becomes more sensitive to train speed when the distance is shorter, the time intervals of the peak-to-peak pressure on the housing for sound emission alleviation gradually decrease when the train speed increases, and the peak-to-peak pressure on the inner surface of the housing for sound emission alleviation becomes lower at higher monitoring points, while that on the outer surface becomes higher. Soper et al. [5] also conducted a series of field experiments to assess the pressure loads acting on housing for sound emission alleviation. They found that the train type has a great influence on the pressure fluctuations acting on the housing for sound emission alleviation and noted that the wind load caused different types of traffic moving on the railway should be taken into consideration.

For enclosed housing for sound emission alleviation, studies on the aerodynamic characteristics of train-induced aerodynamic pressure are rare. As the enclosed housing for sound emission alleviation is similar to a tunnel, the research achievements in the field of tunnel aerodynamics should be instructive for enclosed housing for sound emission alleviation. Many researchers have studied the train-induced aerodynamic pressure inside tunnels through field measurements, dynamic model experiments, and numerical simulations.

Kikuchi et al. [6], Fukuda et al. [7], Liu et al. [8–10], and Ko et al. [11] studied the aerodynamic pressure on a tunnel wall during the passage of trains using field measurements. Kikuchi et al. [6] investigated the wayside low-frequency noise on a tunnel portal created by passing the tunnel of a high-speed train. Their results indicated that the main components of low-frequency noise at the portal of the tunnel are pulsed micro pressure waves and continuous pressure waves. Fukuda et al. [7] investigated the distortion process of a pressure wave. They found that the compression wavefront becomes steep in the early stage and flattened in the later stage of propagation in the tunnel. Liu et al. [8] conducted a series of real vehicle tests on EMUs passing through the tunnel, analyzing pressure changes on the wall of the tunnel, the train wind in the tunnel, and the micro pressure waves at the tunnel entrance with different train speeds. They concluded that the three-dimensional effect of the tunnel entrance is obvious, and when the train is running at 200 km/h, the wind speed caused by the train exceeds the safe wind speed for the human body. Liu et al. [9] studied the influence law of train speed on pressure change, airflow velocity, and micro pressure waves. They found that the pressure change amplitude, airflow velocity, and micro pressure waves are proportional to the square of the train speed, the train speed and the cube of the train speed, respectively. Liu et al. [10] studied the interior pressure variations in high-speed trains passing through tunnels and found that the amplitude and duration of the initial compression effect on the train are independent of the tunnel length when a tunnel is longer than a certain length. Ko et al. [11] measured aerodynamic pressures in tunnels caused by passing high-speed trains and found that the pressure peaks were approximately proportional to the square of the train speed and decreased as the cross-sectional area of the tunnel increased. Field measurement is a direct and reliable method. Through field measurements, a researcher obtained the original data for the train-induced aerodynamic pressure inside tunnels.

Dynamic model experiments are also a reliable method used to investigate the aerodynamic pressure inside tunnels. Iida et al. [12] investigated the characteristics of the pressure wave and the generation process of micro pressure waves by model experiments. The relationship between the amplitude of a micro pressure wave and train speed was obtained. Bellenoue et al. [13] used a 1/77 scale test rig and simulated the first compression wave when a high-speed train enters a tunnel. The experimental results were verified through field observations carried out in the framework of the European Union research project TRANSAERO. Bellenoue and Kageyama [13] investigated the effects of the train/tunnel blockage ratio, the shape of the train nose and the geometry of the entrance hood on the pressure gradient of the compression wave generated by a high-speed train using moving model experiments. They reported that reducing the blocking rate can reduce the amplitude and pressure gradient of the compression wave and that increasing the nose length of the train can reduce the pressure gradient, but these actions have little effect on the amplitude of the pressure wave. Winslow et al. [14] carried out a 1/127 scale moving model experiment to investigate the effect of a scarfed portal on the compression wave generated by a train entering a tunnel. The results showed that optimizing the slope value at the oblique entrance of a tunnel can reduce the pressure gradient by up to 15%. Their results revealed that with the distance from the tunnel entrance, the compression wave becomes a plane wave after it has propagated four times the tunnel diameter inside the tunnel. Iida et al. [15] analyzed the compression wave generated by a train entering a tunnel by performing a model experiment and found that an opening on the sidewall or roof of the tunnel entrance hood can reduce the pressure gradient of the compression wave generated by the entering train. A scaled moving model experiment was conducted to investigate the characteristics of pressure waves induced by entering a high-speed train tunnel [16]. The results revealed that pressure peaks were affected by the train speed and the nose shape; in addition, the initial pressure peak increased slightly with the length of the train. Doi et al. [17] developed a 1/30 scale moving model experimental facility to analyze pressure waves generated by high-speed trains passing through a tunnel. The experimental results agreed well with the field measurements. Heine et al. [18] investigated the effect of tunnel hoods on the pressure waves inside tunnels and concluded that tunnel hoods reduced the pressure gradient by approximately 44%. Yang et al. [19] developed a large-scale ratio of a 1/8 moving model rig that can accelerate a training model to an actual Mach number to achieve an experimental simulation of a train passing through a tunnel or two trains intersecting in a tunnel. Zhang et al. [20] used a 1:20 scale moving model to study the pressures acting on train and tunnel surfaces as well as the effects of train speeds on these surfaces. They found that the pressure amplitude values on the surfaces of trains and tunnels and the micro pressure waves increased sharply with increasing train speed.

With the development of computer technology, numerical simulations have become increasingly popular and reliable. Ozawa et al. [21] simulated the transient flow field and compression wave induced by a high-speed train moving into a tunnel and found that the pressure gradient of the compression wave was related to the train position, the train shape and the tunnel cross-section. Kwon et al. [22] proposed a hybrid dimension method to investigate train-induced tunnel aerodynamics and greatly improved the calculation efficiency. Li et al. [23] obtained the magnitude of the pressure variation on both tunnels and trains when the train passed the tunnel through a numerical simulation. They concluded that the aerodynamic pressure on both the tunnel and train was approximately proportional to the square of the train speed. Uystepruyst et al. [24] investigated the effects of the shape, cross-section and length of the hood on the temporal gradient of the pressure wave generated by a high-speed train entering a tunnel. Chu et al. [25] simulated train-induced aerodynamic pressure in tunnels and found that the maximum and minimum pressures and force coefficients depended on the tunnel length, train length, train speed, and blockage ratio. Rabani et al. [26] concluded that the blockage ratio and train speed were the two main factors affecting the first pressure wave generated when a train enters a tunnel. Liu et al. [27] and Zhou et al. [28] numerically simulated the transient wind load generated by a train passing through a tunnel. Liu et al. [27] found that the pressure waves exhibit good one-dimensional characteristics during propagation in the tunnel. Zhou et al. [28] obtained the pressure change of the train and the tunnel and investigated the flow field around the train.

The above literature review shows that many previous studies have focused on the aerodynamic characteristics of train-induced aerodynamic pressure in tunnels. They help to better understand the aerodynamic characteristics of an enclosed housing for sound emission alleviation when a train passes through. However, because tunnels are rigid structures and are insensitive to dynamic loads, few previous studies have focused on the dynamic characteristics of wind loads. Enclosed housings for sound emission alleviation are always made of steel and are much more flexible and sensitive to dynamic loads. The dynamic characteristics of train-induced aerodynamic pressure on an enclosed housing for sound emission alleviation can act as control factors for the structural design. Consequently, in the present study, the unsteady aerodynamic pressure inside an enclosed housing for sound emission alleviation is numerically simulated when a CRH380 train passes through and exits at a speed of 350 km/h. The dynamic characteristics of the aerodynamic pressure are discussed in terms of wave propagation, impact effect, extreme pressure, duration of extreme pressure, dominant frequency and decay rate.

2. Numerical Methodology

2.1. Geometry

The present study numerically investigates the unsteady aerodynamic pressure of an enclosed housing for sound emission alleviation when a high-speed train passes through. CFD simulations are conducted using ANSYS Fluent software. A geometric model of a high-speed train is simplified as shown in Figure 1a. The details of the train, such as the pantograph, bogies, and wheels, are ignored. As the most common trains, the train model consists of eight coaches, including the head and tail coaches, and six intermediate coaches, with a total length (L_T) of 203 m and a width (W) and height (H) of 3.38 m and 3.7 m, respectively. Figure 1b shows the geometry of the enclosed housing for sound emission alleviation. The model is the preliminary design of a project in a high-speed railway line. The length of the enclosed housing for sound emission alleviation is 840 m, the radius of the cross-section is 6.913 m, the height is 9.637 m, the track spacing is 5.0 m and the cross-sectional area of the enclosed housing for sound emission alleviation is 110.5 m^2. A universal beam (H 500 $\times$ 300 $\times$ 12 $\times$ 25) is set up every two meters inside the housing for sound emission alleviation.

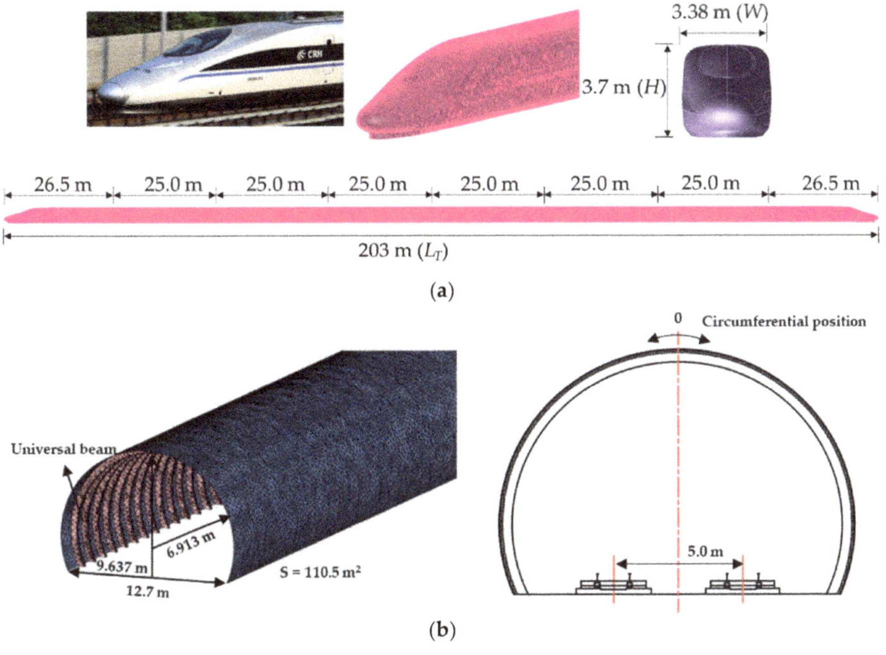

Figure 1. Computational models and surface meshes. (**a**) Train model. (**b**) Enclosed housing for sound emission alleviation.

2.2. Computational Domain and Mesh

Figure 2 shows the overview and cross-section of the computational domain. The computational domain is (550 m + 840 m + 550 m) × 50 m × 50 m in volume and consists of three parts: the two acceleration domains at both ends with a length of 550 m (148.6 H) and the housing for sound emission alleviation domain at the center with a length (L_S) of 840 m. The boundary conditions of the computational domain are defined as shown in Figure 2a. The bottom surface is ground and set as a no-slip solid wall, while the other five surfaces are atmospheric condition and set as outlets with zero pressure [26,29]. The surfaces of the train body and the housing for sound emission alleviation are set as no-slip solid walls. The entire computational model uses a hybrid grid of tetrahedral unstructured grids and prismatic grids. In addition, the meshes on the train body and wall of the housing for sound emission alleviation are refined to improve the calculation accuracy.

To avoid the mesh size effect, a grid independence study was conducted before the formal simulations. Two different meshes with different numbers of cells were divided into two groups: coarse meshes (24 million) and fine meshes (32 million). Figure 3 shows the representative results obtained with the two meshes. When a train travels through the enclosed housing for sound emission alleviation at a speed of 350 km/h, the peak-to-peak aerodynamic pressure at the measuring point located at S2 is 1900.6 Pa and 1849.1 Pa for the two meshes, respectively, and the difference is 2.7%; the mesh is independent of the simulation results. Therefore, the coarse mesh (24 million) is used as the calculation mesh in this paper.

The time step is set as 0.002 s, which is smaller than the 0.0025 s used in Yang et al. [30], and the 0.005 s used in Li et al. [31]. The FVM was selected to discretize the computational domain. The second-order upwind scheme was chosen to discretize the convection and diffusion terms. The second-order implicit scheme was chosen to discretize the time derivative. SIMPLE was chosen as the pressure velocity coupling treatment, which was used by Ferziger and Peric [32], and the convection term applied the second-order upwind form. The $Re = 4.4 \times 10^7$, the y+ around the train surface is approximately 350 and the CFL number is 2. The calculation of each case needs 168 h using 20×24 CPUs for parallel calculations.

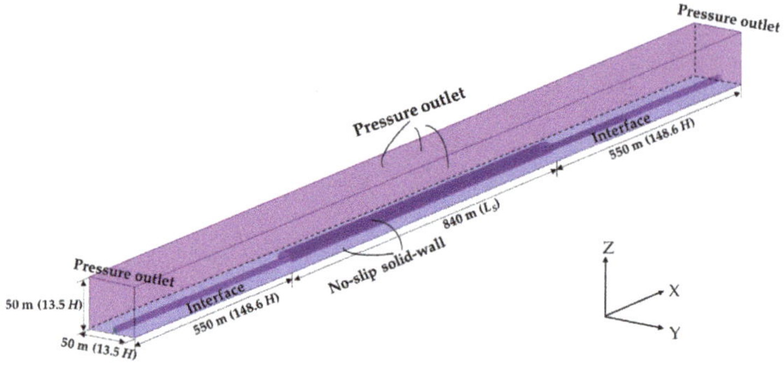

(**a**)

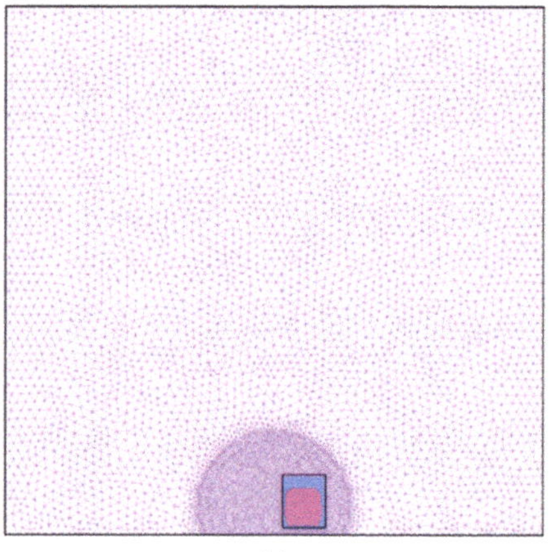

(**b**)

Figure 2. Overview and cross-section of the computational domain. (**a**) Overview of the computational domain. (**b**) Cross-section of the computational domain.

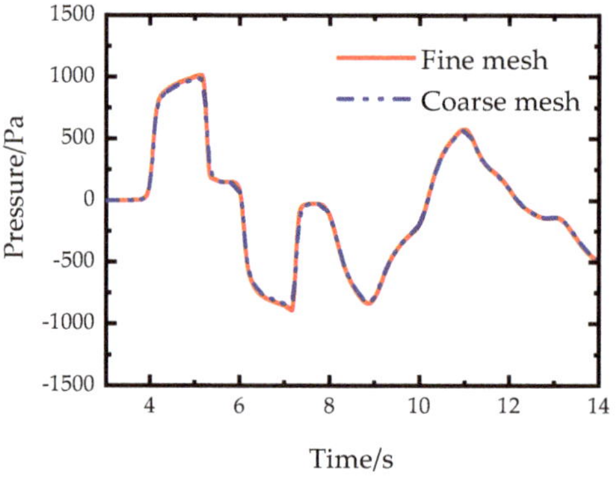

Figure 3. Pressure curves with different mesh resolutions.

The dynamic mesh technique is utilized to simulate the movement of the train model [33–35]. There are three dynamic mesh update methods in FLUENT: smoothing, layering, and remeshing. In the present study, the layering method is adopted to implement the moving train relative to the housing for sound emission alleviation. The computational domain is divided into a stationary region and a moving region, as shown in Figure 4. The train and its surroundings are in the moving region, in which the mesh moves with the train. The movement of the train and its surrounding mesh is defined by a profile file. The outside is a stationary region containing the housing for sound emission alleviation. When the train moves forward, the dynamic layering method adds layers of cells in the field behind the moving region and deletes those in front of the moving region. The two regions exchange flow field information through the interface.

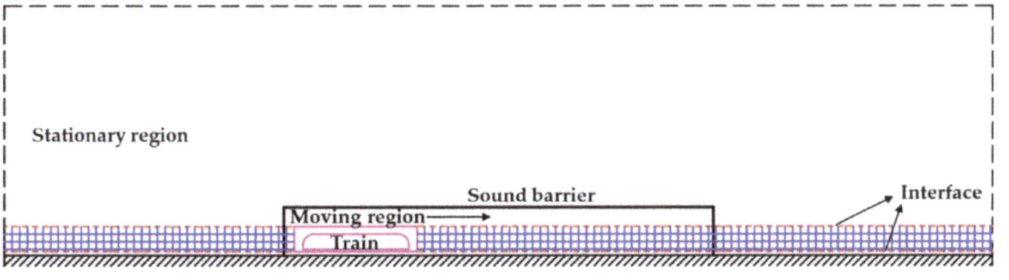

Figure 4. Schematic diagram of the dynamic layering method.

During the simulations, the train model starts to speed up with an acceleration of 48.61 m/s^2 at the location of -250 m (the entrance of the housing for sound emission alleviation is at the location of 0 m) and reaches the design speed at the location of -152.78 m, -152.78 m before the head coach enters the housing for sound emission alleviation, then passes through the housing for sound emission alleviation with a constant speed of 350 km/h, and finally exits the housing for sound emission alleviation. Figure 5 shows the speed strategy of the train model.

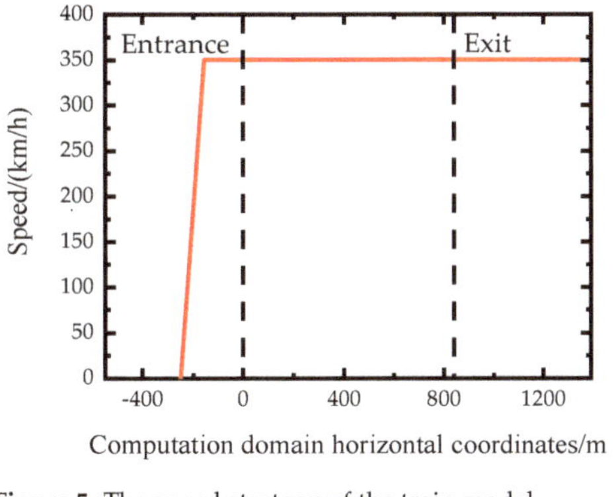

Figure 5. The speed strategy of the train model.

2.3. Numerical Method

When passing through the enclosed housing for sound emission alleviation, the train disturbs the surrounding air and generates strong wind. In the housing for sound emission alleviation, the air is strongly confined by the walls of the housing for sound emission alleviation and the train body; therefore, the flow field is three-dimensional, unsteady, viscous, compressible and turbulent. In the present study, the RANS method proved effective in simulating the flow field around a moving train [36–38], and the

RNG k-ε turbulent model was used to simulate the flow field, similar to Liu et al. [10], Yang et al. [30] and Huang et al. [39]. The number of iteration steps for each time step is preset to 50 steps [29,40–42], and the residual is set to 10^{-6}. This approach considers the influence of vortex factors and low Reynolds number effects in a turbulent flow and adds an extra term to the ε equation of the standard k-ε model to effectively improve the accuracy. The governing equations are presented as below [25,43]:

Continuity Equation:

$$\frac{\partial \rho}{\partial t}+\frac{\partial(\rho u_i)}{\partial x_i}=0 \tag{1}$$

Momentum Equation:

$$\frac{\partial(\rho u_i)}{\partial t}+\frac{\partial}{\partial x_j}(\rho u_i u_j)=-\frac{\partial p}{\partial x_i}+\rho g\delta_{i2}+\frac{\partial}{\partial x_j}[\mu(\frac{\partial u_i}{\partial x_j}+\frac{\partial u_j}{\partial x_i}-\frac{2}{3}\delta_{ij}\frac{\partial u_l}{\partial x_l})]+\frac{\partial}{\partial x_j}(-\rho\overline{u_i' u_j'}) \tag{2}$$

where u and u' refer to the mean velocity and the pulsating velocity, respectively; ρ represents the air density; p is the static pressure; $-\rho\overline{u_i' u_j'}$ is the gradient of Reynolds stress, based on Boussinesq assumption:

$$-\rho\overline{u_i' u_j'}=\mu_t(\frac{\partial u_i}{\partial x_j}+\frac{\partial u_j}{\partial x_i})-\frac{2}{3}(\rho k+\mu_t\frac{\partial u_k}{\partial x_k})\delta_{ij} \tag{3}$$

where μ_t is the turbulent viscosity coefficient, as a function of turbulent kinetic energy k and turbulent dissipation rate coefficient ε:

$$\mu_t=C_\mu\frac{k^2}{\varepsilon} \tag{4}$$

The transport equation of the RNG k-ε turbulence model is similar to the standard k-ε model. Turbulent kinetic energy (k) equation and dissipation rate (ε) equation are:

$$\frac{\partial}{\partial t}(\rho k)+\frac{\partial}{\partial x_i}(\rho k u_i)=\frac{\partial}{\partial x_j}(\frac{\mu_{eff}}{\alpha_k}\frac{\partial k}{\partial x_j})+G_k-\rho\varepsilon-Y_M \tag{5}$$

$$\frac{\partial}{\partial t}(\rho\varepsilon)+\frac{\partial}{\partial x_i}(\rho\varepsilon u_i)=\frac{\partial}{\partial x_j}(\frac{\mu_{eff}}{\alpha_k}\frac{\partial\varepsilon}{\partial x_j})+C_{1\varepsilon}\frac{\varepsilon}{k}G_k-C_{2\varepsilon}^{*}\rho\frac{\varepsilon^2}{k} \tag{6}$$

where μ_{eff} is the effective dynamic viscosity equal to the sum of the molecular and turbulent viscosities, G_k represents the turbulent energy generated by the laminar velocity gradient, the model parameters:$C_{1\varepsilon}=1.42$, $C_{2\varepsilon}^{*}$ is calculated as follows:

$$C_{2\varepsilon}^{*}=C_{2\varepsilon}+\frac{C_\mu\eta^3(1-\eta/\eta_0)}{1+\beta\eta^3} \tag{7}$$

where $C_{2\varepsilon}=1.68$, $\eta_0=4.38$, $\beta=0.012$, $\eta=Sk/\varepsilon$, S is the skewness factor of turbulent velocity.

2.4. Layout of Measurement Points

The unsteady aerodynamic pressure generated by a high-speed train passing through and exiting the enclosed housing for sound emission alleviation was measured by setting pressure monitoring points on the wall of the housing for sound emission alleviation. Figure 6 shows the layout of the monitoring points. The monitoring points were symmetrically set at eleven cross-sections, referred to as S1 to S11, along the longitudinal direction, as shown in Figure 6a. Since the train model was assumed to move from the left side to the right side, section S1 is located at the entrance, and section S11 is located at the exit. The monitoring cross-section was refined at the middle of the housing for sound emission alleviation. There are five measuring points at each monitoring cross-section, as shown in Figure 6b.

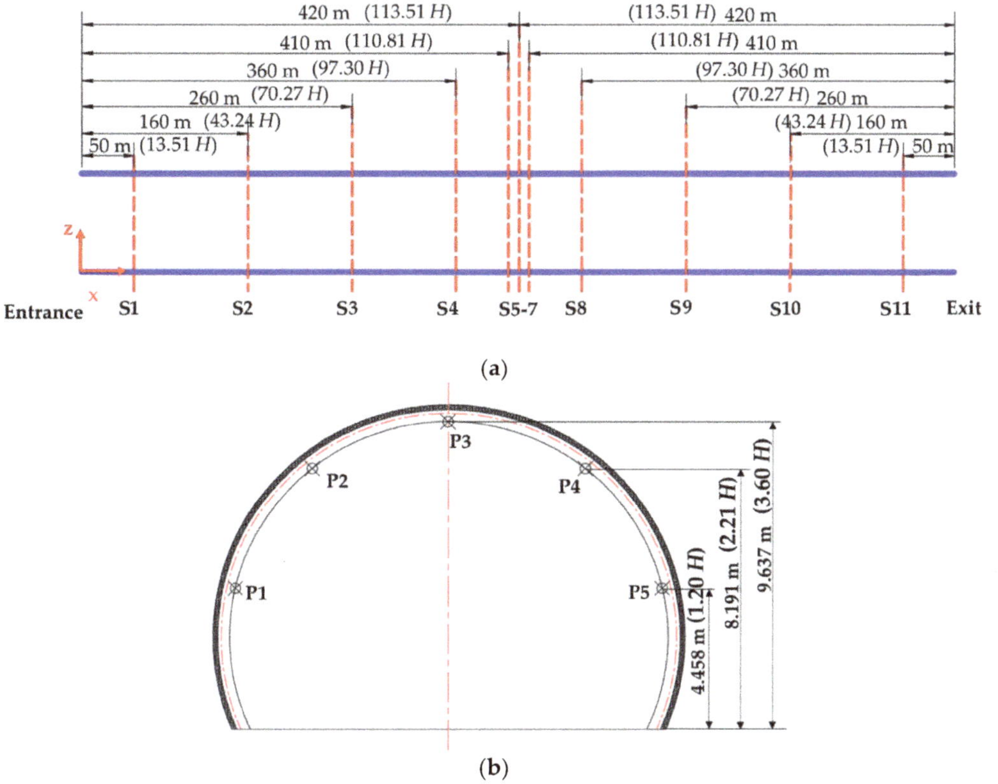

Figure 6. Measuring point distribution on the enclosed housing for sound emission alleviation. (**a**) Schematic diagram of the overall monitoring cross-sectional arrangement. (**b**) Layout of the measuring points on a monitoring cross-section.

3. Validation

To validate the effectiveness of the numerical simulation method, the motion model measurement conducted by Kim et al. [44] was calculated using the above numerical simulation method [45,46], as shown in Figure 7. For detailed information on the tests, please refer to the literature of Kim et al. [44]. Figure 8 compares the numerical and experimental pressure coefficients of PT1, PT2, PT3 and PT4, which are 8.5 m, 15.5 m, 23.5 m and 30.5 m away from the tunnel inlet, respectively. The pressure coefficient is defined as:

$$Cp = \frac{P}{0.5\rho v^2} \tag{8}$$

where P is the experimental or numerical aerodynamic pressure, v is the train speed and ρ is the air density.

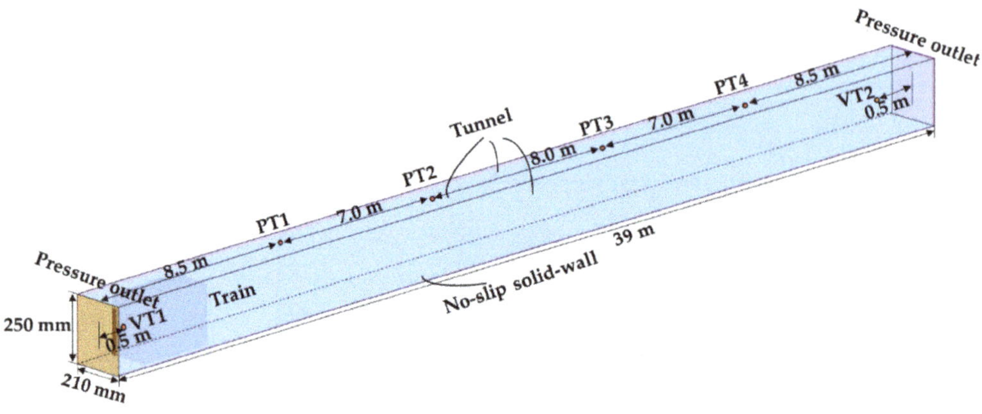

Figure 7. Schematic diagram of the numerical simulation calculation model.

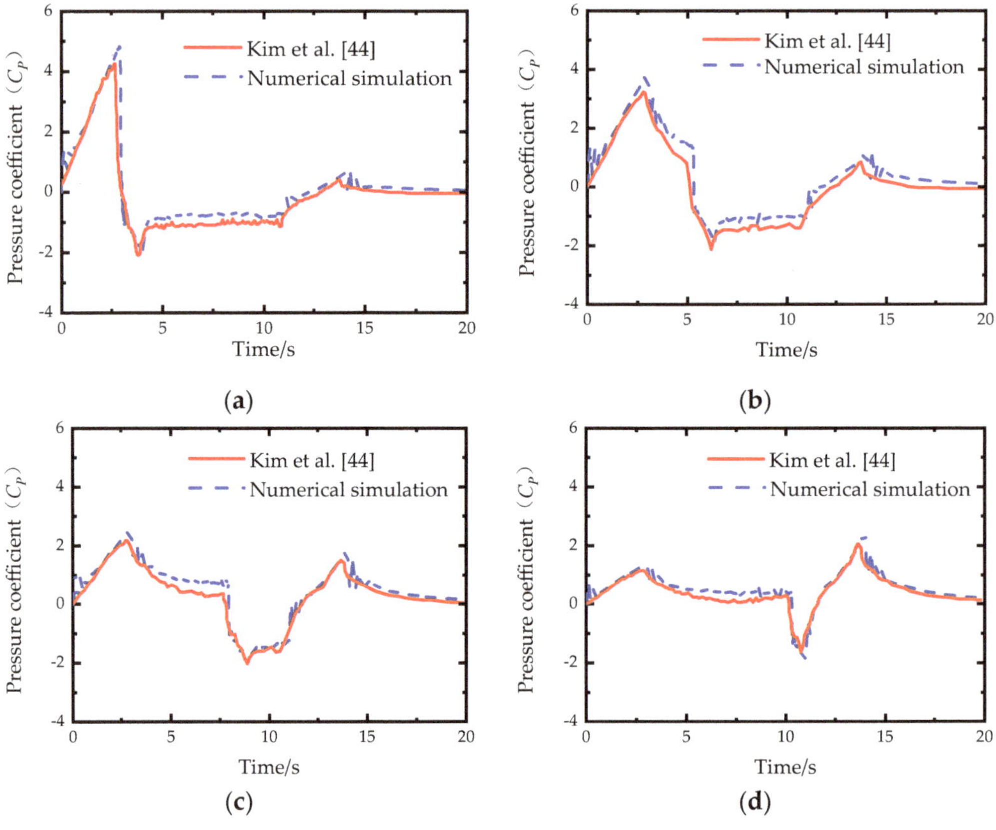

Figure 8. Comparison of Cp variations between the numerical simulation and the experimental results. (**a**) PT1. (**b**) PT2. (**c**) PT3. (**d**) PT4.

Figure 9 compares u/U_{T_MAX} variations between the numerical simulation and the experimental results of VT1 and VT2, which are 0.5 m and 38.5 m away from the tunnel inlet, respectively. U_{T_MAX} is set to as 3.0 m/s. The comparison results show that the numerical calculation and the experimental results are in good agreement. Although the train speed of the validation case is much lower than that of the current work, the comparison results indicate that the numerical method adopted in this research is accurate to simulate the moving object-induced aerodynamic pressure and unsteady flow inside a tunnel structure. In addition, Liu et al. [45] and Izadi et al. [46] had used a similar method to simulate the aerodynamic pressure inside tunnels and obtained reasonable results. It indicates that the numerical method adopted in this research is accurate for obtaining the main characteristics of the flow field and aerodynamic pressure fluctuation when a high-speed train passes the housing for sound emission alleviation.

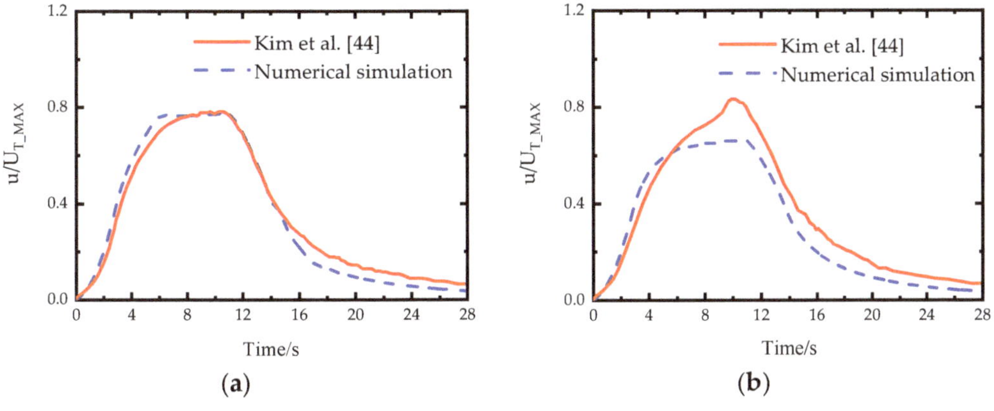

Figure 9. Comparison of u/U_{T_MAX} between the numerical and the experimental results. (**a**) VT1. (**b**) VT2.

4. Results and Discussion

The flow field is simulated using the given method when the train passes through the enclosed housing for sound emission alleviation at a speed of 350 km/h. The time history of the pressure on sections S1, S3, S6, S9 and S11 is monitored and shown in Figure 10. When the aerodynamic pressure is a pushing action, the aerodynamic pressure is defined as positive; conversely, when the aerodynamic pressure is a suction action, the pressure is defined as negative.

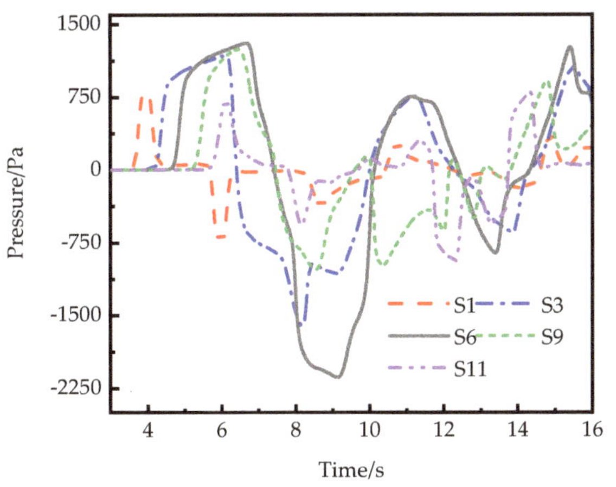

Figure 10. Time history of aerodynamic pressure on measuring points.

The pressure curves show that all sections suffer both positive and negative aerodynamic pressures. The extreme pressures and variation tendencies are different in different sections depending on the distance away from the entrance. In the same section, the pressure curves share a similar tendency; in particular, the pressure curves are very close when the aerodynamic pressure is close to 0 Pa. Meanwhile, the pressure amplitudes at the near-train side are always higher than those at the far-train side when extreme pressure appears. These results indicate that most of the time, the wind flow inside the housing for sound emission alleviation has one-dimensional characteristics; however, when extreme pressure appears, the wind flow should have a three-dimensional structure. These aerodynamic pressure characteristics on the enclosed housing for sound emission alleviation are similar to those for tunnels reported by Liu et al. [10]. However, as most enclosed housings for sound emission alleviation are made of steel structures and their stiffness is much lower than that of tunnels, they are more sensitive to train-induced wind loads, and consequently, the characteristics of the train-induced wind load are more important for the structural design of enclosed housing for sound emission alleviation.

Figure 10 shows that the aerodynamic pressure on the housing for sound emission alleviation is much more complicated than normal dynamic loads. The aerodynamic pressure is not a periodic wave and always rapidly increases or decreases. Therefore, the amplitude and frequency spectra are not enough to describe the dynamic characteristics of the aerodynamic pressure, and the impact effects should be considered. In the following sections, the dynamic characteristics of the aerodynamic pressure will be discussed in terms of wave propagation, impact effects, extreme pressure, duration of extreme pressure, dominant frequency and decay rate.

4.1. Wave Propagation

The aerodynamics inside the enclosed housing for sound emission alleviation are similar to those of tunnels [47,48]. When the train approaches the entrance of the enclosed housing for sound emission alleviation, the train head induces a compression wave, and the train rear motivates an expansion wave. Both compression and expansion waves propagate inside the housing for sound emission alleviation and reflect at both ends in sound velocity. The variation in aerodynamic pressure depends on the wave propagation. Figure 11 shows the details of the propagation of the aerodynamic pressure wave of P3 on section S9 when the train passes through. The black dotted line represents the position of the measuring point on the housing for sound emission alleviation wall, the red solid line (marked by T_N) is the running trajectory of the train head, the green dotted line (marked by T_R) is the running trajectory of the train rear, the blue solid lines (marked by C_W) present the propagation trajectory of the compression waves, and the blue dotted lines (marked by E_W) are the propagation trajectory of the expansion waves. The results show that a compression wave is generated at the instant the train head enters the housing for sound emission alleviation and then propagates inside the housing for sound emission alleviation at sound velocity. When the compression wave reaches section S9 at t_1, the monitored aerodynamic pressure first rapidly and then steadily increases. When the train tail enters the housing for sound emission alleviation, an expansion wave is generated and propagates inside the housing for sound emission alleviation at the sound velocity. When the expansion wave reaches S9 at t_2, the aerodynamic pressure begins to rapidly decrease. When the compression and expansion waves propagate to the exit of the housing for sound emission alleviation, they are reflected back as expansion and compression waves, respectively. At instant t_3, the reflected expansion wave reaches S9, and the aerodynamic pressure continues to rapidly decrease. At instant t_4, the reflected compression wave reaches S9, the aerodynamic pressure increases. When the train head passes section S9 at t_5, the aerodynamic pressure decreases. After that, multiple compression and expansion waves are transmitted to section S6, and the pressure increases and decreases accordingly. The transmission of the aerodynamic pressure waves in the enclosed housing for sound emission alleviation is also consistent with those in tunnels, as reported by Liu et al. [47]

and Ji et al. [48]. Figure 12 shows the pressure distribution on the wall of the housing for sound emission alleviation when t = 6 s. The video of the train transition with pressure shown is given in the attachment.

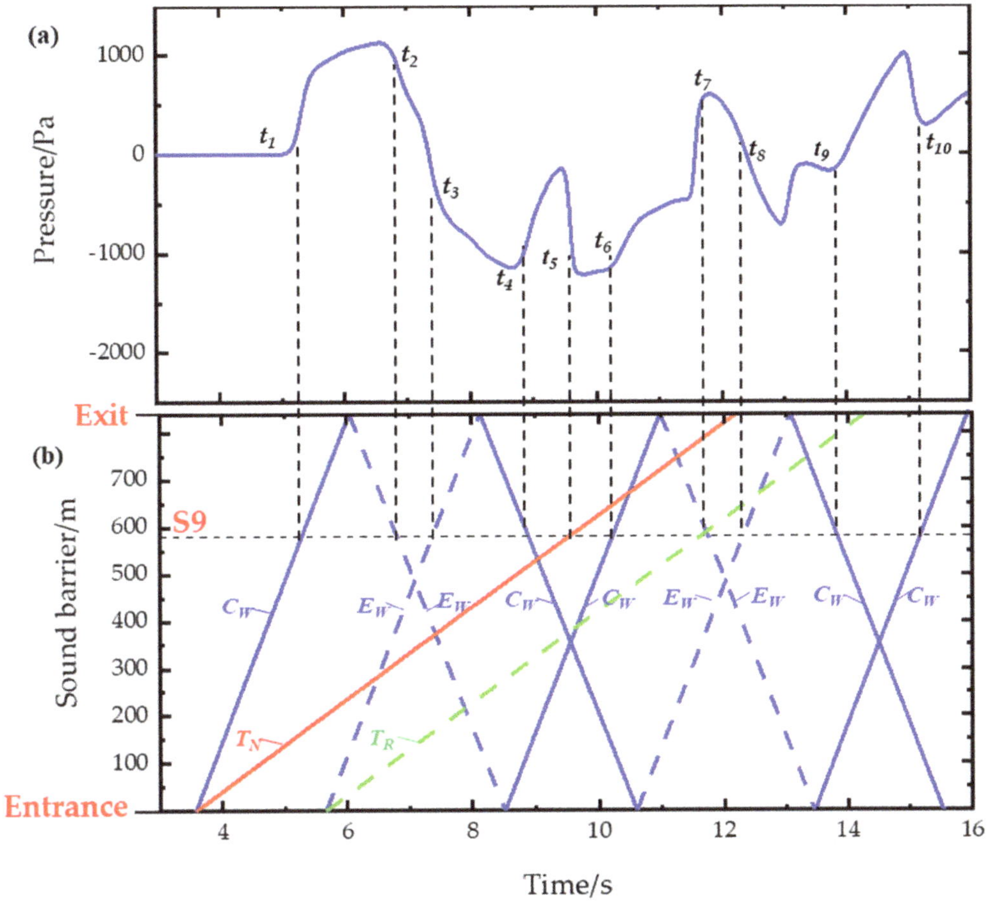

Figure 11. Pressure variations of P3 on the housing for sound emission alleviation section S9 and the propagation process of the pressure waves when the train passing through the housing for sound alleviation: (**a**) Time history of wind pressure of P3 at S9; (**b**) Propagation process of the pressure waves (The black dotted line in (**b**) represents the test section of (**a**), the red solid line is the running trajectory of the train head, marked by T_N; the green dotted line is the running trajectory of the train rear, marked by T_R; the blue solid lines present the propagation trajectory of the compression waves, marked by C_W; and the blue dotted lines are the propagation trajectory of the expansion waves, marked by E_W).

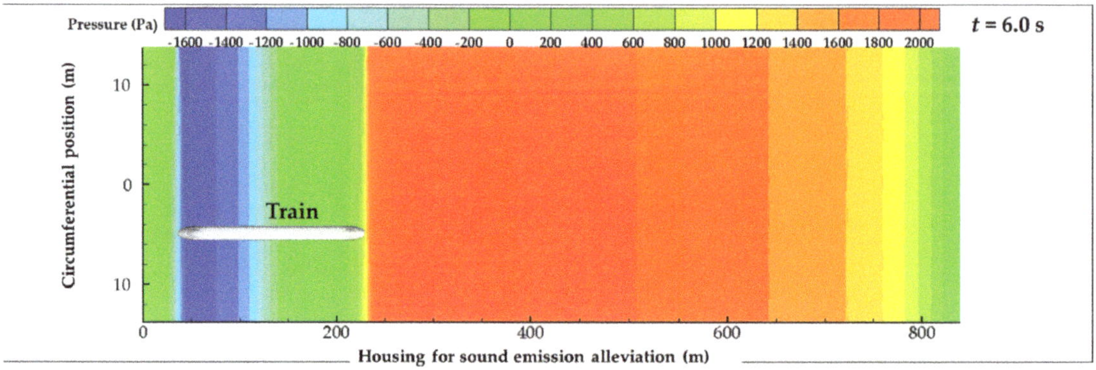

Figure 12. The pressure distribution on the wall of the housing for sound emission alleviation when $t = 6$ s.

4.2. Extreme Pressure, Duration of Extreme Pressure and Impact Effects

As enclosed housings for sound emission alleviations are always made of steel structures, the extreme aerodynamic pressure, duration of the extreme pressure, and impact effects are key factors for the structural design. The extreme pressures are defined as the maximum positive or negative pressure acting on each section when the train passes through at a speed of 350 km/h. Table 1 shows the distributions of the positive extreme value, negative extreme value, and peak-to-peak aerodynamic pressure in the longitudinal direction. P_{pmax} is the positive-extreme pressure, P_{nmax} is the negative-extreme pressure, and the peak-to-peak pressure is the absolute difference of P_{pmax} and P_{nmax} ($P_{pmax} - P_{nmax}$). The results show that the extreme pressures are almost symmetrical against section S6, the middle section. The extreme aerodynamic pressure at the central region is always higher than those close to the entrance/exit. In particular, the negative extreme pressure drastically increases in the range of 150 m to 350 m from the entrance/exit and reaches −2153 Pa at the middle section, more than two times higher than that at the entrance/exit sections. Different from the negative peak, the positive extreme pressure gradually increases from the entrance/exit to the middle section and reaches the maximum value of 1298 Pa at the middle section. The maximum peak-to-peak pressure is calculated as 3451 Pa.

Table 1. Distribution of the extreme pressure in the longitudinal direction of the housing for sound emission alleviation.

Sections	P_{pmax} (Pa)	P_{nmax} (Pa)	Peak-to-Peak (Pa)
1	1090	−1020	2110
2	1047	−1088	2135
3	1145	−1607	2752
4	1237	−2076	3312
5	1285	−2111	3396
6	1298	−2153	3451
7	1277	−2023	3300
8	1227	−1963	3190
9	1170	−1212	2382
10	944	−909	1853
11	799	−936	17

The time history of the train-induced aerodynamic pressure in Figure 10 shows the variation of the aerodynamic pressure is very complicated. It is hard to calculate the wind-induced response using this wind load. In the present study, an ideal model of the aerodynamic pressure is proposed using the duration, T_d, and the change rate of aerodynamic pressure, $\overline{P}$ and P_{pmax} and P_{nmax}, as Figure 13 shows.

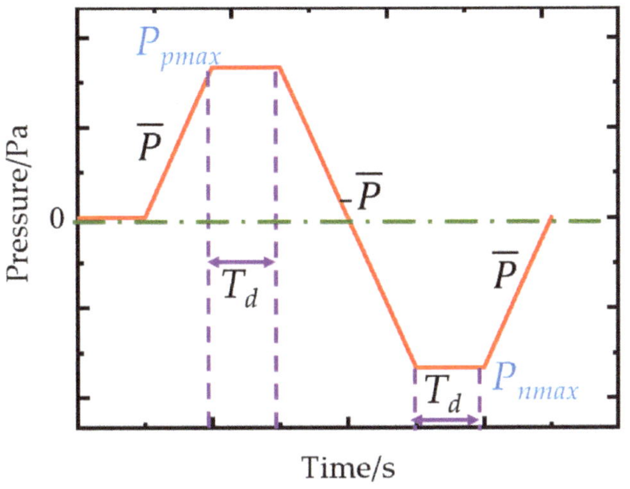

Figure 13. Ideal aerodynamic pressure model.

The duration T_d is defined as the time when the aerodynamic pressure is higher than 85% of extreme pressure and the extreme pressure appears:

$$T_d = T2|_{0.85P_{\max}} - T1|_{0.85P_{\max}} \tag{9}$$

where $T1|_{0.85P_{\max}}$ is the first instant when the pressure reaches $0.85P_{\max}$ before the extreme pressure, $T2|_{0.85P_{\max}}$ is the first instant when the pressure reaches $0.85P_{\max}$ after the extreme pressure, and $P_{\max}$ represents both positive and negative extreme pressures.

The change rate of aerodynamic pressure $\overline{P}$:

$$\overline{P} = \frac{P_{ib} - P_{ih}}{t_{ib} - t_{ih}} \tag{10}$$

where P_{ih} and P_{ib} are the aerodynamic pressures at two adjacent inflection points, t_{ih} and t_{ib} are the corresponding instants. In addition, the absolute difference, $|P_{ih} - P_{ib}|$, should be higher than $0.5P_{\text{pmax}}$.

Table 2 summarizes the T_d and maximum $\overline{P}$ of monitoring point P3 on sections S1, S3, S6, S9, and S11. The sections close to both ends have shorter durations. The middle section S6 has the longest duration of the positive and negative extreme aerodynamic pressures. They are 1.41 s and 1.04 s, respectively. Section S1, at the entrance of the enclosed housing for sound emission alleviation, has the highest positive and negative pressure change rate. They are 9.881 kPa/s and −10.415 kPa/s, respectively. In the other sections, the pressure change rate is relatively close, and the average values are calculated as 5.4 kPa/s and −5.9 kPa/s.

Table 2. The duration of the extreme pressure and the pressure change rate at different sections.

Sections	P_{pmax}		P_{nmax}	
	T_d (s)	$\overline{P}$ (kPa/s)	T_d (s)	$\overline{P}$ (kPa/s)
S1	0.222	9.881	0.154	−10.415
S3	1.379	3.288	0.347	−5.488
S6	1.410	5.780	1.041	−6.569
S9	1.059	6.227	0.719	−5.842
S11	0.356	6.366	0.526	−5.609

Figure 14 compares the ideal aerodynamic pressure of the sections S1, S3, S6, S9 and S11. It shows that the ideal aerodynamic pressures under different sections are significantly different. They are much simpler than the original ones and can be adopted to simulate the structural responses of the housing for sound emission alleviation. However, whether the parameters in the ideal model are reasonable or not needs further study.

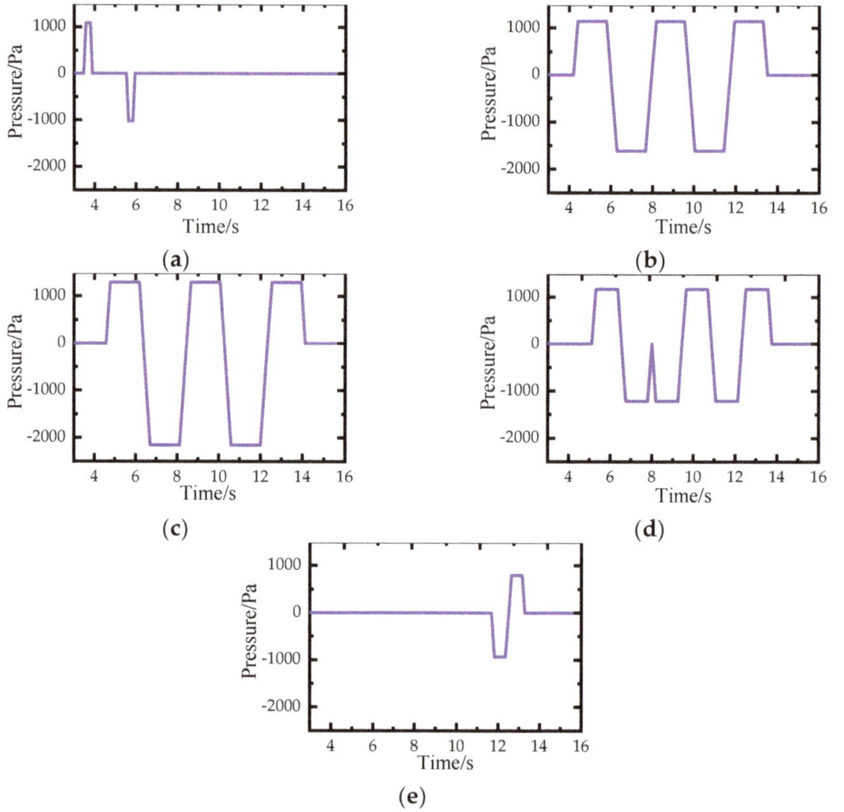

Figure 14. Ideal aerodynamic pressure at different sections. (**a**) S1. (**b**) S3. (**c**) S6. (**d**) S9. (**e**) S11.

As previously noted, when extreme pressures appear, the circumferential distribution of the aerodynamic pressure has obvious no uniformity characteristics. To reveal the circumferential distribution characteristics, the aerodynamic pressure on sections S1 and S6 at the instant when positive/negative extreme pressures appear is investigated and shown in Figures 15 and 16. The baseline values in Figure 15a,b are set as 700 Pa and −600 Pa, respectively. The values in Figure 16a,b are set as 800 Pa and −1300 Pa, respectively. The results confirm that the pressure distribution at both sections S1 and S6 is asymmetrical in the circumferential direction. The pressure at the near side is always higher than that at the far side. However, the circumferential distribution of the aerodynamic pressure at both sections is different.

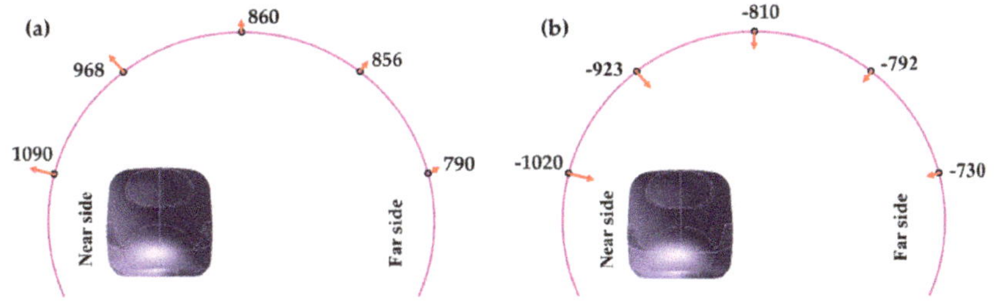

Figure 15. Pressure distribution on S1 when extreme pressure appears.

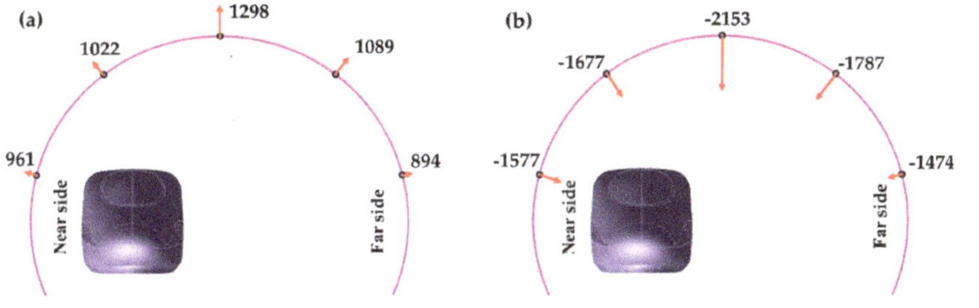

Figure 16. Pressure distribution on S6 when extreme pressure appears.

In section S1, aerodynamic pressure keeps decreasing from the near side to the far side. In particular, the positive and negative pressures on the near side are 37.97% and 39.73% higher than those on the far side, respectively. However, in section S6, the highest aerodynamic pressure appears at the top point P3. The positive and negative extreme pressures at P3 are 45.2% and 46.1% higher than those on the far side, respectively. This obvious no uniformity of the aerodynamic pressure should be carefully considered during the structural design of an enclosed housing for sound emission alleviation.

4.3. Dominant Frequency and Decay Rate of the Aerodynamic Pressure after the Train Exits

After the train exits the enclosed housing for sound emission alleviation, as the pressure wave continues to propagate inside the housing for sound emission alleviation, the aerodynamic pressure does not immediately revert to zero but remains varying. Figure 17 shows the pressure curves of P3 on sections after the train traverses the exit of the housing for sound emission alleviation. The results show that the pressure curves, in particular at the central sections, become periodic waves, and the amplitude gradually decreases, similar to the free vibration decay curves. In addition, the pressure wave is much more periodic, and its amplitude is much higher in the central region than that close to both ends. These characteristics are different from those when the train is passing inside the housing for sound emission alleviation. Consequently, the dominant frequency and decay rate of the pressure curves are analyzed as follows.

The dominant frequency of the aerodynamic pressure at the monitoring points in each section is obtained through spectral analyses. Figure 18 shows the calculated power spectrum density curves of the pressure of P3 on sections after the train traverses the exit of the housing for sound emission alleviation from 14.3–45 s and the sampling frequency is 500 Hz. In all sections, the dominant frequency is equal to 0.2 Hz. In fact, the dominant frequency can be theoretically derived using the propagation law of the pressure wave. When a compression wave reaches the monitoring point, the pressure reaches a positive extreme value; then, the expansion wave is reflected to be an expansion wave at one end of the housing for sound emission alleviation. When the expansion wave reaches the monitoring point, the extreme pressure is negative. After that, the expansion wave is reflected to be a compression wave again at another end of the housing for sound emission alleviation. When the new compression wave reaches the monitoring point, the pressure reaches the next positive extreme value. During this cycle, both compression and expansion waves travel the entire housing for sound emission alleviation once under the sound velocity. Consequently, the dominant frequency can be expressed as:

$$f = \frac{v}{2l} = 0.2\ \text{Hz} \tag{11}$$

where l is the length of the housing for sound emission alleviation and v is the sound velocity. This formula is the same as that proposed by Liu et al. [47] in tunnel aerodynamics.

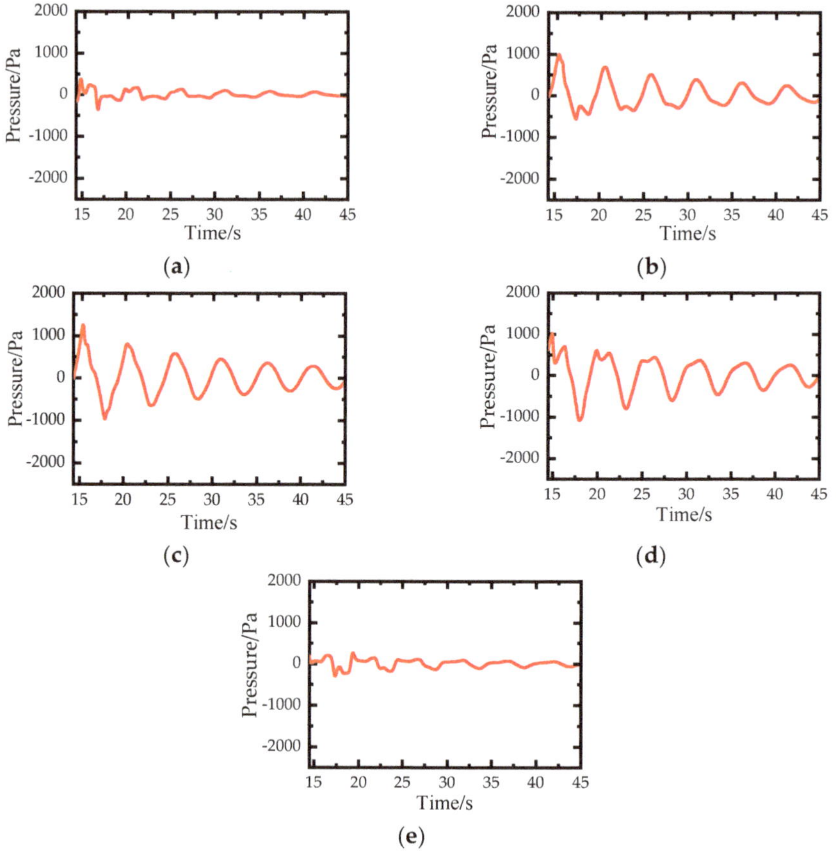

Figure 17. Pressure curves of different measuring points for the train passing through the housing for sound emission alleviation. (**a**) S1. (**b**) S3. (**c**) S6. (**d**) S9. (**e**) S11.

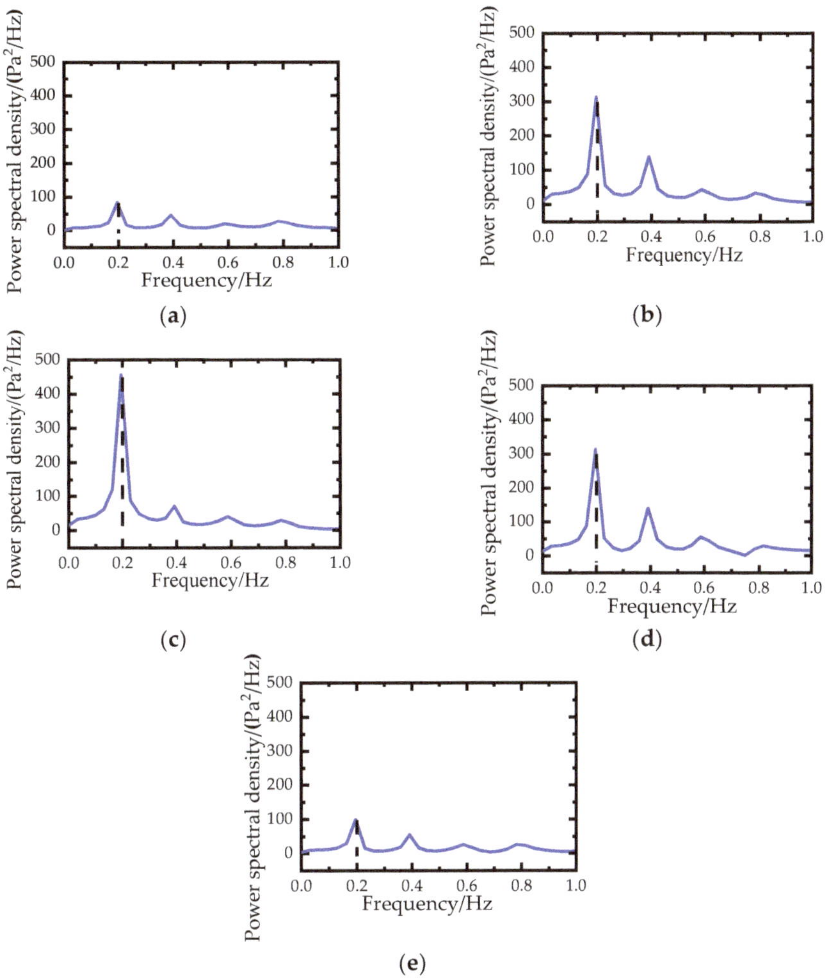

Figure 18. Pressure curves of different measuring points for the train passing through the housing for sound emission alleviation. (**a**) S1. (**b**) S3. (**c**) S6. (**d**) S9. (**e**) S11.

During the propagation, the strength of the aerodynamic pressure decreases due to the friction and refection effects. This dynamic process is important for the structure design. In order to evaluate the attenuation characteristic, the decay rate δ of the aerodynamic pressure is defined as [49]:

$$\delta = \ln(\frac{A_K}{A_{K+1}}) \tag{12}$$

where A_K is a pressure extreme in the process of pressure decay, A_{K+1} is the next pressure extreme after a cycle. The definition of the decay rate of the aerodynamic pressure is similar to that of a damped free vibration.

Figure 19 shows the pressure extremes envelope and the decay rate of the aerodynamic pressure fitted using Equation (11). The decay rate of the aerodynamic pressure on all sections is close, and the average decay rate is calculated as 0.262. The highest decay rate is 0.277 in the middle section. Both decay rate and the dominant frequency are key parameters for the structural design of the enclosed housing for sound emission alleviation because they are necessary when the engineer evaluates the fatigue life.

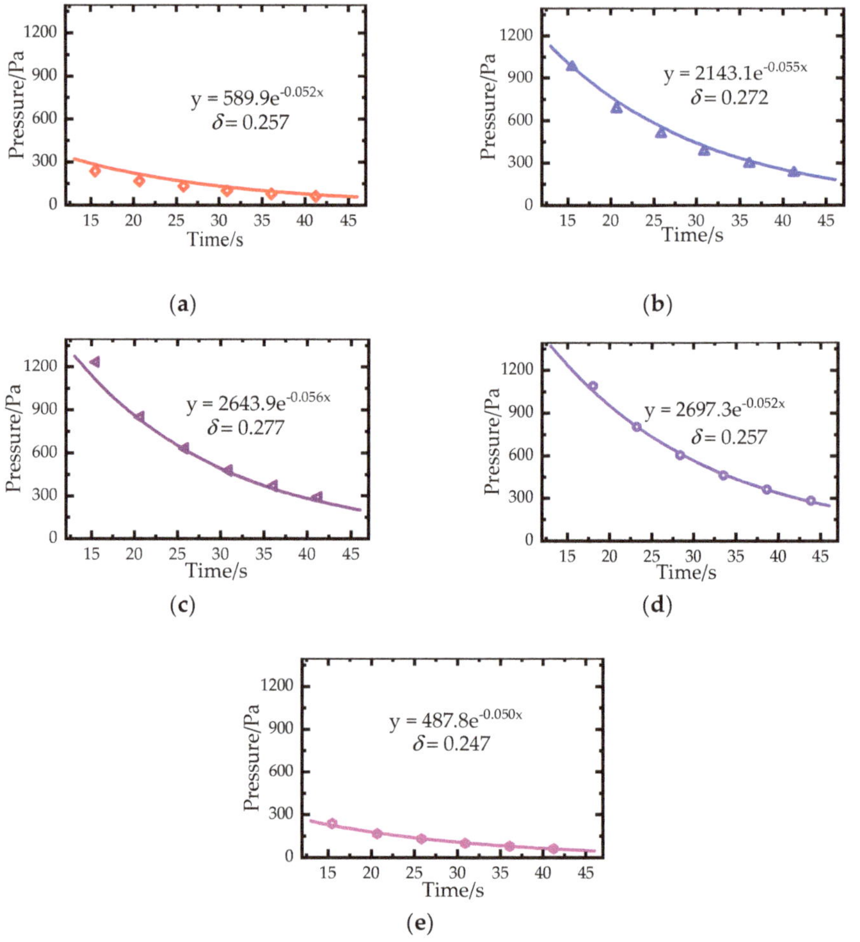

Figure 19. Pressure wave attenuations of different sections. (**a**) S1. (**b**) S3. (**c**) S6. (**d**) S9. (**e**) S11.

5. Conclusions

Consequently, in the present study, the unsteady aerodynamic pressure inside an enclosed noise barrier is numerically simulated when a CRH380 train [29] passes through and exits at a speed of 350 km/h by using the dynamic mesh method. The numerical method is firstly validated by comparing the numerical results with experimental results using a benchmark case. Then, the dynamic characteristics of the aerodynamic pressure are discussed in terms of wave propagation, impact effect, extreme pressure, duration of extreme pressure, dominant frequency and decay rate. The results show that the original reason for the variation of the aerodynamic pressure is the wave propagation process inside the noise barrier, similar to those inside of a tunnel. The extreme aerodynamic pressure at the central region is always higher, and the maximum peak-to-peak pressure reaches 3451 Pa. An ideal model of aerodynamic pressure of enclosed noise barriers is proposed using the duration of the extreme aerodynamic pressure and the pressure change rate, which is a cost-effective model for structure design. After the train exits the housing for sound emission alleviation, the aerodynamic pressure reverts to periodic decay curves. The dominant frequency and decay rate are proposed to express the dynamic characteristics. The main conclusions are summarized as follows:

(1) The unsteady aerodynamic pressure is complicated and aperiodic when the train is running in the enclosed housing for sound emission alleviation. The reasons for the variation in the aerodynamic pressure are clearly caused by the propagation of the aerodynamic pressure wave, similar to that in a tunnel.
(2) The extreme aerodynamic pressure at the central region is always higher than those close to the entrance and exit. In particular, negative and positive extreme pressures appearing in the middle section are −2153 Pa and 1298 Pa, respectively. The maximum peak-to-peak pressure is 3451 Pa.
(3) To further quantify the dynamic characteristics of the train-induced aerodynamic pressure on the housing for sound emission alleviation, an ideal model of aerodynamic pressure is proposed using the duration of the extreme aerodynamic pressure and the pressure change rate. The longest duration of the extreme aerodynamic pressure appears in the middle section, and the highest pressure change rate occurs at the entrance section. In other sections, the pressure change rate is relatively close. The ideal model is much simpler than the original aerodynamic pressure and can be adopted to calculate the structural responses of the housing for sound emission alleviation. The rationality of the parameters in the ideal model needs further studies.
(4) For the special enclosed housing for sound emission alleviation, the longest durations of the positive and negative extreme aerodynamic pressures are calculated as 1.41 s and 1.04 s, respectively. The highest positive and negative pressure change rates are calculated as 9.881 kPa/s and −10.415 kPa/s, respectively. In other sections, the average pressure change rate is calculated as 5.4 kPa/s and −5.9 kPa/s.
(5) After the train exits the housing for sound emission alleviation, the aerodynamic pressure reverts to periodic decay curves. The pressure amplitude at the central region is always higher than those close to the entrance/exit. To better understand the aerodynamic pressure in this process, the dominant frequency and decay rate are proposed to express the dynamic characteristics. For the special enclosed housing for sound emission alleviation, the dominant frequency is identified as 0.2 Hz, and the decay rate is calculated as 0.262.

Author Contributions: All the authors made significant contributions to the work. H.J., X.J. and X.H. conceived this study; H.J., X.J. and X.H. completed the numerical calculations and analyzed the data; X.H. provided advice for the preparation and revision of the paper; H.J., X.J. and X.H. wrote the paper; S.Z., J.Z. and H.Z. reviewed the manuscript for scientific contents. All authors have read and agreed to the published version of the manuscript.

Funding: The numerical calculations in this paper have been done on the supercomputing system of the National Supercomputing Center in Changsha. This research was financially supported by the National Natural Science Foundations of China (No. 52078502, 51925808, U1934209) and the Nat-ural Science Foundations of Hunan Province, China (No. 2019JJ50819).

Institutional Review Board Statement: Not applicable.

Informed Consent Statement: Not applicable.

Data Availability Statement: Data are contained within this article.

Conflicts of Interest: The authors declare no conflict of interest.

References

1. Planning Engineering Consulting + Services China Ltd. *Consultation Report of the Noise Barriers in Chinese Railway Passenger Dedicated Line.* 2007.
2. Baker, C.; Jordan, S.; Gilbert, T.; Quinn, A.; Sterling, M.; Johnson, T.; Lane, J. Transient aerodynamic pressures and forces on trackside and overhead structures due to passing trains. Part 1: Model scale experiments; Part 2: Standards applications. *Proc. Inst. Mech. Eng. Part F J. Rail Rapid Transit* **2014**, *228*, 37–70. [CrossRef]
3. Lü, M.; Li, Q.; Ning, Z.; Ji, Z. Study on the aerodynamic load characteristic of noise reduction barrier on high-speed railway. *J. Wind. Eng. Ind. Aerodyn.* **2018**, *176*, 254–262. [CrossRef]

4. Xiong, X.H.; Li, A.H.; Liang, X.F.; Zhang, J. Field study on high-speed train induced fluctuating pressure on a bridge noise barrier. *J. Wind. Eng. Ind. Aerodyn.* **2018**, *177*, 157–166. [CrossRef]
5. Soper, D.; Gillmeier, S.; Baker, C.; Morgan, T.; Vojnovic, L. Aerodynamic forces on railway acoustic barriers. *J. Wind. Eng. Ind. Aerodyn.* **2019**, *191*, 266–278. [CrossRef]
6. Kikuchi, K.; Iida, M.; Takasaki, T.; Takami, H. Field Measurement of Wayside Low-Frequency Noise Emitted from Tunnel Portals and Trains of High-Speed Railway. *J. Low Freq. Noise Vib. Act. Control* **2005**, *24*, 219–231. [CrossRef]
7. Fukuda, T.; Ozawa, S.; Iida, M.; Takaski, T.; Wakabayashi, Y. Distortion of compression wave propagating through very long tunnel with slab tracks. *JSME Int. J. Ser. B* **2006**, *49*, 1156–1164. [CrossRef]
8. Liu, T.H.; Tian, H.Q.; Jin, X.S. Experimental study of full-scale train on aerodynamics in tunnel. *Acat. Aerodyn. Sin.* **2008**, *26*, 42–46. (In Chinese)
9. Liu, T.H.; Tian, H.Q.; Liang, X.F. Aerodynamic Effects Caused by Trains Entering Tunnels. *J. Transp. Eng.* **2010**, *136*, 846–853. [CrossRef]
10. Liu, T.H.; Chen, X.D.; Li, W.H.; Xie, T.Z.; Chen, Z.W. Field study on the interior pressure variations in high-speed trains passing through tunnels of different lengths. *J. Wind. Eng. Ind. Aerodyn.* **2017**, *169*, 54–66. [CrossRef]
11. Ko, Y.Y.; Chen, C.H.; Hoe, I.T.; Wang, S.T. Field measurements of aerodynamic pressures in tunnels induced by high speed trains. *J. Wind. Eng. Ind. Aerodyn.* **2012**, *100*, 19–29. [CrossRef]
12. Iida, M.; Tanaka, Y.; Kikuchi, K.; Fukuda, T. Pressure waves radiated directly from tunnel portals at train entry or exit. *Q. Rep. RTRI* **2001**, *42*, 83–88. (In Japanese) [CrossRef]
13. Bellenoue, M.; Moriniere, V.; Kageyama, T. Experimental 3-D Simulation of the Compression Wave, due to Train-Tunnel Entry. *J. Fluid Struct.* **2002**, *16*, 581–595. [CrossRef]
14. Winslow, A.; Howe, M.S.; Iida, M. Influence of a Scarfed Portal on the Compression Wave Generated by a High-Speed Train Entering a Tunnel. *J. Low Freq. Nois.* **2009**, *24*, 203–217. [CrossRef]
15. Iida, M.; Kikuchi, K.; Fukuda, T. Analysis and Experiment of Compression Wave Generated by Train Entering Tunnel Entrance Hood. *JSME Int. J. Ser. B* **2006**, *49*, 761–770. [CrossRef]
16. Ricco, P.; Baron, A.; Molteni, P. Nature of pressure waves induced by a high-speed train travelling through a tunnel. *J. Wind. Eng. Ind. Aerodyn.* **2007**, *95*, 781–808. [CrossRef]
17. Doi, T.; Ogawa, T.; Masubuchi, T.; Kaku, J. Development of an experimental facility for measuring pressure waves generated by high-speed trains. *J. Wind. Eng. Ind. Aerodyn.* **2010**, *98*, 55–61. [CrossRef]
18. Heine, D.; Ehrenfried, K. Experimental Study of the Pressure Rise due to Tunnel Entry of a High-Speed Train. In Proceedings of the 18th STAB/DGLR Symposium, Stuttgart, Germany, 1 January 2014.
19. Yang, Q.S.; Song, J.H.; Yang, G.W. A moving model rig with a scale ratio of 1/8 for high speed train aerodynamics. *J. Wind. Eng. Ind. Aerodyn.* **2016**, *152*, 50–58. [CrossRef]
20. Zhang, L.; Yang, M.Z.; Niu, J.Q.; Liang, X.F.; Zhang, J. Moving model tests on transient pressure and micro-pressure wave distribution induced by train passing through tunnel. *J. Wind. Eng. Ind. Aerodyn.* **2019**, *191*, 1–21. [CrossRef]
21. Ogawa, T.; Fujii, K. Numerical investigation of three-dimensional compressible flows induced by a train moving into a tunnel. *Comput. Fluids.* **1997**, *26*, 565–585. [CrossRef]
22. Kwon, H.B.; Kim, T.Y.; Lee, D.H.; Kim, M.S. Numerical simulation of unsteady compressible flows induced by a high-speed train passing through a tunnel. *Proc. Inst. Mech. Eng. Part F J. Rail Rapid Transit* **2003**, *217*, 111–124. [CrossRef]
23. Li, X.H.; Deng, J.; Chen, D.W.; Xie, F.F.; Zheng, Y. Unsteady simulation for a high-speed train entering a tunnel. *J. Zhejiang Univ. Sci. A.* **2011**, *12*, 957–963. [CrossRef]
24. Uystepruyst, D.; William-Louis, M.; Monnoyer, F. 3D numerical design of tunnel hood. *Tunn. Undergr. Space Technol.* **2013**, *38*, 517–525. [CrossRef]
25. Chu, C.R.; Chien, S.Y.; Wang, C.Y.; Wu, T.R. Numerical simulation of two trains intersecting in a tunnel. *Tunn. Undergr. Space Technol.* **2014**, *42*, 161–174. [CrossRef]
26. Rabani, M.; Faghih, A.K. Numerical analysis of airflow around a passenger train entering the tunnel. *Tunn. Undergr. Space Technol.* **2015**, *45*, 203–213. [CrossRef]
27. Liu, T.H.; Chen, Z.W.; Chen, X.D.; Xie, T.Z.; Zhang, J. Transient loads and their influence on the dynamic responses of trains in a tunnel. *Tunn. Undergr. Space Technol.* **2017**, *66*, 121–133. [CrossRef]
28. Zhou, X.S.; Liu, T.H.; Chen, Z.W.; Zou, X.; Liu, D.R. Effect of ambient wind on pressure wave generated by high-speed train entering a tunnel. *J. Cent. South Univ.* **2017**, *24*, 1465–1475. [CrossRef]
29. Li, W.H.; Liu, T.H.; Pedro, M.V.; Chen, Z.W.; Guo, Z.J.; Li, M.; Xia, Y.T.; Liu, H.K. Aerodynamic effects of a high-speed train travelling through adjoining & separated tunnels. *Tunn. Undergr. Space Technol.* **2021**, *113*, 103973.
30. Yang, W.C.; Deng, E.; Lei, M.F.; Zhang, P.P.; Yin, R.S. Flow structure and aerodynamic behavior evolution during train entering tunnel with entrance in crosswind. *J. Wind. Eng. Ind. Aerodyn.* **2018**, *175*, 229–243. [CrossRef]
31. Li, W.H.; Liu, T.H.; Chen, Z.W.; Guo, Z.J.; Huo, X.S. Comparative study on the unsteady slipstream induced by a single train and two trains passing each other in a tunnel. *J. Wind. Eng. Ind. Aerodyn.* **2020**, *198*, 104095. [CrossRef]
32. Ferziger, J.; Peric, M. *Computational Method for Fluid Dynamics*, 3rd ed.; Springer: Berlin/Heidelberg, Germany, 2002.
33. Huang, Y.D.; Gong, X.L.; Peng, Y.J.; Kim, C.N. Effects of the solid curtains on natural ventilation performance in a subway tunnel. *Tunn. Undergr. Space Technol.* **2013**, *38*, 526–533. [CrossRef]

34. Chen, W.L.; Li, H.; Hu, H. An experimental study on the unsteady vortices and turbulent flow structures around twin-box-girder bridge deck models with different gap ratios. *J. Wind. Eng. Ind. Aerodyn.* **2014**, *132*, 27–36. [CrossRef]
35. Li, H.; Chen, W.L.; Xu, F.; Li, F.C.; Ou, J.P. A numerical and experimental hybrid approach for the investigation of aerodynamic forces on stay cables suffering from rain-wind induced vibration. *J. Fluids Struct.* **2010**, *26*, 1195–1215. [CrossRef]
36. Tian, H.Q. *Train Aerodynamics*; China Railway Publishing House: Beijing, China, 2007. (In Chinese)
37. Chen, W.L.; Zhang, Q.Q.; Li, H.; Hu, H. An experimental investigation on vortex induced vibration of a flexible inclined cable under a shear flow. *J. Fluids Struct.* **2015**, *54*, 297–311. [CrossRef]
38. Chen, W.L.; Xin, D.B.; Xu, F.; Li, H.; Ou, J.P.; Hu, H. Suppression of vortex-induced vibration of a circular cylinder using suction-based flow control. *J. Fluids Struct.* **2013**, *42*, 25–39. [CrossRef]
39. Huang, Y.D.; Hong, T.H.; Kim, C.N. A numerical simulation of train-induced unsteady airflow in a tunnel of Seoul subway. *J. Mech. Sci. Technol.* **2012**, *26*, 785–792. [CrossRef]
40. Chen, W.L.; Li, H.; Hu, H. Flow around a circular cylinder with slit. *Exp. Therm. Fluid Sc.* **2017**, *82*, 287–301.
41. Laima, S.; Li, H.; Chen, W.; Li, F. Investigation and control of vortex-induced vibration of twin box girders. *J. Fluids Struct.* **2013**, *39*, 205–221. [CrossRef]
42. Chen, W.L.; Gao, D.L.; Li, H.; Hu, H. Passive jet control of flow around a circular cylinder. *Exp. Fluids.* **2015**, *56*, 1–15. [CrossRef]
43. Xu, F.; Chen, W.L.; Xiao, Y.Q.; Li, H.; Ou, J.P. Numerical study on the suppression of the vortex-induced vibration of an elastically mounted cylinder by a traveling wave wall. *J. Fluid Struct.* **2014**, *44*, 145–165. [CrossRef]
44. Kim, J.Y.; Kim, K.Y. Experimental and numerical analyses of train-induced unsteady tunnel flow in subway. *Tunn. Undergr. Space Technol.* **2007**, *22*, 166–172. [CrossRef]
45. Liu, M.Z.; Zhu, C.G.; Cui, T.; Zhang, H.; Zheng, W.D.; You, S.J. An alternative algorithm of tunnel piston effect by replacing three-dimensional model with two-dimensional model. *Build Environ.* **2018**, *128*, 55–67. [CrossRef]
46. Izadi, T.; Mehrabian, M.A.; Abouali, O.; Ahmadi, G. 3-D numerical analysis of train-induced flow inside four ventilated underground subway stations connecting tunnels. *J. Wind. Eng. Ind. Aerodyn.* **2019**, *193*, 103974. [CrossRef]
47. Liu, F.; Yao, S.; Liu, T.H.; Zhang, J. Analysis on aerodynamic pressure of tunnel wall of high-speed railways by full-scale train test. *J. Zhejiang Univ.* **2016**, *50*, 2018–2024. (In Chinese)
48. Ji, P.; Wang, T.; Wu, F. Calculation grid and turbulence model for numerical simulating pressure fluctuations in high-speed train tunnel. *J. Cent. South Univ.* **2019**, *26*, 2870–2877. [CrossRef]
49. Ray, W.C.; Joseph, P. *Dynamics of Structures*, 2nd ed.; McGraw-Hill: New York, NY, USA, 1993.

applied sciences

MDPI

Article

The Effects of Aerodynamic Interference on the Aerodynamic Characteristics of a Twin-Box Girder

Buchen Wu [1,2,3,4], Geng Xue [1,2], Jie Feng [1,2] and Shujin Laima [1,2,*]

1 Key Lab of Smart Prevention and Mitigation for Civil Engineering Disasters of the Ministry of Industry and Information, Harbin Institute of Technology, Harbin 150090, China; buchenwu@u.nus.edu (B.W.); 20S133174@stu.hit.edu.cn (G.X.); 20b954002@stu.hit.edu.cn (J.F.)
2 Key Lab of Structures Dynamic Behavior and Control of the Ministry of Education, Harbin Institute of Technology, Harbin 150090, China
3 Department of Mechanical Engineering, National University of Singapore, 10 Kent Ridge Crescent, Singapore 119260, Singapore
4 Guangdong Provincial Key Laboratory of Turbulence Research and Applications, Department of Mechanics and Aerospace Engineering, Southern University of Science and Technology, Shenzhen 518055, China
* Correspondence: laimashujin@hit.edu.cn

Abstract: To investigate the aerodynamic characteristics of a twin-box girder in turbulent incoming flow, we carried out wind tunnel tests, including two aerodynamic interferences: leading body-height grid, and leading circular cylinder. In this study, the pressure distribution and the mean and fluctuating aerodynamic forces with the two interferences are compared with bare deck in detail to investigate the relationship between aerodynamic characteristics and the incoming flow characteristics (including Reynolds number and turbulence intensity). The experimental results reveal that, owing to the body-height flow characteristics around the deck interfered with by the body-height grid, the disturbed aerodynamic characteristics of the twin-box girder differ considerably from those of the bare twin-box girder. At the upstream girder, due to the vortex emerging from the body-height grid breaking the separation bubble, pressure plateaus in the upper and lower surface are eliminated. In addition, the turbulence generated by the body-height grid reduces the Reynolds number sensitivity of the twin-box girder. At a relatively high Reynolds number, the fluctuating forces are mainly dominated by turbulence intensity, and the time-averaged forces show almost no change under high turbulence intensity. At a low Reynolds number, the time-averaged forces change significantly with the turbulence intensity. Moreover, at a low Reynolds number, the wake of the leading cylinder effectively forces the boundary layer to transition to turbulence, which reduces the Reynolds number sensitivity of the mean aerodynamic forces and breaks the separation bubbles. Additionally, the fluctuating drag force and the fluctuating lift force are insensitive to the diameter and the spacing ratio.

Keywords: aerodynamic forces; pressure distribution; turbulence intensity; twin-box girder

Citation: Wu, B.; Xue, G.; Feng, J.; Laima, S. The Effects of Aerodynamic Interference on the Aerodynamic Characteristics of a Twin-Box Girder. *Appl. Sci.* **2021**, *11*, 9517. https://doi.org/10.3390/app11209517

Academic Editor: Elsa Caetano

Received: 9 September 2021
Accepted: 11 October 2021
Published: 13 October 2021

1. Introduction

In recent decades, super-long-span bridges have been largely designed using the sharp-edged twin-box girder, due to its superior aerodynamic stability, including the Xihoumen Bridge (main span, 1650 m), the Shanghai Yangtze River Bridge (main span, 730 m), and the Stonecutters' Cable-Stayed Bridge (main span, 1018 m). It is generally acknowledged that the stability of super-long-span bridges is an important indicator that represents the safety of the structures. Super-long-span bridges are often built at sea, where gales often occur, and the aerodynamic forces generated by wind-induced response cannot be neglected. Moreover, the shedding vortices around the box girder induce the vibration behavior, e.g., vortex-induced vibration (VIV). Therefore, the investigation of the aerodynamic performance of the bridge structures by wind tunnel experiment [1–4] or numerical simulation [5–8] is necessary in the pre-research stage of bridge construction. However, the wind tunnel experiments usually examine the bridge section model under uniform inflow conditions, and the aerodynamic characteristics obtained under these conditions are used to represent the aerodynamic characteristics of the bridge, which cannot completely accommodate the dynamic complexity of the structures in the real engineering application environment. The bridge structure is not simply affected by uniform inflow in a natural environment, and the turbulent components usually exist in the incoming flow. For instance, the incoming flow may pass through other structures to reach the bridge deck, so the flow around the windward side of the bridge structure is in the wake of other structures, and is usually unstable, with a large fluctuation velocity component. Therefore, it is necessary to study the flow characteristics of the bridge structure under different incoming flow characteristics.

Recently, the effect of the turbulent components of incoming flow on the aerodynamic performance of bluff bridge sections has fascinated many researchers, and triggered many experimental investigations to discern the effects of the incoming flow characteristics on the bridge. Zhou et al. [9] investigated the effects of the vertical turbulence intensity of incoming flow on the aerodynamic performance of a bridge. The authors claimed that the increase in the vertical turbulence intensity of the incoming flow increases the torsional frequency, and the critical flutter wind speed decreases when the vertical turbulence intensity is 2.84%. Hunt et al. [10] and Sarwar et al. [11] pointed out that the incoming flow with the fluctuation component of long-span bridges is nonlinear with the structural motion, and the turbulence components of incoming flow (such as turbulence intensity and turbulence scale) play a vital role in controlling the aerodynamic characteristics of bridges. Scanlan and Liu [12] experimentally investigated the flutter derivatives of a bridge deck based on the turbulent components of incoming flow, and a new theory was developed, which takes the turbulent components in the incoming flow into account. Haan and Kareem [13] investigated the effects of turbulence on the aero-elastic and self-excited forces of a rectangular prism via experiments. Since turbulence is highly heterogeneous and anisotropic, the self-excited pressure fluctuation, self-excited force, and flutter derivative of a stationary prism are strongly affected by the turbulence. The authors also pointed out that the streamwise position would shift with the increase in turbulence intensity; however, the pressure amplitudes would decrease with a larger turbulence scale. Meanwhile, Matsumoto et al. [14] showed that turbulence can adversely affect the flutter performance of bridges based on experimental results. In recent years, a lot of theoretical models have been developed to evaluate the effects of turbulence on the aerodynamic characteristics of bridges. Chen et al. [15,16] proposed a time-domain approach to predict the aerodynamic response of bridges, and a nonlinear theoretical model was established to analyze the effects of turbulence on the self-excited forces and the flutter performance. Wu and Kareem [17] summarized the latest developments in aerodynamics and aero-elasticity of bluff bodies by turbulent winds. The development of theoretical models of the effects of turbulence on the aerodynamic characteristics of bluff bodies is beneficial in efforts to effectively solve the problem of aerodynamic nonlinear response induced by turbulence. However, to our best knowledge, there is still no effective theoretical model to explain the relationship between the characteristics of incoming flow and the aerodynamic performance of the twin-box girder, because the aerodynamic characteristics of the twin-box girder are highly nonlinear in turbulent flow.

The main objective of this study is to experimentally investigate the influence of different cutting-edge aerodynamic interference methods on the aerodynamic characteristics of a twin-box girder. In the present work, the Xihoumen Suspension Bridge is adopted as a prototype of a twin-box girder to investigate. The turbulence intensity I is adopted to indicate the strength of incoming flow fluctuation, and the aerodynamic characteristics of long-span bridges are generally sensitive to the Reynolds number [18]. Therefore, the comprehensive influence of turbulence intensity and the Reynolds number on the aerodynamic characteristics of the twin-box girder is investigated. The remainder of this paper is organized as follows: In Section 2, the experimental method and geometric details of the twin-box girder are presented. In Section 3, the experimental results of different aerodynamic interferences are described and discussed in detail, including the surface pressure distributions and the time-averaged and fluctuating aerodynamic forces. Finally, conclusions are presented in Section 4.

2. Experimental Setup

The experiments were conducted in a closed-loop wind tunnel (SMC-WT1, Harbin Institute of Technology, Harbin, China). With screens and honeycomb installed before the inlet of the test section, the turbulence intensity was less than 0.4% over the speed range of 4–25 m/s. In the test section, the size of the cross-section was 505 mm × 505 mm.

In this study, the Reynolds number (Re) is the ratio of inertial forces to viscous forces, which is an important dimensionless quantity in fluid mechanics, and is defined as:

$$\mathrm{Re} = \frac{\rho U H}{\mu}, \quad (1)$$

where ρ is the density of the fluid; U is the incoming flow velocity; H is the central height of the twin-box girder, which is adopted as the characteristic length; and μ is the dynamic viscosity of the fluid. Since the present work was conducted in a conventional atmospheric boundary layer wind tunnel, the variation in the Reynolds number was achieved by adjusting the wind speed.

The turbulence intensity is an effective indicator that is associated with the turbulent kinetic energy (TKE), and it can be written as:

$$I = \frac{u'}{\overline{U}} = \frac{\sqrt{\frac{1}{3}\left(u'_x{}^2 + u'_y{}^2 + u'_z{}^2\right)}}{\sqrt{\overline{U_x}^2 + \overline{U_y}^2 + \overline{U_z}^2}}, \quad (2)$$

where $u'_x{}^2$, $u'_y{}^2$, and $u'_z{}^2$ are the root mean square of the turbulence velocity fluctuations in the x, y, and z directions, respectively, $\overline{U_x}$, $\overline{U_y}$, and $\overline{U_z}$ are the mean velocity in the x, y, and z directions, respectively, u' is the root mean square of the turbulence velocity fluctuations, and $\overline{U}$ is the mean velocity.

2.1. Section Model Geometrical Information and Surface Pressure Measurements

The detailed geometrical information of the prototype bridge deck is shown in Figure 1. The twin-box girder has two parallel box girders with a gap of length L = 6 m and width B = 36 m, and the center height of the deck H = 3.51 m. The spanwise length of the section model is L_s = 480 mm, and the geometric scale ratio of the section model is 1:120. Figure 2a shows a 3D sketch of the twin-box girder used in the present study. To obtain the surface pressure distributions, 46 pressure taps with a 0.5 mm radius were installed in the slice, which is 230 mm away from the right end of the bridge deck, as shown in Figure 2b. It should be noted that the experiments on the twin-box girder were stationary in this paper.

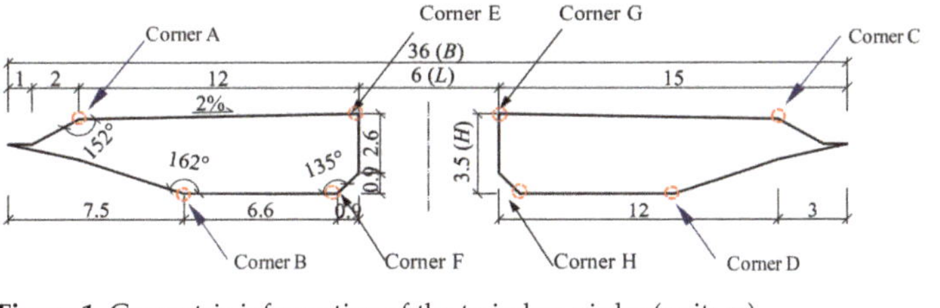

Figure 1. Geometric information of the twin-box girder (unit: m).

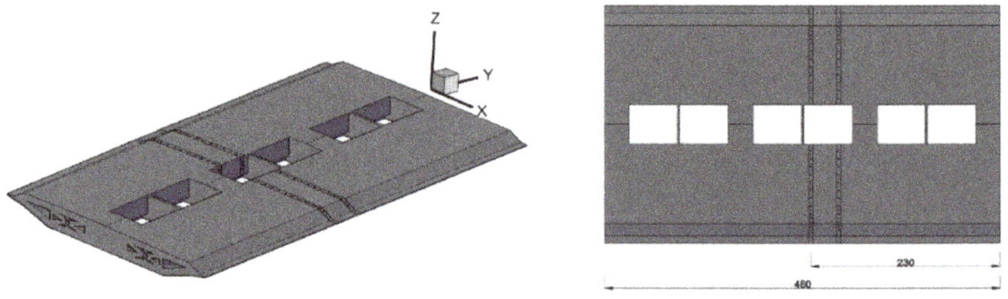

(**a**) (**b**)

Figure 2. Schematic diagrams of the section model: (**a**) 3D sketch of the twin-box girder, and (**b**) spanwise location of the pressure tap (unit: mm).

The distribution of these taps around the section circumference is illustrated in Figure 3. The pressure taps are connected to three pressure scanners (DSA3217, 16 channels for each scanner) with a measuring range of ±2.5 kPa by using connecting tubes (independent polyvinyl chloride (PVC) tubes, internal diameter of 1 mm) with a full length of 500 mm. This system is adopted to measure and record the instantaneous pressure distributions.

Figure 3. Location of the pressure taps in the section circumference.

The Scanivalve system monitors the surface pressure data at a sampling rate of 312.5 Hz, and the time length of one sampling period is 32 s. Based on the correction algorithm [19], the distortion effects brought by the connecting tubes—such as amplification factor and phase shift [20]—are quite small and negligible in the surface pressure measurements of this study. The time-averaged pressure coefficient C_p can be calculated by nondimensionalization of the time-averaged pressure p_m, which can be written as:

$$C_p = \frac{p_m}{\frac{1}{2}\rho_{air}U_\infty^2}, \tag{3}$$

$$p_m = \frac{1}{T}\int_0^T (p_i - p_\infty)dt, \tag{4}$$

where p_i denotes the instantaneous pressure, p_∞ is the pressure of the free stream, ρ_{air} is the density of air, U_∞ is the flow velocity of the free stream, and T is the sample period.

In the present study, the Reynolds number is high (Re > 5 × 10^3); hence, the boundary layer around the solid structure becomes turbulent and the pressure drag force dominates the skin drag force [21]. Thus, the pressure drag force is the dominant component of total drag, and the skin friction drag can be neglected.

Once the instantaneous pressure distributions are obtained, the corresponding aerodynamic forces of the bridge girder can be calculated by standard integration, which can be expressed as follows:

$$F_D = \int \mathbf{P}_{xi}\, ds_i, \tag{5}$$

$$F_L = \int \mathbf{P}_{zi}\, ds_i, \tag{6}$$

$$F_m = \int \mathbf{P}_i \zeta_i\, ds_i, \tag{7}$$

where $\mathbf{P}_i$ is the pressure force, $\mathbf{P}_{xi}$ and $\mathbf{P}_{zi}$ are the pressure force components along the x and z directions, respectively, ds_i denotes the element area, and ζ_i is the moment arm.

It should be noted that, when the gap exists, the pressure forces at pressure taps 9–11 and 32–34 should be considered.

The drag force coefficient C_D, the lift force coefficient C_L, and the moment force coefficient C_m are defined as:

$$C_D = \frac{F_D}{\frac{1}{2}\rho_{air} U_\infty^2 H L_s}, \tag{8}$$

$$C_L = \frac{F_L}{\frac{1}{2}\rho_{air} U_\infty^2 B L_s}, \tag{9}$$

$$C_m = \frac{F_m}{\frac{1}{2}\rho_{air} U_\infty^2 B^2 L_s}, \tag{10}$$

where F_D, F_L, and F_m are the drag, lift, and moment forces, respectively, and B is the width of the twin-box girder.

Then, the six effective indicators that represent the aerodynamic characteristics of the bridge girder can be obtained, including the mean drag force coefficients ($C_{D\,mean}$), the fluctuating drag force coefficients ($C_{D\,rms}$), the mean lift force coefficients ($C_{L\,mean}$), the fluctuating lift force coefficients ($C_{L\,rms}$), the mean moment force coefficients ($C_{m\,mean}$), and the fluctuating moment force coefficients ($C_{m\,rms}$). It should be noted that the root mean square of the specific value is adopted to represent the fluctuation.

2.2. *Aerodynamic Interference Measurements*

2.2.1. Leading Body-Height Grid Aerodynamic Interference

For the practical bridge structures, there exist some other structures at the upstream—for example, in the event that two bridges are close together. It is necessary to investigate the effects of the incompletely developed turbulence from the upstream body on the aerodynamic dynamic characteristics of the bridge. Based on the above consideration, for generality, the turbulence generated by the leading body-height grid is adopted to simulate the wake of the upstream body. With different body-height grids installed at the same height as the section bridge deck at the flow inlet, the turbulence intensity of the incoming flow in the model height range is affected by the geometry of the body-height grid. Therefore, the flow around the windward side of the bridge deck is changed from laminar flow to turbulent flow, with different-sized vortices generated by the interference of the body-height grid. Figure 4 shows the sketch view of the locations of the section model in the test section and the body-height grid in the flow inlet, while Figure 5 presents the detailed geometry of fourteen body-height grids (body-height grids I–XIV).

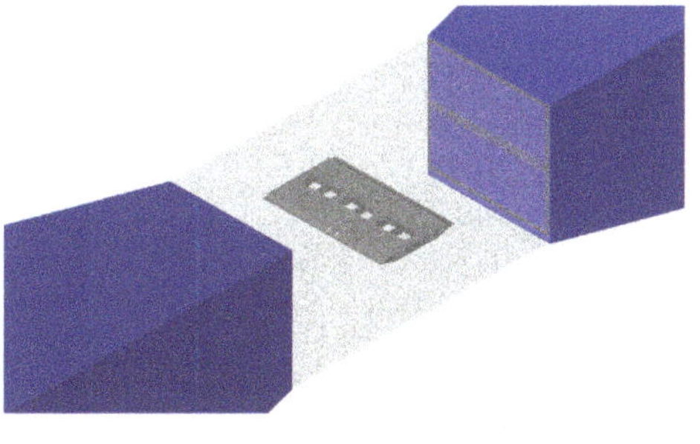

Figure 4. Sketch view of the section model in the test section and the body-height grid in the flow inlet.

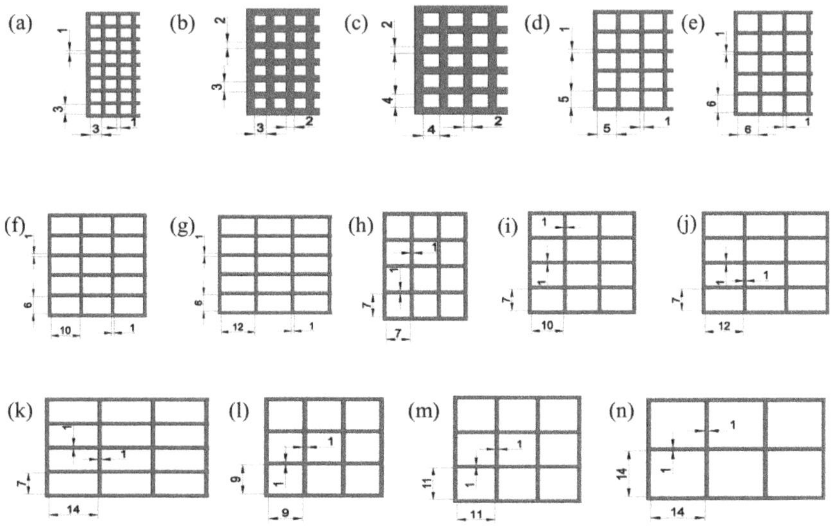

Figure 5. (**a–n**) The detailed geometry of the body-height grids I–XIV, respectively (unit: mm).

2.2.2. Leading Circular Cylinder Aerodynamic Interference

The handrail is an important accessory of the bridge structure, which is usually composed of cylinders, and the cylinders disturb the incoming flow characteristics to change the boundary layer of the box girder. Therefore, with different smooth cylinders (diameters of 3 mm, 4 mm, 5 mm, and 6 mm) installed in front of the essential flow areas of the model (upper edge, middle upper region, leading edge, middle lower region, and lower edge), the effects of different flow characteristics of essential flow regions around the section model on the surface pressure distribution and aerodynamic response of the bridge deck were investigated, and Figure 6 shows the location of cylinders in front of the bridge deck with different spacing ratios. The spacing ratio ε is defined as the ratio of the distance S between the trailing edge of the cylinder and the frontal surface of the twin-box girder to the diameter of cylinder D, $\varepsilon = S/D$. The spacing ratios used in this study were 2, 3, 4, and 5.

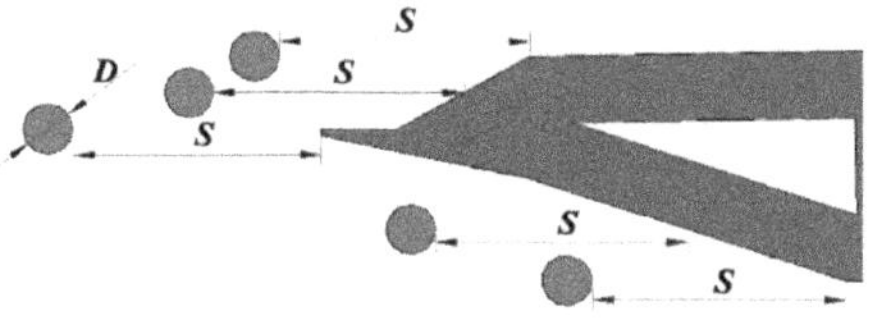

Figure 6. The sketch views of the leading circular cylinder and the twin-box girder (D is the diameter of the circular cylinder).

2.3. Turbulence Intensity Measurement

The turbulence intensity of incoming flow is measured by the Cobra Probe system (including Series 100 Cobra Probe, cabling, and TFI Device Control software). The Cobra Probe is composed of a multi-hole pressure probe, and is able to reconstruct the velocity components along the x, y, and z directions from pressure data. In the present work, the sampling frequency was set as 2 kHz. The Cobra Probe was placed at positions 1, 2 and 3 (see Figure 7) to measure the spread of the turbulent flow. In the present study, the turbulent properties at position 3 were adopted to characterize the incoming flow.

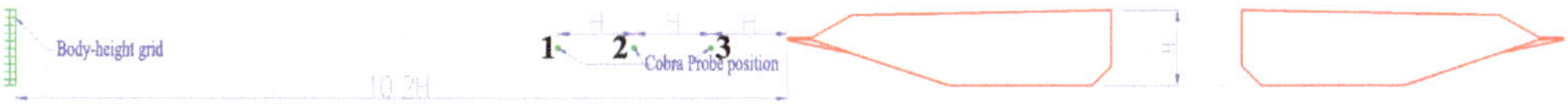

Figure 7. Locations of the Cobra Probe.

3. Results and Discussions

3.1. Undisturbed Surface Pressure Distribution and Aerodynamic Forces

To analyze the effects of the aerodynamic interference methods mentioned above, the aerodynamic characteristics of the bare twin-box girder were experimentally investigated, and the important aerodynamic parameters (including the surface pressure distribution and the aerodynamic forces) were obtained. Figure 8 gives the distribution of the mean pressure coefficient on the surface of the twin-box girder at different Reynolds numbers. At Re = 6.13×10^3, the profile of the surface pressure indicates that the pressure of the upstream girder reaches a peak near the windward corner A, because the incoming flow passes through the windward slope on the upper surface and the incoming flow is affected by the forward pressure gradient, which enhances the flow speed and promotes the continuous increase in suction. After the windward corner, the flow velocity gradually decreases, and the negative pressure also decreases. It should be noted that a short-term pressure plateau is formed after the windward corner on the upper surface of the upstream girder at low Reynolds numbers (e.g., Re = 6.13×10^3), but this pressure plateau will disappear at high Reynolds numbers. The pressure plateau is formed by the separation bubble, because the internal and external fluids of the separation bubble are relatively stable; thus, the pressure in the separation bubble is almost identical. Moreover, the endpoint of the pressure platform could be regarded as the reattachment point. Figure 8 also shows that the amplitude of the surface pressure increases with the Reynolds number, and the pressure on the lower surface of downstream girder has a common tendency to decrease first and then increase at different Reynolds numbers.

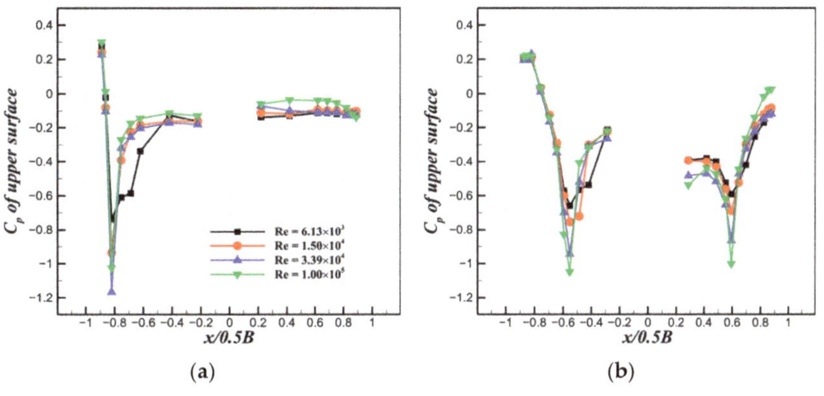

Figure 8. The mean surface pressure distribution of the twin-box girder at different Re: (**a**) the pressure surface distribution of the upper surface, and (**b**) the pressure surface distribution of the lower surface.

Figure 9 shows the mean and fluctuating aerodynamic force coefficients of the twin-box girder at various Reynolds numbers. The aerodynamic force coefficients show a significant change in the Reynolds number range adopted in this study. With the increase in the Reynolds number, the time-averaged drag force coefficient and the fluctuating drag force coefficient both decrease, and the rate of decrease first increases and then decreases; moreover, the maximum rate of decrease is achieved at Re = 9.21×10^3, while the time-averaged lift force coefficient significantly increases first and then flattens with the increase in the Reynolds number. With the Re rising, the time-averaged moment force coefficient first increases rapidly, and then decreases. The fluctuating lift force coefficient and the fluctuating moment force coefficient show a similar tendency to the fluctuating drag coefficient with increasing Re.

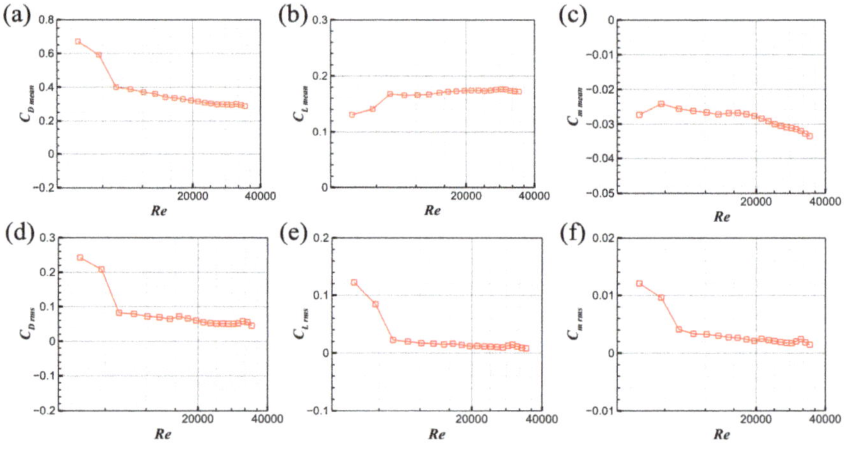

Figure 9. (**a–c**) The time-averaged drag force coefficient, time-averaged lift force coefficient, and time-averaged moment coefficient at different Reynolds numbers, respectively. (**d–f**) The fluctuating drag force coefficient, fluctuating lift force coefficient, and fluctuating moment coefficient at different Reynolds numbers, respectively.

3.2. Modulation of Surface Pressure Distribution by Leading Body-Height Grids

First, the surface pressure distributions with the aerodynamic interference of the leading body-height grid were compared with those of the bare twin-box girder, as shown in Figure 10, and the effects of turbulence generated by different body-height grids on the surface pressure distribution were evaluated and analyzed.

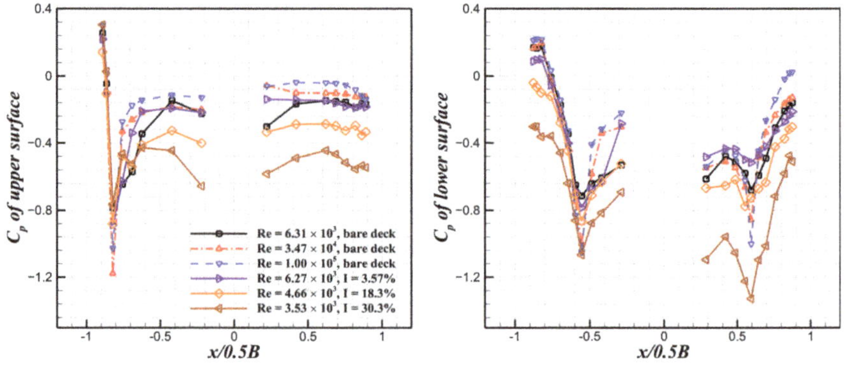

Figure 10. Comparisons of the time-averaged surface pressure distribution of the twin-box girder with body-height grid interference and bare deck.

Figure 10 clearly shows that the turbulence intensity, which is an important characteristic of incoming flow, has a significant influence on the time-averaged surface pressure distribution. Compared with the surface pressure distribution of the bare deck, the characteristic of the pressure distribution with the body-height grid aerodynamic interference changed a lot at low Reynolds numbers (Re < 1.0 × 10^4), because the boundary layer transitioned to the turbulent boundary layer. At low turbulence intensity (I = 3.57–5.15%), the amplitude of the pressure peak was enhanced. It is also noteworthy that, at the upstream girder, the pressure plateaus (appearing in the undisturbed surface pressure distribution) in the upper and lower surfaces were eliminated, because the vortices generated by the body-height grid broke the separation bubbles, and the laminar boundary layer was transformed into turbulence, which cannot maintain a stable pressure to form the separation bubbles. At moderate turbulence intensity (I = 16.9–18.3%), the negative pressures are enhanced when $x/0.5B > -0.7$, indicating that the high turbulence intensity contains more energy to strengthen the flow velocity around the box girder. When the turbulence intensity is further increased (I = 27.9–31.4%), the negative pressures are also increased.

Figure 11 shows the time-averaged surface pressure distribution at different Reynolds numbers when the turbulence intensities are close. It is clearly shown that the turbulence intensity can effectively eliminate the Reynolds number sensitivity of pressure distribution, because the properties of the turbulent boundary layer are dominated by the turbulence intensity of the incoming flows.

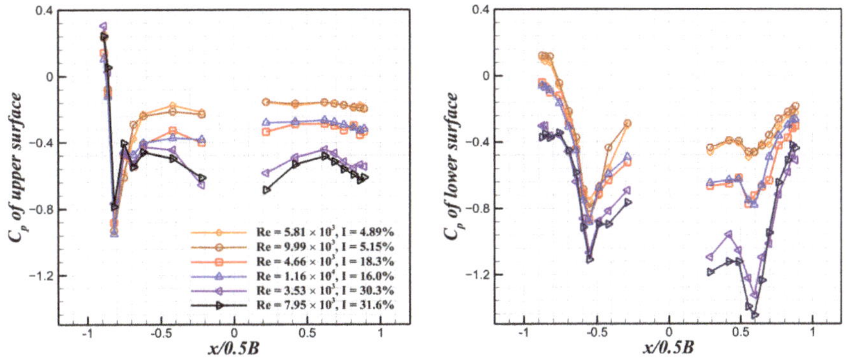

Figure 11. The time-averaged surface pressure distributions of the twin-box girder with body-height grid interference at different Reynolds numbers when turbulence intensities are close.

3.3. Modulation of Surface Pressure Distribution by Leading Circular Cylinders

Second, the time-averaged pressure distributions with cylinder interference were compared with the undisturbed pressure distribution. Figure 12 shows the mean surface pressure distributions with circular cylinder interference (D = 3 mm) at various spacing ratios. As shown in Figure 12, the turbulence generated by the wake of the circular cylinder significantly influences the time-averaged surface pressure distribution at the upstream girder, and the pressure plateaus formed by the separation bubbles are broken by the cylinder wake.

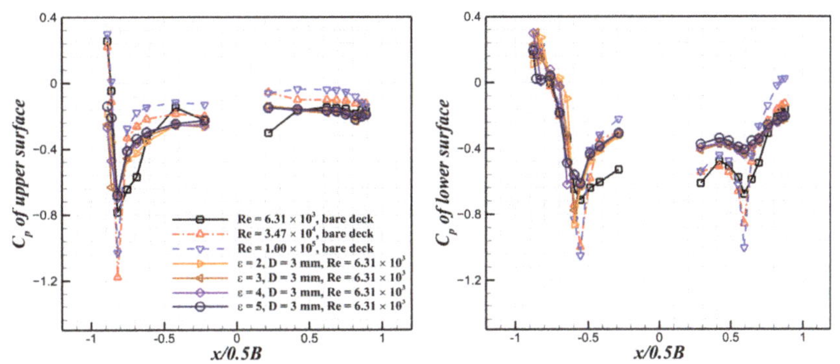

Figure 12. Comparisons of the time-averaged surface pressure distribution of the twin-box girder with circular cylinder interference and bare deck.

Moreover, Figure 12 also shows that the surface pressure on the upper and lower surfaces of the downstream girder exhibit opposite distribution characteristics. The upper surface pressure distribution shows that the negative pressure on the upper surface is lower than the undisturbed pressure at first, and then is higher than the original pressure at the same Reynolds number level. This may be because the flow velocity on the upper surface decreases after passing through the cylinder; however, continuous filling of the surrounding unaffected fluid enhances the flow velocity. The absolute value of minimum pressure on the lower surface is always less than the undisturbed pressure, which demonstrates that the cylinder located in front of the lower surface limits the velocity of the whole lower surface.

The time-averaged surface pressure distributions at different Reynolds numbers are shown in Figure 13. As shown in Figures 12 and 13, it is clear that the pressure distribution at the twin-box girder is insensitive to the spacing ratio and the diameter of the cylinder. Meanwhile, it should be noted that the surface pressure distribution with circular cylinder interference on the downstream girder presents slight Reynolds number sensitivity, indicating that the vortices generated by the flow passing through the leading circular cylinders fully enforce the turbulence on the boundary layer around the twin-box girder.

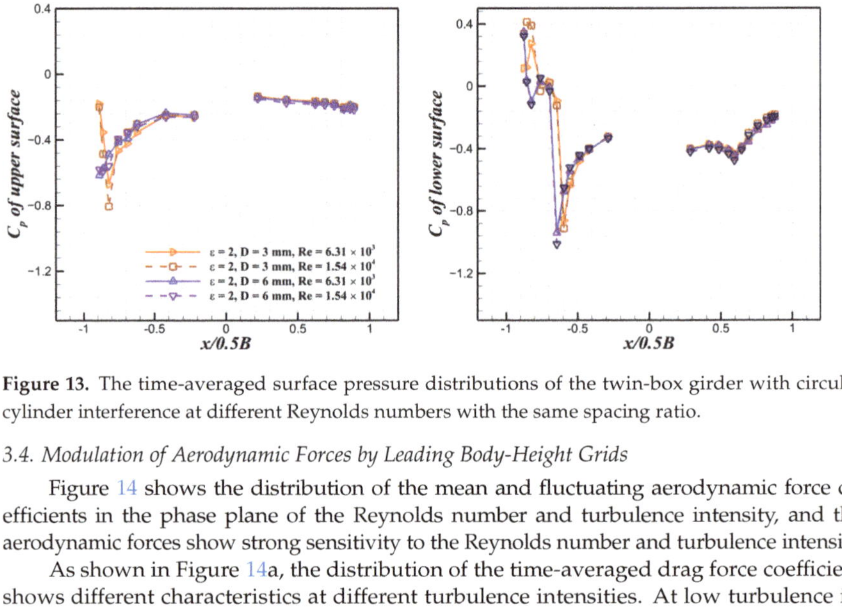

Figure 13. The time-averaged surface pressure distributions of the twin-box girder with circular cylinder interference at different Reynolds numbers with the same spacing ratio.

3.4. Modulation of Aerodynamic Forces by Leading Body-Height Grids

Figure 14 shows the distribution of the mean and fluctuating aerodynamic force coefficients in the phase plane of the Reynolds number and turbulence intensity, and the aerodynamic forces show strong sensitivity to the Reynolds number and turbulence intensity.

As shown in Figure 14a, the distribution of the time-averaged drag force coefficient shows different characteristics at different turbulence intensities. At low turbulence intensity ($I \leq 5\%$), the $C_{D\,mean}$ is sensitive to the Reynolds number when Re $\leq 1.0 \times 10^4$. Moreover, the hypotenuse of the contour at the corner indicates that the turbulence intensity can reduce the Reynolds number effects of $C_{D\,mean}$. When the Reynolds number is further increased, the $C_{D\,mean}$ shows slight Reynolds number sensitivity, which is consistent with the bare deck (see Figure 9a). At moderate turbulence intensity ($5\% \leq I \leq 20\%$), the $C_{D\,mean}$ is enhanced, and the $C_{D\,mean}$ presents slight Reynolds number dependence. It should be noted that there exists a lock-up region of $C_{D\,mean}$. At high turbulence intensity ($I > 20\%$), the $C_{D\,mean}$ is further promoted, and is only associated with the turbulence intensity.

The time-averaged lift force coefficient shows strong sensitivity to the turbulence intensity when Re $\leq 1.0 \times 10^4$ (see Figure 14b), and with the increase in turbulence intensity, the lift force is enhanced, which implies that the strength of turbulent flow can effectively promote the lift force. Meanwhile, at Re > 1e4, the lift force is insensitive to both Reynolds number and turbulence intensity, because the boundary layer transitions to full turbulence.

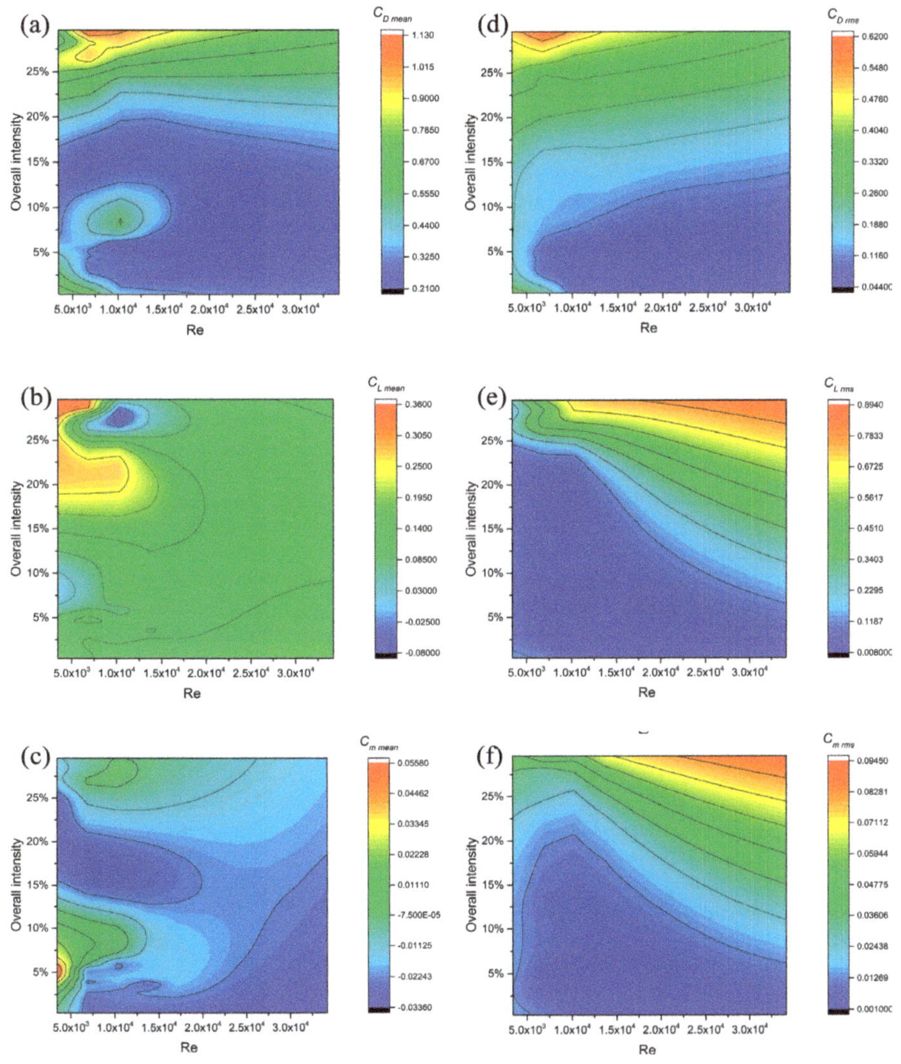

Figure 14. (**a–c**) The distributions of the time-averaged drag, lift, and moment, respectively, in the phase plane of Reynolds number and turbulence intensity. (**d–f**) The distributions of the fluctuating drag, lift, and moment, respectively, in the phase plane of Reynolds number and turbulence intensity.

As for the fluctuating forces, $C_{D\,rms}$ and $C_{L\,rms}$ show different distribution characteristics. At low turbulence intensity ($I < 2.5\%$), $C_{D\,rms}$ shows Reynolds number dependence at low Re, i.e., Re $< 1.0 \times 10^4$. With the increase in turbulence intensity, the $C_{D\,rms}$ is enhanced by the strong fluctuating component of incoming flow, and the Reynolds number sensitivity of the $C_{D\,rms}$ is eliminated (see Figure 14d). However, the $C_{L\,rms}$ depends on the Reynolds number and turbulence intensity, because the increase in Re or turbulence intensity can enhance the vertical fluctuating force.

As shown in Figure 14c,f, when Re $\leq 1.0 \times 10^4$ and turbulence intensity $I \leq 12.5\%$, the mean moment force coefficient increases significantly with the turbulence intensity, which may be due to the increase in the vertical and horizontal fluctuating velocity of the flow field caused by the turbulence. However, in other regions, the mean moment force coefficient does not show strong turbulence intensity sensitivity, because the turbulence mainly affects and increases the horizontal and vertical velocities, but has a small impact on the moment force. The fluctuating moment force coefficient and the fluctuating lift force coefficient exhibit similar trends, indicating that the fluctuating lift force is the dominant component of the fluctuating moment force.

3.5. Modulation of Aerodynamic Forces by Leading Circular Cylinders

Figures 15–20 show the mean and fluctuating aerodynamic force coefficients of the twin-box girder under the aerodynamic interference by leading circular cylinders at different spacing ratios. Figure 15 shows that the time-averaged drag force coefficient of the twin-box girder with cylinder interference decreased significantly compared with the original force coefficient without interference, because the shedding vortices in the wake of the circular cylinder destroyed the original laminar boundary layers; hence, the laminar boundary layers transition to turbulence at the windward corner, and the turbulent boundary layers can be maintained throughout the girder section. Thus, the boundary layer is composed of many small vortices, which effectively reduce the contact area between the fluid boundary layer and the twin-box girder, leading to a decrease in the friction between the turbulent boundary layer and the model. Meanwhile, when the vortices contact the girder, the velocity direction is generally opposite to the incoming flow. In this way, the vortices generate a force that is opposite to the streamwise direction, and the force generated by the interaction between the vortices and the girder reduces the total drag force of the bridge deck. Meanwhile, when the diameter of the cylinder is 3 mm, it clearly shows that the time-averaged drag force coefficient exhibits a periodic trend at different locations from the model, as shown in Figure 15, which indicates that the vortex scale is closer to the diameter. In addition, the time-averaged drag force with cylinder interference at low Re achieves a similar characteristic of undisturbed drag force at high Re.

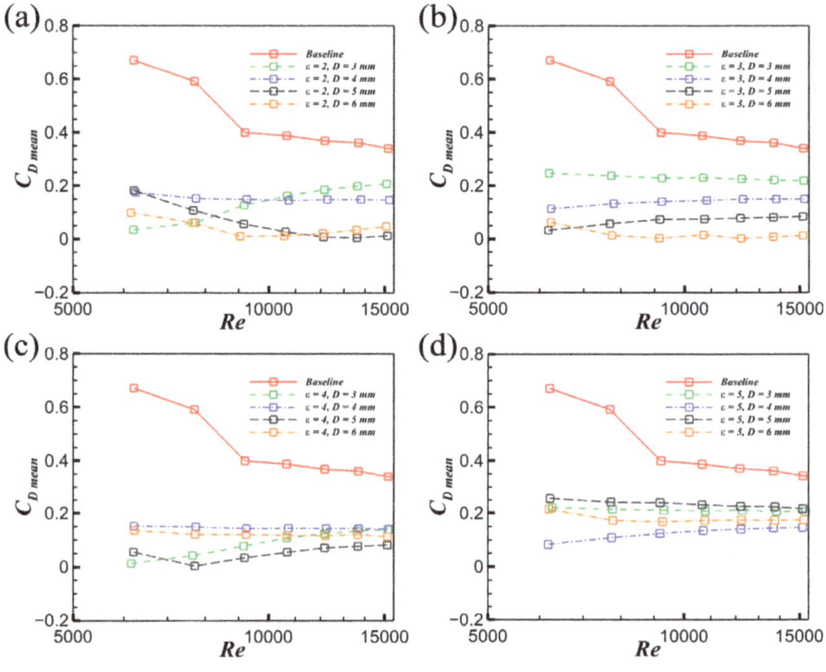

Figure 15. (**a**–**d**) The relationship between the time-averaged drag force coefficient and the Reynolds number at different spacing ratios and cylinder diameters.

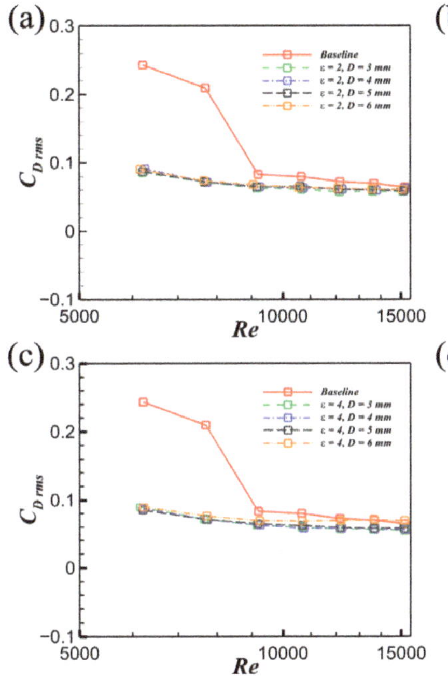

Figure 16. (**a–d**) The relationship between the fluctuating drag force coefficient and the Reynolds number at different spacing ratios and cylinder diameters.

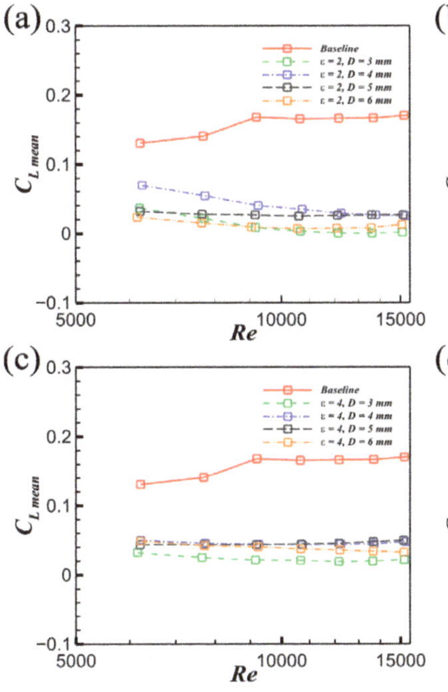

Figure 17. (**a–d**) The relationship between the time-averaged lift force coefficient and the Reynolds number at different spacing ratios and cylinder diameters.

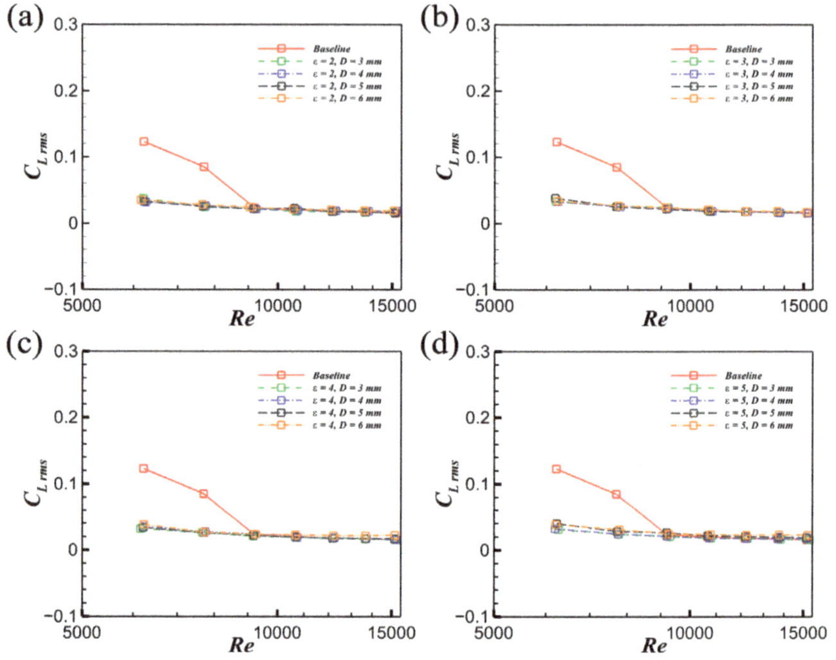

Figure 18. (**a–d**) The relationship between the fluctuating lift force coefficient and the Reynolds number at different spacing ratios and cylinder diameters.

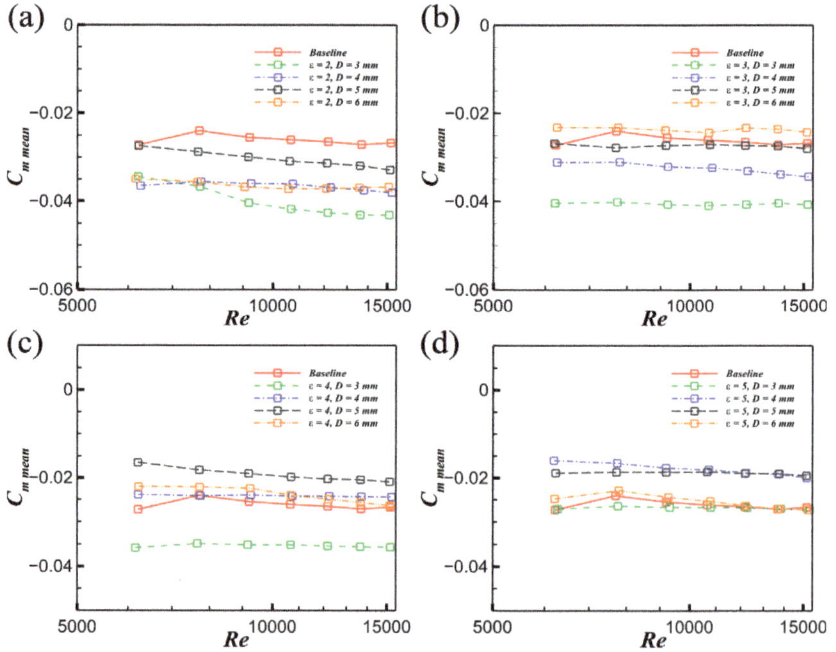

Figure 19. (**a–d**) The relationship between the time-averaged moment force coefficient and the Reynolds number at different spacing ratios and cylinder diameters.

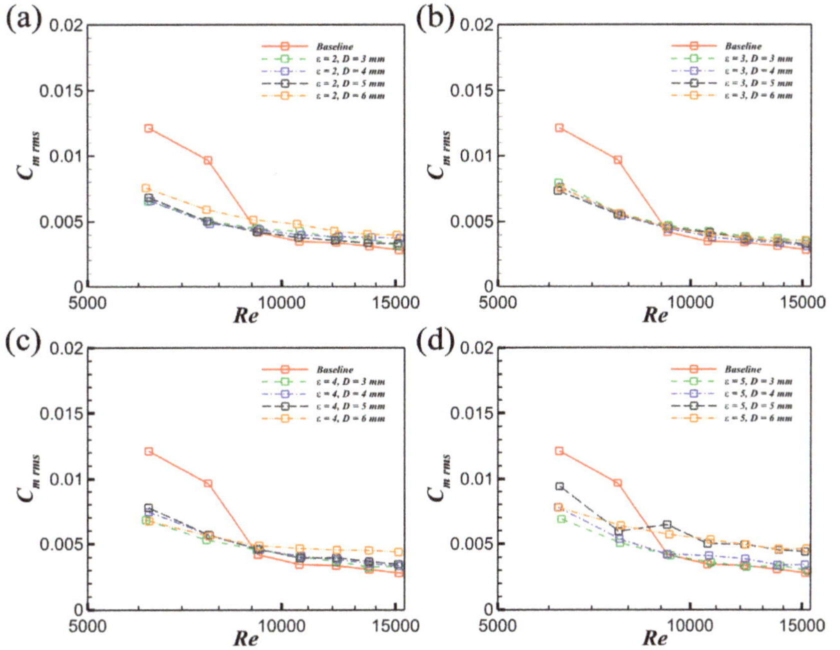

Figure 20. (**a–d**) The relationship between the fluctuating moment force coefficient and the Reynolds number at different spacing ratios and cylinder diameters.

Figure 16 shows that the fluctuating drag force coefficient of the twin-box girder with cylinder interference is much smaller than that of the twin-box girder without interference at Re $\leq 9.0 \times 10^3$. Moreover, the fluctuating drag force is insensitive to the Reynolds number, and the disturbed fluctuating drag force at low Re is closer to that of the undisturbed fluctuating force at high Re. Furthermore, the fluctuating drag force is independent of the spacing ratio ε.

Figure 17 shows that the time-averaged lift force coefficient of the girder with cylinder interference also decreased sharply compared with that of the bare twin-box girder. It should be noted that with the increase in distance between the cylinder and the twin-box girder, the $C_{L\,mean}$ shows less sensitivity to the Reynolds number, because the turbulence in the wake of cylinder has been fully developed, which causes the boundary layer of the body surface to transition to turbulence. When the cylinder is close to the twin-box girder, the wake of the cylinder cannot fully develop into turbulence to change the boundary layer, which explains why the Reynolds number effect still exists. In conclusion, when the spacing ratio $\varepsilon > 2$, the turbulence generated by the wake of the cylinder effectively eliminates the Reynolds number effects on the time-averaged lift force.

Figure 18 shows that the fluctuating lift force coefficient of the twin-box girder with cylinder interference is independent of the Reynolds number and spacing ratio ε. Moreover, the fluctuating lift force is much smaller than that of the undisturbed force at Re $\leq 9.0 \times 10^3$. This may be because the vertical pulsation component in the cylindrical wake is very small and the turbulence generated by the cylinder suppresses the vortex shedding of the twin-box girder to reduce the fluctuation of the lift force.

Compared with the undisturbed time-averaged moment force, Figure 19 shows that the mean moment force coefficient with cylinder interference is large, and is dependent on spacing ratio and diameter. With the decrease in diameter, the time-averaged moment force coefficient $C_{m\,mean}$ is slightly enhanced. Meanwhile, with the increase in the spacing ratio, the absolute value of $C_{m\,mean}$ decreases gradually. When the spacing ratio $\varepsilon = 5$, the

$C_{m\ mean}$ is smaller than the undisturbed mean moment force. This indicates that the fully developed turbulence wake generated by a large spacing ratio can effectively suppress the time-averaged moment force.

Figure 20 shows the fluctuating moment force coefficient $C_{m\ rms}$ of the twin-box girder with cylinder interference. At Re $\leq 9.0 \times 10^3$, the interfered $C_{m\ rms}$ is greater than the undisturbed $C_{m\ rms}$. When the spacing ratio $\varepsilon \leq 3$, the $C_{m\ rms}$ is insensitive to the diameter, and the $C_{m\ rms}$ decreases slightly with the increase in the Reynolds number.

4. Conclusions

In the present work, the effects of two cutting-edge aerodynamic interference measures on the pressure distribution and aerodynamic force of a twin-box girder were investigated. We used the leading body-height grid and leading circular cylinder to increase the turbulence intensity of the incoming flow. The flows gained more energy before reaching the separation point, and the entrainment rate of the flow was enhanced, which not only affects the separation bubbles and reattachment points, but also changes the characteristics of the boundary layer. The conclusions are summarized as follows:

(1) The leading body-height grid generates the turbulent incoming flow, which effectively breaks the separation bubbles and the flow reattachment, and the laminar boundary layer in the undisturbed case at low Re is forced to transition to turbulent flow. Moreover, the characteristics of surface pressure distribution with body-height grid interference are similar to those of bare deck at high Re;

(2) The Reynolds number sensitivity of time-averaged drag force decreases with the increase in turbulence intensity, and the $C_{D\ mean}$ is dominated by the turbulence intensity. While the $C_{L\ mean}$ and $C_{m\ mean}$ are dependent on the Re and turbulence intensity at low Re, at high Re, the $C_{L\ mean}$ and $C_{m\ mean}$ are insensitive to the Re and turbulence intensity. The fluctuating drag force $C_{D\ rms}$ depends on the turbulence intensity, and is insensitive to the Reynolds number, while the $C_{L\ rms}$ and $C_{m\ rms}$ are related to both turbulence intensity and the Reynolds number. In addition, the characteristics of $C_{L\ rms}$ and $C_{m\ rms}$ are similar, indicating that the $C_{L\ rms}$ is the dominant component of the $C_{m\ rms}$;

(3) The coherent turbulence generated by the leading circular cylinders effectively changes the boundary layer of the twin-box girder. The Reynolds number sensitivity of surface pressure distribution is reduced by the interference of cylinders, and it is insensitive to the diameter and the spacing ratio. Moreover, the separation bubbles are also broken by the wake of the cylinder;

(4) The time-averaged drag force $C_{D\ mean}$ is significantly reduced by the interference of the leading cylinder, and its Reynolds number sensitivity is diminished. Moreover, the time-averaged lift force $C_{L\ mean}$ with cylinder interference is drastically decreased, and it is also insensitive to the Reynolds number. With the increase in the spacing ratio, the time-averaged moment force $C_{m\ mean}$ is weakened, and its Reynolds number sensitivity is reduced. In addition, the fluctuating drag force $C_{D\ rms}$ and lift force $C_{L\ rms}$ are both insensitive to the Re, the spacing ratio, and the diameter, while the fluctuating moment force is closely related to these three parameters.

Author Contributions: Conceptualization, B.W. and S.L.; methodology, B.W. and S.L.; experiments, B.W., J.F. and S.L.; software, B.W.; validation, B.W.; formal analysis, B.W. and S.L.; investigation, B.W., G.X. and S.L.; resources, S.L.; data curation, B.W. and S.L.; writing—original draft preparation, B.W.; writing—review and editing, B.W. and S.L.; supervision, S.L.; project administration, S.L.; funding acquisition, S.L. All authors have read and agreed to the published version of the manuscript.

Funding: This study was financially supported by the NSFC under Grant No. 51878230,the Natural Science Foundation of Heilongjiang Province under Grant No. YQ2021E033, and the Postdoctoral Scientific Research Development Fund of Heilongjiang Province under Grant No. LBH-Q20021, along with support by the Heilongjiang Touyan Team and Fundamental Research Funds for the Central Universities.

Institutional Review Board Statement: Not applicable.

Informed Consent Statement: Not applicable.

Data Availability Statement: Data are contained within this article.

Conflicts of Interest: The authors declare no conflict of interest.

References

1. Xu, F.; Ying, X.; Li, Y.; Zhang, M. Experimental explorations of the torsional vortex-induced vibrations of a bridge deck. *J. Bridge Eng.* **2016**, *21*, 04016093. [CrossRef]
2. Chen, W.L.; Li, H.; Hu, H. An experimental study on the unsteady vortices and turbulent flow structures around twin-box-girder bridge deck models with different gap ratios. *J. Wind. Eng. Ind. Aerodyn.* **2014**, *132*, 27–36. [CrossRef]
3. Chen, G.B.; Chen, W.L.; Gao, D.L.; Yang, Z.F. Active control of flow structure and unsteady aerodynamic force of box girder with leading-edge suction and trailing-edge jet. *Exp. Therm. Fluid Sci.* **2021**, *120*, 110244. [CrossRef]
4. Zhang, M.; Xu, F.; Ying, X. Experimental investigations on the nonlinear torsional flutter of a bridge deck. *J. Bridge Eng.* **2017**, *22*, 04017048. [CrossRef]
5. Ying, X.; Xu, F.; Zhang, M.; Zhang, Z. Numerical explorations of the limit cycle flutter characteristics of a bridge deck. *J. Wind. Eng. Ind. Aerodyn.* **2017**, *169*, 30–38. [CrossRef]
6. Chen, W.L.; Li, H.; Ou, J.P.; Li, F.C. Numerical simulation of vortex-induced vibrations of inclined cables under different wind profiles. *J. Bridge Eng.* **2013**, *18*, 42–53. [CrossRef]
7. Liu, J.; Hui, Y.; Wang, J.; Yang, Q. LES study of windward-face-mounted-ribs' effects on flow fields and aerodynamic forces on a square cylinder. *Build. Environ.* **2021**, *200*, 107950. [CrossRef]
8. Xu, L.; Hui, Y.; Yang, Q.; Chen, Z.; Law, S.S. Modeling and modal analysis of suspension bridge based on continual formula method. *Mech. Syst. Signal Process.* **2022**, *162*, 107855. [CrossRef]
9. Zhou, R.; Ge, Y.; Liu, S.; Yang, Y.; Du, Y.; Zhang, L. Nonlinear flutter control of a long-span closed-box girder bridge with vertical stabilizers subjected to various turbulence flows. *Thin-Walled Struct.* **2020**, *149*, 106245. [CrossRef]
10. Hunt, J.C.R.; Kawai, H.; Ramsey, S.R.; Pedrizetti, G.; Perkins, R.J. A review of velocity and pressure fluctuations in turbulent flows around bluff bodies. *J. Wind. Eng. Ind. Aerodyn.* **1990**, *35*, 49–85. [CrossRef]
11. Sarwar, M.W.; Ishihara, T.; Shimada, K.; Yamasaki, Y.; Ikeda, T. Prediction of aerodynamic characteristics of a box girder bridge section using the LES turbulence model. *J. Wind. Eng. Ind. Aerodyn.* **2008**, *96*, 1895–1911. [CrossRef]
12. Scanlan, R.H.; Lin, W.H. Effects of turbulence on bridge flutter derivatives. *J. Eng. Mech. Div.* **1978**, *104*, 719–733. [CrossRef]
13. Haan, F.L., Jr.; Kareem, A. Anatomy of turbulence effects on the aerodynamics of an oscillating prism. *J. Eng. Mech.* **2009**, *135*, 987–999. [CrossRef]
14. Matsumoto, M.; Shiraishi, N.; Shirato, H. Turbulence unstabilization on bridge aerodynamics. In Proceedings of the International Conference Innovation in Cable-Stayed Bridges, Fukuoka, Japan, 18–19 April 1991; pp. 175–183.
15. Chen, X.; Kareem, A. Nonlinear response analysis of long-span bridges under turbulent winds. *J. Wind. Eng. Ind. Aerodyn.* **2001**, *89*, 1335–1350. [CrossRef]
16. Chen, X.; Kareem, A. Aeroelastic analysis of bridges: Effects of turbulence and aerodynamic nonlinearities. *J. Eng. Mech.* **2003**, *129*, 885–895. [CrossRef]
17. Kareem, A.; Wu, T. Wind-induced effects on bluff bodies in turbulent flows: Nonstationary, non-Gaussian and nonlinear features. *J. Wind. Eng. Ind. Aerodyn.* **2013**, *122*, 21–37. [CrossRef]
18. Laima, S.; Wu, B.; Jiang, C.; Chen, W.; Li, H. Numerical study on Reynolds number effects on the aerodynamic characteristics of a twin-box girder. *Wind. Struct.* **2019**, *28*, 285–298.
19. Irwin, H.P.A.H.; Cooper, K.R.; Girard, R. Correction of distortion effects caused by tubing systems in measurements of fluctuating pressures. *J. Wind. Eng. Ind. Aerodyn.* **1979**, *5*, 93–107. [CrossRef]
20. Holmes, J.D.; Lewis, R.E. The dynamic response of pressure-measurement systems. In Proceedings of the 9th AFMC, Australasian Fluid Mechanics Conference, Auckland, New Zealand, 8–12 December 1986; pp. 8–12.
21. Landau, L.D.; Lifshitz, E.M. *Course of Theoretical Physics*; Fluid Mechanics; Pergamon Press: London, UK, 1959; Volume 6.

applied sciences

MDPI

Article

An Experimental Investigation of Passive Jet Control Method on Bridge Tower Wake

Yewei Huang [1,2] and Wenli Chen [1,2,*]

1 Key Lab of Smart Prevention and Mitigation of Civil Engineering Disasters of the Ministry of Industry and Information Technology, Harbin Institute of Technology, Harbin 150090, China; 17b933027@stu.hit.edu.cn
2 Key Lab of Structures Dynamic Behavior and Control of the Ministry of Education, Harbin Institute of Technology, Harbin 150090, China
* Correspondence: cwl_80@hit.edu.cn; Tel.: +86-451-8628-2068

Abstract: In this study, we employed a four-hole cobra probe to measure the wake characteristics of a rec-tangular bridge tower model in a wind tunnel. The scale of the model was 1:30, and the Reynolds number varied from 1.38×10^5 to 2.27×10^5 by changing the yaw angle. A measurement plane with 9×19 measurement points was horizontally set at the middle height behind the model. The wake characteristics of the test model without control, i.e., the baseline case, was first tested in the yaw angle range from 0° to 90°; then, four kinds of passive jet control cases were tested to study their control effects on the bridge tower wake. To evaluate the wake characteristics, three main aspects, i.e., mean velocity, turbulence intensity, and velocity frequency, were investigated. The meas-urement results indicate that the passive jet control method can achieve an effect in suppressing the turbulence of the wake but can slightly modify the mean velocity distribution. The dominant frequency distribution region was eliminated when the yaw angle was small but slightly expanded at a large angle. The differences between cases show a trend that the larger the suction coefficient is, the better the control effects are.

Keywords: passive jet control; tower wake characteristics; cobra probe

Citation: Huang, Y.; Chen, W. An Experimental Investigation of Passive Jet Control Method on Bridge Tower Wake. *Appl. Sci.* **2022**, *12*, 4691. https://doi.org/10.3390/app12094691

Academic Editor: Giovanni Bernardini

Received: 2 March 2022
Accepted: 1 May 2022
Published: 6 May 2022

1. Introduction

The cable system bridge is the most widely used large-span bridge system. With the increasing requirement of the span, a growing number of extra-large cable system bridges have come into service, and the wind resistance performance is the primary limitation to the span of the bridge. As the bridge span grows, the wind effect appears more complex, and unprecedented wind-induced phenomena could occur. The wake flow characteristics of bridge towers is an interesting research branch that could lead to the oscillation of the cable behind the tower. For cable-stayed bridges, some parts are under the influence of the tower wake, particularly the long stay cable. In suspender bridges, the suspenders near the tower are wholly in the influence region of the tower wake. The oscillation motion of these suspenders is much more complex due to the unstable wake flow.

The characteristics of cylinder wake have been sufficiently investigated. Knisely [1] reviewed and further researched the Strouhal numbers of the wake of rectangular cylinders; Norberg [2] also investigated the wake frequencies by using a hot wire. Shimada and Ishihara [3] used a two-layer k-ε model to numerically study the aerodynamic characteristics of an infinite-length rectangular cylinder and gained a great agreement in the distribution of mean pressure, but the fluctuating pressure distribution was underestimated compared with the real model. Assi et al. [4] investigated the wake-induced vibration (WIV) of tandem circular cylinders and considered the vibration as a consequence of the interaction between the wake and the structure, which differs from vortex-induced vibration (VIV). The experiment confirmed the wake displacement theory proposed by Zdravkovich [5]. Carmo et al. [6] performed a numerical study on the phenomenon of WIV to a cylinder at a low Reynolds number (*Re*). Bearman [7] reviewed the research of circular cylinder wakes, regarding wake-induced vibration as a form of VIV. The downstream cylinder is under the collective effect of vortex shedding from both upstream structures and itself. Wang et al. [8] performed a two-dimensional numerical simulation of two tandem cylinders at low *Re* flow and found that the WIV phenomenon comes to aggravation when the size of the cylinder upstream is larger than that downstream. Li et al. [9] conducted a wind tunnel test and a computational fluid dynamics (CFD) simulation to investigate the WIV of the suspenders in wake of a bridge tower. The experiment found that the suspender in the wake of the tower would carry out an elliptical orbit motion at a specific wind speed when the frequency of the wake fits the natural frequency of the suspenders. In another part, a tower with four corners cut decreases the RMS aerodynamic coefficient and facilitates the suppression of the WIV of suspenders.

On the other hand, many studies have been conducted to investigate the suppression of VIV of cylinders. A technology based on suction and jet flow was confirmed to be effective. Amitay et al. [10,11] modified the aerodynamic characteristics of cylinders by using fluidic actuators. The closed recirculating flow regimes generated by jet flow could be equivalent to a virtual surface which can modify the characteristics of aerodynamics. Crook et al. [12] used synthetic jets to delay the separation point on cylinders, with the jet placed near the separation line. Many relevant studies were performed in the 21st century [13–15] that also indicated that the turbulence of wake also can be decreased, and even the vortex street could be eliminated. Hao Feng et al. [16,17] investigated the mode of wake by proper orthogonal decomposition, as the synthetic jet is applied to control the vortex shedding. The particle image velocimetry (PIV) results show that a synthetic jet could generate a vortex pair further downstream, and the proper orthogonal decomposition (POD) analysis indicates that this vortex pair, together with the vorticity shear layers near the cylinder, forms the vortex structures further downstream. Chen et al. [18] performed an experimental investigation on the suction-based flow control method, investigated the pressure distribution on the surface of a circular cylinder with suction slits, and found that the steady suction method exhibits excellent control effectiveness and can distinctly suppress the VIV, the amplitudes of cylinder vibrations, fluctuating pressure coefficients and lift coefficients of the circular cylinder model. They also applied PIV for further study [19], and the results show that the separation points on the test model in the PIV measurement plane are pushed further downstream, and the wake closure length also varies significantly along the span-wise direction of the test model for the cases with the suction flow control. Chen et al. [20,21] first proposed a passive jet apparatus, performed an experimental and numerical study to investigate a passive control method that has passive windward suction and a leeward jet over a circular cylinder, and found that the mean drag and the dynamic wind loads acting on the cylinder are suppressed. This apparatus could manipulate the shedding process of the wake vortices, and the jets could modify the wake stability by displacing the region of absolute instability further downstream. This method was also investigated by Zhang et al. [22] to apply to a bridge girder to control the flow separation. Chen et al. [23] further developed the passive jet method on a rectangular cylinder, taking a tower column as a prototype. The passive jet also has adequate control effects on the aerodynamic forces acting on the column.

The WIV of suspenders was found to be caused by the vortices from the bridge tower shedding into the tower wake [9,24]. Therefore, in this paper, we aim to further apply this passive jet apparatus to the bridge tower to modify the characteristics of the wake. We conducted an experiment to measure the time histories of wind speed at each measurement point. The experiment setup is introduced in Section 2, the results of tower wake characteristics are shown in Section 3, and finally, a discussion and conclusions are presented in Section 4.

2. Experimental Setup

The experiment was conducted in a closed-circuit boundary layer wind tunnel affiliated with the Joint Laboratory of Wind Tunnel and Wave Flume (Joint Lab WTWF), Harbin Institute of Technology. The wind tunnel contains two test sections. The size (width × height × length) of the larger one is 6.0 m × 3.6 m × 50 m, and the smaller one is 4.0 m × 3.0 m × 25 m. Our experiment was performed in the small section, where the wind speed can be adjusted continuously from 3 m/s to 30 m/s, the turbulence intensity of incoming wind speed is no more than 0.46%. The small section equips two turntable systems and a three-dimensional automated traverse measurement system.

2.1. Bridge Tower Model

The bridge tower model was designed using the Xihoumen Bridge as prototype, which is a steel box girder suspension bridge with a main span of 1650 m, located in Zhejiang Province, China. The prototype bridge tower is an overhead gantry type, which can be simplified as two separate independent rectangle cross-section bridge towers when only the wake flow of tower is investigated. We investigate only one tower model, without considering the interaction of the wake of two towers. The cross-section of the model is a 283 mm (B) × 217 mm (D) rectangle, with a side length of 23 mm square corner cut at each corner, and the scale ratio is set to 1:30, as shown in Figure 1. The data presented in this paper correspond to dimensionless distances of B = 283 mm and D = 217 mm in cross-wind and along-wind directions. The height of the column model is 2.9 m, and thus, the remaining 0.1 m to the ceiling can spare space for the clamp device. The model is assembled with thick planks, which provide enough stiffness to limit deformation. The surface of the model is flat and glossy by polishing. A jack system, mounted between the tower model and ceiling, firmly immobilizes the model. The tower model always maintains rigidity and keeps static during the experiment. At the bottom of the tower model, a 20 mm thick wooden turn plate is arranged to adjust the yaw angle to the incoming flow when the jack system is inoperative.

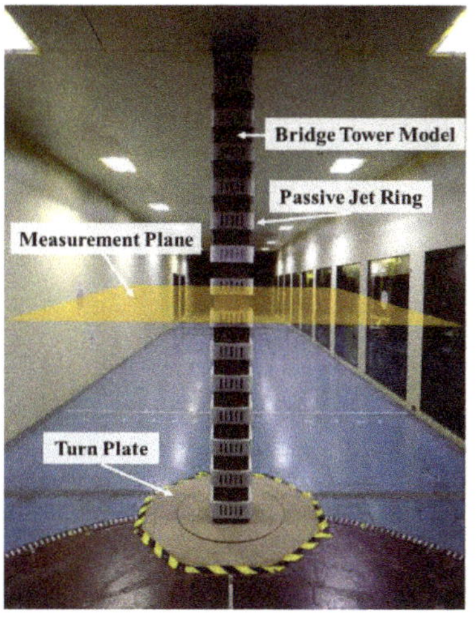

Figure 1. Schematic diagram of test layout.

The yaw angle of model α is set from 0 degrees to 90 degrees, with an interval of 10°. The blocking ratio of the model varies by the yaw angle, reaching a minimum of 5.4% when the yaw angle is 0° and a maximum of 8.9% when the yaw angle is 50°.

2.2. Passive Jet Rings

The present study adopts four kinds of passive jet rings to take control of the tower wake flow, which refers to the design in Chen et al. [23]. The ring is an annular flow channel with slits made by a 1 mm thick plastic plate, as Figure 2 shows. The top view of the ring is a rounded rectangular box, of which the outer size is 323 mm × 257 mm, with a 43 mm radius rounded corner in each corner, and the inner size is 283 mm × 217 mm; thus, it can closely fit the tower model. The intermediate part of 18 mm thickness, considering the thickness of the plate, constitutes the hollow flow channel. The height of the passive jet ring is 100 mm. Several rectangular slits are distributed around the side walls of the ring. The major differences between these four categories of jet ring are the size and the distribution of these slits. The size of slits in the first and third categories is 80 mm × 30 mm, and that in the second and fourth categories is 60 mm × 30 mm. For the first category, there are nine and seven slits on each long and short side, respectively, and for the second category, there are seven and five slits. For the third and fourth categories, the slits of the first and second categories on the long side are sealed, seven and five slits are distributed on each short side, respectively, and no slits are arranged on the long side. The clear spacing between two slits on each side is 30 mm. All the slits adopt axisymmetric distribution on both two principal axes.

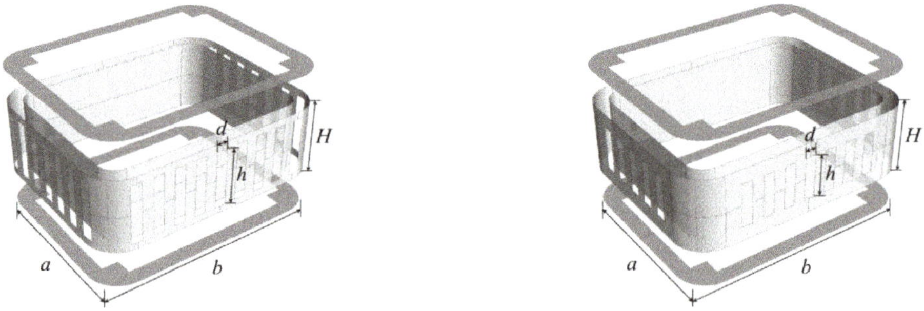

(**a**) Category 1 (**b**) Category 2

Figure 2. Passive jet ring structure schematic diagram.

In this study, we calculate the dimensionless coefficient of suction (Chen et al., 2015):

$$C_{\text{suc}} = \frac{S_{\text{suc}}}{Hl} = \frac{hd(n_{\text{a}} \sin\alpha + n_{\text{b}} \cos\alpha)}{H(a \sin\alpha + b \cos\alpha)}$$

where S_{suc} is the summary area of the suction slits on one passive jet ring projection to the plane perpendicular to the direction of flow; $l = a\sin\alpha + b\cos\alpha$ is the total projection length of the ring perpendicular to the incoming flow; H is the height of the passive jet ring; h is the length of the suction slit; d is the width of the suction slit; a and b are the length of the long and short side of the ring, respectively; n_{a} and n_{b} are the number of the slits distributed on each long and short side, respectively, and α is the yaw angle of the ring.

2.3. Experiment Details and Measurement Point Arrangement

In this experiment, a four-hole pressure probe (Series 100 Cobra Probe, Turbulent Flow Instrumentation Pty Ltd., Victoria, Australia) was used to measure the characteristics of the bridge tower wake flow. There are 172 measurement points in the zone of tower wake, distributed on a 2.4 m × 1.8 m horizontal plane 1.5 m above the ground (half of the wind tunnel section height), as shown in Figure 1. In this plane, nine rows perpendicular to the incoming flow distribute at the interval of 300 mm, and nineteen columns parallel to the incoming flow distribute at the interval of 100 mm. The middle measurement point in the first row is 200 mm away from the back of the tower model. Taking the same scale ratio as the tower model, the realistic locations of four suspenders behind the bridge tower correspond to the middle points in rows 3, 5, 7, and 9. The measurement points distribution can be seen in Figure 3. At each measurement point, the sampling frequency is 625 Hz, and the sampling time is 20 s.

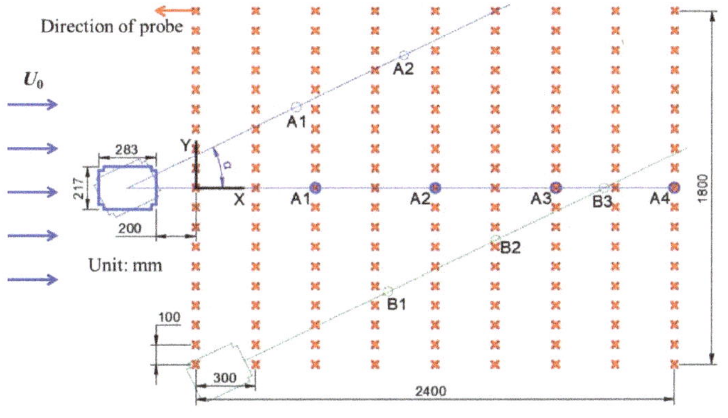

Figure 3. Measurement points arrangement sketch map.

The cobra probe is fixed on the bracket on the three-dimensional traverse measurement system, with its probe head straight towards the incoming flow. The traverse measurement system has a built-in coordinate system, by which we can precisely change the position of the probe. It also has a jack system to firmly fix the whole traverse measurement system when data are collected. The position of the probe is changed to the measurement points mentioned above one by one during the experiment. The incoming wind speed U_0 is set to 10 m/s, corresponding to the range of Reynolds numbers ($Re = \rho U_0 L/\mu$, where ρ is the air density, μ is the dynamic viscosity coefficient of air, L is the projection length of the tower cross-section) from 1.38×10^5 to 2.27×10^5, which varies by the projection length ($L = B\sin\alpha + D\cos\alpha$) when the yaw angle α changes. As the yaw angle changes, the corresponding positions of the suspenders vary, revolving the same degrees around the bridge tower. The locations of the first four imaginary suspenders behind the tower are labeled A1 to A4. When the yaw angle continues to increase, the imaginary suspenders behind the other side tower come into the wake zone of the tower model, the locations of which are labeled B1 to B4. There are five test conditions in this study, recorded as Case 0–Case 4. Case 0 is the baseline condition, in which only a tower model is tested, without passive jet control. Case 1–Case 4, corresponding to the category of the passive jet ring from 1 to 4, respectively, are controlled conditions, in which fourteen passive jet rings are installed on the tower model, distributed at the interval of 100 mm. The ring in the middle is installed 1.5 m from the ground, the same height as the probe. The yaw angle changes from 0 to 90 in Case 0 to Case 2, and from 0 to 30 in Case 3 to Case 4, because few suction slits work at a large angle in Case 3 and 4. The test conditions are shown in Table 1, and the value of suction coefficients at different yaw angles in each case are listed in Table 2.

Table 1. Test conditions of the experiment.

Designation	Type of Passive Jet Ring	Slits Number at Long Side	Slits Number at Short Side	Length of Slits
Case 0	without control	-	-	-
Case 1	Category 1	9	7	80 mm
Case 2	Category 2	7	5	60 mm
Case 3	Category 3	0	7	80 mm
Case 4	Category 4	0	5	60 mm

Table 2. Suction coefficients for each case in the experiment.

Yaw Angle	0°	10°	20°	30°	40°	50°	60°	70°	80°	90°
Case 1	0.654	0.656	0.658	0.660	0.661	0.663	0.664	0.665	0.667	0.669
Case 2	0.350	0.357	0.363	0.367	0.371	0.374	0.378	0.381	0.385	0.390
Case 3	0.654	0.535	0.449	0.379	-	-	-	-	-	-
Case 4	0.350	0.287	0.240	0.203	-	-	-	-	-	-

3. Results and Discussions

3.1. Mean Velocity Distribution

The cobra probe has a natural property called the acceptance ratio. The wind speed data are collected only when the local wind speed is in the range from 2 to 50 m/s; moreover, the direction of the speed vector is contained by the front cone with a 45-degree half-angle [25]. Data history would be placed as zero when the acceptance condition is not satisfied. Zero data history is rejected when the mean velocity distribution is calculated. Only u and v velocity components are taken into consideration due to the two-dimensional model, and the velocity is expressed in non-dimensional form as the local wind speed divides the incoming wind speed. The fourth-order spline interpolation is adopted for each measurement point.

Figure 4 shows the mean dimensionless speed (U/U_0) distribution in each test condition, where the black arrows refer to the local wind velocity vector. The general distribution trend of wind velocity is almost the same, and it can be divided into three regions. A low-value zone is located near the position (0, 0), which can be regarded as the low-speed region. This region is created due to the bridge tower shielding effects and the recirculation zone. Due to the limitation of the cobra probe, the mean velocity is always downstream, which is not tally with the fact; thus, the low-speed region is defined as the location where the data acceptance ratio (the proportion of non-zero data) of the probe is less than 50%. A slow diffusion region is located downstream of the low-speed region, where the speed is approximately half of the incoming wind speed. The high-speed region is distributed on two sides of the Y-axis, where the mean speed is almost not affected by the bridge tower model.

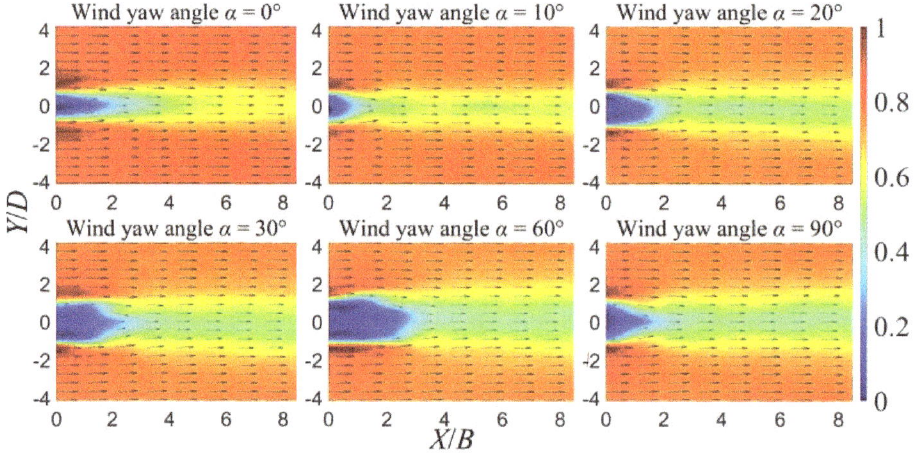

(a). Case 0

Figure 4. *Cont.*

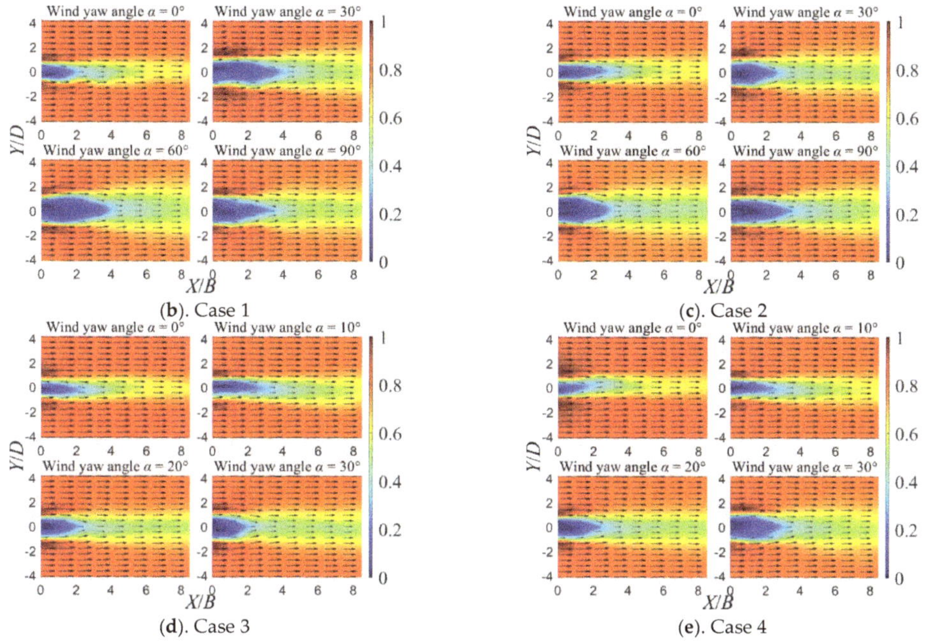

Figure 4. Dimensionless mean speed (U/U_0) distribution of tower wake.

The non-dimensional area (S_1/BD) of the low-speed region is shown in Figure 5, where S_1 is the area where the data acceptance ratio is less than 50% in the tower wake. In Case 0, there is a notable minimum area at 10 degrees equal to 0.386, and the second minimum area is equal to 0.503 at 80 degrees. This indicates that the low-speed area of the rectangular column with corner cuts minimizes not at the position perpendicular to the incoming flow, but at the position of little degree rotation. In Case 1 and Case 2, the low-speed region extends downstream compared with Case 0, every 0–90 degrees; in Case 3 and Case 4, an obvious expanding low-speed region can be observed at 10 and 20 degrees. These phenomena can be explicated by the previous study (Chen et al., 2015), where the PIV results show that the wake vortex is pushed by jet flow and stretched, which would induce the expansion of the low-speed region.

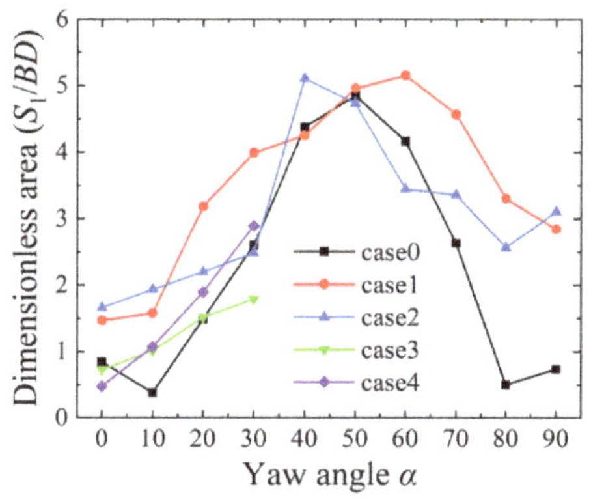

Figure 5. Dimensionless area of low-speed region.

To show the distribution of the velocity on the Y-axis, Figure 6 adopts the velocity of measurement points in the final row (X/B = 8.48) in different cases. The trend of velocity is consistent in all cases. The distribution on the Y-axis shows the shape of a saddle, and the velocity increases first and then decreases as the yaw angle increases. It can be indicated that the passive jet ring significantly modifies the velocity in the near wake but slightly changes the velocity distribution further downstream.

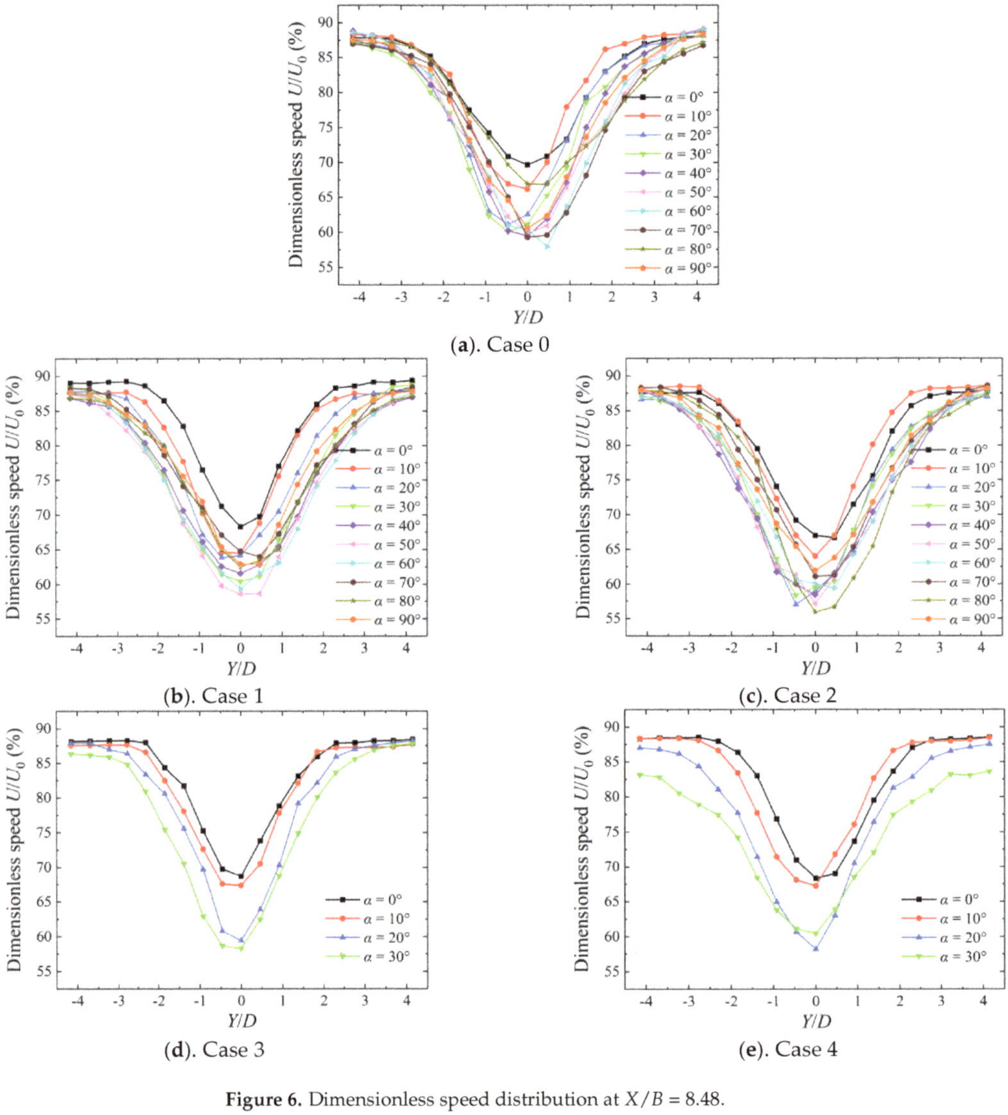

Figure 6. Dimensionless speed distribution at X/B = 8.48.

3.2. Turbulence Intensity Distribution

The turbulence intensity distribution in the wake region is achieved. To the measurement points with zero data, the zeros are rejected in the calculation, the same as velocity. The processing method may lead to some deviation from the true value when the acceptance ratio is low.

The fourth-order spline interpolation is also used to gain the final distribution. Figure 7 shows the distribution of the turbulence intensity distribution. All pictures reveal a universal distribution mode in the tower wake. A core area of the turbulence is located at the back of the tower model, which possesses a high turbulence intensity attaining about 60%. Then, a diffusion zone of the turbulence appears downstream, where the turbulence generally decreases, and the width of the zone increases. Two main differences between uncontrolled and controlled cases must be noticed. First, the core zone of turbulence is enlarged in some controlled cases. The explanation for this phenomenon is that the low-speed region is enlarged in controlled cases as shown in the velocity part, which directly leads to this consequence. Second, there is a low turbulence region inside the core zone in some controlled cases, which may be due to the jet flow from the passive jet ring, as the jet flow would suppress the shedding vortex at the back of the tower model.

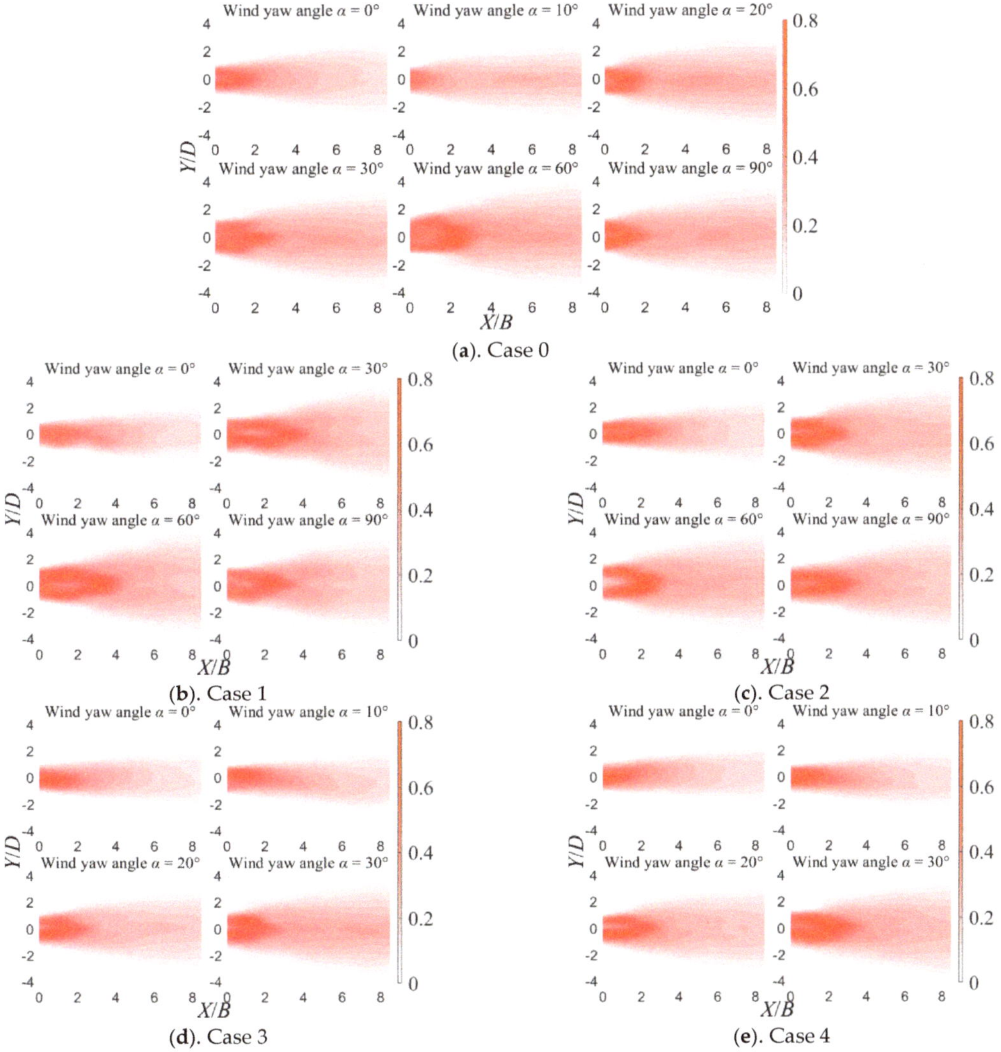

Figure 7. Turbulence intensity distribution of tower wake.

To quantitatively compare the turbulence intensity, Figure 8 shows the turbulence intensity distribution in the row at X/B = 8.48, where the turbulence is totally diffused, and the acceptance ratio is 100%; thus, the deviation can be averted. At 0 degrees, Case 3 shows the best control effect, the peak value reduces by 19.3%, and the width of the wake decreases evidently; other cases show more or fewer control effects. At 10 and 20 degrees, all cases control the turbulence effectively. Case 3 and Case 4 attain the best consequence, which reduces 32.5% and 30.2% at 10 degrees and 17.9% and 20.7% at 20 degrees, respectively. When the yaw angle comes to 30 degrees, Case 3 has no control effect but amplifies the peak value for 7.4%, while Case 1 gains the best effect for a 32.5% reduction. The suction coefficient and flow transmission loss can elucidate this variation trend rationally. When the yaw angle is small (0–20 degrees), the differences in the suction coefficients between Cases 1 and 2 and Cases 3 and 4 are small, but the flow would leak from the side slits in Cases 1 and 2, which induces the abating of the jet flow. Hence, Cases 3 and 4 perform better in small yaw angles. As the yaw angle continuously increases, Cases 1 and 2 obtain larger suction coefficients than Cases 3 and 4; on the other hand, because each side functions as a suction or jet wall, no conception of leakage would exist, so Cases 1 and 2 can perform better at a large yaw angle. This is the reason that Cases 3 and 4 are only conducted at 0–30 degrees. A similar phenomenon also appears in the study of the application to a circular cylinder (Chen et al., 2015), where five pairs of slits perform better in wind fluctuating load suppression than 13 pairs of slits. At 40–90 degrees, Cases 1 and 2 gain a control effect in most degrees, and generally, Case 1 performs better. Case 1 attains the control reduction of [34.1, 26.8, 27.7, 30.2, 0.8, 27.0] percent at 40–90 degrees and Case 2 attains the control reduction of [6.6, 15.9, 10.9, 19.7, −18.5, 18.0] percent. Notably, at 80 degrees, Cases 1 and 2 have no control effects—Case 2 even has a negative reduction. The turbulence of Case 0 has an obvious decrease at 80 degrees, which needs to be further explained. The general tendency indicates that Case 1 would have considerable control effects in different yaw angles because the suction coefficient of Case 1 is the largest. The larger the suction coefficient, the better control effects.

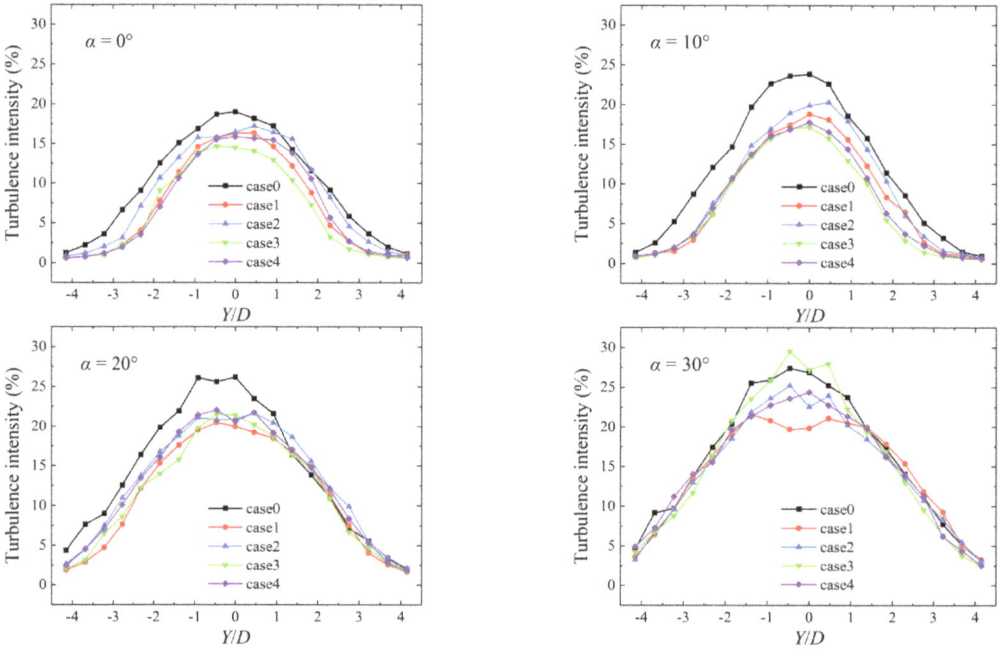

Figure 8. *Cont.*

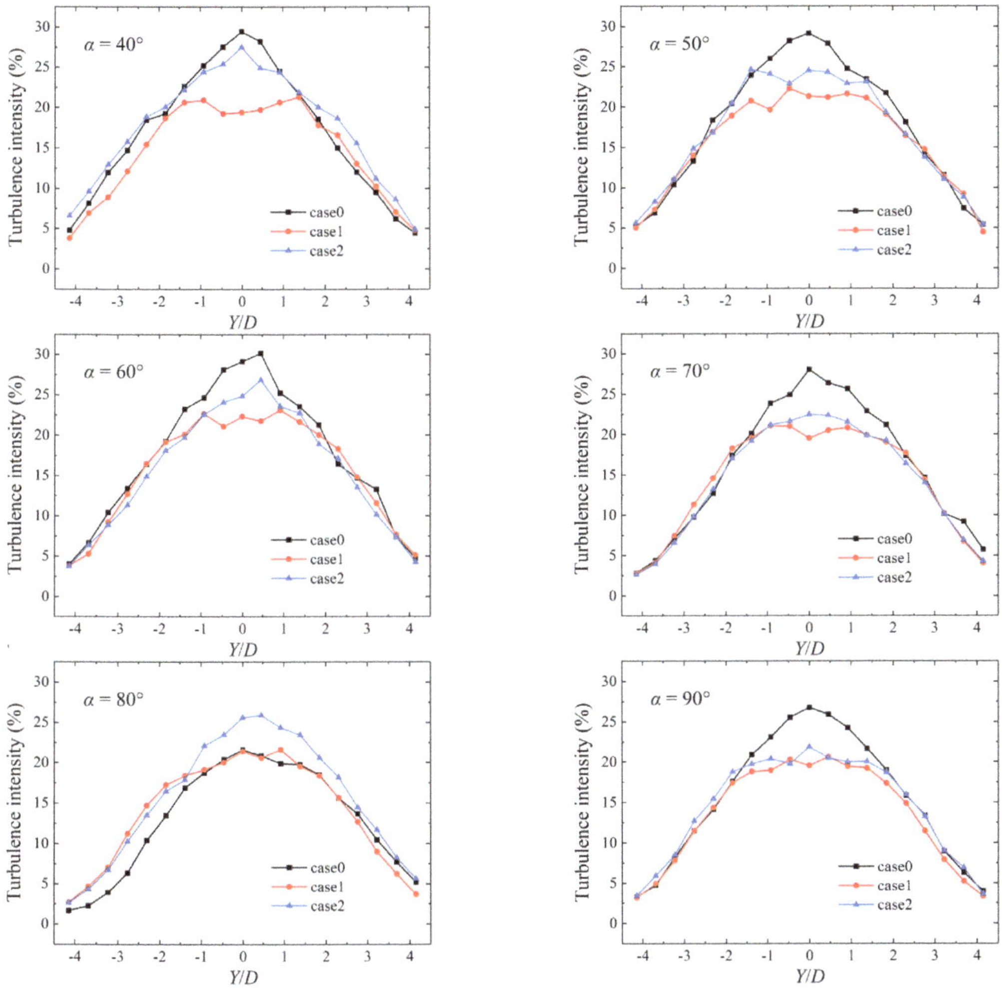

Figure 8. Turbulence intensity distribution at $X/B = 8.48$.

3.3. Dominant Frequency Distribution

Here, we processed the wind speed time history data to gain the power spectrum density. In this part, if there is zero data history, this is also taken into consideration, which means the total length of time history data is the same. We gathered the dominant frequencies of the spectrum in all measurement points, then transformed them into dimensionless frequencies (Strouhal number, $St = f_s L/U_0$, where f_s is the shedding frequency, i.e., the dominant frequency in the wake). The distribution of dimensionless frequencies is obtained, as shown in Figure 9. To eliminate interference from dominant frequencies with small peak values, and for the comparison between cases, a threshold value of power spectral density (PSD) was set. Only the frequencies whose peak value was over the threshold value were counted, and the remaining zones were filled with zeros. If no PSD threshold was set, there is always a dominant frequency anywhere in the wake field, no matter how weak the energy level corresponding to that frequency. The threshold value of PSD in Figure 9 is set as 0.5. In this figure, most St numbers of wake in each case are uniform, but some divergences are mostly located in the middle of wake, especially in Case 0. Comparing the cases, we can find that divergences decline in Cases 1 and 2; especially in Case 1, the St numbers are almost uniform, and the distribution appears clean.

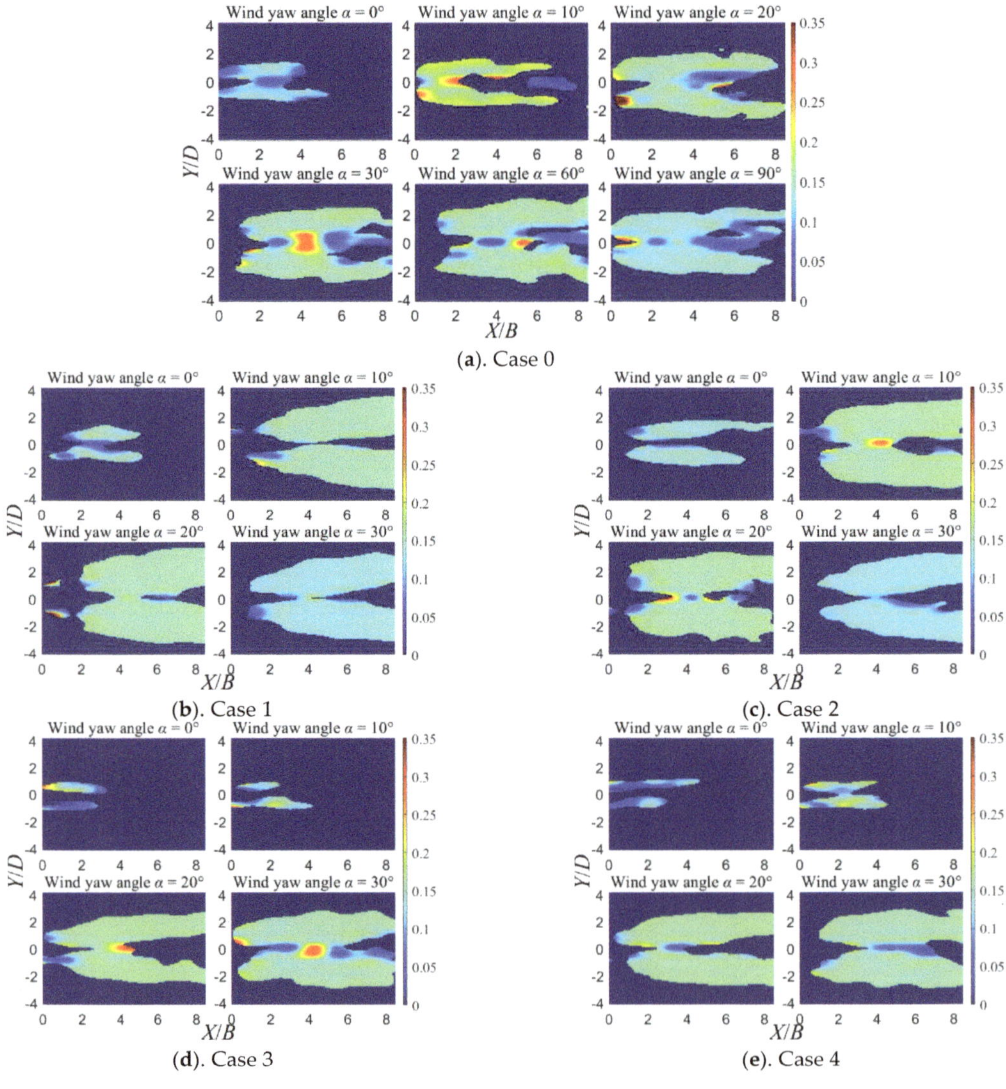

Figure 9. Dimensionless dominant frequency (fD/U_0) distribution of tower wake.

In Cases 3 and 4, the distribution region shrinks to 0 and 10 degrees. For a better comparison of the area change in the distribution region, Figure 10 shows the dimensionless area (S_2/BD) of the region in each case. The threshold value of PSD is also set as 0.5. Two trends in different sections of the yaw angle appear. When the yaw angle is small (0–10 degrees), Cases 1, 3, and 4 all reveal a great control effect. Cases 1–4 attain [36.0, −47.2, 59.1, 50.6] and [63.3, 26.9, 78.0, 61.9] percent reduction of area in 0 and 10 degrees, respectively. However, when the yaw angle increases, the area expands, where Case 1 shows the least expansion, except at 80 degrees, with no more than 4% of the area expanding. As in the turbulence intensity distribution, a drop of the area appears at 80 degrees in Case 0, which causes the amplification of the negative control effects of all cases at 80 degrees. Case 2 has the same trend of the area variation by degrees as Case 1, but the effect is worse—29% amplification of area is the maximum at 30 degrees, without considering 80 degrees. Cases 1 and 3 have larger suction coefficients, so these cases perform better than others. If only a small yaw angle is taken into consideration, Case 3 would be recommended, because the position of the suspenders would come out of the wake region at large yaw angles. For engineering applications, Case 1 would be better for various yaw angles.

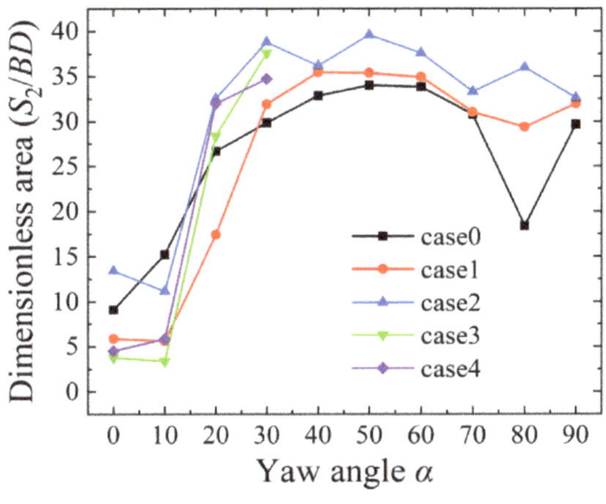

Figure 10. Area of PSD > 0.5 dominant frequency distribution region.

4. Conclusions

In the present study, the characteristics of the wake of a bridge tower were tested with a four-hole cobra probe using the traverse system moving to each measurement point one by one. The cross-section of the tower model is rectangular with corner cuts, and the yaw angle was thus set from 0 to 90 degrees. Overall, the wake region is a diffusion zone, the width increases downstream, and a low-speed region is located at the near back of the tower. As the yaw angle increases, the area of the low-speed region first increases and then decreases, and the maximum appears at 50 degrees. The performance of the cobra probe is evidently restrained in the low-speed region because the acceptance ratio would be extremely low, and deviation is not avoidable. However, the ratio would rapidly reach 100% out of that region. Four controlled cases were set for comparison of the effects of modification in the wake of the model. Two of them were tested at 0–90 degrees, with suction and jet slits distributed on four sides, and the rest were designed for small yaw angles, with slits only distributed on the windward and leeward sides. The suction coefficients of Cases 1 and 2 vary in a small region, from [0.65, 0.35] to [0.67, 0.39], respectively, when the yaw angle changes. The suction coefficients of Cases 3 and 4 sharply decrease when the degree increases. The experimental results indicate that passive jet flow would enlarge the low-speed region; however, velocity distribution further downstream has no significant change. For the turbulence of the wake, the passive jet control method shows the control effects in several cases, where the peak value is suppressed and the width of the high turbulence region is also limited. An imperative trend should be noted: the higher the suction coefficient is, the better the control effects are. From the frequency results, we can conclude that the passive jet method, in some cases, can conspicuously shrink the area of the dominant frequency distribution region with a small yaw angle. However, this conclusion cannot fit the situation with a large yaw angle. The influence of the flow transmission loss is also reflected in the difference between Cases 1 and 3 or Cases 2 and 4. The jet slits distributed at the profile would cause the leak of flow, thus weakening the control effects, but it can contribute to keeping the suction coefficient at a steady level when the yaw angle grows, and thus, a control effect on the turbulence can be maintained. Whether the side jet slits are adopted must take actual demand into consideration.

Similar passive jet rings have been proven to have a good control effect on the aerodynamic force of the bridge tower itself and can effectively suppress the vortex-induced vibration of the bridge tower [23]. In this study, we mainly focused on the effects of the passive jet method on wake characteristics. The results show that the passive jet can change the velocity, turbulence, and main frequency characteristics in the wake region of the bridge tower. It has been proven that the passive jet rings for the bridge tower will effectively control both the around flow field and the wake flow region; however, the control capability on the WIV of the suspenders remains to be further investigated for practical engineering.

Author Contributions: Conceptualization, Y.H. and W.C.; methodology, Y.H. and W.C.; software, Y.H.; validation, W.C.; formal analysis, Y.H.; investigation, Y.H.; resources, W.C.; data curation, Y.H.; writing—original draft preparation, Y.H.; writing—review and editing, W.C.; visualization, Y.H.; supervision, W.C.; project administration, W.C.; funding acquisition, W.C. All authors have read and agreed to the published version of the manuscript.

Funding: This research was funded by the National Natural Science Foundation of China grant number U2106222, 52008140, and 51978222.

Institutional Review Board Statement: Not applicable.

Informed Consent Statement: Not applicable.

Data Availability Statement: Not applicable.

Acknowledgments: This work was funded by the National Natural Science Foundation of China through Grants U2106222, 52008140, and 51978222.

Conflicts of Interest: The authors declare no conflict of interest.

References

1. Knisely, C.W. Strouhal Numbers of Rectangular Cylinders at Incidence: A Review and New Data. *J. Fluids Struct.* **1990**, *4*, 371–393. [CrossRef]
2. Norberg, C. Flow around Rectangular Cylinders: Pressure Forces and Wake Frequencies. *J. Wind Eng. Ind. Aerodyn.* **1993**, *49*, 187–196. [CrossRef]
3. Shimada, K.; Ishihara, T. Application of a Modified K-ε Model to the Prediction of Aerodynamic Characteristics of Rectangular Cross-Section Cylinders. *J. Fluids Struct.* **2002**, *16*, 465–485. [CrossRef]
4. Assi, G.R.S.; Bearman, P.W.; Meneghini, J.R. On the Wake-Induced Vibration of Tandem Circular Cylinders: The Vortex Interaction Excitation Mechanism. *J. Fluid Mech.* **2010**, *661*, 365–401. [CrossRef]
5. Zdravkovich, M.M. REVIEW—Review of Flow Interference Between Two Circular Cylinders in Various Arrangements. *J. Fluids Eng.* **1977**, *99*, 618. [CrossRef]
6. Carmo, B.S.; Sherwin, S.J.; Bearman, P.W.; Willden, R.H.J. Flow-Induced Vibration of a Circular Cylinder Subjected to Wake Interference at Low Reynolds Number. *J. Fluids Struct.* **2011**, *27*, 503–522. [CrossRef]
7. Bearman, P.W. Circular Cylinder Wakes and Vortex-Induced Vibrations. *J. Fluids Struct.* **2011**, *27*, 648–658. [CrossRef]
8. Wang, H.; Yang, W.; Nguyen, K.D.; Yu, G. Wake-Induced Vibrations of an Elastically Mounted Cylinder Located Downstream of a Stationary Larger Cylinder at Low Reynolds Numbers. *J. Fluids Struct.* **2014**, *50*, 479–496. [CrossRef]
9. Li, Y.; Tang, H.; Lin, Q.; Chen, X. Vortex-Induced Vibration of Suspenders in the Wake of Bridge Tower by Numerical Simulation and Wind Tunnel Test. *J. Wind Eng. Ind. Aerodyn.* **2017**, *164*, 164–173. [CrossRef]
10. Amitay, M.; Smith, B.; Glezer, A. Aerodynamic Flow Control Using Synthetic Jet Technology. In Proceedings of the 36th AIAA Aerospace Sciences Meeting and Exhibit, Reno, NV, USA, 12–15 January 1998. [CrossRef]
11. Amitay, M.; Honohan, A.; Trautman, M.; Glezer, A.; Amitay, M.; Honohan, A.; Trautman, M.; Glezer, A. Modification of the Aerodynamic Characteristics of Bluff Bodies Using Fluidic Actuators. In Proceedings of the 28th Fluid Dynamics Conference, Snowmass Village, CO, USA, 29 June–2 July 1997. [CrossRef]
12. Crook, A.; Sadri, A.M.; Wood, N.J. The Development and Implementation of Synthetic Jets for the Control of Separated Flow. In Proceedings of the 17th Applied Aerodynamics Conference, Norfolk, VA, USA, 28 June–1 July 1999; Volume 99, p. 3176. [CrossRef]
13. Fransson, J.H.M.; Konieczny, P.; Alfredsson, P.H. Flow around a Porous Cylinder Subject to Continuous Suction or Blowing. *J. Fluids Struct.* **2004**, *19*, 1031–1048. [CrossRef]
14. Wang, J.; Feng, L.; Xu, C. Experimental Investigations on Separation Control and Flow Structure around a Circular Cylinder with Synthetic Jet. *Sci. China Ser. E Technol. Sci.* **2007**, *50*, 550–559. [CrossRef]

15. Dong, S.; Triantafyllou, G.S.; Karniadakis, G.E. Elimination of Vortex Streets in Bluff-Body Flows. *Phys. Rev. Lett.* **2008**, *100*, 204501. [CrossRef] [PubMed]
16. Feng, L.H.; Wang, J.J. Synthetic Jet Control of Separation in the Flow over a Circular Cylinder. *Exp. Fluids* **2012**, *53*, 467–480. [CrossRef]
17. Feng, L.-H.; Wang, J.-J.; Pan, C. Proper Orthogonal Decomposition Analysis of Vortex Dynamics of a Circular Cylinder under Synthetic Jet Control. *Phys. Fluids* **2011**, *23*, 014106. [CrossRef]
18. Chen, W.-L.; Xin, D.-B.; Xu, F.; Li, H.; Ou, J.-P.; Hu, H. Suppression of Vortex-Induced Vibration of a Circular Cylinder Using Suction-Based Flow Control. *J. Fluids Struct.* **2013**, *42*, 25–39. [CrossRef]
19. Chen, W.-L.; Li, H.; Hu, H. An Experimental Study on a Suction Flow Control Method to Reduce the Unsteadiness of the Wind Loads Acting on a Circular Cylinder. *Exp. Fluids* **2014**, *55*, 1707. [CrossRef]
20. Chen, W.-L.; Wang, X.; Xu, F.; Li, H.; Hu, H. Passive Jet Flow Control Method for Suppressing Unsteady Vortex Shedding from a Circular Cylinder. *J. Aerosp. Eng.* **2017**, *30*, 04016063. [CrossRef]
21. Chen, W.-L.; Gao, D.-L.; Yuan, W.-Y.; Li, H.; Hu, H. Passive Jet Control of Flow around a Circular Cylinder. *Exp. Fluids* **2015**, *56*, 201. [CrossRef]
22. Zhang, L.-Q.; Chen, G.-B.; Chen, W.-L.; Gao, D.-L. Separation Control on a Bridge Box Girder Using a Bypass Passive Jet Flow. *Appl. Sci.* **2017**, *7*, 501. [CrossRef]
23. Chen, W.L.; Huang, Y.W.; Gao, D.L.; Meng, H.; Chen, G.B.; Li, H. Passive Suction Jet Control of Flow Regime around a Rectangular Column with a Low Side Ratio. *Exp. Therm. Fluid Sci.* **2019**, *109*, 109815. [CrossRef]
24. Chen, W.L.; Huang, Y.W.; Meng, H. Wake-Induced Vibration of a Suspender Cable in the Rear of a Bridge Tower. *J. Fluids Struct.* **2020**, *99*, 103166. [CrossRef]
25. Mallipudi, S.; Selig, M.; Long, K. Use of a Four Hole Cobra Pressure Probe to Determine the Unsteady Wake Characteristics of Rotating Objects. In Proceedings of the 24th AIAA Aerodynamic Measurement Technology and Ground Testing Conference, Portland, OR, USA, 28 June–1 July 2004; American Institute of Aeronautics and Astronautics: Reston, VA, USA, 2004; pp. 1–9.

applied sciences

MDPI

Article

Aerodynamic Characteristics of a Square Cylinder with Vertical-Axis Wind Turbines at Corners

Zhuoran Wang [1], Gang Hu [1,2,3,*], Dongqin Zhang [1,*], Bubryur Kim [4], Feng Xu [1] and Yiqing Xiao [1,2,3]

1 School of Civil and Environmental Engineering, Harbin Institute of Technology, Shenzhen 518055, China; 20b954009@stu.hit.edu.cn (Z.W.); xufenghit@hit.edu.cn (F.X.); xiaoyq@hit.edu.cn (Y.X.)
2 Guangdong-Hong Kong-Macao Joint Laboratory for Data-Driven Fluid Mechanics and Engineering Applications, Harbin Institute of Technology, Shenzhen 518055, China
3 Shenzhen Key Laboratory of Intelligent Structure System in Civil Engineering, Harbin Institute of Technology, Shenzhen 518055, China
4 Department of Robot and Smart System Engineering, Kyungpook National University, 80, Daehak-ro, Buk-gu, Daegu 41566, Korea; brkim@knu.ac.kr
* Correspondence: hugang@hit.edu.cn (G.H.); zhangdongqin@hit.edu.cn (D.Z.)

Abstract: A preliminary study is carried out to investigate the aerodynamic characteristics of a square cylinder with Savonius wind turbines and to explain the reason why this kind of structure can suppress wind-induced vibrations. A series of computational fluid dynamics simulations are performed for the square cylinders with stationary and rotating wind turbines at the cylinder corners. The turbine orientation and the turbine rotation speed are two key factors that affect aerodynamic characteristics of the cylinder for the stationary and rotating turbine cases, respectively. The numerical simulation results show that the presence of either the stationary or rotating wind turbines has a significant effect on wind forces acting on the square cylinder. For the stationary wind turbine cases, the mean drag and fluctuating lift coefficients decrease by 37.7% and 90.7%, respectively, when the turbine orientation angle is 45°. For the rotating wind turbine cases, the mean drag and fluctuating lift coefficients decrease by 34.2% and 86.0%, respectively, when the rotation speed is 0.2 times of vortex shedding frequency. Wind turbines installed at the corners of the square cylinder not only enhance structural safety but also exploit wind energy simultaneously.

Keywords: square cylinder; wind turbines; aerodynamic characteristics; vortex shedding

Citation: Wang, Z.; Hu, G.; Zhang, D.; Kim, B.; Xu, F.; Xiao, Y. Aerodynamic Characteristics of a Square Cylinder with Vertical-Axis Wind Turbines at Corners. *Appl. Sci.* **2022**, *12*, 3515. https://doi.org/10.3390/app12073515

Academic Editor: Mohsen Soltani

Received: 21 February 2022
Accepted: 28 March 2022
Published: 30 March 2022

1. Introduction

Techniques to reduce wind-induced vibration of high-rise buildings have practical significance for ensuring structure safety and occupant comfort. In general, there are two traditional strategies, mechanical strategies and aerodynamic modifications, to reduce wind-induced vibrations. The tuned mass damper (TMD) installed inside tall buildings is one of the most effective mechanical strategies. In order to achieve an ideal vibration control effect, the mass of the damper normally needs to reach about 0.5–3% of the building mass. As a result, the damper not only occupies a large amount of building space, but also requires a stronger structure system to support such a heavy device. Furthermore, aerodynamic modification strategies, including rounding corners [1–3] and recessing corners [4–6], have also been studied. Wind tunnel model testing results demonstrate that slotted corners and chamfered corners can cause significant reductions in both along-wind and cross-wind responses [7,8]. Drag forces on a cylinder can also be reduced by adding a splitter plate behind it [9]. Although these strategies are very effective, they sacrifice the valuable space of tall buildings for the sole purpose of reduction in wind loading and wind-induced vibrations.

To further achieve the purpose of reducing drag, many researchers use the moving surface boundary layer control method (MSBC) to inject momentum into the flow field to prevent the formation of boundary layers. In [10], experimental research was conducted on the aerodynamic characteristic of a square cylinder under the momentum injection method. The results show that in the presence of MSBC, the drag of the square cylinder is considerably reduced at all angles of attack. In research reported in [11], a rectangular prism with two rotating cylinders was studied in a wind tunnel, and the effects of different rotation speed and modes on the aerodynamic characteristic were analyzed. Previous scholars mainly relied on the rotation of the small cylinder to inject momentum into the flow field. We consider replacing the small cylinder with wind turbines to increase its possibility and value in engineering applications.

Compared to the sole purpose of aerodynamic modification strategies to suppress wind-induced vibrations of tall buildings described in previous studies, it is desirable to develop effective aerodynamic devices for not only suppressing the wind-induced vibrations but also harvesting wind energy. Extensive studies aimed at making wind energy more efficient have also been carried out. The authors of [12] added two small diameter cylindrical rods on both sides of the cylinder to significantly improve the efficiency of a vortex-induced vibration energy collection system. Wind turbines have been installed in various positions of tall buildings, such as rooftop and through opening and channel between two buildings. The authors of [13] used the scale model to evaluate the wind speed amplifications in the tunnels for wind-power generation through the installation of wind turbines. To obtain optimal environmental benefits, the optimizing design of wind turbines installed on the tall buildings is also discussed [14,15]. Because previous studies have only focused on the energy-saving benefits of installing wind turbines on structures, we consider that installation of the wind turbine at the corners of the square cylinder will not only enhance structural safety but also exploit wind energy simultaneously.

This is a preliminary study to investigate the aerodynamic characteristics of a square cylinder with Savonius wind turbines and attempts to explain the reason why this kind of structure can suppress wind-induced vibrations. Numerical models, including numerical method, computational configuration and mesh arrangement, grid size and time step configuration, and simulation cases are illustrated in Section 2. Mean pressure coefficients, force coefficient and vortex shedding, and flow pattern around the cylinders for both stationary and rotating wind turbines are shown in Section 3. Finally, conclusions are summarized in Section 4.

2. Numerical Simulation Configurations

2.1. Numerical Method

RANS [16–18] treats the vortices of different scales equally in the flow field and smoothens the details of the temporal and spatial variations in pulsation motion through average operation. The turbulence model we used is the SST k–ω.

The continuity equation and momentum equation for impressible flow are expressed as:

$$\nabla \cdot \boldsymbol{U} = 0 \tag{1}$$

$$\frac{\partial \boldsymbol{U}}{\partial t} + \nabla \cdot (\boldsymbol{U}\boldsymbol{U}) = -\frac{1}{\rho}\nabla P + \nabla \cdot (\nu \nabla \boldsymbol{U}) - \nabla \cdot \tau \tag{2}$$

where $\boldsymbol{U}$ and P are the averaged velocity and pressure, respectively; ρ shows air density and ν refers to eddy viscosity; τ is the Reynolds stress which can be expressed as $\tau_{ij} = \nu_t(\partial u_i/\partial x_j + \partial u_j/\partial x_i) - 2/3\rho k \delta_{ij}$ and δ_{ij} is the Kronecker delta.

According to [18,19], the control equations of SST k–ω are expressed as follows:

$$\frac{\partial \rho k}{\partial t} + u_i \frac{\partial \rho k}{\partial x_i} = \frac{\partial}{\partial x_j}\left[\Gamma_k \frac{\partial k}{\partial x_j}\right] + G_k - Y_k \tag{3}$$

$$\frac{\partial \rho \omega}{\partial t} + u_i \frac{\partial \rho k \omega}{\partial x_i} = \frac{\partial}{\partial x_j}\left[\Gamma_\omega \frac{\partial \omega}{\partial x_j}\right] + G_\omega - Y_\omega + D_\omega \tag{4}$$

where $\rho = 1.225\ \text{kg/m}^3$; k is turbulent kinetic energy; ω is the dissipation rate of turbulent kinetic energy; u_i is the velocity component and subscripts i, j = x, y, z; G_w is the term of turbulent kinetic energy generation caused by the velocity gradient; Γ_k and Γ_ω are the convection terms for k and ω, respectively; Y_k and Y_w are the effective diffusion terms of k and w caused by turbulence, respectively; and D_ω is the cross-convection term $(D_\omega = 2(1 - F1)\rho\sigma_{\omega 2}\frac{1}{\omega}\frac{\partial k}{\partial x_j}\frac{\partial \omega}{\partial x_j})$.

The coefficient of eddy viscosity ν_t can be obtained by:

$$\nu_t = a_1 \frac{k}{\max(a_1\omega, \Omega F_2)} \tag{5}$$

$$F_2 = \tanh(\arg_2^2), \arg_2 = \max(2\frac{\sqrt{k}}{0.09\omega y}, \frac{500\nu}{y^2\omega}) \tag{6}$$

where Ω is the absolute value of vorticity, F_2 is a function that is one for boundary-layer flows and zero for free shear layers, and $a_1 = 0.31$ is the empirical coefficient.

The transition from the k–ω model near the wall to the k–ε model far from the wall is controlled by mixed function F_1:

$$F_1 = \tanh(\arg_1^4) \tag{7}$$

$$\arg_1 = \min(\max(2\frac{\sqrt{k}}{0.09\omega y}, \frac{500\nu}{y^2\omega}); \frac{4\rho\sigma_{\omega 2}k}{CD_{k\omega}y^2}) \tag{8}$$

$$CD_{k\omega} = \max(2\rho\sigma_{\omega 2}\frac{1}{\omega}\frac{\partial k}{\partial x_j}\frac{\partial \omega}{\partial x_j}; 10^{-20}) \tag{9}$$

where y is the distance to the next surface, ν_t is the eddy viscosity, and CD_{kw} is the positive portion of the cross-diffusion term. Model coefficient $\sigma_{\omega 2}$ is 0.856.

The open-source computational fluid dynamics package OpenFOAM is used for numerical simulations in this study. The finite volume method is adopted to solve the unsteady incompressible Navier–Stokes equation. The PIMPLE scheme is used to solve the pressure–velocity coupling. The second order implicit scheme is selected for time discretization. Gauss linear upwind scheme is used for spatial discretization. The time step is set as 1.0×10^{-6} s. In terms of time step selection, to assess how the size of the time steps may affect the results, three time steps of 2×10^{-5}, 1.0×10^{-6}, and 1.0×10^{-7} s are tested. At last, 1.0×10^{-6} s is selected to balance the simulation time and the result accuracy. For all simulations, more than 1000 nondimensional time steps, corresponding to approximately 50 vortex-shedding cycles, which is much larger than the number of vortex-shedding cycles adopted in the previous studies [20,21], are taken for assuring reliable results.

2.2. Computational Configuration and Mesh Arrangement

The 2D computational domain, 92 W long and 25 W wide, is illustrated in Figure 1a in which the width (W) of the cross section is 0.1 m and the cylinder centroid is defined as the origin of the coordinate. The lateral boundary is 12.5 W away from the cylinder centroid in terms of the previous studies [22]. In order to allow flow redevelopment behind the wake region, the distances between the cylinder centroid and the inlet and the outlet boundary are 30.5 W and 61.5 W, respectively.

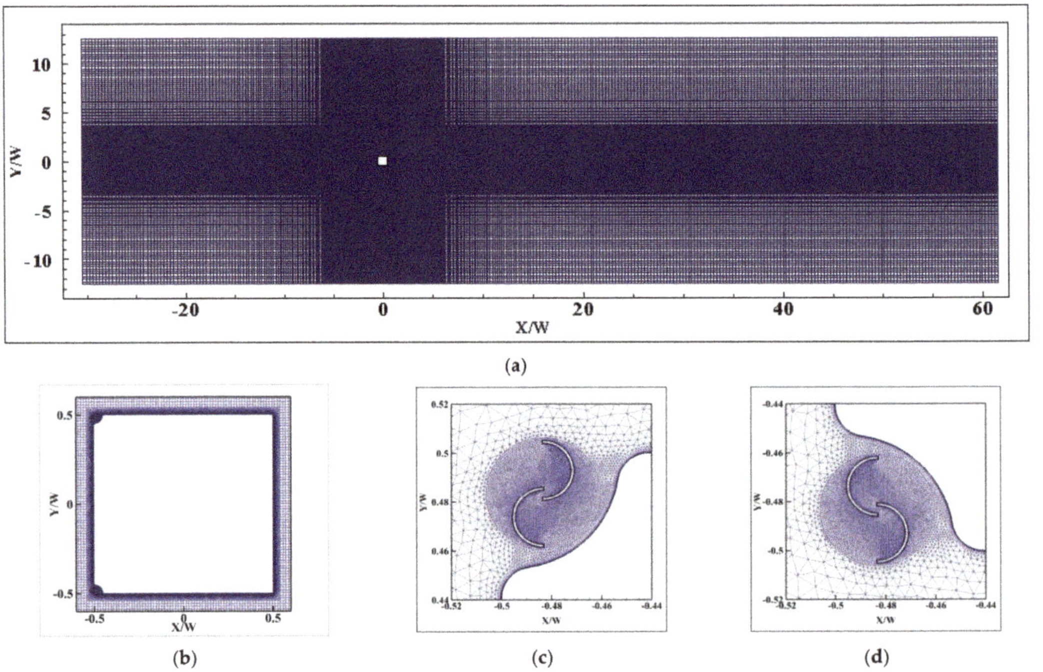

Figure 1. Views of grids for a square cylinder with wind turbines at two leading corners: (**a**) Grid distribution in x–y plane; (**b**) Grid surrounding square cylinder; (**c**) Grid surrounding upper wind turbine; (**d**) Grid surrounding lower wind turbine.

As for the boundary conditions, a smooth flow condition with a wind speed of 9 m/s is used at the inlet ($u = 9m/s$, $\partial p/\partial n = 0$) and the Reynolds number is 6.0×10^4, which is consistent with that in [22]. At the outlet boundary, a zero static pressure is adopted ($p = p_0 = 1.013 \times 10^5$, $\partial u/\partial n = 0$). Two lateral boundaries are defined as a symmetry boundary condition ($v = 0$, $\partial u/\partial n = 0$), which is used to model zero-shear slip walls in viscous flows. An empty condition is enforced on the top and bottom surfaces of the computational domain, which is an ordinary method to solve the 2D problem in OpenFOAM. Two vertical-axis wind turbines are installed at two leading corners of the square cylinder and the nonslip wall boundary condition ($ui = 0$, $\partial p/\partial n = 0$) is applied to surfaces of the cylinder and the wind turbines, where n refers to the normal direction, u and v represent the velocity components in x and y directions, respectively.

2.3. Grid Size and Time Step Configuration

The hybrid grid scheme is used for discretization, as shown in Figure 1. Because of the geometric irregularity of the wind turbines, an unstructured grid is adopted around the turbines in the vicinity of the cylinder. Structural meshes are used for the whole computational domain except the region containing wind turbines. As shown in Figure 2, the diameter of wind turbines is 5/100 W, d is 7/300 W, and e is 1/240 W, which leads to a ratio (e/d) of 1/6. The height of the minimum grid (δ) is 0.004 mm, the maximum $y+$ values for all cases is less than 1. The rotation of the wind turbine is implemented via the dynamic grid model, and a pair of coupling interfaces is set between the rotating region and the stationary region.

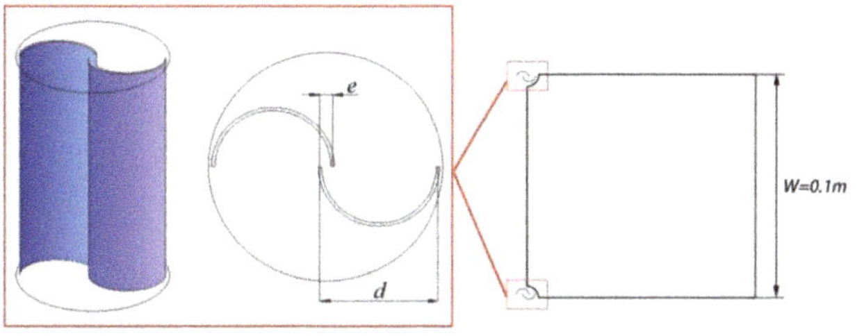

Figure 2. Sketch of Savonius wind turbines.

In order to consider the influence of grid size on numerical simulations, five grids (i.e., G1, G2, G3, G4, and G5) of the plain square cylinder with corner turbines are tested. The drag, lift coefficients, and Strouhal number of the square cylinder obtained from the five grid schemes are compared in Table 1. The mean drag, lift coefficients, and the Strouhal number are defined as:

$$C_d = \frac{F_d}{1/2\rho U^2 W} \tag{10}$$

$$C_l = \frac{F_l}{1/2\rho U^2 W} \tag{11}$$

$$S_t = \frac{f_s W}{U} \tag{12}$$

where F_d and F_l are the drag and lift forces of the cylinder, respectively; f_s is the vortex shedding frequency; W is the width of the cross section; and U is the oncoming flow velocity set at the inlet boundary condition. The specifications of the grids used are given in Table 1 and wind turbine rotation speed is 11.88 rev/s. The Strouhal number and *RMS* lift coefficients C_l^{RMS} of G4 and G5 are identical. Considering accuracy and calculation speed, G4 is selected as the final grid scheme.

Table 1. Grid schemes for grid independence tests.

Grid Schemes	Cell Numbers	y^+(Max)	δ/W	$\overline{C_d}$	C_l^{RMS}	S_t
G1	1.1×10^5	0.66	4×10^{-5}	1.43	0.69	0.136
G2	1.3×10^5	0.52	4×10^{-5}	1.41	0.63	0.136
G3	1.7×10^5	0.55	4×10^{-5}	1.42	0.61	0.136
G4	2.4×10^5	0.57	4×10^{-5}	1.47	0.78	0.134
G5	3.0×10^5	0.61	4×10^{-5}	1.48	0.78	0.134

In addition, the simulation result of a plain cylinder without corner wind turbines is a baseline case to compare to the cases with corner wind turbines. As shown in Figure 3, the mean pressure coefficients on the four faces have favorable agreement with the results in the previous studies [22,23]. Table 2 gives the comparisons of force coefficients between previous studies and present numerical simulations. The definition of δ_{cd} is the relative error of $\overline{C_d}$. As shown in Table 2, the drag coefficient $\overline{C_d}$ for the plain cylinder is 1.99, which agrees well with that of 2.05 for a square cylinder reported in [24]. The relative error δ_{cd} is nearly 5% when compared with all the experimental results. The *RMS* lift coefficient for the plain cylinder is 1.29, which matches well with the *RMS* lift coefficients of 1.37 for a square cylinder reported in [25].

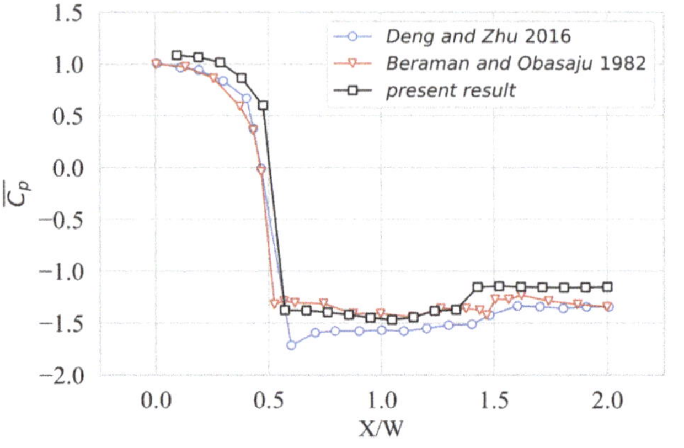

Figure 3. Comparisons of mean pressure distribution between previous studies and present numerical simulations.

Table 2. Comparisons of force coefficients between previous studies and present numerical simulations.

Data from Previous Studies	Re	$\overline{C_d}$	C_l^{RMS}	S_t	δ_{cd}
Experimental results:					
[24]	1.76×10^5	2.05	-	0.122	2.93%
[26,27]	2.14×10^4	2.1	-	0.132	5.24%
[28]	2.20×10^4	2.09	0.95	0.123	4.78%
Simulation results:					
[22]	6.0×10^4	2.26	1.52	0.132	11.9%
[25]	1.9×10^4	2.37	1.37	0.122	16.0%
[29]	2.2×10^4	2.26	1.60	0.136	11.9%
Present simulation results:	6.0×10^4	1.99	1.29	0.123	-

2.4. Simulation Cases

As shown in Figure 4, two types of simulation cases, stationary and rotating wind turbines, are considered. For the stationary wind turbines, their orientation related to the wind flow direction may have a significant effect on aerodynamic characteristics of the cylinder. Similarly, for the rotating wind turbines, it is anticipated that the rotation speed of the turbines is a key factor which affects the aerodynamic characteristics of the cylinder. In order to study the effects of these two factors, 4 different orientations of the stationary wind turbines and 9 different rotation speeds are studied, as shown in Tables 3 and 4. The orientation of the stationary turbine is denoted by θ, which represents the angle of rotation of the wind turbines. Two wind turbines rotate inward relative to the square cylinder. V_t is the rotation speed of the wind turbines and f_s is the vortex shedding frequency of the square cylinder calculated according to Strouhal number of 0.132 for the square cylinder provided by [22]. The rotation speed ranges from 0 to 4 times that of f_s, and a rotation speed of 0 refers to a stationary case. In this study, the turbine rotates independently from the wind speed. According to [30], the rotation speed of the Savonius wind turbine is about 19.5 (r/s), which is closest to the rotation speed of R6. Thus, R6 is considered as an actual case. In order to conduct a comparative study on the influence of different rotational speeds on the aerodynamic characteristics of the square cylinder, we created 8 other working conditions.

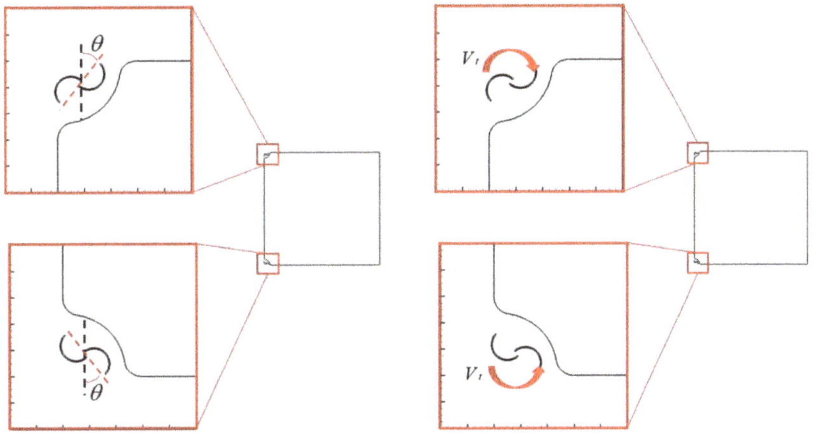

(**a**) Stationary cases (**b**) Rotating cases

Figure 4. Simulation cases: (**a**) Stationary cases; (**b**) Grid surrounding square cylinder.

Table 3. Rotating test cases.

Cases	Plain	*R0*	*R1*	*R2*	*R3*	*R4*	*R5*	*R6*	*R7*	*R8*
V_t/f_s	-	0	0.2	0.4	0.6	0.8	1	2	3	4
V_t (r/s)	-	0	2.38	4.75	7.13	9.50	11.88	23.76	35.64	47.52

Table 4. Stationary test cases.

Cases	Plain	*S1*	*S2*	*S3*	*S4*
θ	-	0°	45°	90°	135°

3. Results and Discussions

3.1. Mean Pressure Coefficients

3.1.1. Stationary Wind Turbines at Cylinder Corners

Comparisons of mean pressure coefficients ($\overline{Cp}$) around the cylinder surface among the four cases with stationary wind turbines at the cylinder corners and the plain cylinder are provided in Figure 5.

On the windward face, $\overline{Cp}$ at the stagnation point is approximately 1.0 for all cases. $\overline{Cp}$ gradually decreases from the stagnation point to two corners and $\overline{Cp}$ on two side faces are symmetrically distributed as expected. Because of the flow separation, the value of $\overline{Cp}$ on two side faces is negative. Compared with the cases with turbines at corners, $\overline{Cp}$ on the side faces of the plain cylinder has the lowest value. It means that the existence of stationary wind turbines in the corners reduces the suction on the side faces. Furthermore, $\overline{Cp}$ on the side faces has the largest value when θ is $\pm 45°$ (*S2*), which means that the suction on the side faces of *S2* is lower than the other cases.

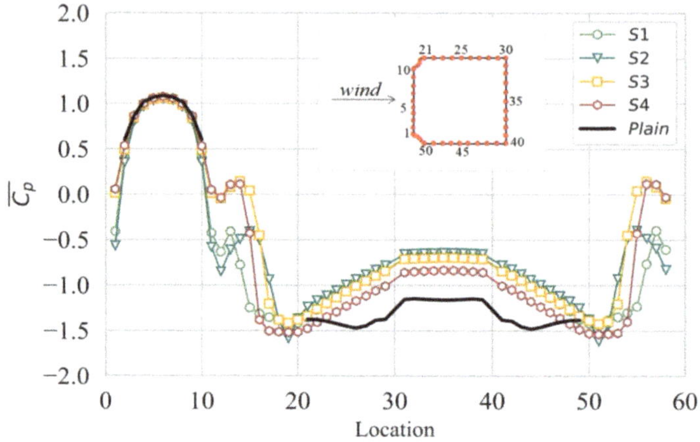

Figure 5. Mean pressure coefficients around the cylinder for the four stationary cases and plain cylinder.

On the leeward face, the suctions for cases with stationary wind turbines are significantly lower than that of the plain cylinder. Similar to the observations on the side face, the stationary wind turbines weaken the suction on the leeward face and it becomes more obvious when θ is $\pm 45°$, i.e., *S2*.

3.1.2. Rotating Wind Turbines at Cylinder Corners

Mean pressure coefficients ($\overline{Cp}$) around the cylinder surfaces of nine rotating cases and the plain cylinder are compared in Figure 6. Figure 6a presents the cases when V_t/f_s is less than 1 while Figure 6b shows the cases when V_t/f_v is equal to or larger than 1.

Figure 6. Mean pressure coefficients around the cylinder for the nine rotating cases and the plain cylinder: (**a**) $V_t/f_s < 1$; (**b**) $V_t/f_s \geq 1$.

On the two side faces, $\overline{Cp}$ of the plain cylinder has the lowest value among the nine cases. Therefore, the existence of rotating wind turbines in the corner reduces the suction on the two side faces. As shown in Figure 6a, the suction on the two sides increases as the rotation speed increases when the rotation speed of wind turbines is less than f_S. As shown in Figure 6b, the suction on the two sides has an obvious reduction when the wind turbine rotation speed increases from f_S (*R5*) to $2f_S$ (*R6*). Lastly, the suction will not be lower than *R6* as the rotation speed further increases.

On the leeward face, the suction of cases with the rotating wind turbines is lower than the plain cylinder. Similar to the observations on the side faces, it is believed that the rotating wind turbine is able to weaken the suction on the leeward face. It is also noted that the suction on the leeward face has the lowest value, as shown in Figure 6b, when the rotating speed is 2fs (*R6*).

3.2. *Force Coefficients and Vortex Shedding Characteristics*

3.2.1. Stationary Wind Turbines at Cylinder Corners

$\overline{C_d}$, C_l^{RMS}, and S_t of four stationary cases and the plain cylinder are listed in Table 5. The percentage of the values relative to the plain cylinder is shown in Figure 7.

Table 5. Force coefficient for four stationary cases and plain cylinder.

Case	Cell Numbers	y^+(Max)	δ/W	$\overline{C_d}$	C_l^{RMS}	S_t
Plain cylinder	2.0×10^4	1.4	2×10^{-4}	1.99	1.29	0.123
S1	2.4×10^5	0.41	4×10^{-5}	1.26	0.19	0.144
S2	2.4×10^5	0.43	4×10^{-5}	1.24	0.12	0.145
S3	2.4×10^5	0.44	4×10^{-5}	1.35	0.21	0.141
S4	2.4×10^5	0.59	4×10^{-5}	1.48	0.43	0.143

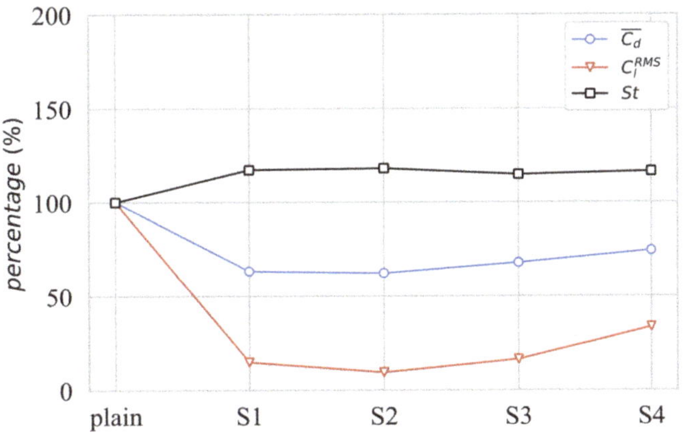

Figure 7. Comparisons of *RMS* lift coefficients, mean drag coefficients, and Strouhal number for the four stationary cases and plain cylinder.

As shown in Table 5 and Figure 7, four cases with stationary wind turbines have lower $\overline{C_d}$ than the plain cylinder since the suction on the two sides and leeward face is lower than the plain cylinder, as shown in Figure 5.

As shown in Figure 7, $\overline{C_d}$ is reduced 37.7% compared to the plain cylinder when θ is 45° (*S2*), which is the most significant for all cases. The lowest suction on the leeward face shown in Figure 5 contributes to the lowest $\overline{C_d}$ of *S2*. The variation in C_l^{RMS} is similar to that of $\overline{C_d}$ and the *RMS* lift coefficients has a 90.7% reduction when θ is 45° (*S2*). That is to say, the fluctuation of C_l was suppressed effectively. There is still a drop of 66.7% of C_l^{RMS} compared with the plain cylinder when θ is 135° (*S4*). Figure 8a,b presents the comparison of time history of *Cl* for *S2* and *S4*. The red curve represents the time history of *Cl* of the plain cylinder and the black curves represent the time history of the typical cases. It can be found that the fluctuation of *Cl* for *S2* has a greater inhibition than *S4*.

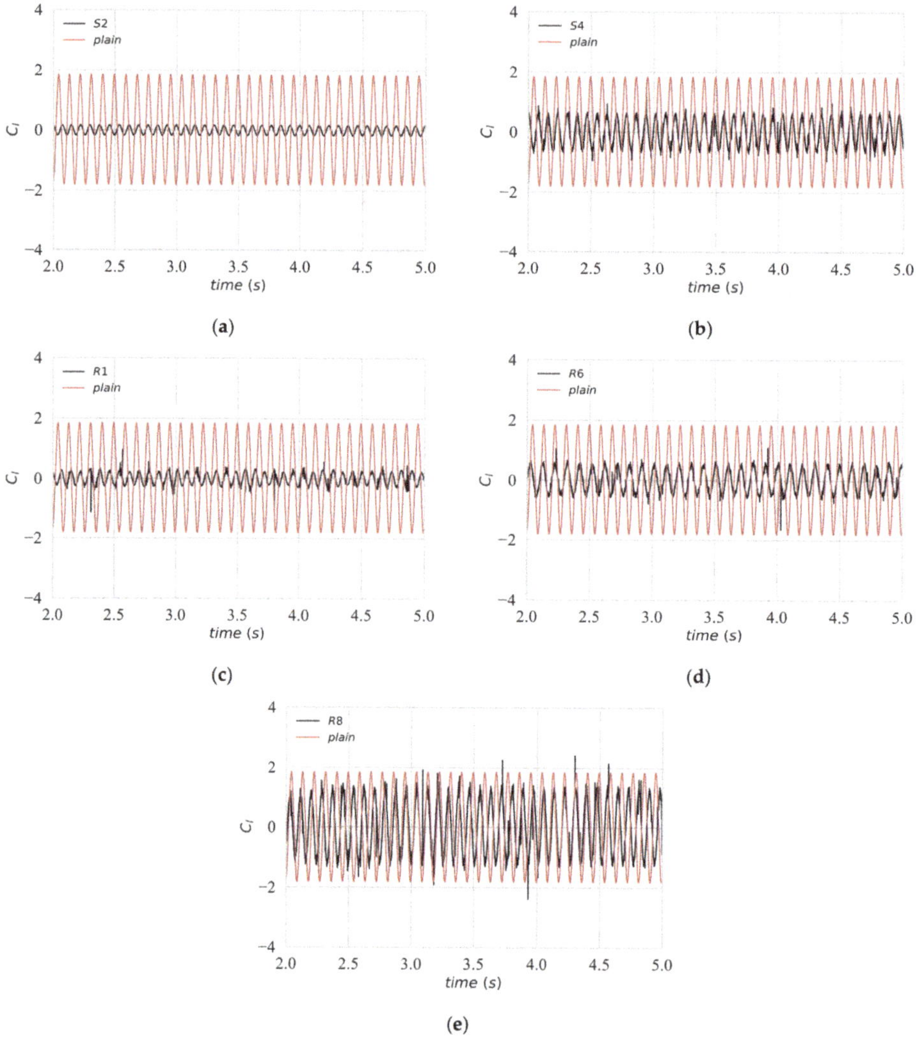

Figure 8. Comparison of time history of *Cl*: (**a**) *S2*; (**b**) *S4*; (**c**) *R1*; (**d**) *R6*; (**e**) *R8*.

Figure 9 illustrates the power spectrum of the lift force for the four stationary cases and plain cylinder; the vortex shedding energy is represented by the peak value of the spectrum. The frequency corresponding to the maximum value of the power spectrum is the vortex shedding frequency, and then S_t can be calculated by $f_s W/U$ (W is the width of the cylinder and U is the wind speed). As shown in Figure 9, the spectrum of the lift coefficient of the plain cylinder fits well with that in [31]. Four stationary cases have a much larger Strouhal number, approximately 20% larger than that of the plain cylinder. The vortex shedding energy presented by the peak value of the spectrum of four stationary cases is evidently smaller than that of the plain cylinder. Even though the stationary wind turbines installed at two leading corners increase the frequency of vortex shedding of the square cylinder, they can effectively reduce the energy of vortex shedding. Furthermore, the vortex shedding energy of the cylinder with corner wind turbines exhibits the lowest value when θ is 45° (*S2*).

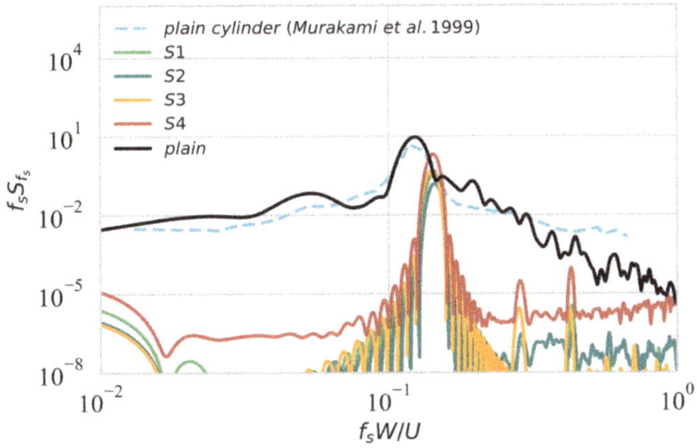

Figure 9. Power spectra of lift forces for the four stationary cases and plain cylinder.

3.2.2. Rotating Wind Turbines at Cylinder Corners

$\overline{C_d}$, C_l^{RMS}, and S_t of the nine rotating cases and the plain cylinder are listed in Table 6. Ratios of these values to that for the plain cylinder are shown in Figure 10. All cases of the square cylinder with rotating corner wind turbines have lower mean drag coefficients ($\overline{C_d}$) than the plain cylinder. Compared with the plain cylinder, when the rotation speed is $0.2f_s$ (*R1*), $\overline{C_d}$ of the cylinder with wind turbines dramatically decreases by about 34.2%, and then the drop rate gradually decreases as the rotating speed increases before reaching f_s. When the rotation speed increases to $2f_s$ (*R6*), the drop rate of $\overline{C_d}$ again increases to 34.2%. Similarly, all cases of the square cylinder with rotating corner wind turbines have lower C_l^{RMS} than the plain cylinder. As rotation speed increases, the variation in the drop rate of C_l^{RMS} is similar to that of $\overline{C_d}$, but the reduction in C_l^{RMS} is more significant than $\overline{C_d}$. Compared with the plain cylinder, C_l^{RMS} of the cylinder with corner wind turbines dramatically decreases by about 86.0% as the wind turbine rotation speed is $0.2f_s$(*R1*) and the drop rate becomes 71.3% when the rotating speed increases to $2f_s$ (*R6*). When the rotation speed is $4f_s$(*R8*), both $\overline{C_d}$ and C_l^{RMS} have the least reduction. It is worth noting that even though $\overline{C_d}$ and C_l^{RMS} have a significant reduction when the wind turbine has a rotating speed of $0.2f_s$ and $2f_s$, it is still smaller than the reduction in the case with the stationary wind turbines (*R0*). Figure 8c–e present the comparison of time history of *Cl* of *R1*, *R6*, and *R8*. It can be found that the fluctuation of *Cl* of *R1* is smallest, which corresponds to the largest drop rate of C_l^{RMS} of *R1*. Figure 8d,e shows that the fluctuation of *Cl* of *R6* is also effectively suppressed, but the effect of fluctuation inhibition of *R8* is not obvious.

Table 6. Force coefficient for the nine rotating cases and plain cylinder.

Case	Cell Numbers	y^+(Max)	δ/W	$\overline{C_d}$	C_l^{RMS}	S_t
Plain cylinder	2.0×10^4	1.4	2×10^{-4}	1.99	1.29	0.123
R0 (*S2*)	2.4×10^5	0.43	4×10^{-5}	1.24	0.12	0.145
R1	2.4×10^5	0.46	4×10^{-5}	1.31	0.18	0.142
R2	2.4×10^5	0.49	4×10^{-5}	1.34	0.25	0.142
R3	2.4×10^5	0.53	4×10^{-5}	1.37	0.33	0.147
R4	2.4×10^5	0.50	4×10^{-5}	1.39	0.52	0.127
R5	2.4×10^5	0.57	4×10^{-5}	1.47	0.78	0.134
R6	2.4×10^5	0.46	4×10^{-5}	1.31	0.37	0.114
R7	2.4×10^5	0.51	4×10^{-5}	1.45	0.58	0.129

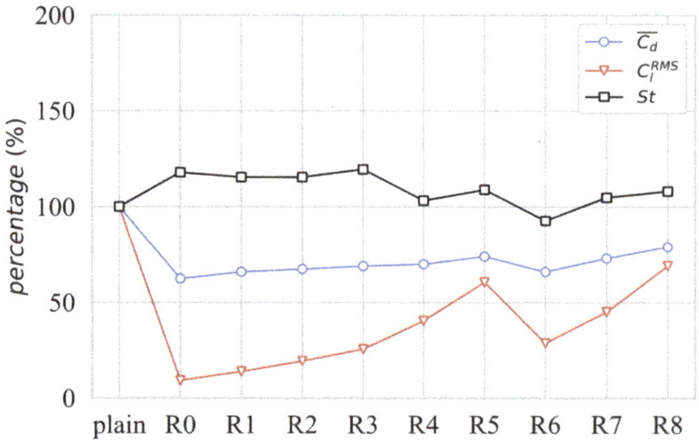

Figure 10. Comparisons of *RMS* lift coefficients, mean drag coefficients, and Strouhal number for the nine rotating cases and the plain cylinder.

Figure 11 presents the power spectrum of the lift force for the nine rotating cases and plain cylinder. Figure 11a refers to the cases with V_t/f_s less than 1 and Figure 11b shows the cases with V_t/f_s equal to or larger than 1. For Strouhal number (S_t) shown in Figure 11, all of the rotating cases have larger Strouhal numbers than the plain cylinder, except for *R6*. The vortex shedding energy represented by the peak value of the spectrum of the nine rotating cases is evidently smaller than that of the plain cylinder. Moreover, the vortex shedding energy of the cylinder with corner wind turbines has the lowest value when the wind turbine rotation speed is $2f_s$ (*R6*).

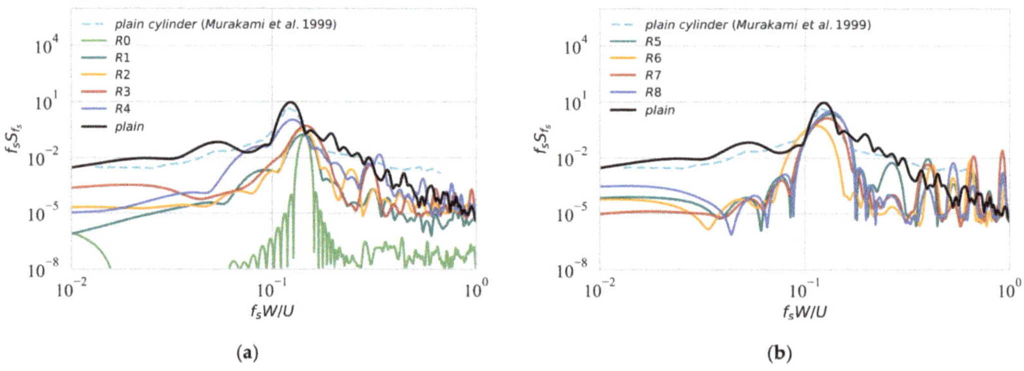

Figure 11. Power spectra of lift forces for the nine rotating cases and plain cylinder: (**a**) $V_t/f_s < 1$; (**b**) $V_t/f_s \geq 1$.

3.3. Flow Pattern around the Cylinders

The considerable effects of stationary and rotating wind turbines on the aerodynamic characteristics of a square cylinder are explained above. It can be found that there is a significant difference between aerodynamic characteristics of a plain square cylinder and a square cylinder with stationary or rotating wind turbines at the corners. To clarify the related mechanism, the flow patterns are discussed in this section.

3.3.1. Stationary Wind Turbines at Cylinder Corners

Time−averaged pressure coefficient distribution around the plain cylinder and the cylinder of the two stationary cases (*S2* and *S4*) are shown in Figure 12a–c. As for the stationary cases, *S2* has the best effect of drag reduction while *S4* has the worst result, thus the flow pattern of these two cases is discussed herein. As for *S2*, the pressure distribution on the windward face is consistent with the plain cylinder, but the negative pressures coefficient of 0.68 on the leeward face is much smaller than 1.4 of the plain cylinder, which leads to a significant decrease in $\overline{C_d}$, as presented in Figure 7. The time−averaged pressure coefficient on the leeward face of *S4* is about −0.84, leading to a 25.6% decrease in $\overline{C_d}$.

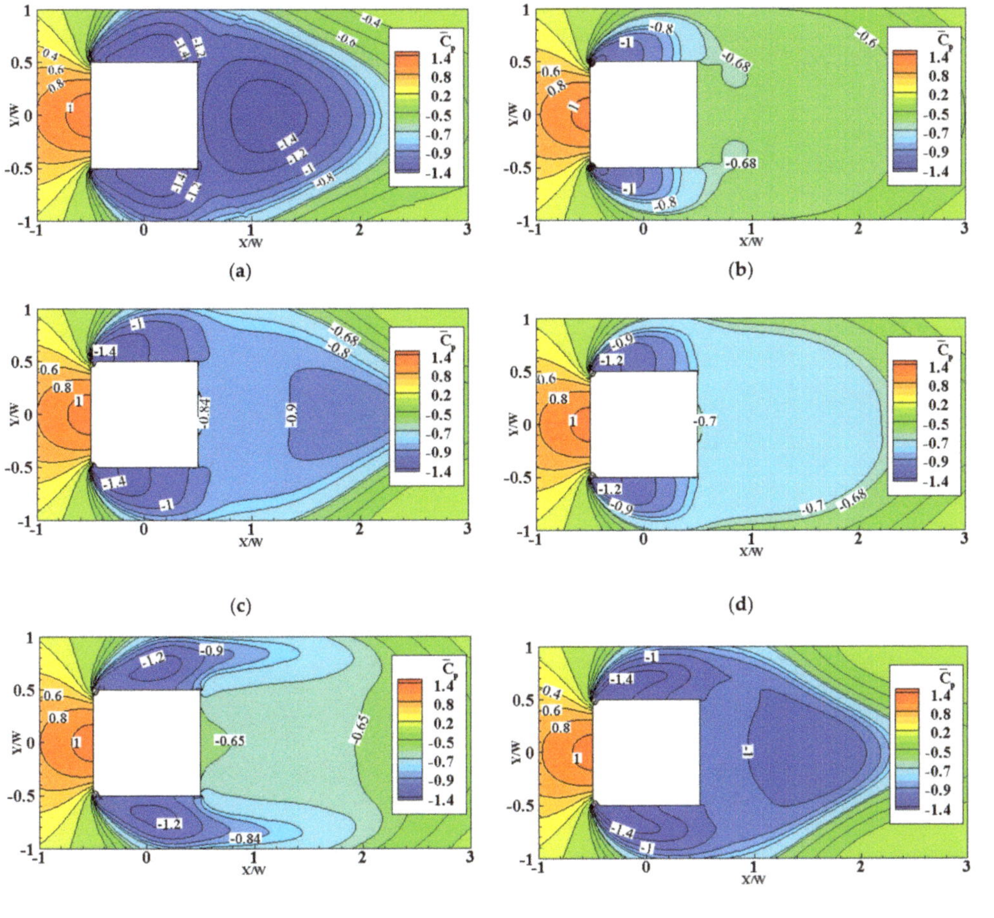

Figure 12. Time-averaged pressure coefficient distribution: (**a**) plain cylinder; (**b**) *S2*; (**c**) *S4*; (**d**) *R1*; (**e**) *R6*; (**f**) *R8*.

Time-averaged streamlines around the plain cylinder and the cylinder of the two stationary cases (*S2* and *S4*) are shown in Figure 13a–c. For the plain cylinder, two recirculation zones beside the side faces and a pair of recirculation zones which are symmetrically distributed in the near wake can be found. The color of the streamline represents the x component of the velocity (*U*). It can be found that the wind flow has an increasing speed effect at the corner. The length of the recirculation zones is defined as the length between the plain cylinder center and the saddle point of the time-averaged streamtraces along the centerline of the test models [15,32,33]. It has been reported that the drag coefficient is inversely proportional to the length of the recirculation zones in studies on drag reduction for square cylinders with cut-corners at the front edges [4] and the bluff body fitted with a splitter plate [34–37]. Less fluctuating lift coefficient can also be attributed to the elongated wake recirculation zone [29,38,39]. Apparently, the wake recirculation zone is dramatically elongated when θ is 45° (*S2*) compared with the plain cylinder and *S4*. Therefore, lower $\overline{C_d}$ and C_l^{RMS} for *S2* shown in Table 5 can be inferred. By contrast, the elongation of the length of the recirculation zone in *S4* is shorter than *S2*, leading to only a 25.6% decrease in $\overline{C_d}$ compared with the plain cylinder, as presented.

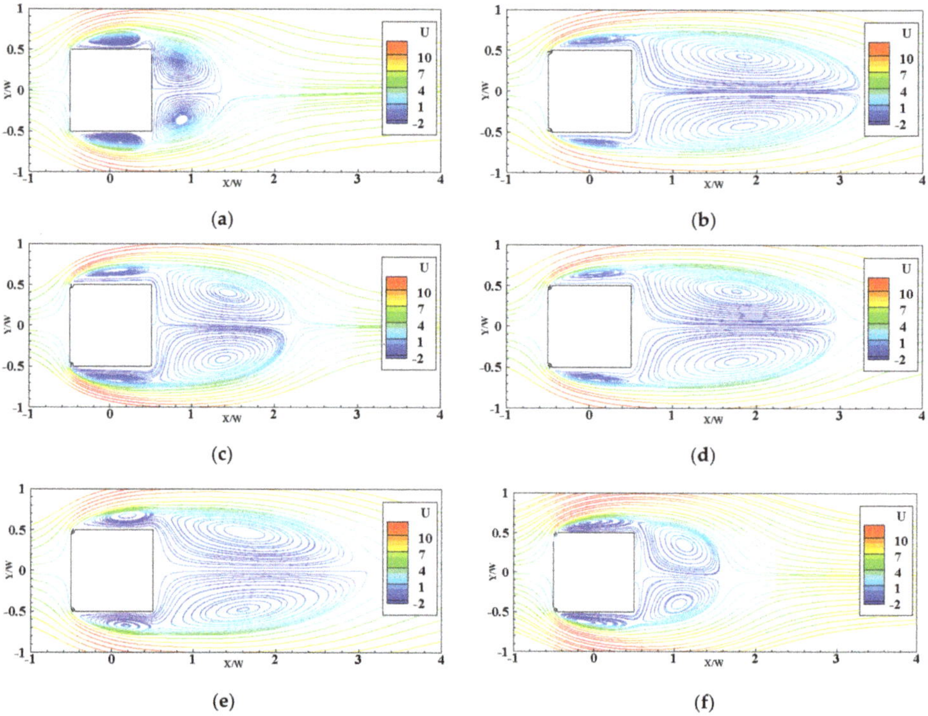

Figure 13. Time-averaged streamlines around the cylinder: (**a**) plain cylinder; (**b**) *S2*; (**c**) *S4*; (**d**) *R1*; (**e**) *R6*; (**f**) *R8*.

The suction on the side faces and leeward face depends on the curvature of the separated shear layer, the suction increases as the curvature of the separated shear increases [29]. Figure 14a–c shows the normalized time-averaged vorticity distributions of the plain cylinder and *S2* and *S4*. When θ is 45° (*S2*), the contour of normalized time-averaged vorticity distributions becomes thinner than that of the plain cylinder, which means the separated shear layer has smaller curvature, resulting in the lower suction that is shown in Figure 5. However, the contour of the normalized time-averaged vorticity distributions of the plain cylinder is stubbier, which means the curvature of the separated layer of the plain cylinder is larger. This may cause the increase in suction on the side faces and leeward faces of the plain cylinder, which may further lead to the increase of $\overline{C_d}$ in the structure.

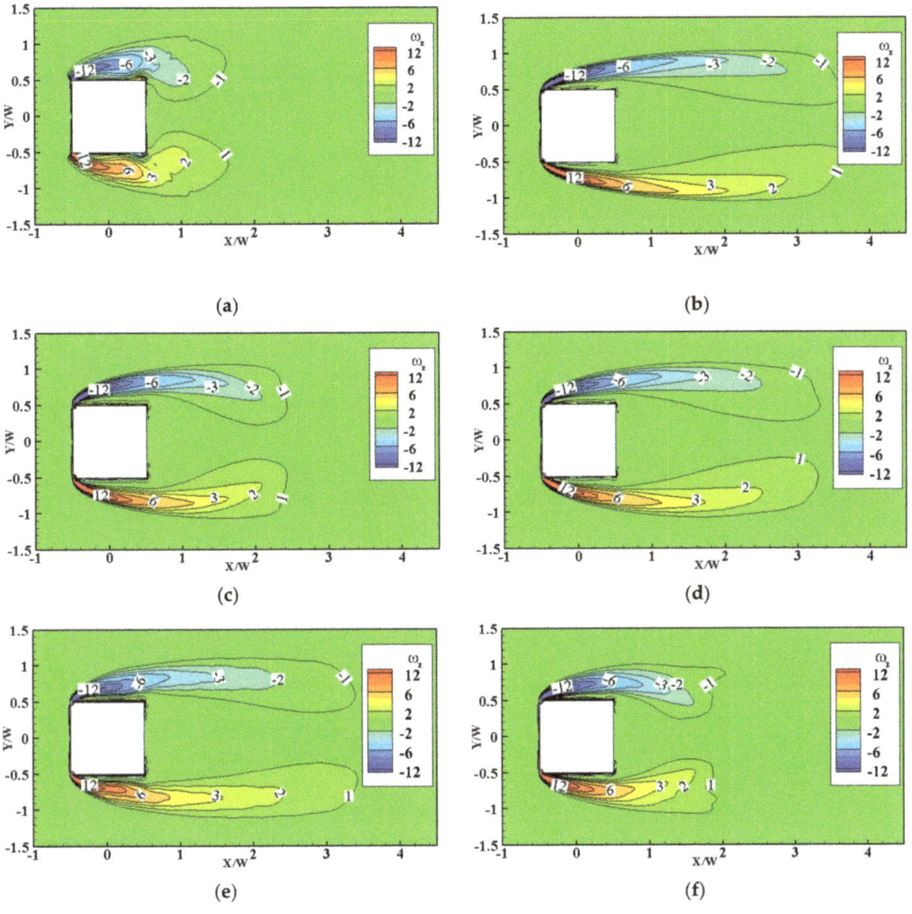

Figure 14. Normalized time-averaged vorticity distributions: (**a**) plain cylinder; (**b**) *S2*; (**c**) *S4*; (**d**) *R1*; (**e**) *R6*; (**f**) *R8*.

Compared with the instantaneous spanwise vorticity distributions of the plain cylinder as shown in Figure 15a, the normalized instantaneous spanwise vorticity distributions in one period of *S2* in Figure 15b indicate that the stationary wind turbines push the separated shear layer away and the vortices form farther downstream compared with the plain cylinder, which weakens its interaction with the side faces [38,40–43]. The weaker interaction causes a lower spectrum peak in the lift force spectrum in Figure 8 and lower C_l^{RMS}, as shown in Figure 7. Overall, the stationary wind turbines installed in the corner can reduce the mean drag force and the fluctuation of lift remarkably.

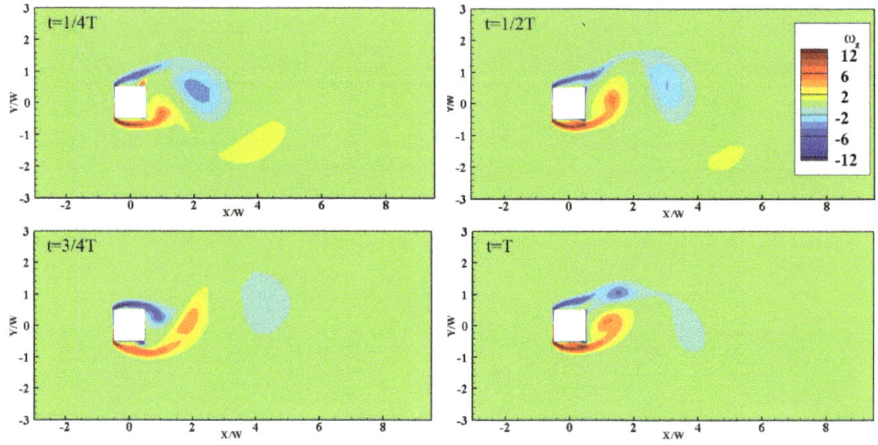

(a)

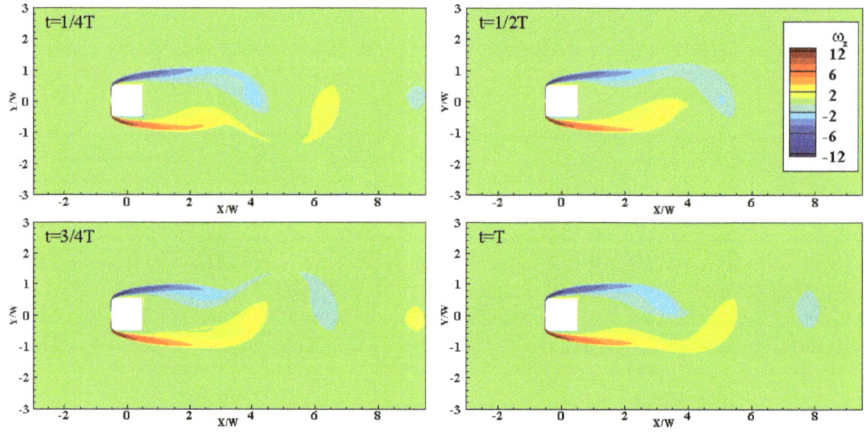

(b)

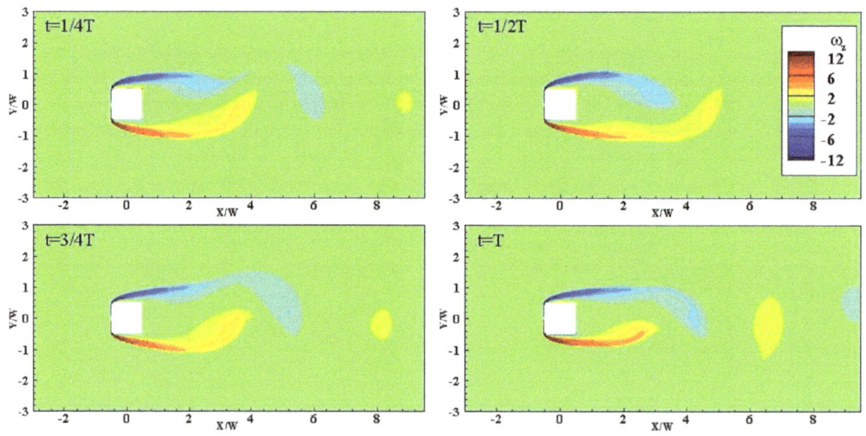

(c)

Figure 15. *Cont.*

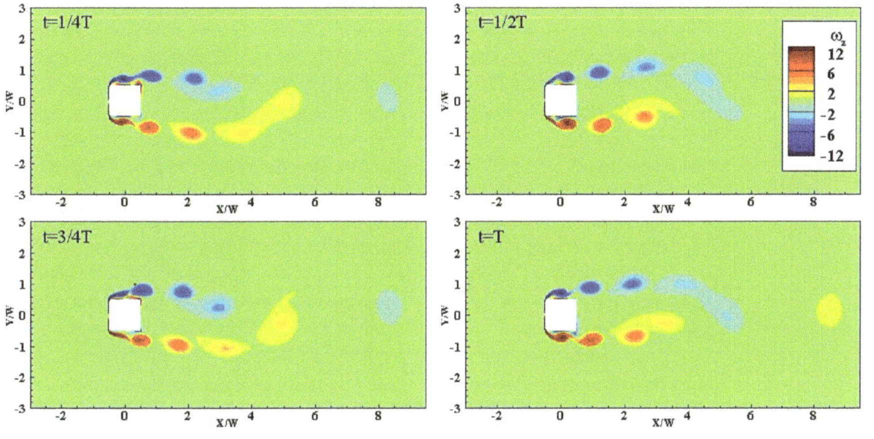

(d)

Figure 15. Normalized instantaneous vorticity distributions during one period: (**a**) plain cylinder; (**b**) *S2*; (**c**) *R1*; (**d**) *R6*.

3.3.2. Rotating Wind Turbines at Cylinder Corner

Time-averaged pressure coefficient distributions of the three rotating cases are shown in Figure 12d–f. (*R1* and *R6* have excellent effects of drag reduction while *R8* has the worst result). The pressure distributions on the windward face of *R1* and *R6* are consistent with the plain cylinder, but *R1* and *R6* have lower negative pressure coefficients (0.7 and 0.65) on the leeward face, leading to about a 35% decrease in drag coefficient, as presented in Table 6.

Time-averaged streamlines around the cylinder of the three rotating cases and the plain cylinder are shown in Figure 13d–f. Apparently, compared with the plain cylinder, the wake recirculation zones are dramatically elongated when the rotation speeds are 0.2*fs*(*R1*) and 2*fs*(*R6*). Thus, 86.0% and 71.3% decrease in C_l^{RMS} of *R1* and *R6* can be observed. On the contrary, the wake recirculation zone of *R8* is shorter than that of *R6*, which results in a worse effect on drag reduction.

Figure 14d–f shows the normalized time-averaged vorticity distributions of *R1*, *R6*, and *R8*. When the rotating speed is 2*fs*(*R6*), the contour of normalized time-averaged vorticity distributions becomes more elongated than the plain cylinder, and the separated shear layer has smaller curvature due to the lower suction, as shown in Figure 6. As for *R1*, the situation is similar. However, the curvature of the separated shear layer of *R8* is distinctly larger, so the suction of *R8* is greater than *R1* and *R6*.

In Figure 15d, the symmetrical vortex shedding is observed in the normalized instantaneous vorticity distributions during one period of *R6*. The typical Karman-type vortex shedding is replaced by the symmetric shedding modes, which can lead to the large suppression of lift fluctuations shown in Figure 9. That is why the *RMS* lift coefficient (C_l^{RMS}) of *R6* is significantly lower than that of the plain cylinder. As for *R1*, when the rotation speed is $0.2f_s$, the normalized instantaneous vorticity distribution during one period, shown in Figure 15c, is similar to *S2*, which means that *R1* has a similar mechanism of drag reduction as *S2*.

4. Concluding Remarks

This is a preliminary study to investigate the aerodynamic characteristics of a square cylinder with Savonius wind turbines and to explain the reason why this kind of structure can suppress wind-induced vibrations. The results indicate that the turbine orientation and the turbine rotation speed are the key factors that affect aerodynamic characteristics of the square cylinder. The numerical simulation results show that presence of either the stationary or the rotating wind turbines has a significant effect on wind forces acting on the square cylinder. The stationary wind turbines and rotating wind turbines installed at the leading corners can reduce both C_l^{RMS} and $\overline{C_d}$ of the plain cylinder. For the stationary wind turbine cases, when the turbine orientation angle is 45°, the stationary wind turbines push the separated shear layer away and the vortices form farther away from the square cylinder compared with the plain cylinder, which weakens its interaction with the side faces, thus causing 37.7% and 90.7% reduction in the mean drag and fluctuating lift coefficients, respectively. For the rotating wind turbine cases, the mean drag and fluctuating lift coefficients are reduced by 34.2% and 86.0%, respectively, when the rotation speed is 0.2 times of vortex shedding frequency. When the rotation speed is twice the vortex shedding frequency, the typical Karman-type vortex shedding is replaced by the symmetric shedding modes, which leads to a large suppression of lift fluctuations (71.3%) and mean drag (34.2%).

This study investigated the effect of stationary or rotating Savonius wind turbines with two blades on the aerodynamic characteristics of a square cylinder. For the next stage, other Savonius wind turbines with more blades will be considered to investigate the effects of blades on the aerodynamic performance of a square cylinder.

Author Contributions: Conceptualization, G.H. and Y.X.; methodology, G.H., F.X. and D.Z.; software, Z.W.; validation, Z.W., G.H. and D.Z.; formal analysis, Z.W. and D.Z.; investigation, Z.W. and F.X.; resources, G.H. and Y.X.; data curation, Z.W. and B.K.; writing—original draft preparation, Z.W.; writing—review and editing, G.H., D.Z., B.K., F.X. and Y.X.; visualization, Z.W.; supervision, G.H. and Y.X.; project administration, G.H.; funding acquisition, G.H. and Y.X. All authors have read and agreed to the published version of the manuscript.

Funding: This study was supported by National Key R&D Program of China (2021YFC3100702), National Natural Science Foundation of China (52108451, 52078175), Shenzhen Science and Technology Innovation Commission (GXWD20201230155427003-20200823230021001, GXWD20201230155427003-20200823134428001), Shenzhen Key Laboratory Launching Project (ZDSYS20200810113601005), and Guangdong–Hong Kong–Macao Joint Laboratory for Data-Driven Fluid Mechanics and Engineering Applications (2020B1212030001).

Institutional Review Board Statement: Not applicable.

Informed Consent Statement: Not applicable.

Conflicts of Interest: The authors declare no conflict of interest.

References

1. Carassale, L.; Freda, A.; Marre-Brunenghi, M. Experimental investigation on the aerodynamic behavior of square cylinders with rounded corners. *J. Fluids Struct.* **2014**, *44*, 195–204. [CrossRef]
2. Tamura, T.; Miyagi, T. The effect of turbulence on aerodynamic forces on a square cylinder with various corner shapes. *J. Wind. Eng. Ind. Aerodyn.* **1999**, *83*, 135–145. [CrossRef]
3. Tamura, T.; Miyagi, T.; Kitagishi, T. Numerical prediction of unsteady pressures on a square cylinder with various corner shapes. *J. Wind. Eng. Ind. Aerodyn.* **1998**, *74*, 531–542. [CrossRef]
4. He, G.S.; Li, N.; Wang, J.J. Drag reduction of square cylinders with cut-corners at the front edges. *Exp. Fluids* **2014**, *55*, 1745. [CrossRef]
5. Kurata, M.; Ueda, Y.; Kida, T.; Iguchi, M. Drag reduction due to cut-corners at the front-edge of a rectangular cylinder with the length-to-breadth ratio being less than or equal to unity. *J. Fluids Eng.* **2009**, *131*, 064501. [CrossRef]
6. Tse, K.; Hitchcock, P.A.; Kwok, K.C.; Thepmongkorn, S.; Chan, C.M. Economic perspectives of aerodynamic treatments of square tall buildings. *J. Wind. Eng. Ind. Aerodyn.* **2009**, *97*, 455–467. [CrossRef]
7. Kwok, K.; Wilhelm, P.A.; Wilkie, B.G. Effect of edge configuration on wind-induced response of tall buildings. *Eng. Struct.* **1988**, *10*, 135–140. [CrossRef]
8. Kwok, K.C.; Bailey, P.A. Aerodynamic devices for tall buildings and structures. *J. Eng. Mech.* **1987**, *113*, 349–365. [CrossRef]
9. Bearman, P.W.; Trueman, D.M. An investigation of the flow around rectangular cylinders. *Aeronaut. Q.* **1972**, *23*, 229–237. [CrossRef]

10. Modi, V.J.; Deshpande, V.S. Fluid dynamics of a cubic structure as affected by momentum injection and height. *J. Wind. Eng. Ind. Aerodyn.* **2001**, *89*, 445–470. [CrossRef]
11. Munshi, S.R.; Modi, V.J.; Yokomizo, T. Aerodynamics and dynamics of rectangular prisms with momentum injection. *J. Fluids Struct.* **1997**, *11*, 873–892. [CrossRef]
12. Hu, G.; Tse, K.T.; Wei, M.; Naseer, R.; Abdelkefi, A.; Kwok, K. Experimental investigation on the efficiency of circular cylinder-based wind energy harvester with different rod-shaped attachments. *Appl. Energy* **2018**, *226*, 682–689. [CrossRef]
13. Li, Q.S.; Chen, F.B.; Li, Y.G.; Lee, Y.Y. Implementing wind turbines in a tall building for power generation: A study of wind loads and wind speed amplifications. *J. Wind. Eng. Ind. Aerodyn.* **2013**, *116*, 70–82. [CrossRef]
14. Jafari, S.; Kwok, K.C.; Hassanli, S. Integration of wind turbines in tall buildings for wind power generation. In Proceedings of the 8th International Colloquium on Bluff Body Aerodynamics and Applications, Boston, MA, USA, 7–11 June 2016.
15. Cochran, B.C.; Damiani, R.R. Harvesting wind power from tall buildings. In *Wind Power*; CPP, Inc.: Mountain View, CA, USA, 2008.
16. Li, T.; Qin, D.; Zhang, J. Effect of RANS turbulence model on aerodynamic behavior of trains in crosswind. *Chin. J. Mech. Eng.* **2019**, *32*, 85. [CrossRef]
17. Lu, Y.; Yang, M.; Qian, B. The influence of reduced cross-section on pressure transients from high-speed trains intersecting in a tunnel. *J. Wind. Eng. Ind. Aerodyn.* **2020**, *201*, 104–161. [CrossRef]
18. Deng, E.; Yang, W.; He, X.; Zhu, Z.; Wang, H.; Wang, Y. Aerodynamic response of high-speed trains under crosswind in a bridge-tunnel section with or without a wind barrier. *J. Wind. Eng. Ind. Aerodyn.* **2021**, *210*, 104–502. [CrossRef]
19. Wang, T.; Wu, F.; Yang, M.; Ji, P.; Qian, B. Reduction of pressure transients of high-speed train passing through a tunnel by cross-section increase. *J. Wind. Eng. Ind. Aerodyn.* **2018**, *183*, 235–242. [CrossRef]
20. Rodi, W.; Ferziger, J.H.; Breuer, M.; Pourquie, M. Status of large eddy simulation: Results of a workshop. *Trans.-Am. Soc. Mech. Eng. J. Fluids Eng.* **1997**, *119*, 248–262. [CrossRef]
21. Voke, P.R. Flow past a square cylinder: Test case LES2. In *Direct and Large-Eddy Simulation II*; Kleiser, L., Ed.; Springer: Dordrecht, The Netherlands, 1997; Volume 5, pp. 355–373.
22. Deng, Y.; Zhu, Z. Numerical simulation of two dimensional flow around square column at full wind Angle. *J. Railw. Sci. Eng.* **2016**, *13*, 1796–1802.
23. Bearman, P.W.; Obasaju, E.D. An experimental study of pressure fluctuations on fixed and oscillating square-section cylinders. *J. Fluid Mech.* **1982**, *119*, 297–321. [CrossRef]
24. Lee, B.E. The effect of turbulence on the surface pressure field of a square prism. *J. Fluid Mech.* **1975**, *69*, 263–282. [CrossRef]
25. Wang, Q.; Jiang, Q.; Hu, G.; Chen, X.; Li, C.; Xiao, Y. Aerodynamic characteristics of a square cylinder with corner fins. *Adv. Bridge Eng.* **2021**, *2*, 20. [CrossRef]
26. Lyn, D.A.; Einav, S.; Rodi, W.; Park, J.H. A laser-Doppler velocimetry study of ensemble-averaged characteristics of the turbulent near wake of a square cylinder. *J. Fluid Mech.* **1995**, *304*, 285–319. [CrossRef]
27. Lyn, D.A.; Rodi, W. The flapping shear layer formed by flow separation from the forward corner of a square cylinder. *J. Fluid Mech.* **1994**, *267*, 353–376. [CrossRef]
28. Hangan, H.M. Wake Aerodynamics for 2D Bluff Bodies. Ph.D. Thesis, The University of Western Ontario, London, ON, Cnada, 1997.
29. Xu, J. *Effects of Vertical Clearance on Aerodynamic Characteristics of Long-Span Bridge Girders*; Hunan University: Changsha, China, 2014.
30. Wenehenubun, F.; Saputra, A.; Sutanto, H. An experimental study on the performance of Savonius wind turbines related with the number of blades. *Energy Procedia* **2015**, *68*, 297–304. [CrossRef]
31. Murakami, S.; Iizuka, S.; Ooka, R. CFD analysis of turbulent flow past square cylinder using dynamic LES. *J. Fluids Struct.* **1999**, *13*, 1097–1112. [CrossRef]
32. Gao, D.; Chen, G.; Chen, W.; Huang, Y.; Li, H. Active control of circular cylinder flow with windward suction and leeward blowing. *Exp. Fluids* **2019**, *2*, 60. [CrossRef]
33. Konstantinidis, E.; Balabani, S.; Yianneskis, M. The effect of flow perturbations on the near wake characteristics of a circular cylinder. *J. Fluids Struct.* **2003**, *18*, 367–386. [CrossRef]
34. Gao, D.; Chen, G.; Huang, Y.; Chen, W.; Li, H. Flow characteristics of a fixed circular cylinder with an upstream splitter plate: On the plate-length sensitivity. *Exp. Therm. Fluid Sci.* **2020**, *117*, 110135. [CrossRef]
35. Gao, D.; Huang, Y.; Chen, W.; Chen, G.; Li, H. Control of circular cylinder flow via bilateral splitter plates. *Phys. Fluids* **2019**, *31*, 057105. [CrossRef]
36. Bearman, P.W. Investigation of the flow behind a two-dimensional model with a blunt trailing edge and fitted with splitter plates. *J. Fluid Mech.* **1965**, *21*, 241–255. [CrossRef]
37. Chen, W.; Xin, D.; Xu, F.; Li, H.; Ou, J.; Hu, H. Suppression of vortex-induced vibration of a circular cylinder using suction-based flow control. *J. Fluids Struct.* **2013**, *42*, 25–39. [CrossRef]
38. Gao, D.; Meng, H.; Huang, Y.; Chen, G.; Chen, W. Active flow control of the dynamic wake behind a square cylinder using combined jets at the front and rear stagnation points. *Phys. Fluids* **2021**, *33*, 047–101. [CrossRef]
39. Lam, K.; Lin, Y.F. Large eddy simulation of flow around wavy cylinders at a subcritical Reynolds number. *Int. J. Heat Fluid Flow* **2008**, *29*, 1071–1088. [CrossRef]

40. Gao, D.; Chen, W.; Li, H.; Hu, H. Flow around a circular cylinder with slit. *Exp. Therm. Fluid Sci.* **2017**, *82*, 287–301. [CrossRef]
41. Chen, W.; Gao, D.; Yuan, W.; Li, H.; Hu, H. Passive jet control of flow around a circular cylinder. *Exp. Fluids* **2015**, *11*, 56. [CrossRef]
42. Gao, D.; Chen, W.; Chen, G.; Li, H. Effects of steady wake-jets on subcritical cylinder flow. *Exp. Therm. Fluid Sci.* **2019**, *102*, 575–588. [CrossRef]
43. Gao, D.; Deng, Z.; Yang, W.; Chen, W. Review of the excitation mechanism and aerodynamic flow control of vortex-induced vibration of the main girder for long-span bridges: A vortex-dynamics approach. *J. Fluids Struct.* **2021**, *105*, 103348. [CrossRef]

applied sciences

MDPI

Article

Effect of the Extended Rigid Flapping Trailing Edge Fringe on an S833 Airfoil

Hongtao Yu and Zifeng Yang *

Department of Mechanical and Materials Engineering, Wright State University, Dayton, OH 45435, USA; yu.41@wright.edu
* Correspondence: zifeng.yang@wright.edu

Abstract: A 2D numerical simulation was conducted to investigate the effect of an extended rigid trailing edge fringe with a flapping motion on the S833 airfoil and its wake flow field, as an analogy of an owl's wing. This study aims to characterize the influence of the extended flapping fringe on the aerodynamic performance and the wake flow characteristics downstream of the airfoil. The length (L_e) and flapping frequencies (f_e) of the fringe are the key parameters that dominate the impact on the airfoil and the flow field, given that the oscillation angular amplitude is fixed at 5°. The simulation results demonstrated that the airfoil with an extended fringe of 10% of the chord at a flapping frequency of f_e = 110 Hz showed a substantial effect on the pressure distribution on the airfoil and the flow characteristics downstream of the airfoil. An irregular vortex street was predicted downstream, thus causing attenuations of the vorticities, and shorter streamwise gaps between each pair of vortices. The extended flapping fringe at a lower frequency than the natural shedding vortex frequency can effectively break the large vortex structure up into smaller scales, thus leading to an accelerated attenuation of vorticities in the wake.

Keywords: flapping fringe; CFD simulation; vortex attenuation; aerodynamics enhancement

Citation: Yu, H.; Yang, Z. Effect of the Extended Rigid Flapping Trailing Edge Fringe on an S833 Airfoil. *Appl. Sci.* **2022**, *12*, 444. https://doi.org/10.3390/app12010444

Academic Editor: Giangiacomo Minak

Received: 13 October 2021
Accepted: 29 December 2021
Published: 3 January 2022

Publisher's Note: MDPI stays neutral with regard to jurisdictional claims in published maps and institutional affiliations.

1. Introduction

For decades, the quiet flight of owls has drawn interest from researchers in the aerodynamic and aeroacoustics fields. Owl-wing-inspired modifications on either leading-edge or trailing-edge designs have attracted plenty of research interest aiming to improve the aerodynamic performance and/or, in the meanwhile, suppress the aerodynamic noise from airfoils or turbine/fan blades. Multiple bio-inspired airfoil designs have been evaluated, including the trailing-edge or leading-edge serrations, trailing-edge flaplets, flapping wings, the compliant surface, and the flexible trailing-edge fringe, etc. [1–9]. It has been found that the soft trailing edge fringe can attenuate the primary noise source generated from the trailing edge through the modification of the coherent scattering mechanism. Consequently, aeroacoustic noise above 2 kHz can be effectively attenuated by the compliant surface covered by the down feathers through a bypass dissipation mechanism [10,11]. In the last decade, the use of an owl-wing-inspired trailing edge design as a passive control approach for an airfoil model to minimize drag and improve lift coefficients has gained attention [12–16].

The brush-like trailing edge has been utilized as the trailing edge extension to reduce both the narrowband bluntness noise and broadband turbulent boundary-layer trailing-edge noise, as it contains the characteristics of both porosity and flexibility as the natural fringe [17–21]. A remarkable noise reduction was observed for an airfoil with a flexible trailing edge equipped to the suction side of the airfoil experimentally [21]. The attachment of flaplets on the trailing edge of an airfoil has clearly shown reductions in tonal noise, primarily resulting from the generation of small-scale vortices by the flaplets [22].

Recently, Murayama et al. (2021) carried out an experiment to study the effects of flexible flaps with various gaps on the aerodynamic characteristics of a fixed-wing [23].

The results indicated that the adjustable flaps enabled the suppression of the large-scale vortex shedding, and thus, reduced the fluctuation of aerodynamic forces. This finding is consistent with our experimental findings of an airfoil with soft feathers installed on the trailing edge [24]. Our previous experiment demonstrated that the real bird feather enabled the large-scale vortex to break up into small-scale turbulence. It is illustrated that the vibrating porous feather fringe produces a significant vertical velocity in the flow field around the trailing edge. As a result, it suppressed the leading-edge vortices approaching the flexible tail, and accelerated the shedding vortices' dissipation. Additionally, Yu et al. (2020) performed computational fluid dynamics (CFD) simulations on an airfoil with an extended trailing-edge fringe [25]. The effects of the fringe lengths and flapping frequencies on shedding vortices have been investigated. The flapping behavior with a certain length and frequency can attenuate the strength of the vortices, and improve the aerodynamics in terms of reducing the drag coefficients, and increasing the lift coefficients. The abovementioned studies aimed at improving the aerodynamics of an airfoil, and reducing the noise. The flapping trailing edge fringe can be applied to wind turbine blades and airplanes to improve noise reduction from the aerodynamic perspective.

In this study, to leverage the influence of the extended flapping trailing edge fringe on the aerodynamic performance of the airfoil, CFD simulations are performed for the S833 airfoil with an extended flapping trailing edge fringe. This numerical simulation model is different from our previous experimental study. First, the experimental model was equipped with real feathers at the trailing edge of the airfoil as a passive flow control without any controlled flapping motion. Due to the complexity of the real feather structure, the porosity, and the biological material properties, such numerical studies haven't been conducted yet. In this work, a simplified bare fringe without any porosity in a 2D fashion, together with specific flapping motions working as an active control, is adopted in the numerical modeling. In the simulation, three fringe lengths (L_e) at four different flapping frequencies (f_e) are evaluated thoroughly. The analysis focuses on the flapping-induced changes on the Q-criterion and velocity distributions within the flow field, and the aerodynamic pressure distributions over the airfoil.

2. Materials and Methods

2.1. Computational Model

The NREL's S833 airfoil has been extensively used to design wind turbine blades [26]. However, with the rapid development of wind farms, noise emission has become a significant issue. The ultimate goal of this study is to explore a novel airfoil design to reduce the noise generation associated with the shedding vortices from wind turbine blades. Thus, the S833 airfoil is selected to investigate the aerodynamic interactions between the vortex shedding in the wake and the flapping fringe installed at the trailing edge. Two-dimensional CFD simulations are performed for the airfoil model. The dimensions and boundary conditions applied in simulations are shown schematically in Figure 1.

The rectangular area of the outside domain is 20C by 10C (C: the chord length of the airfoil), with 5C from the inlet to the leading-edge of the airfoil. The flow boundary conditions applied in the simulation are adopted from the parameters in our previous experimental study [24]. The upstream velocity is set at 6 m/s, and the density and viscosity of the air are assumed as 1.23 kg/m^3 and 1.84 $\times$ 10^{-5} Pa s, respectively, to match the air property at room temperature. This boundary condition corresponds to a Reynolds number (Re) of 40,000, which is the same as the Re number evaluated in the experiment. This Re number falls within the range of the Re number of owls' flight. The outlet boundary condition is defined as zero static pressure, and surfaces on the airfoil, as well as the top and bottom walls, are defined as non-slip wall conditions. The airfoil's angle of attack is set at 9 degrees, which is also consistent with the experimental study [24]. The fringe length (L_e) ranges from 8% to 12% of the chord, with a step increment of 2%. The flapping frequency (f_e) ranges from 80 Hz to 170 Hz, with a step increment of 30 Hz. The initial f_e is determined as the shedding vortex frequency (f) of the baseline bare airfoil model, which is 140 Hz, obtained via the power spectral density analysis of the velocity in the downstream wake [25]. The dynamic motion of the trailing edge fringe can be expressed as $\theta = A\sin(2\pi ft)$, where A is the amplitude flapping angle ($A = 5°$). The connection point of the flapping motion is fixed at the tip of the trailing edge.

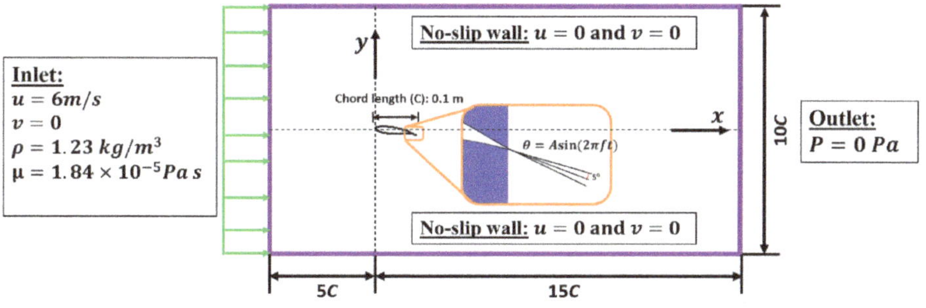

Figure 1. The 2-dimensional computational domain and boundary conditions for the CFD simulation. C: chord length of the airfoil; u: velocity in the x-direction; v: velocity in the y-direction; ρ: density of the fluid; μ: viscosity of the fluid; P: pressure; θ: dynamic motion of the trailing edge fringe; A: amplitude of the flapping angle and set at 5 degrees; f: frequency of the flapping motion.

2.2. CFD Simulations

In this work, three fringe lengths at four different flapping frequencies of the extended fringe were evaluated. These variables resulted in 12 simulation cases, and a bare airfoil model was also simulated as the baseline case. Simulations were performed using the CFD software Cradle SC/Tetra 2021 (Dayton, OH, USA). The computational software solves the N–S equations by using the finite volume method with hybrid computational grids, wherein hexahedral grids were generated as prism layers around the airfoil to capture the boundary layer behavior, and tetrahedral grids were generated in the rest of the region. A local mesh refinement technique (grids become coarser along the downstream direction) was adopted to resolve the flow around the flapping fringes installed at the trailing edge to save computational effort.

The turbulence model LKE K-KL-ω was adopted to evaluate the turbulence region over the airfoil. This model is one of the Reynolds-Averaged Navier–Stokes (RANS)-based approaches to predict the transitional flow [25,27,28]. In the LKE model, the energy of the disturbances in the pre-transitional region of a boundary layer is expressed as "Laminar Kinetic Energy", whereas the turbulence energy is defined as k, and the transport equation of KL is solved with two equations of the fully turbulent model. The flow over the airfoil is governed by the Navier–Stokes equation for two-dimensional, viscous, incompressible flow. The continuity equation and the Reynolds-averaged N–S equations can be expressed as:

$$\frac{\partial(\rho\overline{u_i})}{\partial(x_i)} = 0 \tag{1}$$

$$\rho\frac{D(\rho\overline{u_i})}{Dt} = \frac{\partial(\overline{P})}{\partial(x_i)} + \frac{\partial}{\partial(x_i)}(\mu\left(\frac{\partial(\overline{u_i})}{\partial(x_j)} + \frac{\partial(\overline{u_i})}{\partial(x_i)}\right) - \rho\overline{u_i'u_j'} \tag{2}$$

where i and j are coordinate indices; u is the velocity; x is the flow direction; ρ is the density of the fluid; μ is the dynamic viscosity of the fluid.

The dynamic motion of the fringe was fulfilled using the stretching mesh technique, as it enables the expansion and shrinking of the mesh elements to accommodate the movement of the solid boundary, which can avoid the pressure or velocity discontinuity/fluctuations at interfaces [25]. The CFD simulation, in terms of the turbulence model, grid sensitivity, and the dynamic moving mesh technique, has been validated statically by comparing static lift and drag coefficients with experiments, and dynamically, by comparing pressure contours for an oscillatory plunging motion of an airfoil with the literature reported in our previous study [25].

The numerical simulations were performed on a computational cluster containing 3.4 GHz AMD Phenom II X4 965 Quad-cores and 8 GB RAMs. Each case was run with 32 processors. In the current study, a total physical time period of 1.5 s of the flow was simulated with a time step of 5×10^{-5} s. The mean computational time for each simulation case was about 23 h.

3. Results and Discussion

The impact of the extended flapping fringe on the airfoil's aerodynamics were studied primarily through the alterations in the distribution of the Q-criterion, velocities in the flow field, and pressure coefficients over the airfoil model. The variable Q-criterion quantifying the swirling vortex strength is defined as

$$Q = \frac{1}{2}\left[\left(\frac{\partial u}{\partial x}\right)^2 + \left(\frac{\partial v}{\partial y}\right)^2\right] - \frac{\partial u}{\partial y}\frac{\partial v}{\partial x} \quad (3)$$

where u is the instantaneous velocity in the x-direction, and v is the instantaneous velocity in the y-direction.

3.1. Effect of the Flapping Motion on the Q-Criterion and Velocity Distributions

The instantaneous (at t = 1 s) and time-averaged Q-criterion and velocity distributions with streamlines of the bare airfoil model (baseline) are presented in Figure 2. The quasi-steady-state of the simulation was reached by checking the periodicity of the flow field. The vortices were shed evenly in the wake, with an identical magnitude of the Q-criterion for each pair, indicating typical Karman vortex street characteristics. The time-averaged Q-criterion shows very weak vortices downstream in the wake. It implies that the rotational direction of two vortices for each pair was opposite, and thus, after averaging, the opposite vorticities counteracts the magnitude of the Q-criterion. A separation bubble was predicted around the middle section on the upper surface of the airfoil, as shown in the instantaneous velocity distribution panel in Figure 2. In the time-averaged flow field, the detachment and reattachment point for the separation bubble can be clearly identified by the streamlines surrounding the circulation region. The separation bubble has a length of about 25% of the chord. In fact, the shedding vortices downstream of the airfoil were initiated by the separation bubble.

The Q-criterion (around the airfoil and in the wake) and velocity distributions for the airfoil model with the fringe length of 8% of the chord at four different f_e are presented in Figures 3–5. The large-scale vortices initiated over the upper surface of the airfoil were broken up into smaller scale vortices compared to the shed vortices of the bare airfoil model. The flapping motion at the flapping frequency of 80 Hz creates an irregular pattern of vortices compared to the cases with higher frequencies (110–170 Hz), as shown in Figure 3. The time-averaged Q-criterion distributions demonstrate an accelerated decay of vortices by the extended fringe with a low flapping frequency (80 Hz), resulting from the unstructured small-scale vortices being shed irregularly. This observation might be induced by the velocity "offset effect" between induced vortices and the shed vortices from upstream. With the increased flapping frequency, the vortices decayed slowly. This might be attributed to the fact that the contribution from the flapping fringe to the overall swirling strength becomes more dominant with the increased flapping frequency.

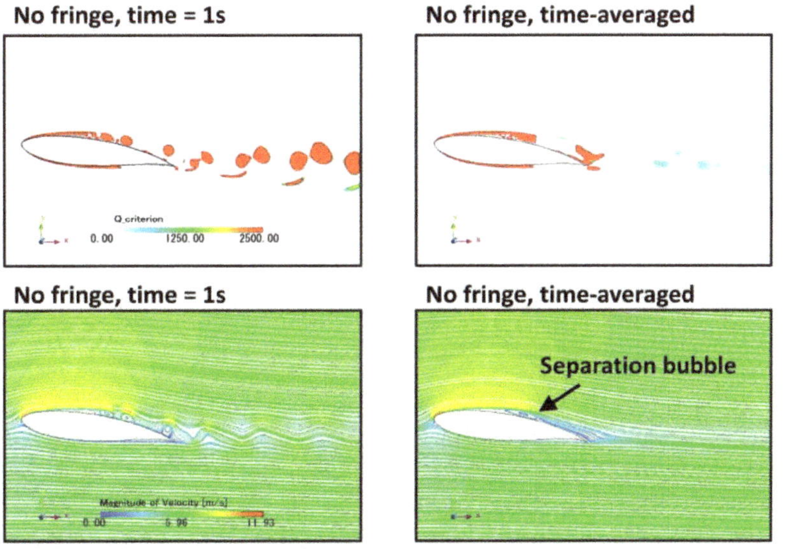

Figure 2. The simulated instantaneous (time = 1 s) and time-averaged Q-criterion and velocity distributions with streamline of the bare airfoil model.

In the downstream wake, the airfoil model with f_e = 110, 140, or 170 Hz shows evenly shed vortices, as shown in Figure 4. The increased flapping frequency seems to shorten the gap between each pair of vortices, which is directly correlated to the flapping frequency of the fringe. From the comparison of the four flapping frequencies, it is suggested that when the flapping frequency f_e is 60 Hz (40%) lower than f (natural vortex shedding frequency of the bare airfoil model, 140 Hz), the regular vortex shedding is interfered with, and results in an irregular vortex distribution with non-uniform scales and shorter gaps between each pair. As can be observed, the flapping trailing edge fringe with f_e = 80 and 100 Hz alters the coherent structure of the large-scale vortices in the wake, and at the same time, reduces the vorticity in the flow field (Figure 4).

However, it is difficult to differentiate the influence of the fringe tip position (fringe tip at the top or bottom positions) on the distribution of the Q-criterion. Thus, the instantaneous and time-averaged flow fields around the airfoil were plotted to illustrate the effects of the flapping fringe on the flow field, as shown in Figure 5. The flapping motion induces movement of the separation bubble towards the leading edge of the airfoil for all cases (Figure 5). The existence of the separation bubble caused fluctuated pressure forces on the upper surface, reflecting the unstable pressure distribution over the airfoil model, as shown in Figure 6. Higher flapping frequencies tend to increase the velocity over the top surface of the airfoil, and, as a consequence, it results in a decrease in the aerodynamic pressure. Subsequently, the pressure drop could lead to an increase in the lift coefficient, and a decrease in drag coefficient. It is indicated that the pressure acting on the top surface tends to be small when the flapping fringe tip is at the bottom position, in contrast to that when the fringe tip is at the top position.

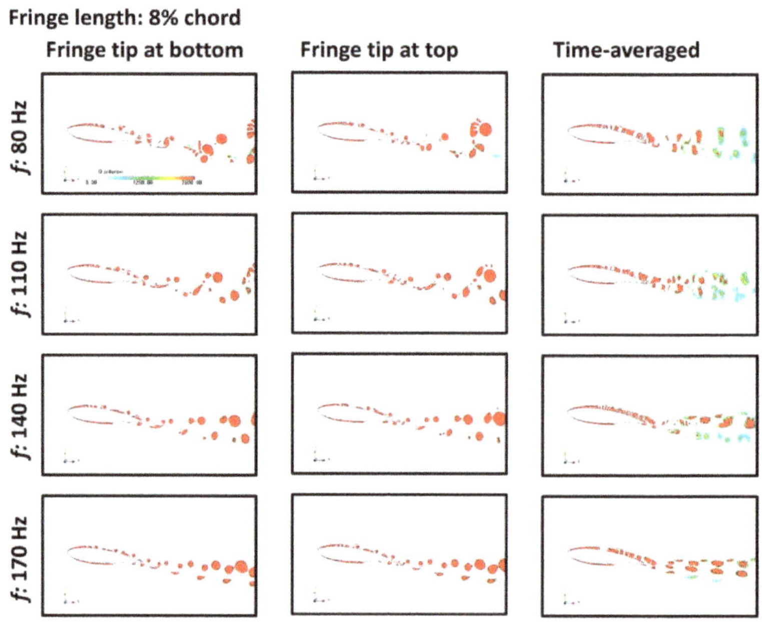

Figure 3. Q-criterion distributions around the airfoil model with the fringe length of 8% chord at different flapping frequencies. Left column: instantaneous Q-criterion distributions at the fringe tip moved to the top; Middle column: instantaneous Q-criterion distributions at the fringe tip moved to the bottom; Right column: time-averaged Q-criterion distribution.

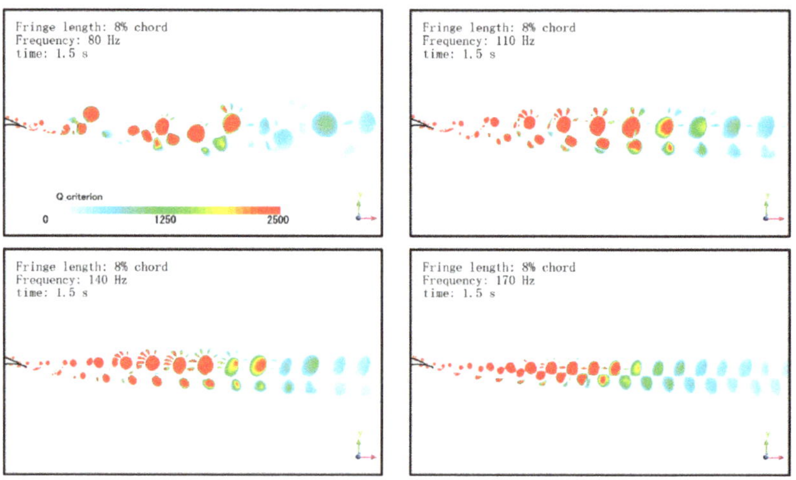

Figure 4. Instantaneous Q-criterion distributions in the wake of the airfoil model with the fringe length of 8% chord at different flapping frequencies at t = 1.5 s.

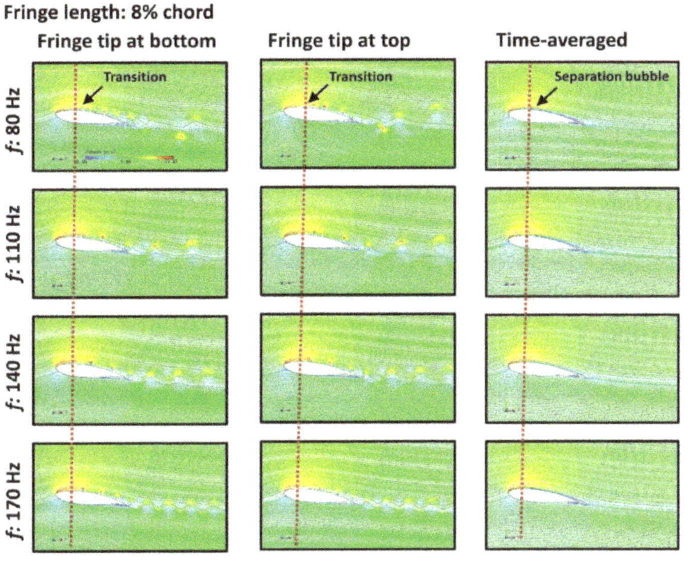

Figure 5. The velocity distribution around the airfoil model with the fringe length of 8% chord at different flapping frequencies. **Left column**: instantaneous velocity distributions when the fringe tip moved to the top; **Middle column**: instantaneous Q-criterion distributions when the fringe tip moved to the bottom; **Right column**: time-averaged Q-criterion distribution.

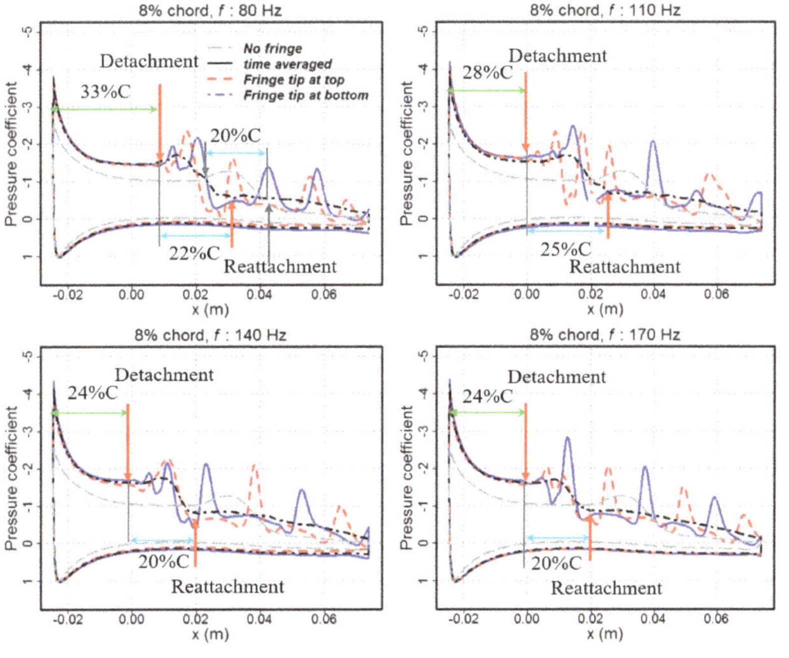

Figure 6. Pressure coefficient distribution over the airfoil model with the fringe length of 8% chord at four flapping frequencies.

The pressure coefficient distributions over the airfoil model with the fringe length of 8%C at four flapping frequencies are compared, as shown in Figure 6. The instantaneous pressure coefficient distribution shows a significant fluctuation corresponding to the shedding vortex over the upper surface as compared to the averaged pressure distribution over the baseline airfoil model. An obvious phase difference of the pressure oscillations can be observed between the case with the fringe tip at the bottom position and that with the fringe tip at the top position. The phase difference is more complicated in the region of the separation bubble compared to the region downstream of the separation bubble. By comparing the time-averaged pressure distributions, one can observe that higher flapping frequencies tend to move the separation bubble forward (towards the leading ledge). The distance from the leading edge to the take-off point of the separation bubble decreases from 33%C to 24%C as the flapping frequency increases from 80 Hz to 170 Hz. The separation bubble length is about 22%C, 25%C, 20%C, and 20%C for the case of f_e = 80 Hz, f_e = 110 Hz, f_e = 140 Hz, and f_e = 170 Hz, respectively. Overall, higher flapping frequencies tend to generate a lower time-averaged pressure coefficient over the upper surface, which would result in an increase in the lift coefficient.

Figures 7–10 present the instantaneous Q-criterion distribution over the airfoil, the Q-criterion distribution in the wake, the velocity, and the pressure coefficient distributions over the airfoil model, respectively, with the fringe length of 10% at different flapping frequencies. Figures 11–14 present the instantaneous Q-criterion distribution over the airfoil, the Q-criterion in the wake, the velocity, and the pressure coefficient distributions over the airfoil model, respectively, with the fringe length of 12% chord at various flapping frequencies. Similarly, the vortex shedding characteristics are similar to the airfoil model with the fringe length of 8% chord. The flapping motion with a relatively lower f_e (<140 Hz) can break the large-scale vortex into small-scale vortices at the trailing edge, and, subsequently, alter the coherent structure of the vortex shedding. The irregularly shed vortices and the increased gap between each pair of vortices for cases with a lower f_e accelerated the vortex decay compared to those with a higher f_e. As the flapping frequency goes up to 170 Hz, the shedding frequency of the vortex is dominated by the flapping motion, which results in a regulated pattern of the shedding vortices. The time-averaged Q-criterion distributions show that the resultant time-averaged vortex strength has been elevated by the flapping fringe. It is indicated that the extended flapping fringe did not always facilitate the reduction of vorticity; conversely, it could promote the vortex domination in the wake, its interaction with the blunt trailing edge, and, possibly, the consequent noise emission from the trailing edge of the airfoil.

With the increased fringe length, the scale of the vortex became smaller, as shown in Figures 7 and 11. Correspondingly, the overall Q-criterion distribution over the top surface of the airfoil is decreased, as presented in Figures 8 and 12. These alterations in the flow field around the airfoil led to a decrease in the pressure coefficient over the upper surface of the airfoil model, as shown in Figures 10 and 14. Compared with the flapping frequency, the effect of the fringe length is less significant in terms of slight changes in both the vortex strength and the pressure distribution.

3.2. *Effect of the Extended Flapping Fringe on the Lift and Drag Coefficients*

The extended flapping trailing edge fringe not only reduced the swirling strength in the wake of the airfoil, but also decreased the drag coefficient, and increased the lift coefficient, surprisingly, as shown in Figure 15. It should be noted that in the integration calculation of the lift and drag, only the pressure distribution over the airfoil surface, excluding the trailing edge fringe, was considered. The drag coefficient essentially decreased with the increase in fringe length, and always decreased with the increase in flapping frequency, whereas the lift coefficient was essentially proportional to these two factors. The airfoil model with L_e = 12%C and f_e = 170 Hz generated the minimum drag coefficient, i.e., 22.5% lower than the baseline, and the maximum lift coefficient, i.e., 63.2% higher than the baseline, as can be observed in Figure 15. However, this combination would not alleviate the swirling strength of vortices shedding in the wake. For this case, the flapping fringe serves more like a propeller tail, such as the tail of a fish, and thus, reduces the drag, and improves the lift. The drag coefficients of the models with f_e = 80 Hz are slightly lower than that of the baseline model, whereas the lift coefficients are elevated dramatically. An f_e = 140 Hz, which is the natural shedding frequency for the bare airfoil, did not generate drastically different results. The difference in the lift and drag generation due to the fringe length is the minimum for this case.

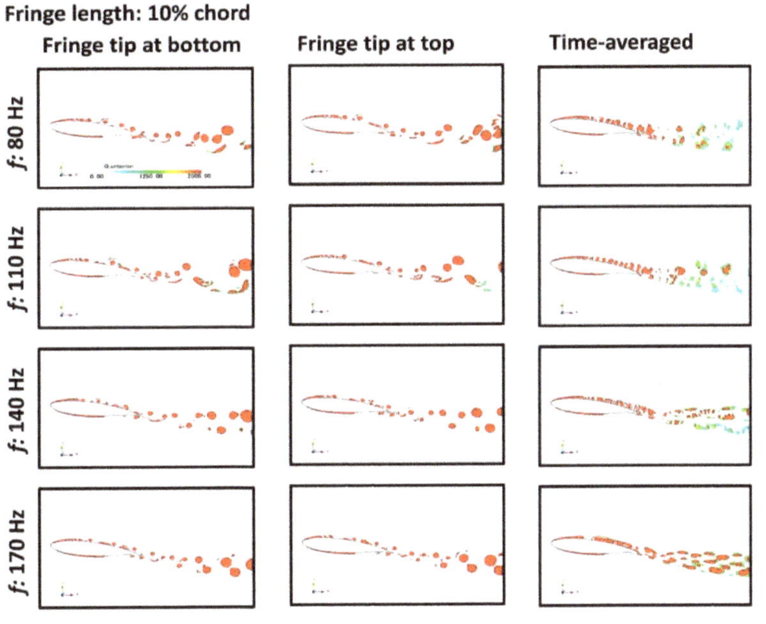

Figure 7. Q-criterion distributions around the airfoil model with the fringe length of 10% chord at different flapping frequencies.

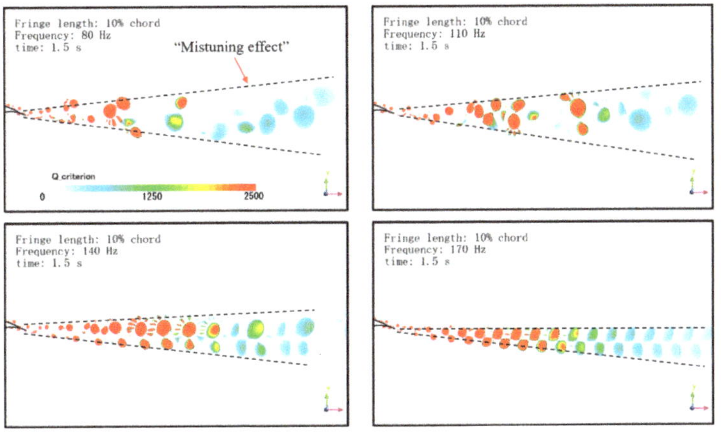

Figure 8. Instantaneous Q-criterion distributions in the wake of the airfoil model with the fringe length of 10% chord at different flapping frequencies at t = 1.5 s.

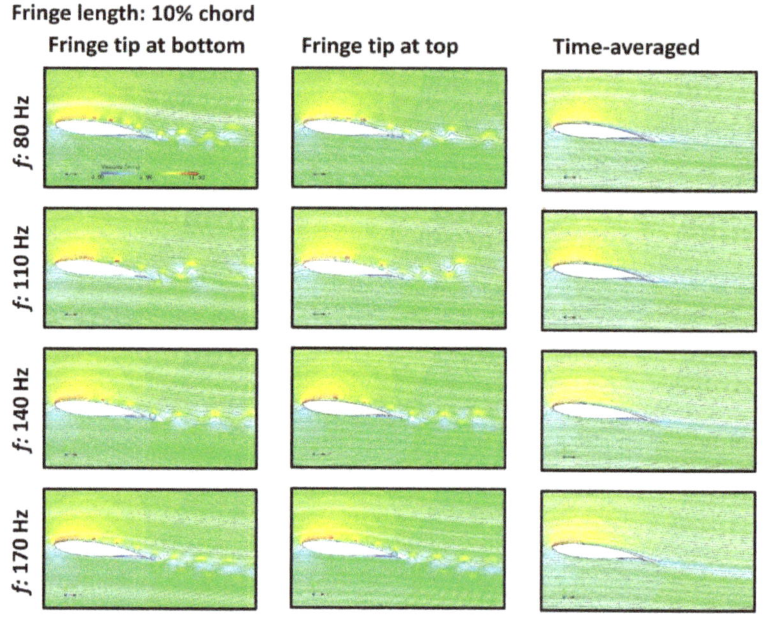

Figure 9. The velocity distribution around the airfoil model with the fringe length of 10% chord at different flapping frequencies.

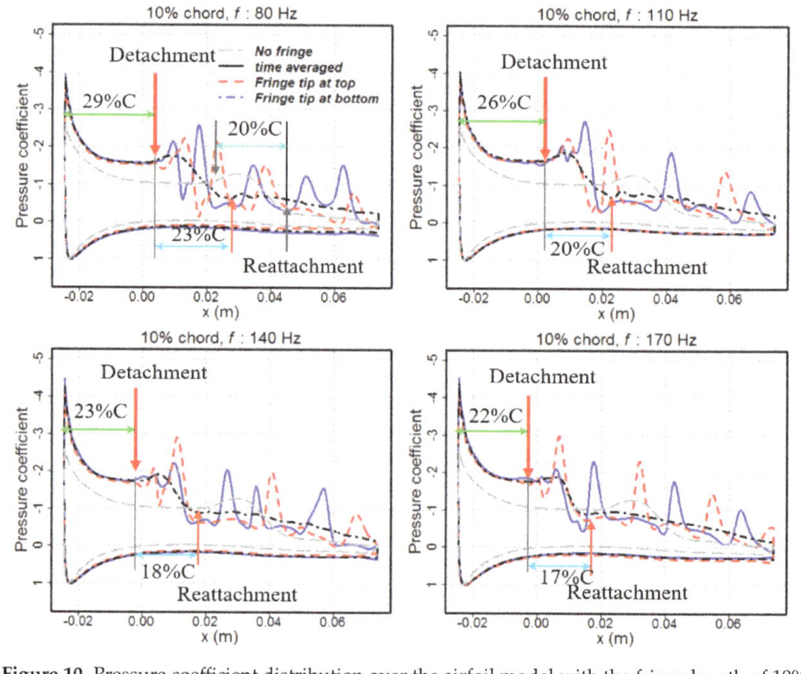

Figure 10. Pressure coefficient distribution over the airfoil model with the fringe length of 10% chord at different flapping frequencies.

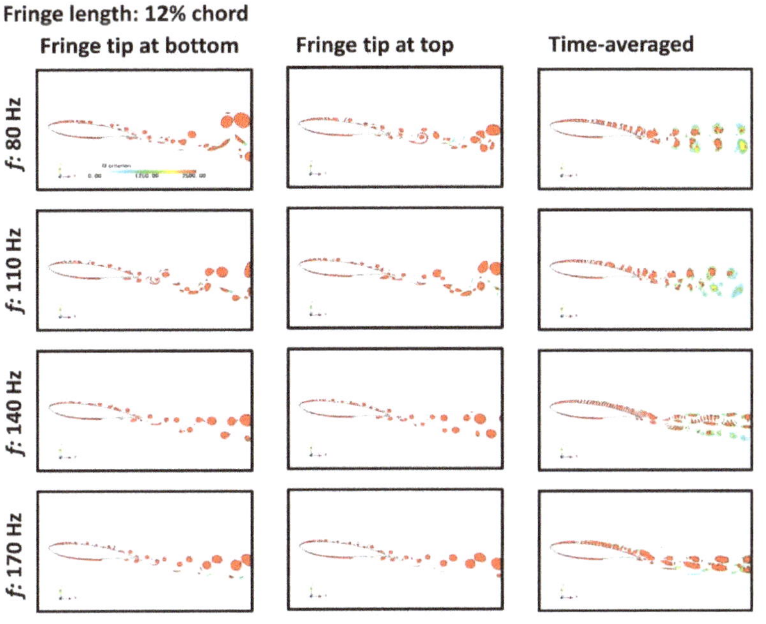

Figure 11. Q-criterion distributions around the airfoil model with the fringe length of 12% chord at different flapping frequencies.

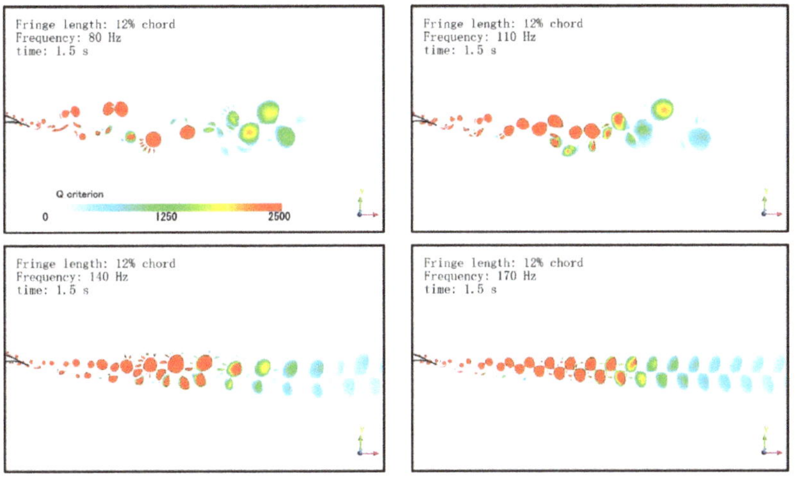

Figure 12. Instantaneous Q-criterion distributions in the wake of the airfoil model with the fringe length of 12% chord at different flapping frequencies at t = 1.5 s.

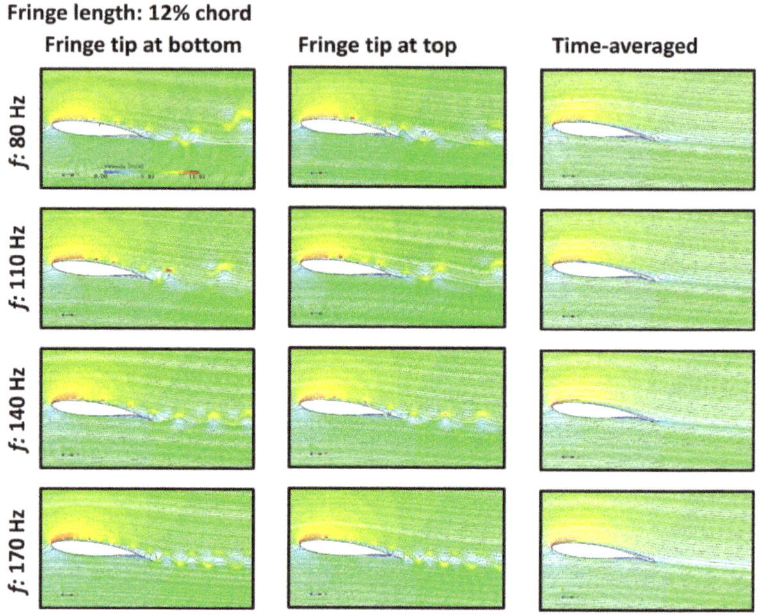

Figure 13. The velocity distribution around the airfoil model with the fringe length of 12% chord at different flapping frequencies.

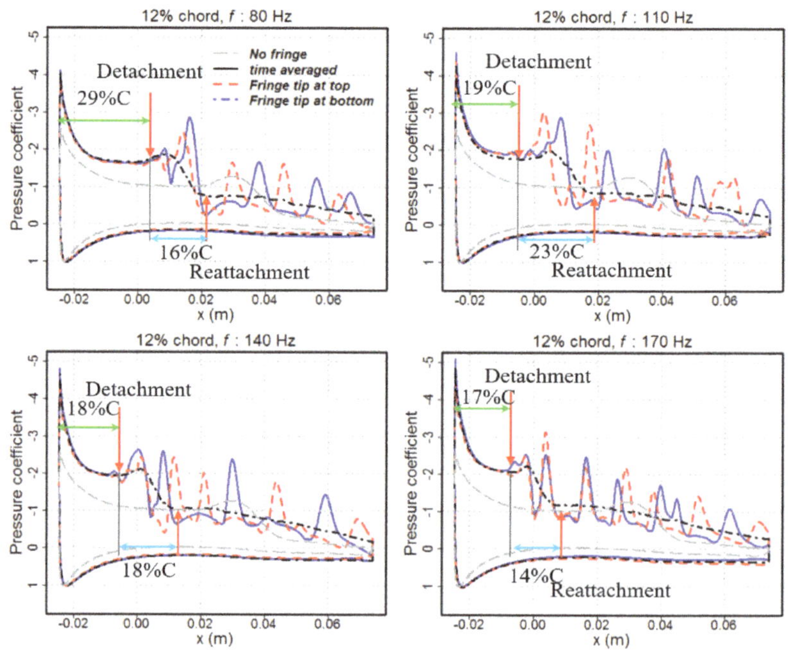

Figure 14. Pressure coefficient distribution over the airfoil model with the fringe length of 12% chord at different flapping frequencies.

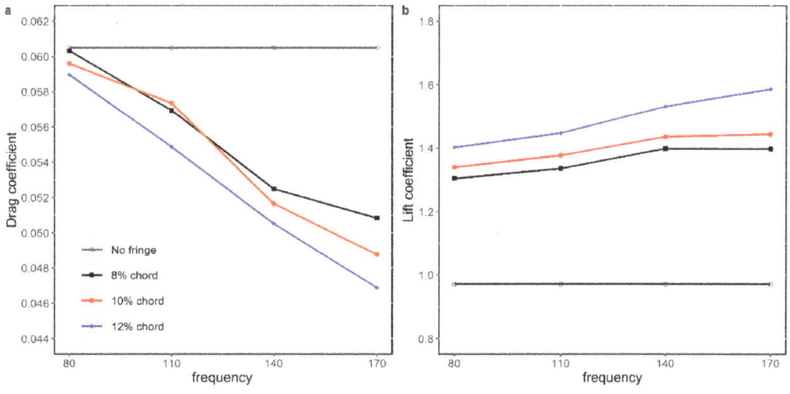

Figure 15. Time-averaged force coefficients for airfoil models with different fringe lengths at different flapping frequencies: (**a**) Drag coefficient; (**b**) Lift Coefficient.

In summary, the effects of the fringe length and flapping frequency on the vortex wake and aerodynamic performance are investigated numerically. It is found that a favorable f_e should be determined according to the nature shedding frequency, f, of the bare airfoil model. The structure of the shedding vortices or their strength would be either maintained or elevated by the vibration of the fringe at a higher f_e ($>f$). A higher flapping frequency (f_e) tends to serve as a booster, rather than a disturber, to increase the local velocity at the trailing edge, resulting in the elevation of the swirling strength, and a shrinking of the dimension of vortices. However, most importantly, a lower f_e was able to destruct the shedding vortices' coherent structure due to the disruption of the fringe on the original shedding pattern from the bare airfoil. Consequently, the swirling strength of vortices became weak, and the irregular distribution made it easy to dissipate.

3.3. Limitation of the Current Study

There are several limitations of this study. First, a RANS turbulence model was used to conduct the simulation. This would cause less accurate simulation results on the structure of shedding vortices. Due to the lack of comparable experimental data, direct validation of the simulation results could not be made, but, qualitatively, the predicted vortex reduction is similar to what we observed in our previous experiment [24]. In addition, to simplify the computational effort, a 2D model geometry was used, and thus, the span-wise porosity effect of feather fringes, as used in the experiment, could not be studied. To improve the understanding, 3D simulations on porous fringes attached at the trailing edge will be investigated in future work. To better mimic the feather fringe of an owl's wing, a soft material with a biological property should be adopted as well. The flapping motion will then be determined by the fluid-structure interaction as a passive flow control.

4. Conclusions

The airfoil S833 equipped with an extended flapping fringe is used to investigate the effects of two factors, i.e., the trailing fringe length and the flapping frequency, on the vortex shedding characteristics and the aerodynamic performance of the airfoil. The extended fringe with a flapping frequency significantly below the natural shedding frequency (~140 Hz), such as 80 Hz, can alter the coherence structure. Likewise, an irregular vortex structure in the wake can be generated, and thus, cause decrements in the swirling strength, and faster decay of the vortices. The flapping motion can decrease the pressure coefficient over the upper surface of the airfoil, leading to increases in the lift coefficient about 40% relative to the bare airfoil model. Out of the cases in the present study, the model with L_e = 0.01 m (10% of the airfoil chord) at a flapping frequency of f_e = 110 Hz outperforms other cases, and shows an overall substantially better aerodynamic performance of the airfoil, as well as more favorable vortex characteristics downstream of the airfoil. In spite of the limitations of the study, we can observe the aerodynamic benefits of using an extended flapping fringe at the trailing edge. The pressure alteration around the airfoil would potentially reduce the noise generation for wind turbines especially, which will be studied in our future work as well.

Author Contributions: Conceptualization, H.Y. and Z.Y.; methodology, H.Y. and Z.Y.; software, H.Y. and Z.Y.; validation, H.Y. and Z.Y.; formal analysis, H.Y. and Z.Y.; investigation, H.Y. and Z.Y.; resources, H.Y. and Z.Y.; data curation, H.Y. and Z.Y.; writing—original draft preparation, H.Y. and Z.Y.; writing—review and editing, H.Y. and Z.Y.; visualization, H.Y. and Z.Y.; supervision, Z.Y.; project administration, Z.Y.; funding acquisition, Z.Y. All authors have read and agreed to the published version of the manuscript.

Funding: This research received no external funding.

Institutional Review Board Statement: Not applicable.

Informed Consent Statement: Not applicable.

Data Availability Statement: The analysis data are available from the corresponding author on reasonable request.

Conflicts of Interest: The authors declare no conflict of interest.

References

1. Zhao, M.; Cao, H.; Zhang, M.; Liao, C.; Zhou, T. Optimal design of aeroacoustic airfoils with owl-inspired trailing edge serrations. *Bioinspir. Biomim.* **2021**, *16*, 056004. [CrossRef] [PubMed]
2. Oerlemans, S.; Fisher, M.; Maeder, T.; Kögler, K. Reduction of wind turbine noise using optimized airfoils and trailing-edge serrations. *AIAA J.* **2009**, *47*, 1470–1481. [CrossRef]
3. Li, D.; Liu, X.; Hu, F.; Wang, L. Effect of trailing-edge serrations on noise reduction in a coupled bionic aerofoil inspired by barn owls. *Bioinspir. Biomim.* **2019**, *15*, 16009. [CrossRef] [PubMed]
4. Talboys, E.; Geyer, T.F.; Prüfer, F.; Brücker, C. A parametric study of the effect of self-oscillating trailing-edge flaplets on aerofoil self-noise. *Appl. Acoust.* **2021**, *177*, 107907. [CrossRef]
5. Geissler, W.; van der Wall, B.G. Dynamic stall control on flapping wing airfoils. *Aerosp. Sci. Technol.* **2017**, *62*, 1–10. [CrossRef]
6. Xinyu, L.; Bifeng, S.; Wenqing, Y.; Wenping, S. Aerodynamic performance of owl-like airfoil undergoing bio-inspired flapping kinematics. *Chin. J. Aeronaut.* **2021**, *34*, 239–252.
7. Moreau, D.J.; Doolan, C.J. Noise-reduction mechanism of a flat-plate serrated trailing edge. *AIAA J.* **2013**, *51*, 2513–2522. [CrossRef]
8. Winzen, A.; Klaas, M.; Schröder, W. PIV measurements comparing natural and model owl wings. In Proceedings of the 17th International Symposium on Applications of Laser Techniques to Fluid Mechanics, Lisbon, Portugal, 7–10 July 2014.
9. Winzen, A.; Klän, S.; Klaas, M.; Schröder, W. Flow field analysis and contour detection of a natural owl wing using PIV measurements. In *Nature-Inspired Fluid Mechanics*; Springer: Berlin/Heidelberg, Germany, 2012; pp. 119–134.
10. Lilley, G. A study of the silent flight of the owl. In Proceedings of the 4th AIAA/CEAS Aeroacoustics Conference, Toulouse, France, 2–4 June 1998; p. 2340.
11. Sarradj, E.; Fritzsche, C.; Geyer, T. Silent owl flight: Bird flyover noise measurements. *AIAA J.* **2011**, *49*, 769–779. [CrossRef]
12. Kerho, M.; Hutcherson, S.; Blackwelder, R.F.; Liebeck, R.H. Vortex generators used to control laminar separation bubbles. *J. Aircr.* **1993**, *30*, 315–319. [CrossRef]
13. Klan, S.; Bachmann, T.; Klaas, M.; Wagner, H.; Schröder, W. Experimental analysis of the flow field over a novel owl based airfoil. In *Animal Locomotion*; Springer: Berlin/Heidelberg, Germany, 2010; pp. 413–427.
14. Miao, J.M.; Ho, M.H. Effect of flexure on aerodynamic propulsive efficiency of flapping flexible airfoil. *J. Fluids Struct.* **2006**, *22*, 401–419. [CrossRef]
15. Shan, H.; Jiang, L.; Liu, C.; Love, M.; Maines, B. Numerical study of passive and active flow separation control over a NACA0012 airfoil. *Comput. Fluids* **2008**, *37*, 975–992. [CrossRef]
16. Veldhuis, L.L.M.; Jansen, D.P.; El Haddar, J.; Correale, G. Novel Passive and Active Flow Control for High Lift. In Proceedings of the 28th International Congress Aeronautucal Science, Brisbane, Australia, 23–28 September 2012.
17. Finez, A.; Jacob, M.; Jondeau, E.; Roger, M. Broadband noise reduction with trailing edge brushes. In Proceedings of the 16th AIAA/CEAS Aeroacoustics Conference, Stockholm, Sweden, 7–9 June 2010; p. 3980.
18. Herr, M.; Dobrzynski, W. Experimental Investigations in Low-Noise Trailing Edge Design. *AIAA J.* **2005**, *43*, 1167–1175. [CrossRef]
19. Das, C. An Experimental Investigation of Flow-Induced Noise Mechanism of a Flexible Flat-Plate Trailing-Edge. 2015. Available online: https://www.semanticscholar.org/paper/AN-EXPERIMENTAL-INVESTIGATION-OF-FLOW-INDUCED-NOISE-Das-Mimani/6b0724fa4ad87b65336240d253d30b5a3b1c4176#citing-papers (accessed on 10 October 2021).

20. Schlanderer, S.C.; Sandberg, R.D. DNS of a compliant trailing-edge flow. In Proceedings of the 19th AIAA/CEAS Aeroacoustics Conference, Berlin, Germany, 27–29 May 2013; p. 2013.
21. Kamps, L.; Brücker, C.; Geyer, T.F.; Sarradj, E. Airfoil self noise reduction at low Reynolds numbers using a passive flexible trailing edge. In Proceedings of the 23rd AIAA/CEAS Aeroacoustics Conference, Denver, CO, USA, 5–9 June 2017; p. 3496.
22. Talboys, E.; Geyer, T.F.; Brücker, C. An aeroacoustic investigation into the effect of self-oscillating trailing edge flaplets. *J. Fluids Struct.* **2019**, *91*, 102598. [CrossRef]
23. Murayama, Y.; Nakata, T.; Liu, H. Flexible flaps inspired by avian feathers can enhance aerodynamic robustness in low Reynolds number airfoils. *Front. Bioeng. Biotechnol.* **2021**, *9*, 374. [CrossRef] [PubMed]
24. Yang, Z.; He, Z.; Chen, F. Study on the Vortex Wake of an Airfoil Equipped with Flexible Trailing Edge Fringe. In Proceedings of the 53rd AIAA Aerospace Sciences Meeting, Kissimmee, FL, USA, 5–9 January 2015; p. 1707. [CrossRef]
25. Yu, H.; Yang, Z. A numerical simulation on the airfoil s833 equipped with flapping trailing edge fringes. *J. Appl. Fluid Mech.* **2020**, *13*, 571–582. [CrossRef]
26. Somers, D.M. *S833, S834, and S835 Airfoils: November 2001–November 2002*; National Renewable Energy Lab. (NREL): Golden, CO, USA, 2005.
27. Walters, D.K.; Leylek, J.H. A new model for boundary layer transition using a single-point RANS approach. *J. Turbomach.* **2004**, *126*, 193–202. [CrossRef]
28. Walters, D.K.; Cokljat, D. A three-equation eddy-viscosity model for Reynolds-averaged Navier–Stokes simulations of transitional flow. *J. Fluids Eng.* **2008**, *130*, 121401. [CrossRef]

applied sciences

MDPI

Article

Study on Traveling Wave Wall Control Method for Suppressing Wake of Flow around a Circular Cylinder at Moderate Reynolds Number

Xin Liu [1], Weifeng Bai [2] and Feng Xu [1,*]

[1] School of Civil and Environmental Engineering, Harbin Institute of Technology (Shenzhen), Shenzhen 518055, China; 21b954008@stu.hit.edu.cn
[2] Khoury College of Computer Sciences, Northeastern University, Vancouver, BC V6B 1Z3, Canada; bai.wei@northeastern.edu
* Correspondence: xufenghit@hit.edu.cn

Abstract: In the present paper, the computational fluid dynamics (CFD) numerical simulation was utilized to investigate the effectiveness of the transverse traveling wave wall (TWW) method with the expectation of inhibiting the vortex shedding from a fixed circular cylinder. We mainly focused on the variations of four kinds of wave propagation directions, five different maximum wave amplitudes and ten different wave velocities for suppressing vortices shedding and aerodynamic forces. The aerodynamic coefficients and vortex structures under different propagation directions, wave amplitudes, wave numbers and wave velocities were investigated in detail. The results demonstrate that the alternate wake behind the cylinder can be effectively eliminated resorting to the "Downstream" propagating TWW. The mean drag coefficient is positively associated with wave velocity. Drag and lift coefficients remain relatively stable at different wave amplitudes. When the velocity ratio (wave velocity divided by incoming velocity) is 1.5, the lift coefficient fluctuation decreases to the minimum. In contrast, the optimal combination of control parameters under the present Reynolds number is concluded with "Downstream" propagating direction, maximum wave amplitude ratio of 0.02, and velocity ratio of 1.5.

Keywords: flow control; traveling wave wall; circular cylinder; numerical simulation; CFD

Citation: Liu, X.; Bai, W.; Xu, F. Study on Traveling Wave Wall Control Method for Suppressing Wake of Flow around a Circular Cylinder at Moderate Reynolds Number. *Appl. Sci.* **2022**, *12*, 3433. https://doi.org/10.3390/app12073433

Academic Editor: Artur Tyliszczak

Received: 14 February 2022
Accepted: 23 March 2022
Published: 28 March 2022

1. Introduction

Fluid flow around a cylinder is a universal phenomenon in engineering practice, i.e., heat exchanger tubes, marine cables, high-rises, and civil engineering structures. Vortex shedding behind bluff cylinders is the cause of vortex-induced vibration (VIV) which can result in structural damage under certain unfavorable conditions. In addition, there have been a variety of investigations about vortex-induced vibration from different perspectives in recent years [1,2]. Eliminating the alternating shedding vortices in the cylinder wake and suppressing the cylinder vibration are of important significance.

Fluid has a huge influence on bluff body structures, and there are many relative research [3–5]. Studies on flow control of bluff body have specified a new orientation for the flow control problems, and they have gained more and more concerns in recent years. Controlling methods can be mainly divided into two types: passive control and active control. Passive flow control could change the flow conditions to achieve the goal of flow control without consuming any external energy [6–10]. Another flow control method is active flow control, which requires external energy infused into the flow field. Moreover, active flow control could introduce the proper perturbation to change the inner flow mode. Several active flow control methods have been investigated to suppress vibration as well as improve aerodynamic performance, i.e., drag reduction, using wall vibration [11,12], suction and injection [13–16], momentum injection [17–20], bionic control method of traveling wave wall, etc.

The TWW is one of the bionic based flow control methods and it has been developing rapidly in recent years. The boundary layer near the rigid wall comes into existence when the viscous fluid flows pass the wall. However, making the cylinder surface to be flexible or movable may help to weaken the boundary layer.

The "fluid roller bearing" (FRB) effects of the axisymmetric TWW was used by Yang and Wu [21] to sharp into a series of vortex rings. The main flow was separated from the near-wall flow accompanied by friction drag and pressure drag significantly reduced. Wu et al. [22] carried out a numerical simulation to find out the proper parameter (wave amplitude, wavelength and ratio of wave velocity to incoming velocity) for the FRB effects on infinite and finite two-dimensional TWW. Large flow separation was eliminated after applying TWW on airfoils, and the vortex shedding of streamline airfoils with a large wind attack angle was restrained to provide more lift. Wu et al. [23] proposed a transverse traveling wave (TTW) control method to manipulate the unsteady flow around a circular cylinder at Re = 500. A fluid FRB was introduced in this method, and the vortex shedding was eliminated because the global flow remained attached to the surface. Xu et al. [24] simulated the rear section of a cylinder with TWW, and the cylinder was elastically mounted with two degrees of freedom. The simulation took fixed wave amplitude, wave number and ratio of wave velocity to incoming velocity. In addition, the whole process was completely simulated starting with flow around the fixed cylinder, to the oscillating cylinder, and finally to the oscillating cylinder with TWW.

In the present paper, TWW was placed on the rear section of a fixed cylinder, and the CFD numerical simulation was adapted to investigate the influencing factors for suppressing the cylinder oscillating wake. The TWW was activated when the alternating shedding vortex behind the fixed circular could be stably observed in the present numerical simulation. There are several steps we need to comply with as per control variate technique. Firstly, fixed velocity ratio and maximum wave amplitude were used to measure the control effect of four kinds of TWW propagation direction for suppressing the oscillating wake. Then, the most effective propagation direction was selected, based on which we forward studying the controlling effectiveness of 5 different wave amplitudes, 4 different wave numbers and 10 different wave velocities. The control effect under various influencing factors was elaborated by comparing the characteristic values, aerodynamic force time histories and the wake flow patterns.

2. Numerical Model and Validation

2.1. Governing Equations and TWW

The governing equations of two-dimensional incompressible flow in a Cartesian coordinate system can be written as follows.

$$\frac{\partial u_i}{\partial x_i} = 0 \tag{1}$$

$$\frac{\partial u_i}{\partial t} + u_j \frac{\partial u_i}{\partial x_j} = -\frac{1}{\rho}\frac{\partial p}{\partial x_i} + \frac{\mu}{\rho}\frac{\partial^2 u_i}{\partial x_j^2} \tag{2}$$

where u_i is the velocity components, i.e., u_1 is the velocity component at the inline flow direction and u_2 is the velocity component at the cross-flow direction, μ denotes the kinematic viscosity coefficient, ρ is the fluid density, and p denotes the pressure in flow field.

As shown in Figure 1, The transverse traveling wave at the rear section of cylinder was established, i.e., the transverse traveling wave can propagate from both point B and C to point A at the same time.

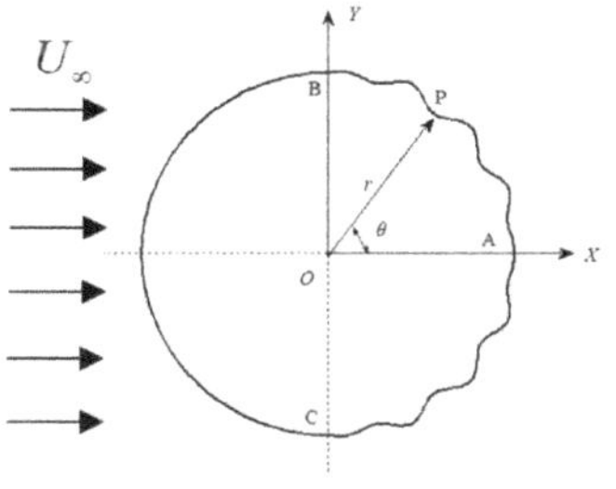

Figure 1. Schematic of fixed circular cylinder with TWW.

The origin *O* is positioned at the center of the plane polar coordinate system (*r*, *θ*), and the wave equations of the TWW can be written as,

$$\begin{cases} x = r\cos\theta \\ y = r\sin\theta \\ r = r_0 + A(l)\cos[k(l-ct)] \end{cases}, -\pi/2 \le \theta \le \pi/2 \quad (3)$$

$$A(l) = \begin{cases} \hat{A}\frac{l}{\lambda}, 0 \le l \le \lambda \\ \hat{A},\ \lambda \le l \le (N-1)\lambda \\ \hat{A}\frac{(N\lambda-l)}{\lambda}, (N-1)\lambda \le l \le N\lambda \end{cases} \quad (4)$$

where $r_o = D/2$ is the radius of the standard circular cylinder without any traveling wave, D is the diameter of the cylinder, $A(l)$ is the vibration amplitude of each point on the TWW, $k = 2\pi/\lambda$ is the wave number, l is the arc length from point A on the cylinder rear edge to any point P on the TWW, c is the wave velocity, t is the time, λ is the wave length, $\hat{A}$ is the maximum wave amplitude, N is the wave number in 1/4 circle. In the first wave, the amplitude increases linearly from 0; In the end wave, the amplitude decreases linearly to 0, while other intermediate waves have the same amplitude as the complete waveform, as shown in Equation (4). The TWW connect smoothly with the non-moving cylinder surface in this configuration.

2.2. Computational Domain and Boundary Conditions

Figure 2 shows the grid distribution and the computational domain. The computational domain is a rectangle area with a length of 60*D* and a width of 40*D*, and the center of the cylinder is at the coordinate origin. The cylinder diameter is 0.12 m. The upstream inlet is 20*D* ahead of the coordinate origin, and the downstream outlet is 40*D* after the coordinate origin. The distances between either of the upside and downside and the coordinate origin are 20*D*. Unstructured grid is adopted to discretize the computational domain into four layers. The grids in the regions with large gradients of flow parameters are locally refined, i.e., the region near cylinder surface and wake region. The geometric model and meshing are performed by pre-processing software ICEM.

The boundary conditions are set up as follows, the inlet is set as the velocity-inlet with a uniform velocity U_∞, the outlet is set as the pressure-outlet with the relative pressure of 0, the upper and lower sides are set as the symmetry boundary, and the cylinder surface is set as the no-slip wall.

The numerical calculations are carried out with a Reynold number of 4.1×10^4, and the turbulence flow is calculated by SST *k*-*ω* model. SIMPLE algorithm is adopted to calculate the coupling between the pressure and velocity fields. The pressure interpolation format is set as "Standard". The second-order upwind scheme is used for the momentum discretizing because of its accuracy and stability. During the solution process, the convergence residual standard of continuity equation is 1.0×10^{-6}, and the convergence residual standard of momentum equation and turbulence parameters is 3.0×10^{-7}.

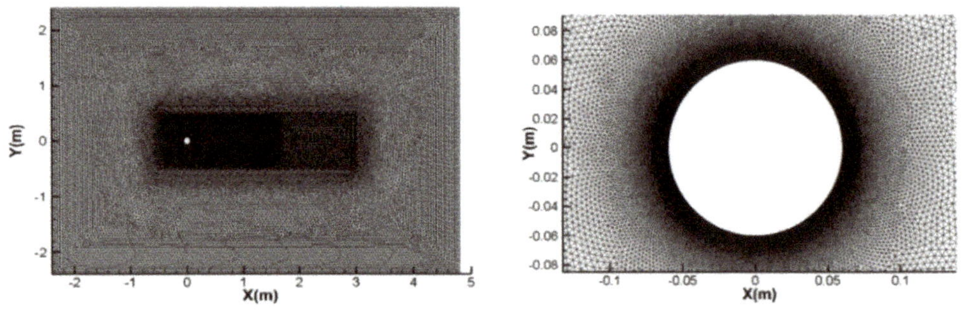

Figure 2. Grid distribution and computational domain near the cylinder surface.

The CFD general software Fluent is used to calculate the flow around a fixed cylinder. Firstly, the governing Equations (1) and (2) are solved to obtain the flow characteristics around the cylinder, and then the aerodynamic coefficient of the cylinder is obtained. When the amplitude of the aerodynamic coefficient is stable (t = 1.8 s), TWW is activated. The wave Equation (3) of traveling wave is solved by User Defined Function (UDF) in Fluent, and the radial velocity at any point P in Figure 1 can be obtained according to Equation (3), and then the radial velocity is decomposed into the x direction velocity and the y direction velocity. The DEFINE_GRID_MOTION macro of UDF is used to assign the velocity for all grid points on the rear semi-cylindrical surface, and the motion of cylinder surface grids are realized by the dynamic mesh technique. When the grids are updated to the specified position, the flow field is calculated to converge with this shape as the boundary, and then the calculation of the next time step begins. This loop calculation ends until the TWW achieves a significant wake control effect.

2.3. Validity Investigation

The grid and time step independence verification were carried out because the mesh size and the time step determine the accuracy of CFD computational results. The comparisons of the global parameters for the grid and time step independences studies are shown in Tables 1 and 2. In the tables, N_c is the number of nodes for 1/4 circle, N_{mesh} is the total number of the mesh, and Δt is the time step. The global parameters include the fluctuating lift coefficient C_l', the mean drag coefficient $\overline{C}_d$, the fluctuating drag coefficient C_d', the Strouhal number S_t, and the minimum, mean and maximum values y^+_{min}, $\overline{y}^+$ and y^+_{max}, respectively. In order to guarantee the correctness of the turbulence simulation results and sufficient amount of mesh in the viscous sublayer, the y^+_{max} should be less than 11.63 which is the demarcation point between the log-law region and the viscous sublayer recommended by Versteeg and Malalasekera [25].

Table 1. Comparisons of the global parameters for the grid independence study.

N_c	N_{mesh}	$\Delta t(s)$	$\overline{C}_d$	C_d'	C_l'	S_t	y^+_{min}	$\overline{y}^+$	y^+_{max}
100	82,456	2.5×10^{-4}	1.0999	0.0873	0.8329	0.267	0.745	8.331	16.232
150	92,304	2.5×10^{-4}	1.3675	0.0949	0.9927	0.249	0.442	5.589	11.405
200	101,504	2.5×10^{-4}	1.3420	0.0884	0.9702	0.243	0.334	4.116	8.389
250	111,600	2.5×10^{-4}	1.2674	0.0812	0.9211	0.243	0.247	3.187	6.902
300	120,594	2.5×10^{-4}	1.2224	0.0822	0.8956	0.243	0.184	2.580	5.750

Table 2. Comparisons of the global parameters for the time step independence study.

Δt (s)	N_c	N_{mesh}	$\overline{C}_d$	$C_d^{'}$	$C_l^{'}$	S_t
1.0×10^{-3}	250	111,600	1.1433	0.0591	0.7928	0.226
5.0×10^{-4}	250	111,600	1.2378	0.0745	0.9003	0.237
2.5×10^{-4}	250	111,600	1.2674	0.0812	0.9211	0.243
1.0×10^{-4}	250	111,600	1.2819	0.0853	0.9351	0.243

Through comparison and analyzation of the global parameter results, $N_c = 250$ and $\Delta t = 2.5 \times 10^{-4}$ s were adopted in the final results discussion. Current results accompanied with some previous simulation and experimental results [26–36] for the flow around a single cylinder are shown in Figure 3. The mean drag coefficient $\overline{C}_d$ versus Reynolds is shown in Figure 3a, and it can be observed that $\overline{C}_d$ is very close to previous results at similar Re range. The S_t in the previous 3D flow experiment is less than that in this paper at similar Re range. In the present 2D numerical simulation, the computational domain height is extraordinarily small, so the 3D flow in spanwise direction cannot be fully developed. Thus, the present S_t results are closer to the 2D flow results [28,34,35], as shown in Figure 3b. From the previous comparison, it can be concluded that the present numerical model and grids are reliable and have good reference value for further research of TWW flow control.

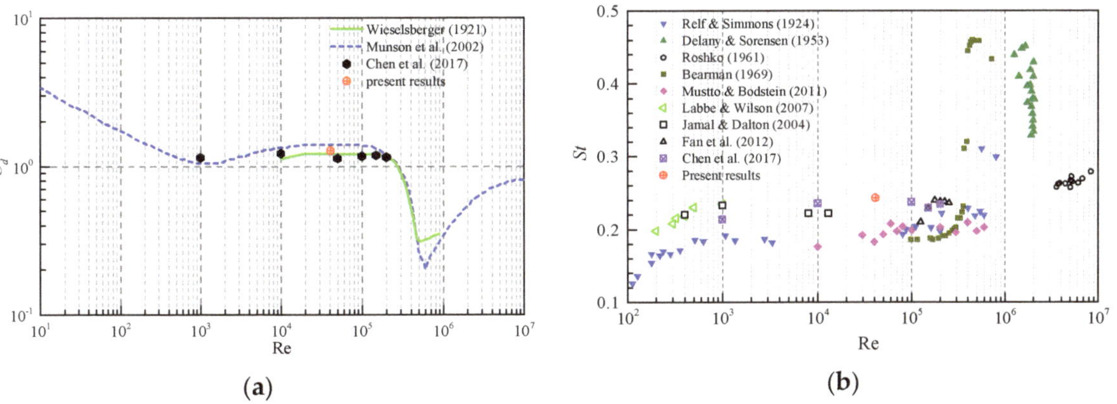

Figure 3. Comparisons of aerodynamic coefficient and the Strouhal number with previous results. (**a**) $\overline{C}_d$ versus Re. (**b**) S_t versus Re.

3. Results and Discussion

3.1. Influence of Different Propagation Directions

The control effectiveness of 4 types of TWW propagation should be considered first. The four propagation directions include the "Downstream", the "Upstream", the "Corotating-Clockwise" and the "Corotating-Counterclockwise", as shown in Figure 4. During calculation process, the effects of the TWW propagation directions are considered under fixed velocity ratio $c/U_\infty = 2.0$, wave number $N = 4$, and maximum wave amplitude $\hat{A}/D = 0.02$. This numerical simulation begins with the flow around a fixed cylinder (FR). The TWW is activated to control the flow field after the flow field is brought into stability with alternating shedding vortices. The aerodynamic coefficients and vortex shedding patterns before and after activating the TWW could serve as a good measurement for the control effect of TWW control method.

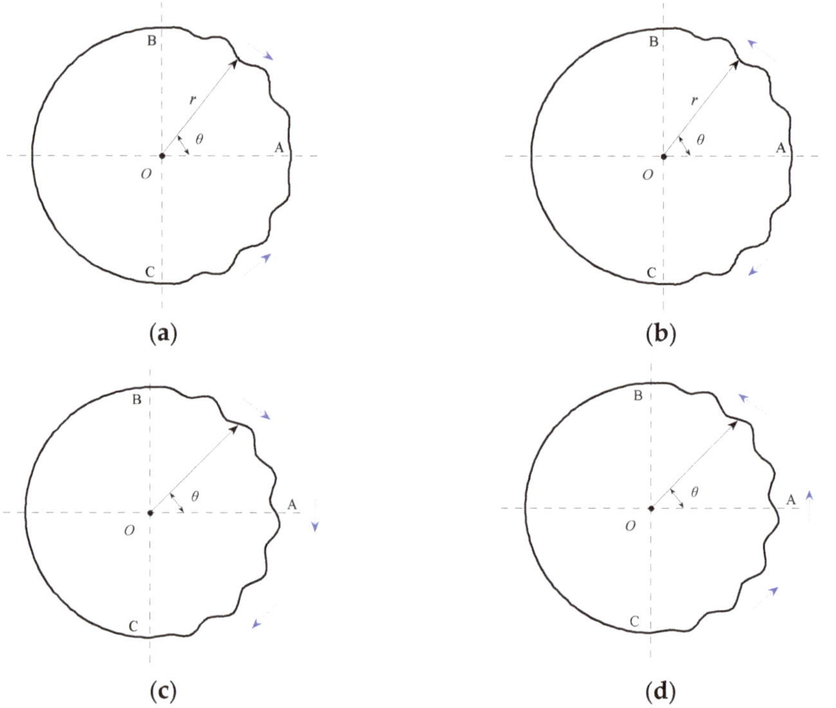

Figure 4. Four types of propagation direction of TWW. (**a**) Downstream. (**b**) Upstream. (**c**) Corotating-Clockwise. (**d**) Corotating-Counterclockwise.

The time histories of lift and drag coefficient for the TWW cylinder with the four different propagation directions are shown in Figure 5. When $t < 1.8$ s, TWW was not activated, the result of FR was obtained. When $t \geq 1.8$ s, TWW was activated, the suppression effect of TWW on the vortex shedding and wake under the four different propagation directions can be obtained.

Figure 5 shows that the lift coefficient fluctuation reduces dramatically only when traveling wave propagates in the downstream direction. It indicates that the TWW propagating downstream has the effect of eliminating Vortex Street. In addition, the mean drag coefficient decreases at this propagation direction, while the fluctuating drag coefficient increases dramatically, as shown in Figure 5a. Figure 5b shows that the fluctuating lift coefficient and mean drag coefficient increase significantly when traveling wave propagates in the upstream direction. It illustrates that the traveling wave propagating upstream enhances the wake behind the cylinder. The drag coefficient fluctuation increases, and the mean lift coefficient varies significantly when traveling wave co-propagates along the cylinder rear surface. When traveling wave Co-rotates clockwise, the mean lift coefficient increases significantly to a positive value, as shown in Figure 5c. When traveling wave Co-rotates counterclockwise, the mean lift coefficient decreases significantly to a negative value, as shown in Figure 5d. It indicates that there is a negative pressure zone on the upper or lower surface with these two propagation directions, and the oscillating wake of traveling wave cylinder cannot be eliminated.

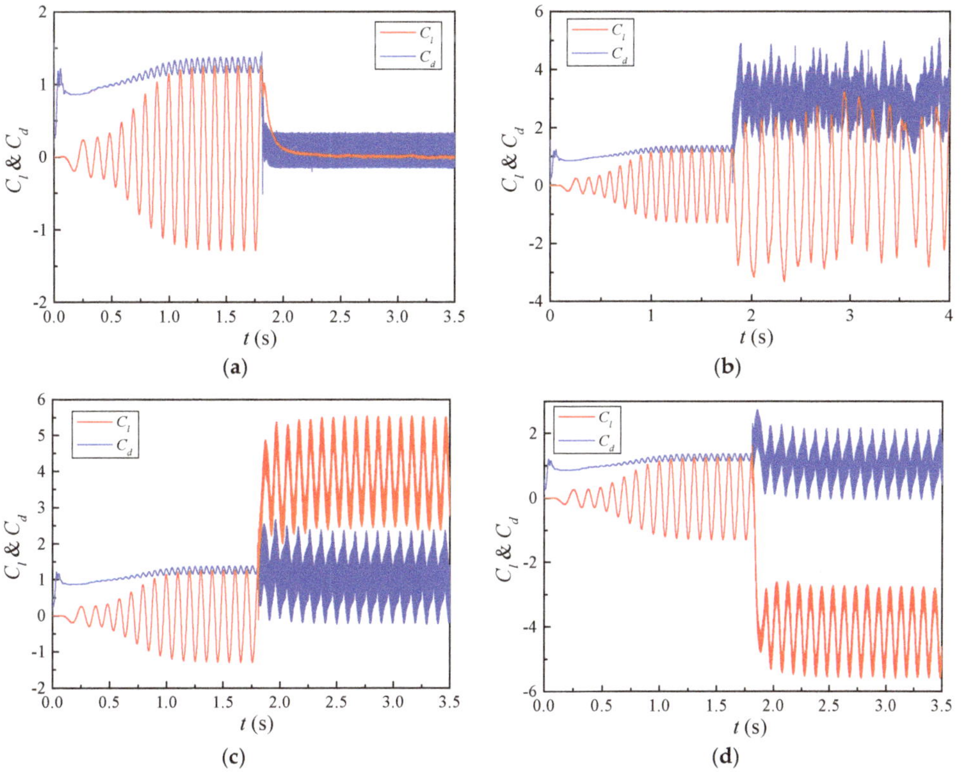

Figure 5. Lift and drag coefficient time history curves under various propagation directions. (**a**) Downstream. (**b**) Upstream. (**c**) Corotating-Clockwise. (**d**) Corotating-Counterclockwise.

The mean value reflects the variant characteristics within a period, and the fluctuation value (root-mean-square, RMS) reflects the discrete degree of the variant. Before activating the TWW, the mean lift coefficient $\overline{C}_l$ approximates 0 and C_l' is 0.921, the mean drag coefficient $\overline{C}_d$ is 1.267 and C_d' is 0.081. The time history curves of lift and drag coefficients for uncontrolled cylinder are taken for reference in the following parts to measure the influence of different control parameters on the control effect. Figure 6 shows the lift and drag coefficients characteristics with different traveling wave propagation directions, while the dashed curves inside the same figure demonstrate the results of the standard circular cylinder. As shown in Figure 6a, $\overline{C}_l$ approaches to 0 for "Downstream" and "Upstream" propagation direction, while the $\overline{C}_l$ increases dramatically to 3.917 and −4.057 for co-propagations. For the "Downstream" propagation, the C_l' approaches to 0, which indicates that the cylinder oscillating wake is eliminated by TWW. For the "Upstream" propagation, C_l' increases dramatically to 1.825. In addition, C_l' for the co-propagations is close to that of the standard circular cylinder. These illustrate that the other three propagations are not able to suppress the cylinder wake vortex except "Downstream". Figure 6b shows that the fluctuating drag coefficients C_d' vary between 0.172 and 0.763 in various propagation directions, and all of them are larger than that of the standard circular cylinder, which means that TWW leads to a larger amplitude fluctuating drag. For "Downstream" propagation, $\overline{C}_d$ decreases dramatically to −0.152, meaning the cylinder is subjected to reverse thrust due to TWW. For "Upstream" propagation, $\overline{C}_d$ increases dramatically to 2.809, i.e., the inline flow drag increases dramatically. The $\overline{C}_d$ of other two propagation directions equal to 0.905 and 0.902, respectively, which are slightly lower than the standard circular cylinder. Compared with the standard circular cylinder results, C_l' and

$\overline{C}_d$ of "Downstream" propagation decreases by 99.48% and 112%, respectively. It illustrates the cylinder wake oscillation in FR stage can be eliminated by the TWW propagating downstream.

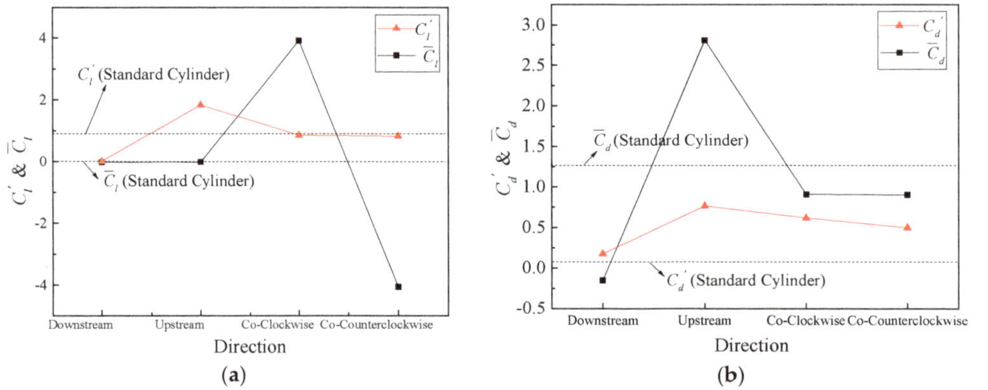

Figure 6. Characteristic values of lift and drag coefficient for different propagation directions. (**a**) C_l' & $\overline{C}_l$. (**b**) C_d' & $\overline{C}_d$.

The vorticity contours surrounding TWW cylinder with "Downstream" propagation are shown in Figure 7. In the interval $t < 1.8$ s, the traveling wave was not activated, and the vortex sheds from both side of the cylinder alternately, as shown in Figure 7a. Figure 7b shows that the TWW activated shortly has little effects, and the separated free shear layer still rolls into vortices. Form Figure 7c–g, as time advanced, the control effect of TWW became more obvious, and the vortex street appeared in uncontrolled cylinder was eliminated. The shear layer detaches from the cylinder surface, and then enters the wake region without rolling into vortices due to the "fluid roller bearing" (FRB) effect of TWW, and the vorticity is distributed steadily and symmetrically in the cylinder wake region, which means that the TWW cut off the energy source for generating vortex. When $t = 2.97$ s, the TWW eliminates oscillating wakes of the cylinder successfully.

In addition to the "Downstream" propagation, the rest three kinds of propagations are less effective in eliminating the cylinder oscillating wake, which can be seen from the vorticity contours obtained after activating TWW, as shown in Figures 8 and 9. In Figure 8, the vortices in the wake formed by FR are replaced by stronger vortices, so that the fluctuating lift coefficient and the mean drag coefficient increase dramatically. In Figure 9, an overall downward vortex street is formed in the cylinder wake region because of the "Corotating-Clockwise" traveling wave. In addition, the wake for the "Corotating-Counterclockwise" traveling wave induces an overall upward vortex street, which is opposite to that in Figure 9. For both clockwise and counterclockwise propagation directions, the TWW is activated at the same time (t = 1.8 s), and the initial state of the flow field controlled by the TWW propagating in the two directions is exactly the same. If the clockwise and counterclockwise TWWs are activated at opposite phases of the lift coefficient oscillations, it is expected to obtain a mirror-symmetrical flow field and exactly equal aerodynamic results. The choice of TWW activating phase is an important reason for the slight difference in the aerodynamic statistic values of the circular cylinder with the two different propagation directions of TWW, as shown in Figure 6.

3.2. *Influence of Different Wave Amplitudes*

The "Downstream" propagation direction is therefore selected as the base to study the influence of traveling wave amplitude, the fixed TWW velocity ratio c/U_∞ is taken as 2.0, and the wave number per quarter circle is taken as 4. The aerodynamic coefficients and vortex shedding pattern after activating TWW are chosen to access the control effectiveness.

The lift and drag coefficient time histories for given $\hat{A}/D$ of 0.01, 0.02, 0.03, 0.04, and 0.05 are shown in Figure 10. The TWW is activated at $t = 0$ throughout the following calculation process. The mean lift coefficient decreases slightly, and the fluctuating lift coefficient remains almost the same. With an increase of wave amplitude, the fluctuating drag coefficient increases significantly and the mean drag coefficient decreases to negative values. It can be found that the lift coefficient is less affected by wave amplitude, but wave amplitude has significant influence on drag coefficient, as shown in Figure 10.

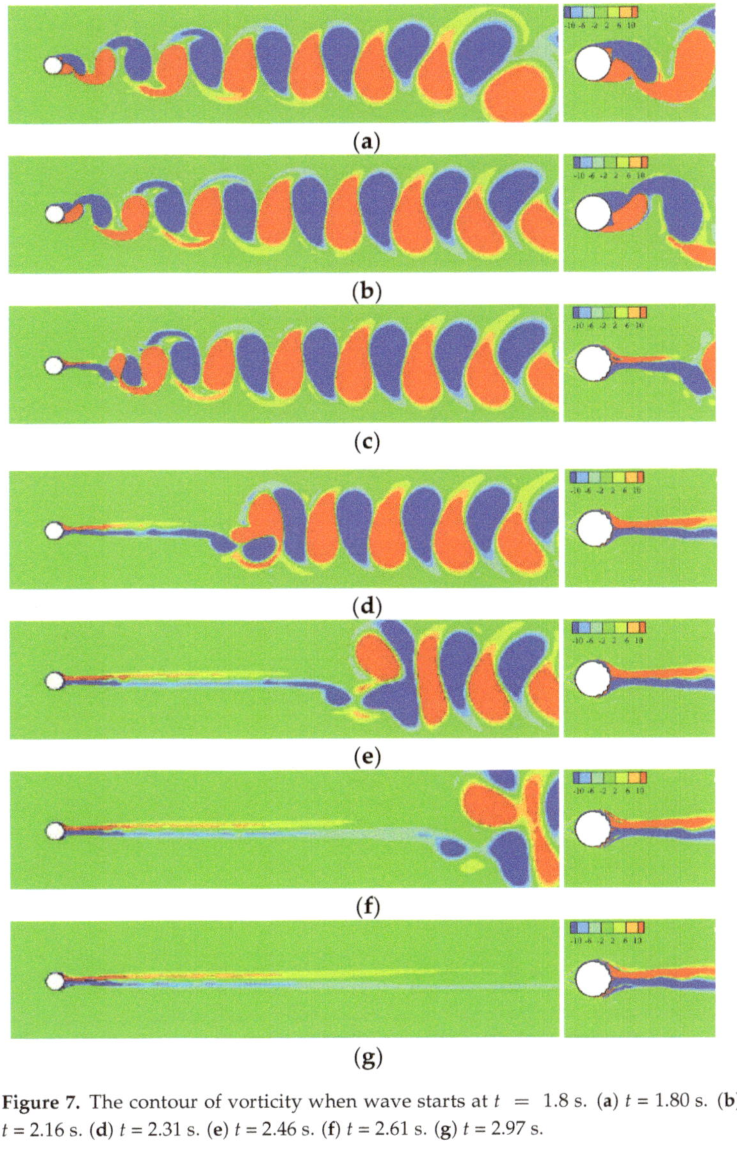

Figure 7. The contour of vorticity when wave starts at $t = 1.8$ s. (**a**) t = 1.80 s. (**b**) t = 1.96 s. (**c**) t = 2.16 s. (**d**) t = 2.31 s. (**e**) t = 2.46 s. (**f**) t = 2.61 s. (**g**) t = 2.97 s.

Figure 11 shows the lift and drag coefficient characteristics under different wave amplitudes, including the mean and RMS values of lift and drag coefficients. Form Figure 11a, with the increase of $\hat{A}/D$, $\overline{C}_l$ decreases slightly, which is smaller than the standard circular cylinder result. Overall, there is less difference in $\overline{C}_l$ compared to that in standard cylinder.

When amplitude ratio $\hat{A}/D = 0.02$, C_l' dramatically decreases to 0.005 and reaches the minimum, which drops by 99.48% from that of standard circular cylinder. When $\hat{A}/D$ increases from 0.02 to 0.05, C_l' shows a slightly increasing trend. C_l' equals to 0.012 at $\hat{A}/D = 0.05$, which dropped by 98.7% from that of standard circular cylinder. In Figure 11b, with the increase of $\hat{A}/D$, $\overline{C}_d$ decreases from -0.010 to -0.456, and C_d' increases from 0.085 to 0.385, which is almost 4.7 times the standard circular cylinder result. It's easily observed that different wave amplitudes have little impact on C_l' and $\overline{C}_l$. However, C_d' and $\overline{C}_d$ change obviously as $\hat{A}/D$ increases from 0.01 to 0.05. To make TWW perform an ideal functionality in eliminating the aerodynamic forces, $\hat{A}/D = 0.02$ should be chosen, which can eliminate lift without extra significant drag oscillation.

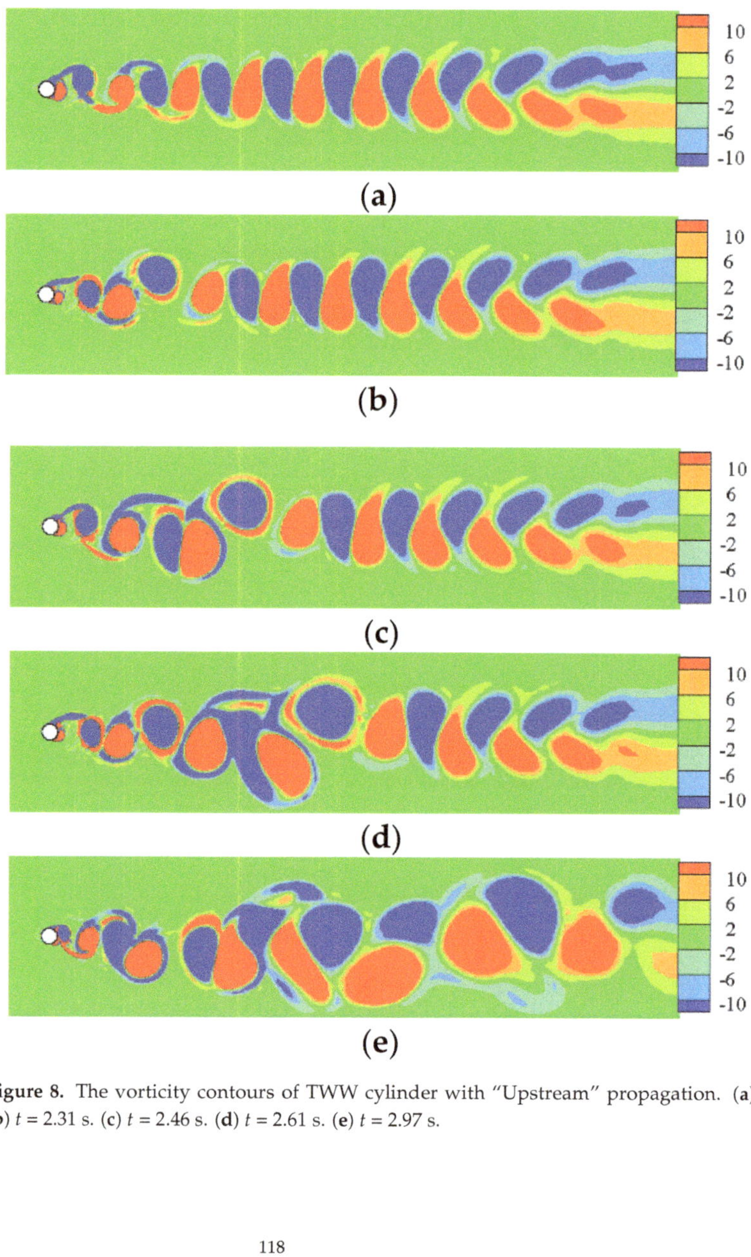

Figure 8. The vorticity contours of TWW cylinder with "Upstream" propagation. (**a**) t = 2.16 s. (**b**) t = 2.31 s. (**c**) t = 2.46 s. (**d**) t = 2.61 s. (**e**) t = 2.97 s.

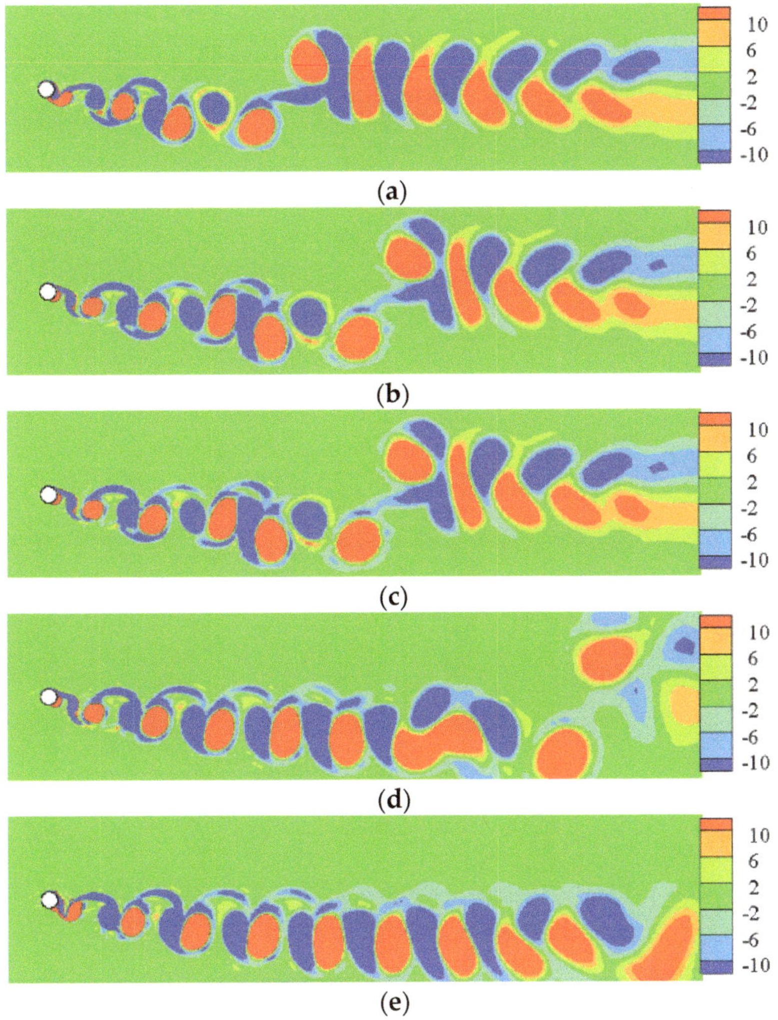

Figure 9. The vorticity contours of TWW cylinder with "Corotating-Clockwise" propagation. (**a**) t = 2.16 s. (**b**) t = 2.31 s. (**c**) t = 2.46 s. (**d**) t = 2.61 s. (**e**) t = 2.97 s.

3.3. Influence of Different Wave Numbers

In this part of simulation, the wave number is, respectively, set as 3, 4, 5 and 6 in rear section of 1/4 cylinder. As TWW propagated downstream, the amplitude ratio $\hat{A}/D$ equals to 0.02 and the velocity ratio c/U_∞ equals to 2.0. The lift and drag coefficient time histories under different wave numbers are shown in Figure 12. The results indicate that the TWW with different wave numbers can suppress the cylinder wake. On the other hand, the mean and fluctuating lift coefficients keep almost unchanged with the increase in wave number. In the meantime, as wave number changes from 3 to 6, there is little change in fluctuating drag coefficient, while the mean drag coefficient decreases to negative values.

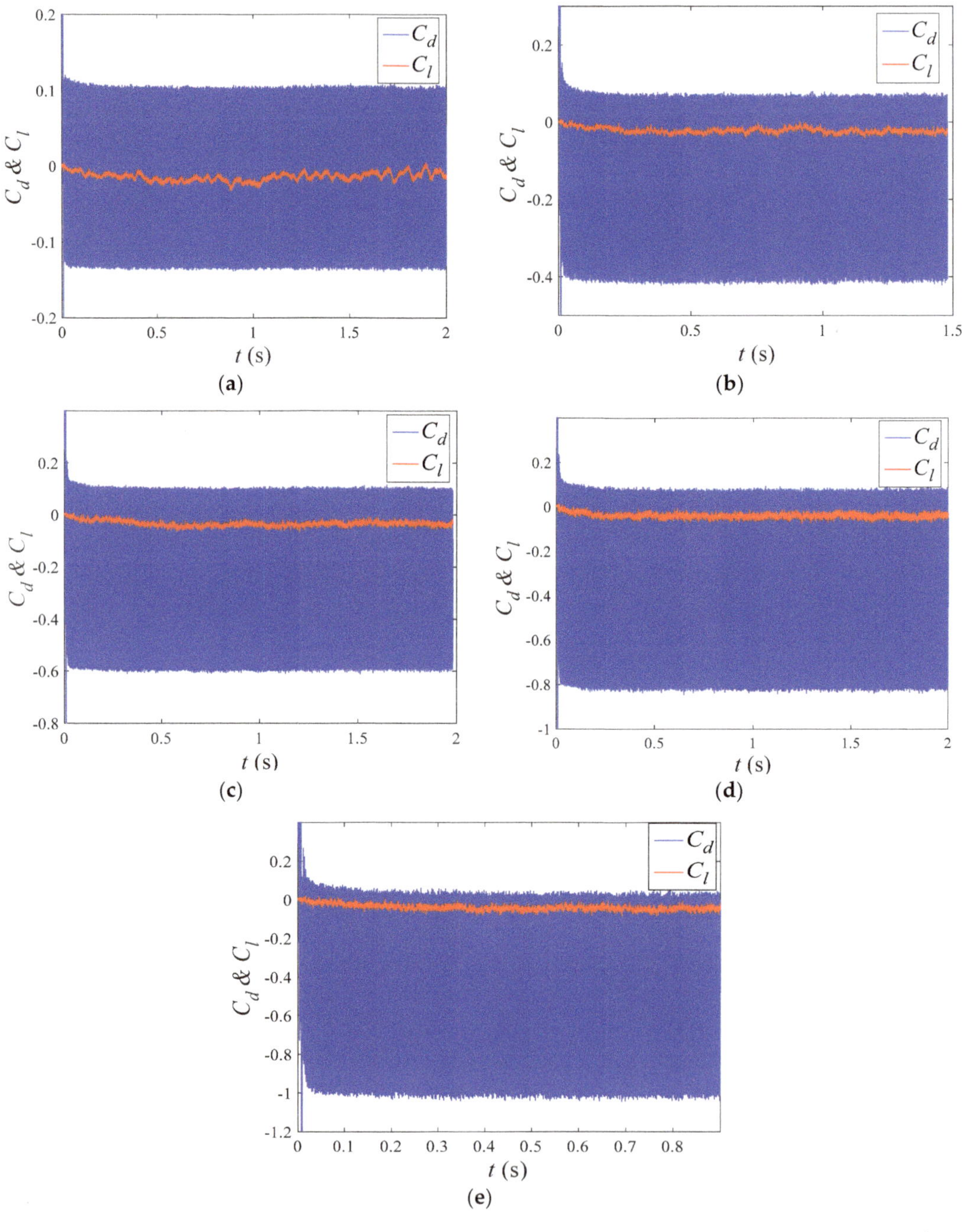

Figure 10. Lift and drag coefficients of TWW cylinder under various wave amplitudes. (**a**) $\hat{A}/D = 0.01$. (**b**) $\hat{A}/D = 0.02$. (**c**) $\hat{A}/D = 0.03$. (**d**) $\hat{A}/D = 0.04$. (**e**) $\hat{A}/D = 0.05$.

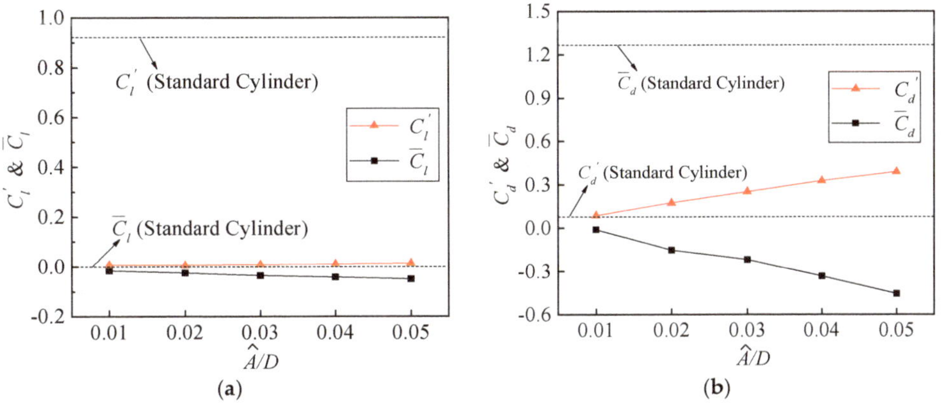

Figure 11. Characteristic values of lift and drag coefficient under different wave amplitudes. (**a**) C_l' & $\overline{C}_l$. (**b**) C_d' & $\overline{C}_d$.

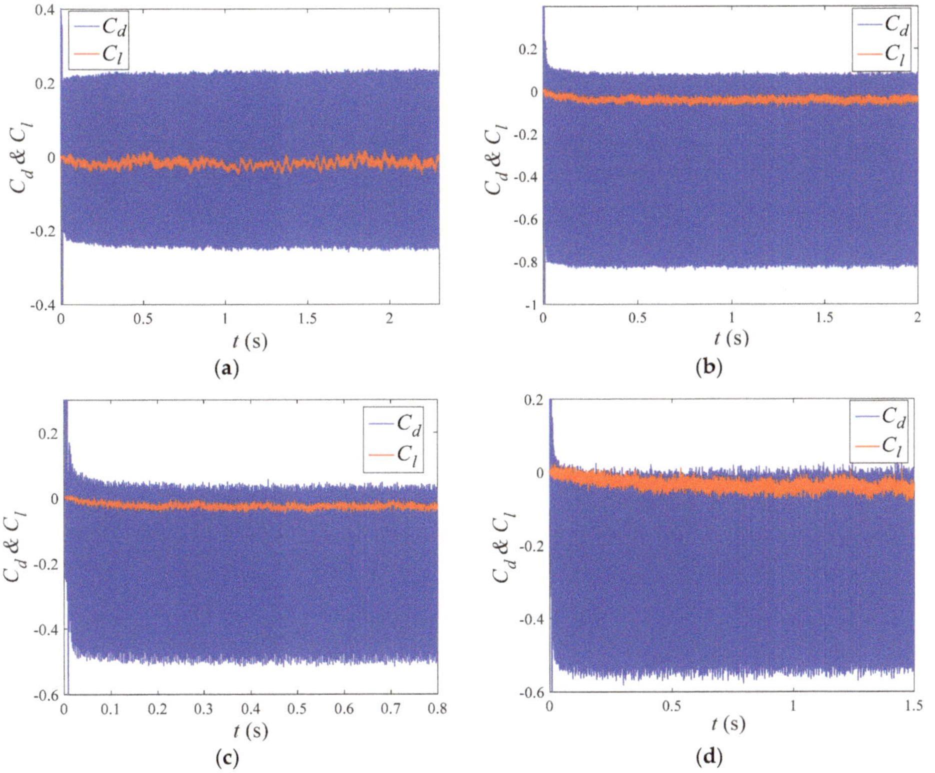

Figure 12. Lift and drag coefficients under different wave numbers. (**a**) N = 3. (**b**) N = 4. (**c**) N = 5. (**d**) N = 6.

Figure 13 shows the characteristic values of lift and drag coefficient under different wave numbers, including the mean and RMS values of lift and drag coefficient. In Figure 13a, with the increase in N, $\overline{C}_l$ decreases slightly from −0.019 to −0.039. C_l' almost hardly changes, whose values are between 0 and 0.01. It is observed that the fluctuating lift coefficient has been tremendously suppressed, however, different wave numbers have little impact on fluctuating lift coefficient after activating TWW, in comparison with the standard circular cylinder. Figure 13b shows that $\overline{C}_d$ decreases to negative values and C_d' changes from 0.166 to 0.195 with the increase in N. It indicates that the cylinder's reverse thrust increases gradually with the increase of N, and the wave number has some effect on $\overline{C}_d$.

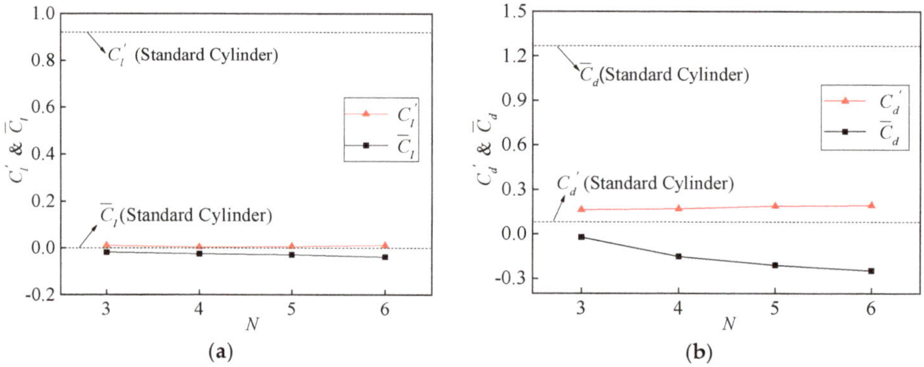

Figure 13. Characteristic values of lift and drag coefficient under different wave numbers. (**a**) C_l' & $\overline{C}_l$. (**b**) C_d' & $\overline{C}_d$.

3.4. Influence of Different Wave Velocities

The "Downstream" propagation direction of TWW is selected to further study the effect of wave velocity. The wave number is set as 4, and the amplitude ratio $\hat{A}/D$ is 0.02. Figure 14 shows the lift and drag coefficient time histories under given velocity ratios of 0.5, 1.0, 1.5, 2.0, 2.5, 3.0, 3.5, 4.0, 4.5, and 5.0. From the results of the lift coefficient fluctuation, the cylinder wake is suppressed in different degrees, but the fluctuation of lift coefficient increases slightly after $c/U_\infty > 2$. The mean lift coefficient decreases to negative values as c/U_∞ increases. With the increase in wave velocity, the mean drag coefficient decreases dramatically but the fluctuating drag coefficient increases gradually, as shown in Figure 14.

Figure 15 shows the characteristic value analysis of lift and drag coefficients under different wave velocities, including the mean and RMS values of lift and drag coefficients. In Figure 15a, with the increase of c/U_∞, $\overline{C}_l$ changes from a result larger than that of standard circular cylinder to a result less than that of standard circular cylinder. As for the overall time, $\overline{C}_l$ has little difference with that of standard circular cylinder in the range of the whole velocity ratios. As c/U_∞ increases from 0.5 to 1.5, C_l' decreases dramatically from 0.115 to 0.003. In addition, C_l' reaches the minimum at $c/U_\infty = 1.5$, whose value decreases by 99.63% than that of standard circular cylinder. C_l' increases slightly from 0.005 to 0.036 as c/U_∞ increases from 2.0 to 5.0. In addition, C_l' equals to 0.036 at $c/U_\infty = 5.0$, which achieves 96.09% reduction from the result of standard circular cylinder. In Figure 15b, with the increase of c/U_∞, $\overline{C}_d$ decreases from 0.761 (60.04% of standard cylinder) to -1.8774, and C_d' increases from 0.014 to 1.037 (almost 12.8 times the result of standard circular cylinder). From the analysis results above, it is indicated that the velocity ratio $c/U_\infty = 1.5$ is the optimal wave velocity for TWW flow control with respect to the control effect and energy input.

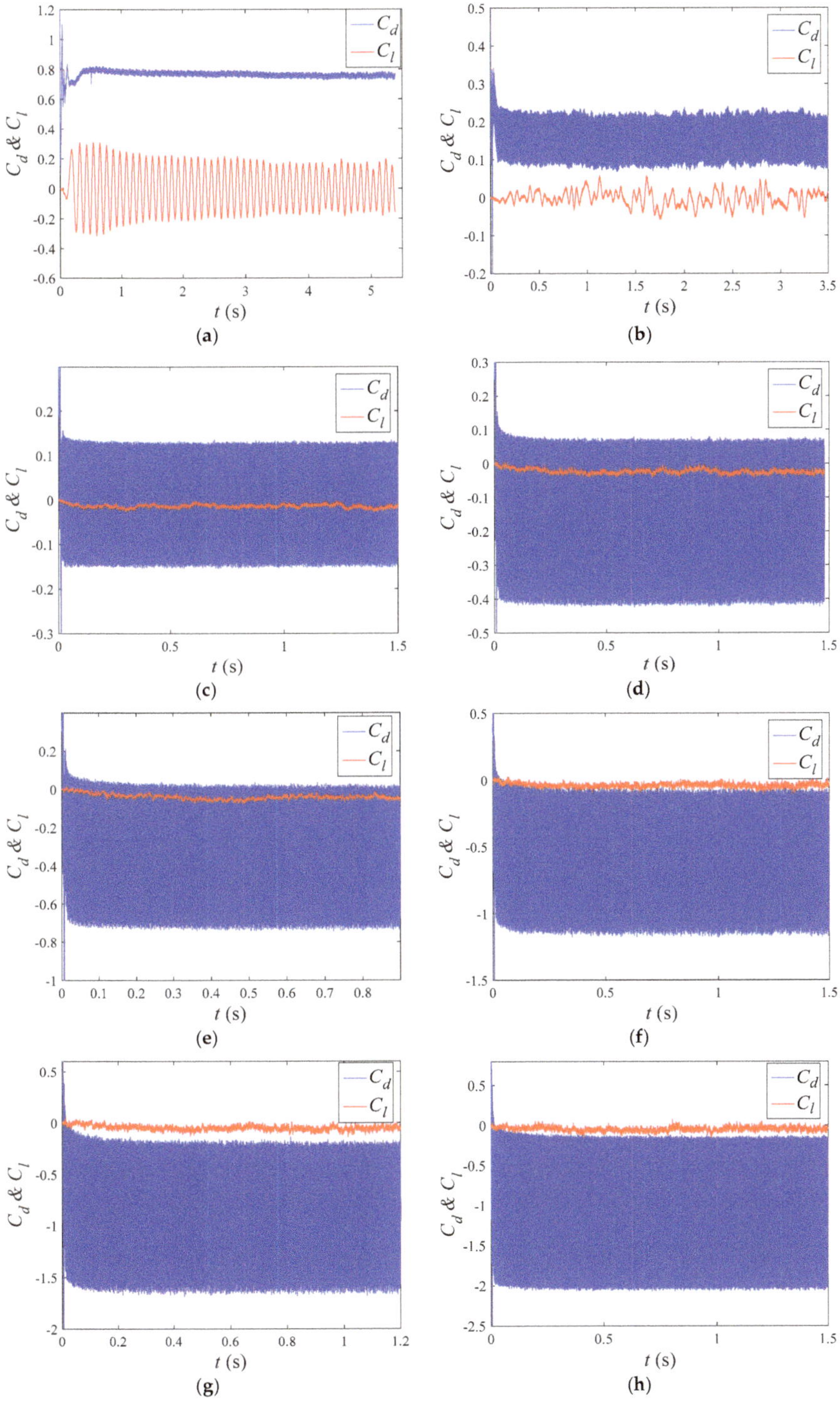

Figure 14. *Cont.*

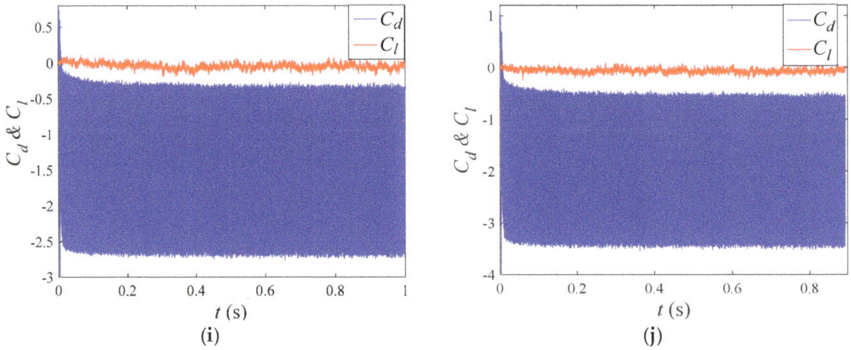

Figure 14. Lift and drag coefficient under different wave velocities. (**a**) $c/U_\infty = 0.5$. (**b**) $c/U_\infty = 1.0$. (**c**) $c/U_\infty = 1.5$. (**d**) $c/U_\infty = 2.0$. (**e**) $c/U_\infty = 2.5$. (**f**) $c/U_\infty = 3.0$. (**g**) $c/U_\infty = 3.5$. (**h**) $c/U_\infty = 4.0$. (**i**) $c/U_\infty = 4.5$. (**j**) $c/U_\infty = 5.0$.

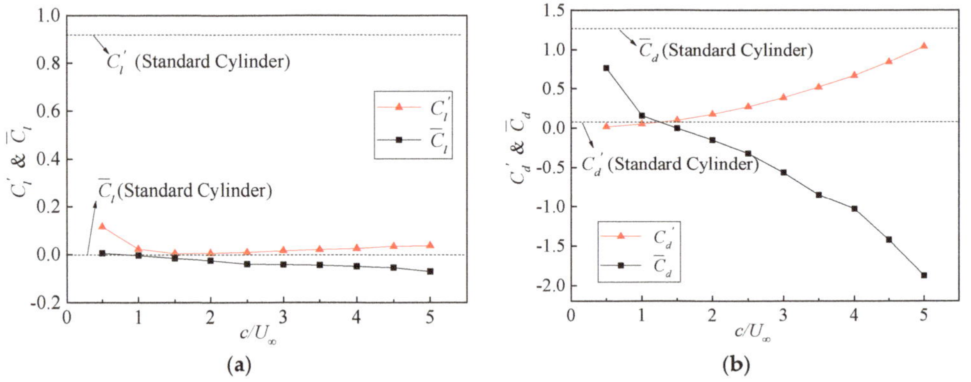

Figure 15. Characteristic values of lift and drag coefficient under different wave velocities. (**a**) C_l' & $\overline{C}_l$. (**b**) C_d' & $\overline{C}_d$.

4. Conclusions

In the present research, the CFD numerical simulation was employed to investigate the oscillating wake of flow around a fixed cylinder with the traveling wave wall at a high Reynolds number. The influence factors of TWW flow control method were investigated, i.e., the TWW propagation direction, wave amplitude, wave number and wave velocity. The main conclusions are as follows.

The TWW with "Downstream" propagation direction can successfully eliminate the alternating shedding vortex behind the cylinder under the condition of $c/U_\infty = 2.0$, and C_l' of TWW cylinder decreases by 99.48% from that of standard circular cylinder. The TWW cylinder in "Upstream" propagation direction enhances the cylinder wake. Neither "Corotating-Clockwise" nor "Corotating-Counterclockwise" could eliminate the cylinder oscillating wake.

Wave amplitude should be taken as $0.02D$ to obtain the optimal control effect. At the wave amplitude ratio $\hat{A}/D$ of 0.02, C_l' decreases dramatically to 0.005. When $\hat{A}/D$ increases from 0.02 to 0.05, C_l' increases slightly, while C_l' dropped by 98.7% from the result of standard circular cylinder at $\hat{A}/D = 0.05$. With the increase in $\hat{A}/D$, $\overline{C}_d$ decreases from -0.010 to -0.456, and C_d' increases to almost 4.7 times the result of standard circular cylinder.

Wave number N has relatively slighter effect on lift and drag coefficients. The $\overline{C}_l$, C_l' and C_d' show little changes with the increase of wave number N. $\overline{C}_d$ is less than 0, and the absolute value of $\overline{C}_d$ increases as wave number N increases, which indicates that the reverse thrust applied to the cylinder increases gradually with the increase in wave number N.

The fluctuating lift coefficient C_l' reaches the minimum at $c/U_\infty = 1.5$, which drops by 99.63% from the result of standard cylinder. When $c/U_\infty = 2.0 \sim 5.0$, C_l' takes a slow increasing tendency and increases to 3.91% of the result of standard circular cylinder. With the increase of c/U_∞, $\overline{C}_d$ decreases gradually and finally approaches to zero at $c/U_\infty = 1.5$. However, C_d' increases gradually with increasing c/U_∞ and finally reaches to about 12.8 times the result of standard circular cylinder at $c/U_\infty = 5.0$. Considering the control effect and energy input under the Reynolds number in this study, $c/U_\infty = 1.5$ is the optimal wave velocity for TWW flow control in this paper.

Author Contributions: Conceptualization, F.X.; investigation, F.X. and X.L.; writing—original draft preparation, F.X.; writing—review and editing, X.L. and W.B.; visualization, W.B. All authors have read and agreed to the published version of the manuscript.

Funding: This research was funded by the National Natural Sciences Foundation of China (NSFC) (52078175, 51778199 and U1709207), the Natural Science Foundation of Guangdong Province (2019A1515012205), the fundamental research funds of Shenzhen Science and Technology plan (JCYJ20190806144009332, JCYJ20180306172123896), and the Stability Support Program for colleges and universities in Shenzhen (GXWD20201230155427003-20200823134428001).

Institutional Review Board Statement: Not applicable.

Informed Consent Statement: Not applicable.

Data Availability Statement: Data are contained within this article.

Conflicts of Interest: The authors declare no conflict of interest.

References

1. Laima, S.; Li, H.; Chen, W.-L.; Li, F. Investigation and control of vortex-induced vibration of twin box girders. *J. Fluids Struct.* **2013**, *39*, 205–221. [CrossRef]
2. Chen, W.-L.; Zhang, Q.-Q.; Li, H.; Hui, H. An experimental investigation on vortex induced vibration of a flexible inclined cable under a shear flow. *J. Fluids Struct.* **2015**, *54*, 297–311. [CrossRef]
3. Chen, W.-L.; Li, H.; Hu, H. An experimental study on the unsteady vortices and turbulent flow structures around twin-box-girder bridge deck models with different gap ratios. *J. Wind Eng. Ind. Aerodyn.* **2014**, *132*, 27–36. [CrossRef]
4. Li, H.; Chen, W.-L.; Xu, F.; Li, F.-C.; Ou, J.-P. A numerical and experimental hybrid approach for the investigation of aerodynamic forces on stay cables suffering from rain-wind induced vibration. *J. Fluids Struct.* **2010**, *26*, 1195–1215. [CrossRef]
5. Gao, D.-L.; Chen, W.-L.; Li, H.; Hu, H. Flow around a circular cylinder with slit. *Exp. Therm. Fluid Sci.* **2017**, *82*, 287–301. [CrossRef]
6. Bearman, P.W.; Owen, J.C. Reduction of bluff-body drag and suppression of vortex shedding by the introduction of wavy separation lines. *J. Fluids Struct.* **1998**, *12*, 123–130. [CrossRef]
7. Bechert, D.W.; Bruse, M.; Hage, W.; van der Hoeven, J.G.T.; Hoppe, G. Experiments on drag-reducing surfaces and their optimization with an adjustable geometry. *J. Fluid Mech.* **1997**, *338*, 59–87. [CrossRef]
8. Choi, K.-S. Near-wall structure of a turbulent boundary layer with riblets. *J. Fluid Mech.* **1989**, *208*, 417–458. [CrossRef]
9. Lee, S.-J.; Jang, Y.-G. Control of flow around a NACA 0012 airfoil with a micro-riblet film. *J. Fluids Struct.* **2005**, *20*, 659–672. [CrossRef]
10. Owen, J.C.; Bearman, P.W.; Szewczyk, A.A. Passive control of VIV with drag reduction. *J. Fluids Struct.* **2001**, *15*, 597–605. [CrossRef]
11. Choi, J.I.; Xu, C.X.; Sung, H.J. Drag reduction by spanwise wall oscillation in wall bounded turbulent flows. *AIAA J.* **2002**, *40*, 842–850. [CrossRef]
12. Choi, K.-S.; Clayton, B.R. The mechanism of turbulent drag reduction with wall oscillation. *Int. J. Heat Fluid Flow* **2001**, *22*, 1–9. [CrossRef]
13. Chen, W.-L.; Xin, D.-B.; Xu, F.; Li, H.; Ou, J.-P.; Hui, H. Suppression of vortex-induced vibration of a circular cylinder using suction-based flow control. *J. Fluids Struct.* **2013**, *42*, 25–39. [CrossRef]
14. Chen, W.-L.; Li, H.; Hui, H. An experimental study on a suction flow control method to reduce the unsteadiness of the wind loads acting on a circular cylinder. *Exp. Fluids* **2014**, *55*, 1–20. [CrossRef]
15. Chen, W.-L.; Cao, Y.; Li, H.; Hui, H. Numerical investigation of steady suction control of flow around a circular cylinder. *J. Fluids Struct.* **2015**, *59*, 22–36. [CrossRef]

16. Chen, W.-L.; Gao, D.-L.; Yuan, W.-Y.; Li, H.; Hui, H. Passive jet control of flow around a circular cylinder. *Exp. Fluids* **2015**, *56*, 1–15. [CrossRef]
17. Modi, V.J. Moving surface boundary-layer control: A review. *J. Fluids Struct.* **1997**, *11*, 627–663. [CrossRef]
18. Kubo, Y.; Yukoku, E.; Modi, V.J.; Yamaguchi, E.; Kato, K.; Kawamura, S.-I. Control of flow separation from leading edge of a shallow rectangular cylinder through momentum injection. *J. Wind Eng. Ind. Aerodyn.* **1999**, *83*, 503–514. [CrossRef]
19. Mittal, S. Control of flow past bluff bodies using rotating control cylinders. *J. Fluids Struct.* **2001**, *15*, 291–326. [CrossRef]
20. Korkischko, I.; Meneghini, J.R. Suppression of vortex-induced vibration using moving surface boundary-layer control. *J. Fluids Struct.* **2012**, *34*, 259–270. [CrossRef]
21. Yang, Z.; Wu, J.Z. Drag reduction by axisymmetric travelling wavy wall. *J. Univ. Sci. Tech.* **2005**, *35*, 471–479.
22. Wu, C.J.; Xie, Y.Q.; Wu, J.Z. "Fluid Roller Bearing" effect and flow control. *Acta Mech. Sin.* **2003**, *19*, 476–484.
23. Wu, C.-J.; Wang, L.; Wu, J.-Z. Suppression of the von Kármán vortex street behind a circular cylinder by a travelling wave generated by a flexible surface. *J. Fluid Mech.* **2007**, *574*, 365–391. [CrossRef]
24. Xu, F.; Chen, W.-L.; Xiao, Y.-Q.; Li, H.; Ou, J.-P. Numerical study on the suppression of the vortex-induced vibration of an elastically mounted cylinder by a traveling wave wall. *J. Fluids Struct.* **2014**, *44*, 145–165. [CrossRef]
25. Versteeg, H.K.; Malalasekera, W. *An Introduction to Computational Fluid Dynamics: The Finite Volume Method*; Wiley: New York, NY, USA, 1995.
26. Wieselsberger, C. Neuere Feststellungen über die Gesetze des Flüssigkeits-und Luftwiderstands. *Phys. Z.* **1921**, *22*, 321–328. (In German)
27. Munson, B.R.; Young, F.D.; Okiis, T.H. *Fundamentals of Fluid Mechanics*, 5th ed.; Wiley: New Delhi, India, 2002.
28. Chen, W.L.; Wang, X.L.; Xu, F.; Li, H.; Hu, H. Passive jet flow control method for suppressing unsteady vortex shedding from a circular cylinder. *J. Aerospace Eng.* **2017**, *30*, 04016063. [CrossRef]
29. Relf, E.F.; Simmons, E.F.G. *The Frequency of Eddies Generated by the Motion of Circular Cylinders through a Fluid*; Taylor & Francis: Abingdon, UK, 1924.
30. Delany, N.K.; Sorensen, N.E. *Low-Speed Drag of Cylinders of Various Shapes*; Technical Note 303; National Advisory Committee for Aeronautics: Washington, DC, USA, 1953.
31. Roshko, A. Experiments on the flow past a circular cylinder at very high Reynolds number. *J. Fluid Mech.* **1961**, *10*, 345–356. [CrossRef]
32. Bearman, P.W. On vortex shedding from a circular cylinder in the critical Reynolds number régime. *J. Fluid Mech.* **1969**, *37*, 577–585. [CrossRef]
33. Mustto, A.A.; Bodstein, G.C.R. Subgrid-Scale Modeling of Turbulent Flow Around Circular Cylinder by Mesh-Free Vortex Method. *Eng. Appl. Comput. Fluid Mech.* **2011**, *5*, 259–275. [CrossRef]
34. Labbé, D.F.L.; Wilson, P.A. A numerical investigation of the effects of the spanwise length on the 3-D wake of a circular cylinder. *J. Fluids Struct.* **2007**, *23*, 1168–1188. [CrossRef]
35. Al-Jamal, H.; Dalton, C. Vortex induced vibrations using Large Eddy Simulation at a moderate Reynolds number. *J. Fluids Struct.* **2004**, *19*, 73–92. [CrossRef]
36. Fan, J.J.; Tang, Y.G.; Zhang, R.Y. Numerical simulation of viscous flow around circular cylinder at high Reynolds numbers and forced oscillating at large ratio of amplitude. *J. Hydrodyn.* **2012**, *27*, 24–32.

applied sciences

MDPI

Article

Moving Surface Boundary-Layer Control on the Wake of Flow around a Square Cylinder

Te Song [1], Xin Liu [2] and Feng Xu [2,*]

[1] Engineering Management Department, Guangzhou Communications Investment Group Co., Ltd., Guangzhou 510290, China; estimos@163.com
[2] School of Civil and Environmental Engineering, Harbin Institute of Technology (Shenzhen), Shenzhen 518055, China; 21b954008@stu.hit.edu.cn
* Correspondence: xufenghit@hit.edu.cn

Abstract: In this paper, the entire process of the flow around a fixed square cylinder and the moving surface boundary-layer control (MSBC) at a low Reynolds number was numerically simulated. Two small rotating circular cylinders were located in each of the two rear corners of the square cylinder, respectively, to transfer momentum into the near wake behind the square cylinder. The rotations of the two circular cylinders were realized via dynamic mesh technology, when the two-dimensional incompressible Navier–Stokes equations for the flow around the square cylinder were solved. We analyzed the effects of different rotation directions, wind angles θ, and velocity ratios k (the ratio of the tangential velocity of the rotating cylinder to the incoming flow velocity) on the wake of flow around a square cylinder to evaluate the control effectiveness of the MSBC method. In the present work, the aerodynamic forces, the pressure distributions, and the wake patterns of the square cylinder are discussed in detail. The results show that the high suction areas near the surfaces of the rotating cylinders can delay or prevent the separation of the shear layer, reduce the wake width, achieve drag reduction, and eliminate the alternating vortex shedding. For a wind angle of 0°, the inward rotation of the small circular cylinders is the optimal arrangement to manipulate the wake vortex street behind the square cylinder, and $k = 2$ is the optimal velocity ratio between the control effectiveness and external energy consumption.

Keywords: wake control; drag reduction; MSBC; square cylinder; numerical simulation

Citation: Song, T.; Liu, X.; Xu, F. Moving Surface Boundary-Layer Control on the Wake of Flow around a Square Cylinder. *Appl. Sci.* **2022**, *12*, 1632. https://doi.org/10.3390/app12031632

Academic Editor: Zifeng Yang

Received: 4 November 2021
Accepted: 30 January 2022
Published: 4 February 2022

1. Introduction

The development of complex structures has led to the construction of structures with reduced stiffness, such as large-span bridges and high-rise buildings, and wind-induced forces have become the main controlling factors in the design of these types of structures. The suppression of vortex shedding can lead to the lessening of the unsteady forces acting on the bluff bodies, thus resulting in less wind-induced vibrations. In terms of the flow field, the flow control changes the fluid-induced loads by altering the turbulence structures and the flow characteristics, whereupon the purpose of suppressing structural vibrations is achieved. The flow control may be accomplished by manipulating the boundary-layer separation or the structure of the shear layer in the wake.

A passive flow control does not consume external energy to realize its control purpose [1–4]. Laima et al. [5] and Chen et al. [6] studied the turbulence vortex structures, vortex-induced vibration characteristics, and passive flow control methods of the twin-box-girder bridge section model through wind tunnel tests. From the perspective of vortex dynamics, Gao et al. [7] presented a selective review of recent progress on the mechanism of VIV (vortex-induced vibration) occurred in long-span bridges and proposed several passive and active flow control methods to manipulate the surrounding flow patterns around the girder. Li et al. [8] and Chen et al. [9] studied the wind-induced vibration characteristics and wind–rain excitation mechanism of cylindrical cable structure by wind tunnel test and numerical simulation, which laid a foundation for the development of new flow control methods. To realize the control purpose, an active flow control channels the external energy into the flow field and injects the proper perturbation so that it may interact with the inner mode of the flow. Unlike the wall vibration, the bubble method, injection and suction [10–12], or the traveling wave wall [13–15], the moving surface boundary-layer control (MSBC) method is another type of active flow control.

The passive wake control behind a circular cylinder in a uniform flow for a Reynolds number (hereafter denoted as Re) ranging from 80 to 300 has been studied by Kubo et al. [16]. In their study, the vortex street behind the main cylinder still existed; however, the fluctuating lift and the form drag on the main cylinder significantly and monotonously decreased when Re increased from 80 to 300. Hwang and Yang [17] placed two splitter plates with the same length as the cylinder diameter along the horizontal centerline on the upstream and downstream of the cylinder and analyzed the aerodynamic forces and vortex shedding modes of the cylinder by adjusting the distances between the two plates and the cylinder. The study found that the maximum value of drag reduction was achieved 38.6% at a certain set of gap ratios. Assi et al. [18] investigated the suppression of cross-flow and in-line vortex-induced vibrations of a circular cylinder with a low combined mass-and-damping parameter of up to 0.014 by using two-dimensional control plates. The results showed that the maximum drag reduction was approximately 38% and occurred when using parallel plates. Koca and Genç [19,20] revealed the relationship between aerodynamic performance and vortex shedding from suction surface and wake of different types of wind turbine blades at a low Reynolds number by using the wind tunnel test, which provided a theoretical basis for further flow control research. In order to improve the aerodynamic performance and power output of wind turbine blades, Genç et al. [21,22] and Koca et al. [23] achieved the purpose of suppressing or eliminating laminar separation bubbles and reducing aerodynamic force by changing the surface roughness and arranging flexible membrane material at appropriate positions on the suction surface or on both suction and pressure surfaces of wind turbine blades. Bayramoğlu and Genç [24] replaced the flexible membrane material with piezoelectric material, which can achieve the dual purpose of flow control and energy harvesting. Malekzadeh and Sohankar [25] used a numerical simulation method to study the reduction effect of the aerodynamic forces and heat transfer of a control plate on a square cylinder at $\mathrm{Re} = 50 - 200$ based on the width of the square cylinder (W). This study shows that the optimum position and width for the control plate are a distance of $3W$ away from the cylinder and a width of $0.5W$ at $\mathrm{Re} = 160$, where the cylinder drag and the total drag (cylinder and plate) have a reduction of 86% and 37%, respectively, and the rms lift and drag coefficients on the cylinder have a 92% and 90% reduction compared to that of an isolated cylinder. Blowing and suction are also an effective way to control the wake of bluff bodies. Turhal and Çuhadaroğlu [26] investigated some aerodynamic parameters of the flow around perforated-surface square, horizontal, and diagonal cylinders at three different Reynolds numbers in a wind tunnel. The experimental results showed that the surface injections through the top-rear, rear, and all surfaces of a diagonal square cylinder reduce the drag coefficient for all the Reynolds numbers, while the injection through all surfaces only reduces the drag coefficient of a horizontal square cylinder. Sohankar et al. [27] numerically investigated the effects of uniform suction and blowing through the surfaces of a square cylinder on the vortex shedding, wake flow, and heat transfer at low Reynolds numbers. The results show that the lift and drag fluctuations for the optimum configuration decay, and the maximum reduction on the drag force are 61%, 67%, and 72% for Re = 70, 100, 150, and respectively. Chen et al. [28] and Gao et al. [29] realized the passive jet flow control by attaching hollow opening pipes on the cylinder surface and opening a slit in the middle of the cylinder section. Both methods have achieved the purpose of suppressing or delaying the alternating shedding vortices in the wake of the cylinder.

Based on the wind tunnel tests, the theoretical prediction, and the flow visualization, Munshi et al. [30] applied the MSBC to two-dimensional rectangular prisms to achieve drag reduction and suppression of flow-induced vibrations. The results indicated that the existence of the MSBC effectively eliminated the vortex resonance and galloping-type instabilities and that the maximum momentum injection corresponding to a velocity ratio of 3 was sufficient to delay boundary-layer separation and to achieve a minimum drag condition. On the basis of the existing research methods, Munshi et al. [31] conducted numerical simulations and analyzed the drag of two-dimensional flat plates and rectangular prisms with MSBC. Both the numerical and the flow visualization results showed narrowing of the wake as the momentum injection increased, leading to eventual suppression of the vortex shedding. Kubo et al. [32] studied the suppression of the aerodynamic response of square-sectioned tall structures with MSBC. The results of their study clearly showed that the boundary-layer control through the use of rotating cylinders at the leading edges of a two-dimensional square prism were effective in suppressing the vortex-excited and galloping oscillations. The rotor length and its position on the structure were the main factors affecting the effectiveness of MSBC. Kubo et al. [33] conducted extensive wind tunnel experiments to study the role of the MSBC in the suppression of torsional flutter of a shallow rectangular prism, which was one of the typical fundamental configurations of bridge deck sections. The results that they obtained from measurements of surface pressure distributions and aeroelastic vibration responses showed that the proposed boundary-layer control method through momentum injection was effective in suppressing the torsional flutter of a shallow rectangular prism. Modi and Deshpande [34] conducted wind tunnel tests to study the aerodynamics of a cubic structure in the presence of MSBC, using rotating circular cylindrical elements at the two adjacent vertical edges. The results revealed the significant effect of the MSBC, which changes the height of the boundary-layer separation and reattachment, thus affecting the fluid dynamical parameters.

The moving surface boundary-layer control (MSBC) is also called angular momentum injection. Through numerical simulation, Patnaik and Wei [35] proposed an angular momentum injection strategy to control the wake dynamics and turbulence behind a fixed square cylinder at Re = 200. This study observed a new zone formation of a recirculation-free zone (RFZ) behind the cylinder at $\zeta = 1.25$. With the increase of angular momentum injection, the zones of absolute and convective instabilities (AIZ and CIZ) are diminished, and the recirculation-free zone is enhanced. Finally, the wake turbulence is completely suppressed at $\zeta = 1.5$. Muddada and Patnaik [36] compared the wake characteristics of a flow past an isolated circular cylinder, with momentum injection in the subcritical Re range. The numerical flow visualization through streaklines and streamlines clearly demonstrated the effectiveness of the rotating-type actuators in modifying the turbulent vortex structures. As the ratio of the momentum input became higher, the flow tended to be smoother. Mittal [37] investigated the flow past a bluff body with two rotating control cylinders using two-dimensional numerical simulation. When the control cylinders rotated with a high speed, in a manner that the tip speed was five times the free-stream speed, the flow at Re = 100 reached a steady state. For Re = 10^4, although the flow remained unsteady, the wake was much more organized and narrower compared to the one without control. Korkischko and Meneghini [38] verified the efficiency of the MSBC in suppressing vortex-induced vibrations (VIV); the flow control results in a mean drag of almost 60% compared to that of the plain cylinder, reduction at $U_c/U = 5$. The wake is highly organized and narrower compared to the one observed in cylinders without control.

In the present work, the entire process involving the flow around a fixed square cylinder and the MSBC was numerically simulated. The present numerical simulation started from the flow around a fixed square cylinder; when the alternating vortex shedding appeared and became stable, the MSBC was then activated to control the wake. Two small rotating circular cylinders were employed, which were located in the rear corners of the square cylinder, similar to Patnaik and Wei [35], to achieve momentum injection into the flow field. In this work, we mainly study the control effectiveness of the rotation direction of the small circular cylinders, the wind angle, and the velocity ratio on the oscillating wake of the square cylinder. The aerodynamic force time histories and their statistics, the mean pressure distribution, and the vortex shedding pattern were studied to demonstrate the control effectiveness of the MSBC for different influencing factors and to explain the mechanism of the MSBC flow control method.

2. Numerical Calculation Model

2.1. Governing Equations of the Fluid Flow

For a two-dimensional incompressible flow, the governing equations, including the continuity equation and the Navier–Stokes equation, in the Cartesian coordinate system can be written as:

$$\frac{\partial u_i}{\partial x_i} = 0, \tag{1}$$

$$\frac{\partial u_i}{\partial t} + u_j \frac{\partial u_i}{\partial x_j} = -\frac{1}{\rho}\frac{\partial p}{\partial x_i} + \frac{\mu}{\rho}\frac{\partial^2 u_i}{\partial x_j^2}, \tag{2}$$

where u_i is the velocity components in the i direction, ρ is the fluid density, μ is the kinematic viscosity coefficient, and p is the pressure in the flow field.

2.2. Numerical Model and Solution Setting

The computational domain is a rectangular region of $40L \times 20L$, where the side length of the square cylinder $L = 0.054$ m; the center of the square cylinder is located at the origin of the coordinates, as shown in Figure 1. The distances from the upstream inlet, the downstream outlet, and the two sides to the cylinder center are $10L$, $30L$, and $10L$, respectively. The selection of the above size parameters of the computational domain refers to the research results of Sohankar et al. [39] and Cheng et al. [40]. It meets the requirement that the blocking ratio is less than or equal to 5%, and the distance of $30L$ from model to outlet can ensure the full development of the downstream flow of the square cylinder. Two rotating circular cylinders with a diameter of D are set on the rear corners of the square cylinder, and the distance between the rotating cylinders and the square cylinder is represented by h. The upper and lower sides of the square cylinder are tangential to the two circular cylinder surfaces.

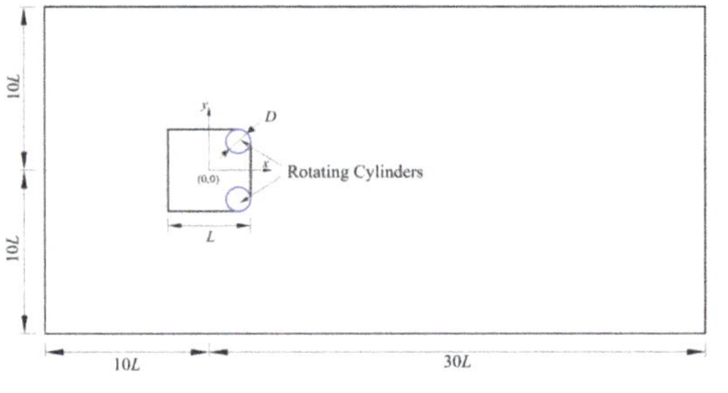

Figure 1. Schematic diagram of physical model and computational domain.

The computational domain is discretized using unstructured grids and is divided into two regions, namely Zone 1 and Zone 2, as shown in Figure 2. Zone 1 is the region around the two circular cylinders where the circular grids can rotate with the circular cylinders. The remaining region is Zone 2, and the fixed grids near the square cylinder surface are locally refined. The interface between the rotation grid (Zone 1) and fixed grid (Zone 2) is set as a slip boundary.

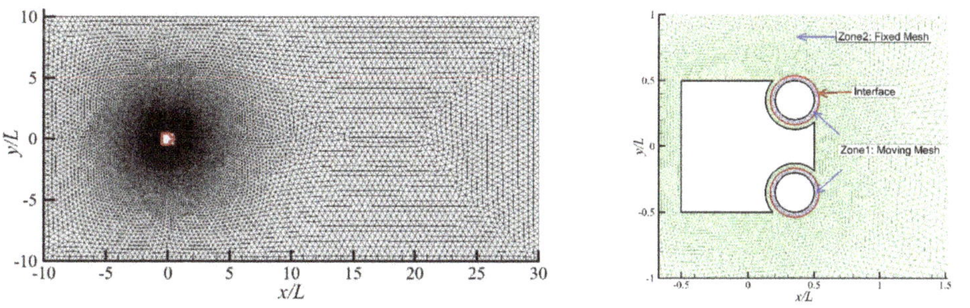

(**a**) (**b**)

Figure 2. Computational mesh arrangement and domain partition, (**a**) Computational mesh, (**b**) Computational domain partition and local refined mesh.

The wind angle, rotation directions of the two circular cylinders, and arrangement of the pressure monitoring points for $h/D = 1/4$ and $D/L = 0.3$ are shown in Figure 3. The rotation directions of the two circular cylinders were designed as in the following three schemes: rotating in the downstream direction, rotating in the upstream direction, and co-rotating in the clockwise direction; each of the three schemes is represented in Figure 3a–c, respectively. The wind angle θ is defined as in the following: when the direction of the incoming airflow and the x-axis are the same, the wind angle is set to 0°, and the flow along the clockwise direction is considered as the positive direction. When the wind angle is 0°, the two rotating circular cylinders are placed at the two rear corners of the square cylinder. A total of 32 monitoring points were arranged on the model surface, out of which 18 points were evenly arranged on the straight edges of the square cylinder, and 7 points were evenly located on the quarter arc of each rotating circular cylinder. The main purpose is to obtain the pressure distribution characteristics on each straight edge of the square cylinder and on the two quarter arcs tangent to the surface of the square cylinder. In the calculation process, the pressure coefficient of each point was monitored to reveal the underlying mechanism of the MSBC method for manipulating the wake behind the square cylinder.

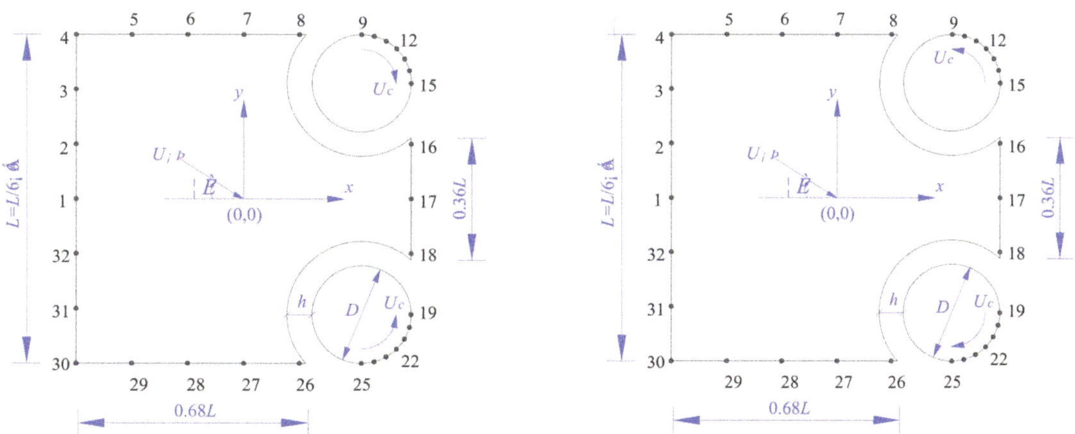

(**a**) (**b**)

Figure 3. *Cont.*

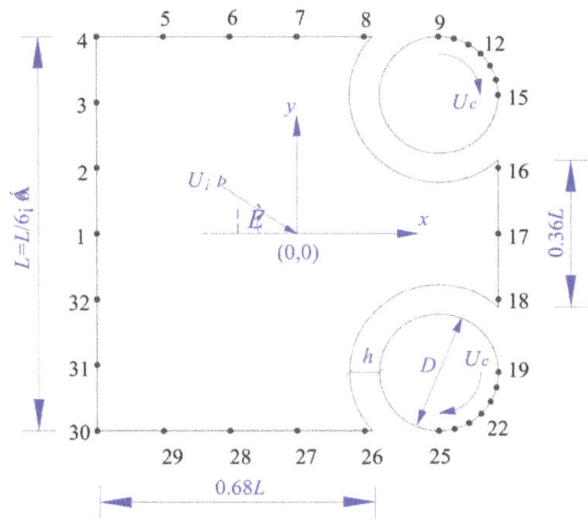

(c)

Figure 3. Wind angle, rotation direction and pressure monitoring point layout ($h/D = 1/4$, $D/L = 0.3$), (**a**) Inward rotation, (**b**) Outward rotation, (**c**) Clockwise co-rotation.

The velocity ratio k ($k = U_c/U_\infty$) is the ratio of the tangential velocity of the rotating circular cylinder surface U_c to the uniform incoming velocity U_∞. The velocity ratio was adjusted by changing the rotation speed of the small cylinders, while the incoming airflow remained fixed. When $k = 0$, the two small circular cylinders were stationary; as the velocity ratio became higher, the momentum that the small cylinders delivered to the flow field became greater.

The flow direction was from the left to the right, and the following boundary conditions were set. The left side was a velocity inlet with a uniform velocity U_∞ of 0.054 m/s; the right side was a pressure outlet, where the relative pressure was set to 0; and the upper and lower sides were set as symmetry boundaries. In addition, the square cylinder and the rotating circular cylinder surfaces were set as no-slip walls, and the interfaces between Zone 1 and Zone 2 were set as slip boundaries. The rotation of the circular cylinder and the surrounding mesh was realized through a dynamic mesh technique.

The numerical simulations were carried out at Re = 200, and the laminar model was employed. The finite volume method (FVM) was used to discrete the flow field around a square cylinder. The SIMPLE algorithm was used for the calculation of the coupling between the pressure and the velocity fields. The format of the pressure interpolation was selected to be 'Standard'. The second-order upwind scheme was utilized for the momentum discretization because of its stability and veracity. The time steps were changed according to the numerical simulation conditions. At the beginning, the dimensionless time step was set to 0.05. When the flow became fully developed and the vortex shedding occurred in the wake of the square cylinder, the MSBC became activated; then, the dimensionless time step was automatically reduced to 0.01.

2.3. Validity Investigation

The square cylinder with $\theta = 0^\circ$ and $k = 0$ is defined as the 'standard square cylinder', whereas the square cylinder without rotating cylinder is referred to as a 'single square cylinder'. First, a mesh independence study was conducted in order to quantify the effect of mesh density on computational results. Simulations of flow around the single square cylinder were then performed at Re = 200 using four different mesh densities as listed in Table 1, where N_c, N_{mesh}, and N_{nodes} represent, respectively, the node number on each side length of the square cylinder, the total number of grids, and nodes in the computational domain. The minimum grid size of each side of the single square cylinder $l_{min} = L/N_c$, and the grids grow into the computational domain with each side of the square cylinder as source and the growth rate of 1.04, and the maximum grid size $l_{max} = 0.4L$. The comparison shown in Table 1 includes the fluctuation of lift coefficient ($C_l\prime$), the mean and fluctuation of drag coefficient ($\overline{C}_d$, $C_d\prime$), and the Strouhal number (S_t). As shown in Table 1, as the mesh density increases, the $C_d\prime$ and S_t of different mesh schemes do not change significantly, and both $\overline{C}_d$ and $C_l\prime$ show a decreasing trend. The differences in highlighted aerodynamic coefficients $\overline{C}_d$ and $C_l\prime$ between the normal (Scheme 3) and dense (Scheme 4) meshes are 0.6 % and 4.2 %, and that between the coarse (Scheme 2) and dense (Scheme 4) meshes are 2.3 % and 12.8 %. Therefore, the normal mesh of 'Scheme 3' is appropriate in the following simulations of this study.

Table 1. Effect of mesh refinement on the calculation result of the flow around a single square cylinder at Re = 200.

Mesh Density	N_c	N_{mesh}	N_{nodes}	$\overline{C_d}$	$C_d^{\prime}$	$C_l^{\prime}$	S_t
Scheme 1, coarsest	40	30,344	15,387	1.523	0.024	0.446	0.146
Scheme 2, coarse	55	36,208	18,349	1.489	0.022	0.404	0.148
Scheme 3, normal	70	42,092	21,323	1.465	0.020	0.373	0.148
Scheme 4, dense	85	48,590	24,602	1.456	0.019	0.358	0.149

Then, the mesh independent results of flow around the single square cylinder are compared with the existing research results to verify the accuracy of numerical simulation, and the results from the flow around the standard square cylinder are compared with those from the single square cylinder. The minimum grid size, growth ratio, and maximum grid size of each side of the standard square cylinder and the surfaces of the two rotating small cylinders are kept consistent with the parameters of the 'Scheme 3' meshing scheme of the single square cylinder. Figure 4 shows the time histories of the lift and drag coefficients of the single and standard square cylinders. The dimensionless lift coefficient by a spectral analysis is converted to the Strouhal number (S_t). The dimensionless time is $t^* = tU_\infty/L$. The alternating shedding vortex wake is the main cause of the periodic variation in the lift and drag coefficients, as shown in Figure 5. The figure shows the vorticity contours in the z direction (perpendicular to the x-o-y plane). The red color in the figure represents the positive vorticity of counterclockwise rotation, and the blue color represents the negative vorticity of clockwise rotation. The comparison between the results from the present numerical simulation and the experimental and numerical results from other previous work is presented in Table 2. The statistical results of aerodynamic parameters in the Table 2 are analyzed by using the time histories of lift and drag coefficients in the time period of $t^* = 80 \sim 150$, shown in Figure 4a. In the subsequent calculation cases, all pressure and aerodynamic coefficient time histories are calculated to the amplitude stable state, and the results of at least five vortex shedding characteristic periods are taken for statistical analysis. The simulation results of the aerodynamic forces and the Strouhal number are consistent with the previous results, and it indicates that the present grid accuracy, time step, and numerical settings can accurately capture the characteristics of the flow around a square cylinder. The mean drag coefficient $\overline{C}_d = 1.015$ of the standard square cylinder is 30.7% lower than that of the single square cylinder, the fluctuation drag coefficient $C_d\prime = 0.0128$ is 36.0% lower than that of the single square cylinder, and the fluctuation lift coefficient $C_l\prime = 0.324$ is 13.1% lower than that of the single square cylinder. The Strouhal number of the standard square cylinder ($S_t = 0.15$) is slightly higher than that of the single square cylinder. It is shown that a small change in the shape of the rear corner of the square cylinder can effectively reduce the aerodynamic force. However, the vortex shedding frequency will be slightly increased. In the following sections, the numerical results will be compared with the results of the flow around the standard square cylinder.

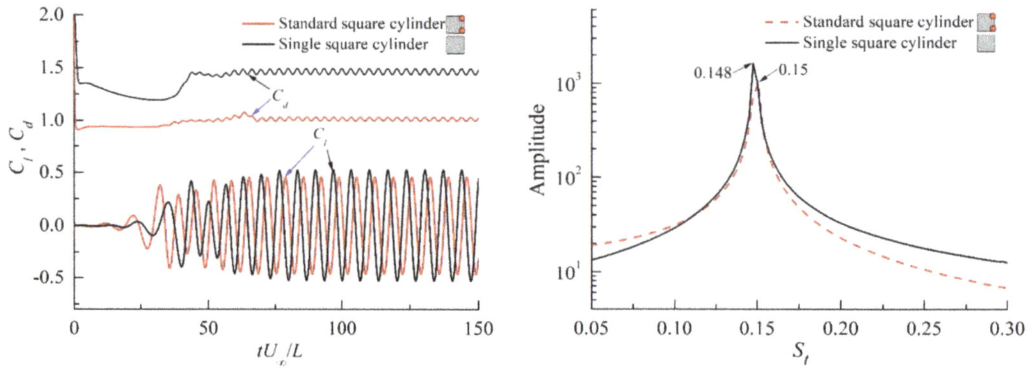

(a) (b)

Figure 4. Comparisons of results of flow around a standard square cylinder and a single square cylinder, (**a**) Time histories of lift and drag coefficients, (**b**) Spectral analysis of C_l.

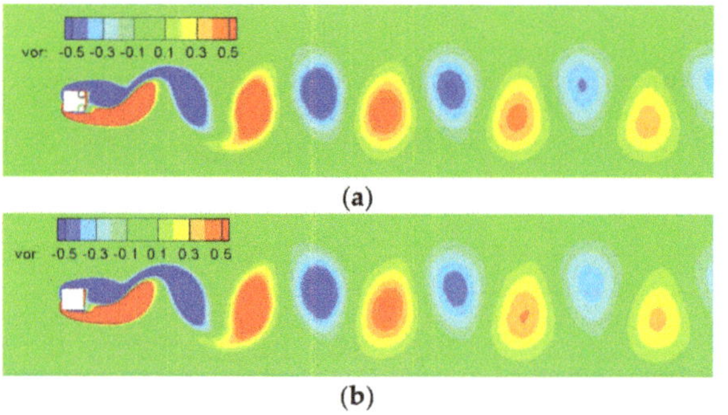

Figure 5. Vortex shedding in the wake of a standard square cylinder and a single square cylinder, (**a**) Standard square cylinder, (**b**) Single square cylinder.

Table 2. Comparison of the force coefficients statistics and Strouhal number of flow around a single square cylinder at Re = 200.

Investigation	$\overline{C_d}$	C_d'	C_l'	S_t
Okajima [41], Experimental	1.45	-	-	0.14~0.148
Sohankar et al. [39], Numerical, 2D	1.462	-	0.377	0.15
Cheng et al. [40], Numerical, 2D	1.45	-	0.372	0.15
Jan and Sheu [42], Numerical, 2D	-	-	-	0.148
Abograis and Alshayji [43], Numerical, 2D	1.488	0.027	0.332	0.153
Present, Numerical, 2D ($N_c = 70$)	1.465	0.020	0.373	0.148

3. Results and Discussion

3.1. Analysis of Influence of D/L and h/D Parameters

The diameter (D) of the small rotating circular cylinders and the spacing (h) between the rotating circular cylinders and the controlled square cylinder are important parameters that affect the control effect on the wake of the square cylinder. When wind angle $\theta = 0°$, velocity ratio $k = 2$, and the rotation direction of the circular cylinders is fixed to 'inward' rotation, as shown in Figure 3a, the influences of D/L and h/D on the statistical values and frequency characteristics of the aerodynamic coefficients of the square cylinder are mainly analyzed. The two dimensionless parameters $E_{C_l} = (C_l{\prime} - C_{l\_c}{\prime})/C_l{\prime}$ and

$E_{C_d} = (\overline{C}_d - \overline{C}_{d\_c})/\overline{C}_d$ are defined, which represent the relative variation in the RMS value of the lift coefficient and the mean value of the drag coefficient, respectively. These two parameters are used to evaluate the control effectiveness of the MSBC on the wake of flow around the square cylinder and determine the optimal parameters of the D/L and h/D. When E_{C_l} and E_{C_d} are close to 1.0, the suppression effect on the wake is optimal. In contrast, when E_{C_l} and E_{C_d} are close to 0, no control exists.

Figure 6 shows the aerodynamic statistical parameters and Strouhal number (S_t and S_{t_c}) of flow around the uncontrolled ($k = 0$) and controlled ($k = 2$) square cylinder versus $D/L = 0.05 \sim 0.4$ when $h/D = 1/4$, together with the control effectiveness. The specific values corresponding to Figure 6 are shown in Table 3, where the aerodynamic statistical parameters include the RMS value of the lift coefficient ($C_l{\prime}$ and $C_{l\_c}{\prime}$), the mean value of the drag coefficient ($\overline{C}_d$ and $\overline{C}_{d\_c}$), and the RMS value of the drag coefficient ($C_d{\prime}$ and $C_{d\_c}{\prime}$). For different D/L cases, the centers of the rotating circular cylinders can ensure that the side extended lines of the square cylinder are tangent to the circumference of the rotating circular cylinders. For the uncontrolled standard square cylinder, $C_l{\prime}$ is slowly decreased by 8%, $\overline{C}_d$ and $C_d{\prime}$ are decreased by 29% and 50%, respectively, while S_t is kept unchanged at 0.15 with the increase of D/L from 0.05 to 0.4. When there are two nonrotating cylinders at the corners of the square cylinder, it is equivalent to the rounded treatment of the leeward sharp corners of the square cylinder. This has less influence on the fluctuation degree of the lift coefficient in the cross-flow direction and vortex shedding frequency but can reduce mean value and fluctuation of the drag coefficient in the along-wind direction.

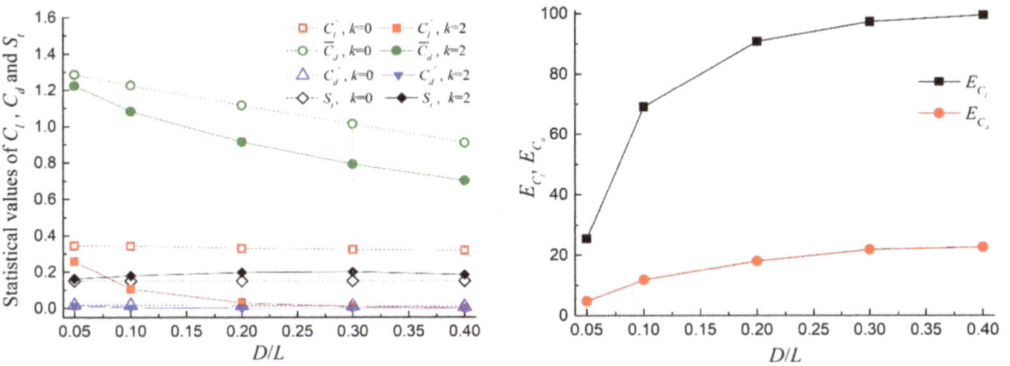

(**a**) (**b**)

Figure 6. Statistical values of lift and drag coefficients, Strouhal number, and control effects under different D/L values, (**a**) Statistical values of C_l, C_d, and S_t, (**b**) Control effects of C_l, C_d.

Table 3. Lift and drag coefficient statistics, Strouhal number, and control effects of flow around an uncontrolled ($k = 0$) and controlled ($k = 2$) square cylinder with $h/D = 1/4$.

D/L	k=0				k=2					
	C_l'	$\overline{C_d}$	C_d'	S_t	C_{l_c}'	$\overline{C_{d\_c}}$	C_{d_c}'	S_{t_c}	E_{C_l} (%)	E_{C_d} (%)
0.05	0.3466	1.2873	0.0186	0.15	0.2584	1.2255	0.0141	0.1625	25.45	4.80
0.1	0.3433	1.2274	0.0180	0.15	0.1063	1.0828	0.0051	0.1786	69.04	11.78
0.2	0.3309	1.1159	0.0155	0.15	0.0306	0.9144	0.0012	0.1961	90.75	18.06
0.3	0.3238	1.0145	0.0128	0.15	0.0089	0.7927	0.0008	0.2000	97.25	21.86
0.4	0.3189	0.9091	0.0093	0.15	0.0018	0.7027	0.0011	0.1832	99.44	22.70

When the two circular cylinders are rotating, the $C_{l\_c}{\prime}$, $\overline{C}_{d\_c}$, and $C_{d\_c}{\prime}$ of the controlled square cylinder are lower than those of the uncontrolled square cylinder, whereas the S_{t_c}

of the controlled square cylinder is higher than that of the uncontrolled square cylinder. With the increasing D/L, the control effect E_{C_l} on the RMS value of the lift coefficient is gradually increased from 25.45% to 99.44%, and the control effect E_{C_d} on the mean value of the drag coefficient is gradually increased from 4.8% to 22.7%. When $D/L = 0.3$, the RMS value of drag coefficient plummets to its minimum value, which is 93.7% lower than that of the uncontrolled square cylinder, and the E_{C_l} and E_{C_d} at this time are 97.25% and 21.86%, respectively. When $D/L = 0.05 \sim 0.3$, the S_{t_c} is gradually increased from 0.1625 to 0.2; in turn, when the D/L is increased to 0.4, the S_{t_c} is decreased slightly to 0.1832. It can be seen that the rotating circular cylinders play the role of controlling the wake of the square cylinder. The mean value of the drag coefficient can be reduced by up to 22.7%, and the fluctuations of the lift and drag coefficients are basically completely suppressed, while the vortex shedding frequency in the wake of the square cylinder is increased from 8.3% to 33.3%.

When $D/L = 0.3$, the aerodynamic statistical parameters, Strouhal number (S_t and S_{t_c}), and control effectiveness of flow around the uncontrolled ($k = 0$) and controlled ($k = 2$) square cylinder versus h/D, the range of which varies from 1/8 to 3/8 with an interval of 1/16, are shown in the Figure 7 and Table 4. With gradually increasing h/D, the $C_l{\prime}$ of the uncontrolled square cylinder does not change significantly between 0.3226 and 0.3388, and the S_t is also kept unchanged at 0.15, while $\overline{C}_d$ and $C_d{\prime}$ are decreased by 13% and 23%, respectively. When $k = 2$, the $C_{l\_c}{\prime}$ and $\overline{C}_{d\_c}$ of the controlled square cylinder are gradually increased with increase of h/D, and the corresponding control effects of E_{C_l} and E_{C_d} are gradually decreased with the increasing h/D. The $C_{d\_c}{\prime}$ of controlled square cylinder changes from 0.0006 to 0.0039, which is about 5.2% to 26% of the results of uncontrolled square cylinder, and the S_{t_c} of controlled square cylinder is between 0.18 and 0.2, which is still greater than the result of the uncontrolled square cylinder.

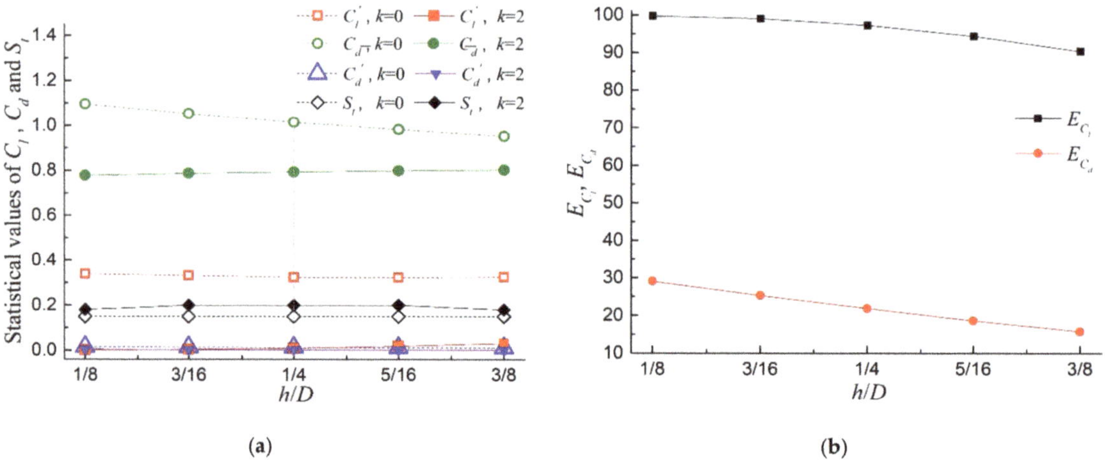

Figure 7. Statistical values of lift and drag coefficients, Strouhal number, and control effects under different h/D values, **(a)** Statistical values of C_l, C_d, and S_t, **(b)** Control effects of C_l, C_d.

Table 4. Lift and drag coefficient statistics, Strouhal number, and control effects of flow around an uncontrolled ($k = 0$) and controlled ($k = 2$) square cylinder with $D/L = 0.3$.

D/L	k=0				k=2					
	C_l'	$\overline{C_d}$	C_d'	S_t	C_{l_c}'	$\overline{C_{d\_c}}$	C_{d_c}'	S_{t_c}	E_{C_l} (%)	E_{C_d} (%)
1/8	0.3388	1.0956	0.0150	0.15	0.0009	0.7770	0.0039	0.18	99.73	29.08
3/16	0.3318	1.0534	0.0140	0.15	0.0032	0.7870	0.0020	0.20	99.04	25.29
1/4	0.3238	1.0145	0.0128	0.15	0.0089	0.7927	0.0008	0.20	97.25	21.86
5/16	0.3226	0.9821	0.0121	0.15	0.0181	0.7996	0.0013	0.20	94.39	18.58
3/8	0.3254	0.9524	0.0115	0.15	0.0314	0.8026	0.0006	0.18	90.35	15.73

Through comparative analysis, we can see that the smaller the spacing (h) between the rotating circular cylinders and the square cylinder, the better the control effect of the MSBC method on the wake of the square cylinder. As the spacing ratio h/D decreases from 3/8 to 1/8, the results of E_{C_l} and E_{C_d} are increased by 9.38% and 13.35%, respectively, which are less than the increasing extents of E_{C_l} and E_{C_d} with increase of D/L and are especially far less than the increasing extent of 73.99% for E_{C_l}, as shown in Table 2. It can be seen that the influence of the diameter of the rotating circular cylinders on the control effect of the wake of the square cylinder is greater than the influence of the variation of the spacing between the rotating circular cylinders and the square cylinder on the control effect.

Compared with the control effects of $D/L = 0.3$ and $h/D = 1/4$, when D/L continues to increase from 0.3 to 0.4, the improvements of the control effects E_{C_l} and E_{C_d} of the controlled square cylinder are not obvious, only increased by 2.19% and 0.84%. As the h/D is decreased from 1/4 to 1/8 continually, the E_{C_l} is only increased from 97.25% to 99.73%, and the E_{C_d} is only increased by 7.22% at the same time. On the contrary, the fluctuation of the drag coefficient $C_{d\_c}\prime$ is increased from 6.3% to its maximum of 26% of $C_d\prime$ for the uncontrolled standard square cylinder. From the above analysis, it can be concluded that $D/L = 0.3$ and $h/D = 1/4$ are the better combination of the parameters to achieve a better control effect on the wake of flow around the square cylinder.

3.2. Detailed Results and Analysis of $D/L = 0.3$ and $h/D = 1/4$

The present numerical simulation starts from the flow around a fixed square cylinder; when the alternating vortex shedding street has been formed behind the square cylinder, the MSBC becomes activated. The rotation directions of the circular cylinders include the following three forms: inward rotation, outward rotation, and clockwise co-rotating, as shown in Figure 3. The wind angle range was $0^\circ \sim 180^\circ$ and the angle interval was 30°. For the inward rotation case, an extra wind angle of 15° was added. The velocity ratio range was $0 \sim 4.0$. The aerodynamic coefficients of the square cylinder under different wind angles and velocity ratios were calculated. Then, the statistical and frequency characteristics of the aerodynamic coefficients were obtained. In the following sections, the pressure coefficient and the vortex shedding mode are also presented.

3.2.1. Aerodynamic Statistics and Frequency Characteristics

Figures 8–10 show the variations in the lift and drag coefficient statistics of the square cylinder for the wind angle (θ) and the velocity ratio (k) for each of the three rotation directions, respectively. The results include the mean value $\overline{C}_l$ and the fluctuating value $C_l\prime$ of the lift coefficient, as well as the mean value $\overline{C}_d$ and the fluctuating value $C_d\prime$ of the drag coefficient. When $k > 0$, the small rotating cylinders transfer the momentum into the flow field. The results for $k = 0$ given in the abovementioned figures correspond to the results from the flow around a standard square cylinder. As shown in Figure 8a, when $\theta = 0^\circ$ and $\theta = 180^\circ$, the square cylinder is symmetrically placed in the flow direction. Therefore, the value of $\overline{C}_l$ is near zero for different k values at these two wind angles. When $15^\circ < \theta < 90^\circ$ and $k > 0$, the absolute value of $\overline{C}_l$ is greater than the absolute value of $\overline{C}_l$ when $k = 0$. When $\theta = 15^\circ$ and $k = 4$, the value of $\overline{C}_l$ increases 5.36 times compared with its value when

$k = 0$. When $\theta \geq 120^{\circ}$, the values of $\overline{C}_l$ are similar among them for various k values. As illustrated in Figure 8b, when $\theta \leq 60^{\circ}$ and $k > 0$, $C_l\prime$ is lower than the value of $C_l\prime$ when $k = 0$. It may be observed that for smaller wind angles, the rate of decrease of $C_l\prime$ becomes more obvious. The alternate shedding vortex behind the square cylinder is responsible for the lift coefficient fluctuation; the results show that the vortex shedding of the square cylinder should be suppressed at small wind angles. When $k = 4$ and θ increases from 0° to 15°, $C_l\prime$ decreases by 98% and 87%, respectively, compared with the value of $C_l\prime$ when $k = 0$. When $\theta = 60^{\circ}$ and $k = 4$, the value of $C_l\prime$ is slightly higher than the value of $C_l\prime$ when $k = 2$. This instance suggests that the increment of momentum that is transferred to the flow field may lead to the diminution of the control effect. When $\theta > 60^{\circ}$, $C_l\prime$ increases with the increase in k; this means that the MSBC is involved in the enhancement of the wake. As shown in Figure 8c, the variation of $\overline{C}_d$ at $k = 0$ is similar to that of $C_l\prime$ at $k = 0$ in Figure 8b. When $k = 4$ and θ increases from 0° to 15° and then increases to 30°, the values of $\overline{C}_d$ decrease by 36%, 54%, and 47%, respectively, compared with the value of $\overline{C}_d$ when $k = 0$. When $\theta = 60^{\circ}$, the value of $\overline{C}_d$ at $k = 2$ decreases to the lowest point. When $\theta = 90^{\circ}$ and $\theta = 120^{\circ}$, the values of $\overline{C}_d$ at $k = 4$ are 1.94 and 1.38 times the values of $\overline{C}_d$ when $k = 0$, respectively. When $\theta = 150^{\circ}$ and $k > 0$, the values of $\overline{C}_d$ are near the value that corresponds to $k = 0$. When $\theta = 180°$, the values of $\overline{C}_d$ at $k = 2,\ 4$ increase by 37% and 44%, respectively. As shown in Figure 8d, when $\theta \geq 30^{\circ}$, $C_d\prime$ increases with the increase in k, and the change is similar to the results when $k = 0$. When $\theta = 15^{\circ}$, the values of $C_d\prime$ for $k > 0$ can decrease by 25% compared to the value of $C_d\prime$ for $k = 0$.

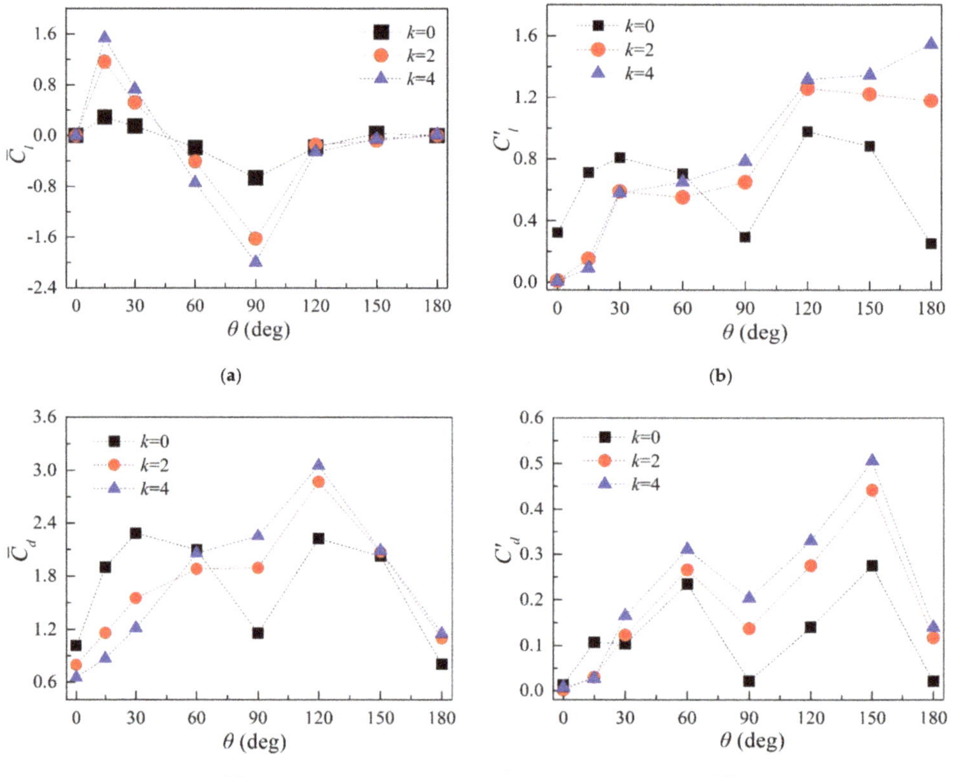

Figure 8. Aerodynamic coefficient statistics of the square cylinder for the wind angles (θ) and the velocity ratios (k) under inward rotation, **(a)** $\overline{C}_l$, **(b)** $C_l\prime$, **(c)** $\overline{C}_d$, **(d)** $C_d\prime$.

Through the above analysis, it may be observed that the control method may help to improve the wake stability of a square cylinder when $\theta \leq 60^{\circ}$, and it is unfavorable to the stability of the wake for the remaining wind angles. The control effect on the oscillating wake of the square cylinder is optimal when $\theta = 0^{\circ}$. When $\theta = 15^{\circ}$, this particular control method resulted in a great decrease in $\overline{C}_d$ and a great increase in $\overline{C}_l$ of the square cylinder. From this viewpoint, the optimum effect of 'drag reduction and lift increment' is achieved at this particular wind angle.

In Figure 9, the variation of the statistical parameters of the aerodynamic forces with wind angles is similar for $k = 2$ and $k = 4$; the difference lies only in terms of the individual angle. When $k = 0$, the aerodynamic parameters of the square cylinder (outward rotation) are consistent with the results in Figure 8. When $\theta = 0^{\circ}$, the values of $\overline{C}_l$ for different k values are near 0; however, the value of $C_l\prime$ increases from 0.32 at $k = 0$ to 0.97 at $k = 4$. In a similar manner, the mean and fluctuating values of the drag coefficient have increased significantly, thus suggesting that the momentum generated by the rotating cylinder leads to a more intense vortex shedding in the wake of the square cylinder. When $\theta = 30^{\circ}$ and $\theta = 60^{\circ}$, $\overline{C}_l$ is not equal to 0, and $C_l\prime$, $\overline{C}_d$, and $C_d\prime$ present growth trends as the k value incrementally increases; this illustrates that the wake vortex pattern has changed, and the MSBC presents an effect of 'lift and drag increment'. When $\theta = 90^{\circ}$ and $k = 4$, $C_l\prime$ and $C_d\prime$ increase slightly compared with their values when $k = 0$. $\overline{C}_d$ remains unchanged, and $\overline{C}_l$ slightly decreases. This indicates that the momentum injection does not improve the flow structure. When $\theta = 120^{\circ}$ and $\theta = 150^{\circ}$, $\overline{C}_l$ increases significantly with the increase in k; however, $C_l\prime$, $\overline{C}_d$, and $C_d\prime$ present a significantly decreasing trend. This suggests that the oscillating wake of the square cylinder has been suppressed and that the MSBC presents an effect of 'drag reduction and lift increment'. When $\theta = 120^{\circ}$ and $k = 4$, $\overline{C}_d$ decreases by 90% compared with the value of $\overline{C}_d$ when $k = 0$, and $C_d\prime$ decreases by 60%; therefore, the 'drag reduction' effect is obvious. In addition, the value of $\overline{C}_l$ becomes six times higher than the value of $\overline{C}_l$ when $k = 0$, whereas $C_l\prime$ decreases by 75%; thus, the 'lift increment' effect is obvious. When $\theta = 180^{\circ}$, $\overline{C}_d$ increases slightly with the increase in k; the values of the remaining parameters do not present significant differences with their respective values when $k = 0$. The results show that the MSBC method has no obvious effect on the oscillating wake for this particular wind direction.

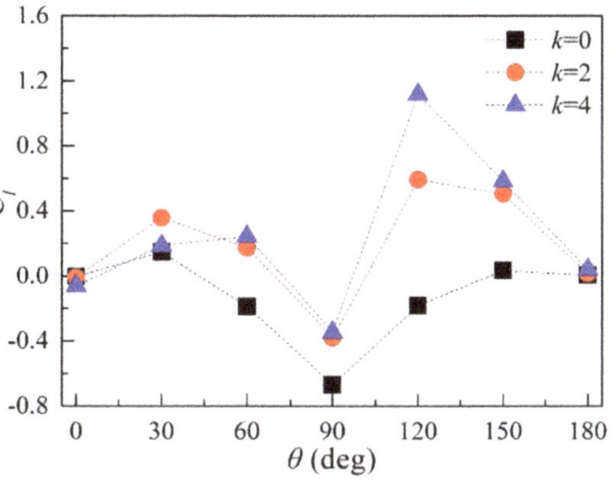

(a)

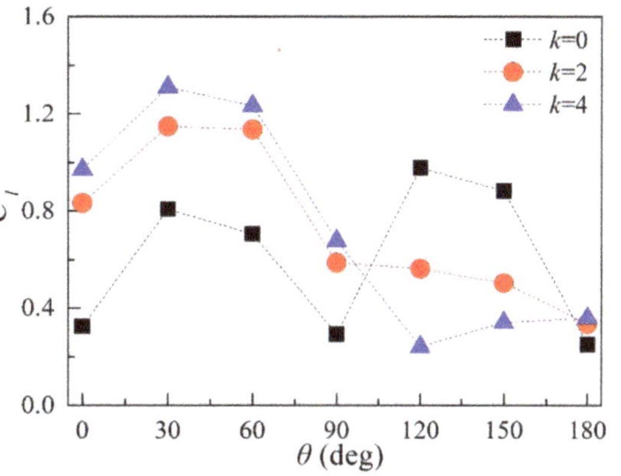

(b)

Figure 9. *Cont.*

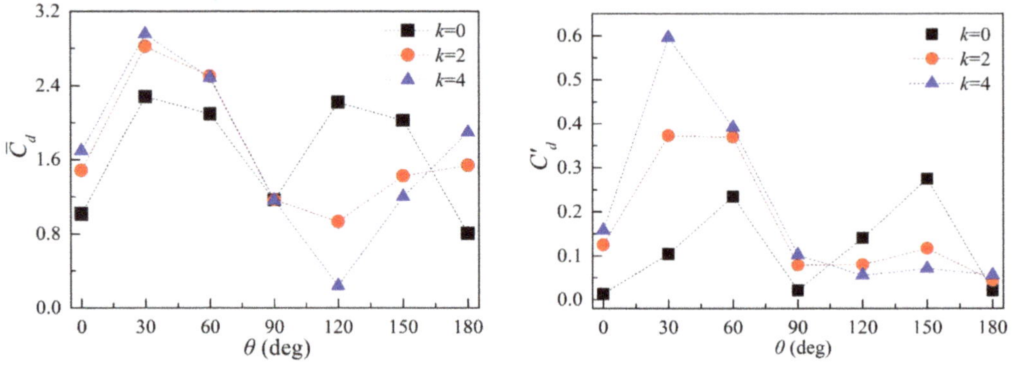

(c) (d)

Figure 9. Aerodynamic coefficient statistics of the square cylinder for the wind angles (θ) and the velocity ratios (k) under outward rotation, **(a)** $\overline{C}_l$, **(b)** $C_l\prime$, **(c)** $\overline{C}_d$, **(d)** $C_d\prime$.

As shown in Figure 10, when $k > 0$, the aerodynamic statistics of the square cylinder present a similar change for different wind angles. With the increase in k from 0 to 4, the momentum of the rotation that the circular cylinders transfer to the flow field gradually increases. When $\theta = 0^\circ$, $\overline{C}_l$ increases from 0 to 1.6, and $C_l\prime$ increases approximately by 100%; $\overline{C}_d$ remains almost unchanged, whereas $C_d\prime$ slightly increases. When $\theta = 30^\circ$ and $\theta = 60^\circ$, $\overline{C}_l$ at $k = 4$ increases to 2.1 and 1.1, respectively, whereas $C_l\prime$ decreases by 36% and 90%, respectively. This reveals that the oscillating wake of the square cylinder can be suppressed to different levels for these particular two wind angles. Although the fluctuation of the lift coefficient has been eliminated, the mean of the lift coefficient has increased. Simultaneously, the value of $\overline{C}_d$ decreased by 54% and 58% for $\theta = 30^\circ$ and $\theta = 60^\circ$, respectively. The values of $C_d\prime$ at $\theta = 30^\circ$ slightly decrease with the increase in k, whereas $C_d\prime$ at $\theta = 60^\circ$ decreases by 87%. When $\theta = 90^\circ$, $C_l\prime$ at $k = 2$ and $k = 4$ decreases to 0; this reveals that the alternate shedding vortices in the wake of the square cylinder have been eliminated. When $\theta = 120^\circ$ and $\theta = 150^\circ$, $\overline{C}_l$ still presents a growth trend, whereas the remaining statistical parameters present lower values than their corresponding values when $k = 0$; therefore, the input of momentum presents an effect of 'drag reduction and lift increment'. When $\theta = 180^\circ$, $\overline{C}_l$ presents a negative increment, whereas the remaining statistics present higher values than their corresponding values when $k = 0$. This demonstrates that the input of momentum causes the wake vortex shedding to become enhanced in this particular wind direction.

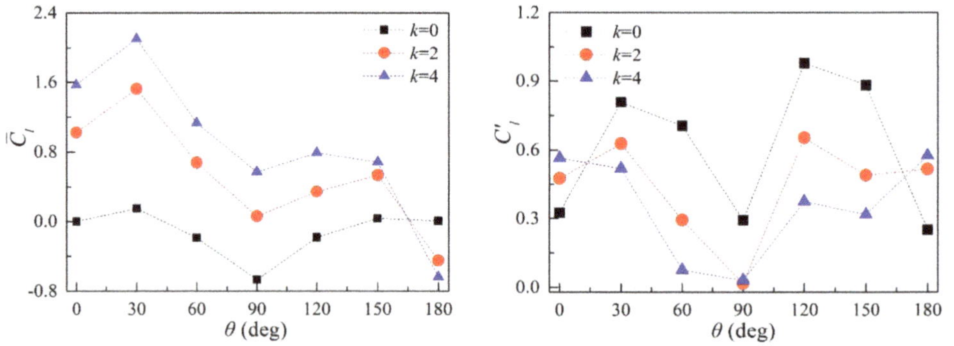

(a) (b)

Figure 10. *Cont.*

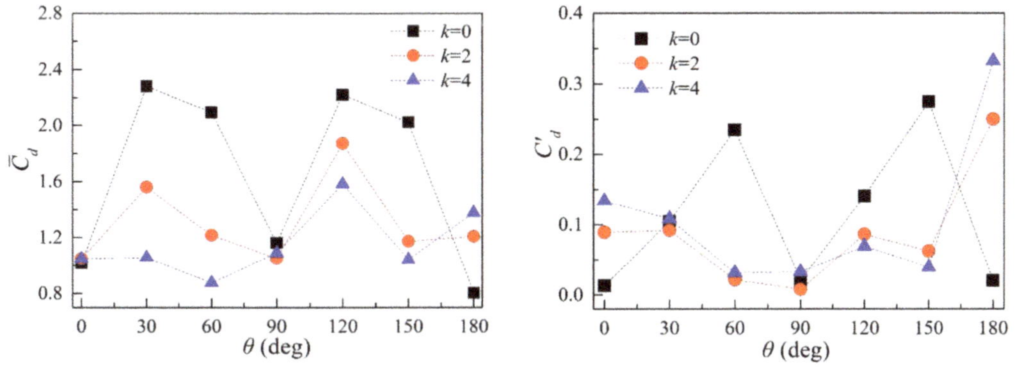

(c) (d)

Figure 10. Aerodynamic coefficient statistics of the square cylinder for the wind angles (θ) and the velocity ratios (k) under co-rotation, (**a**) $\overline{C}_l$, (**b**) $C_l{'}$, (**c**) $\overline{C}_d$, (**d**) $C_d{'}$.

Figures 11–13 show the dimensionless amplitude spectra of the lift coefficients of the square cylinder for different velocity ratios and different wind angles for the cases of inward rotation, outward rotation, and clockwise co-rotation, respectively. The two horizontal axes represent the wind angle and the Strouhal number, respectively; the vertical axis represents the amplitude of the fast Fourier transform (FFT). The control effect of the MSBC method on the oscillating wake of the square cylinder can be observed through the maximum amplitude for different k values; simultaneously, the high harmonics with small amplitudes caused by the input of momentum can be observed, as well. When the circular cylinders start to rotate, the amplitude of certain spectral analysis curves decreases, or the peak disappears completely. This shows that the alternate shedding vortex street of the square cylinder wake is suppressed or eliminated under this wind direction.

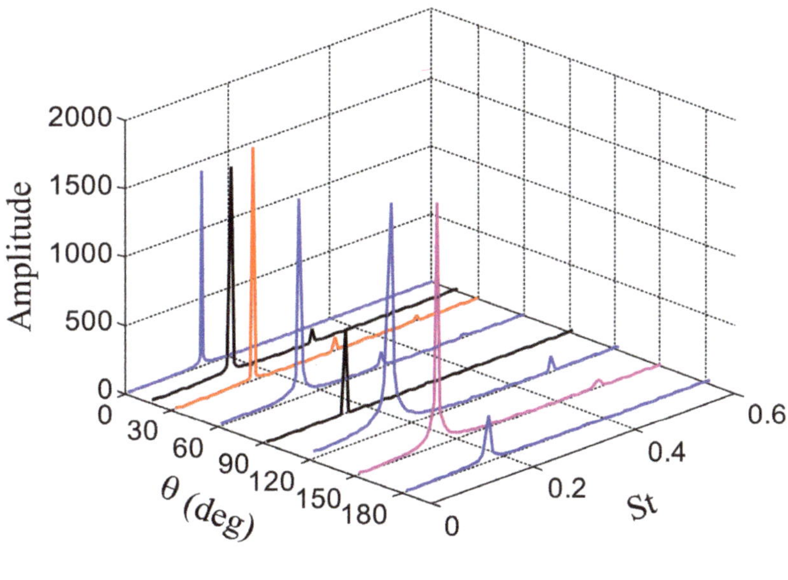

(a)

Figure 11. *Cont.*

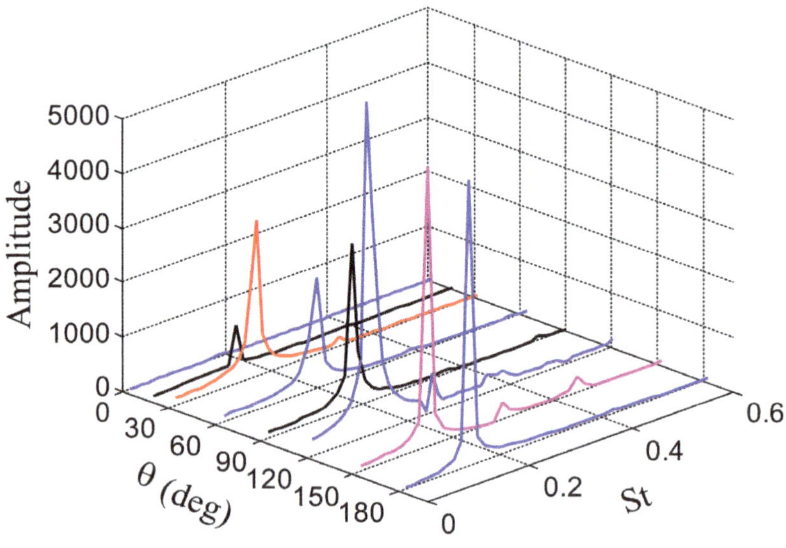

(b)

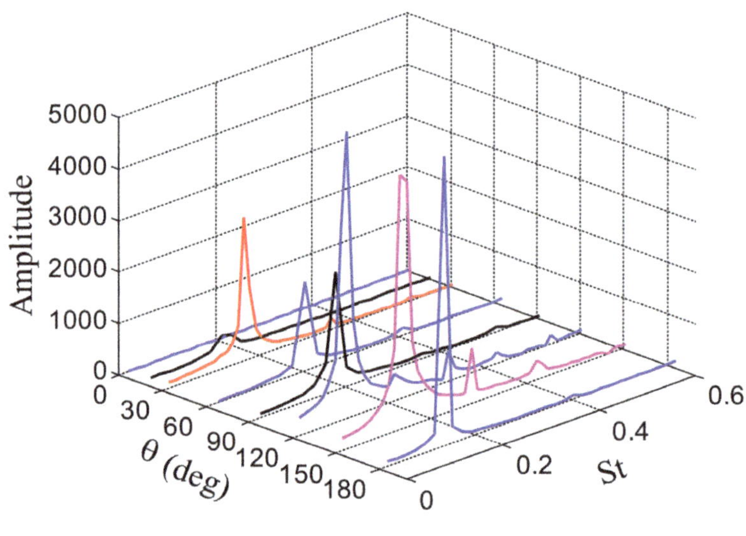

(c)

Figure 11. Amplitude spectra of Strouhal number under different wind angles with the change of velocity ratios (inward rotation), (**a**) $k = 0$, (**b**) $k = 2$, (**c**) $k = 4$.

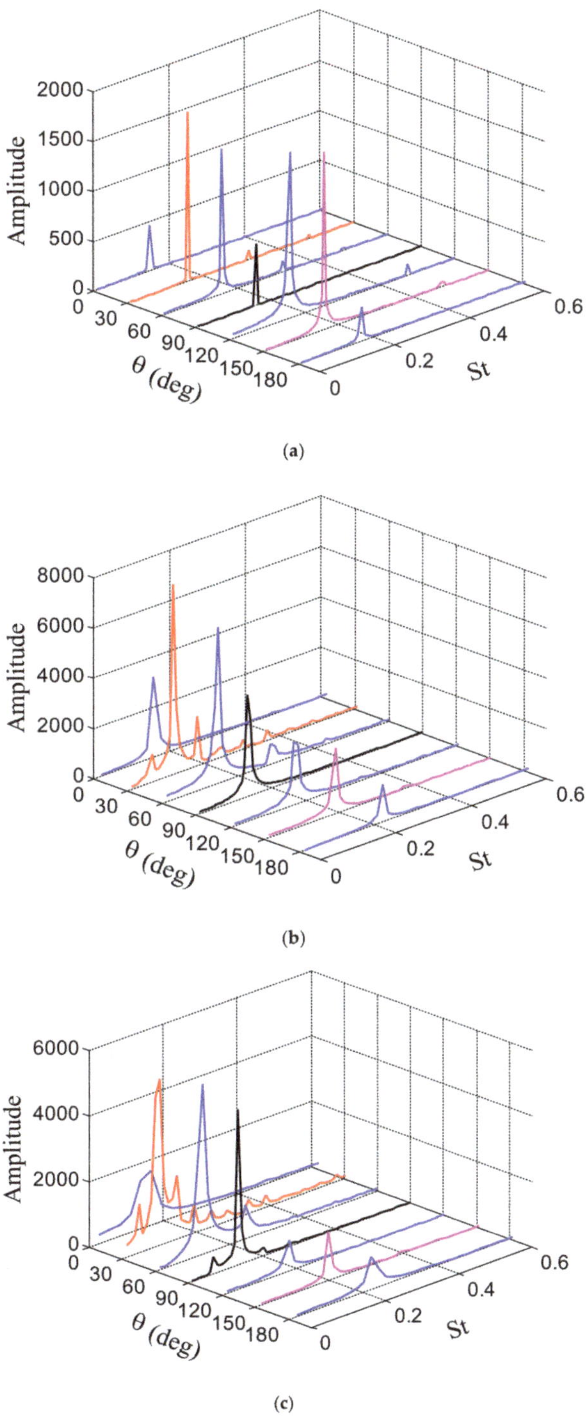

Figure 12. Amplitude spectra of Strouhal number under different wind angles with the change of velocity ratios (outward rotation), (**a**) $k = 0$, (**b**) $k = 2$, (**c**) $k = 4$.

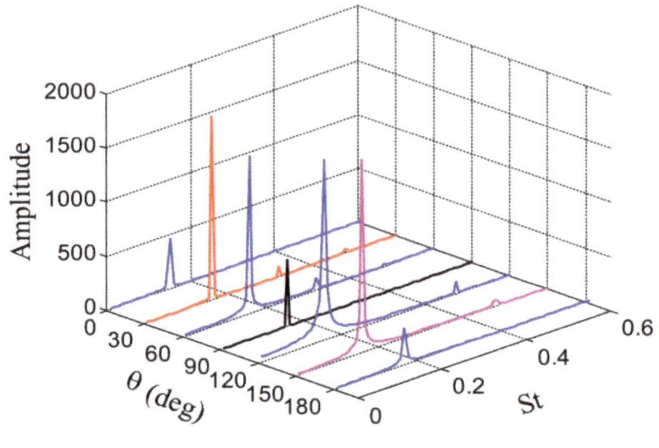

(**a**)

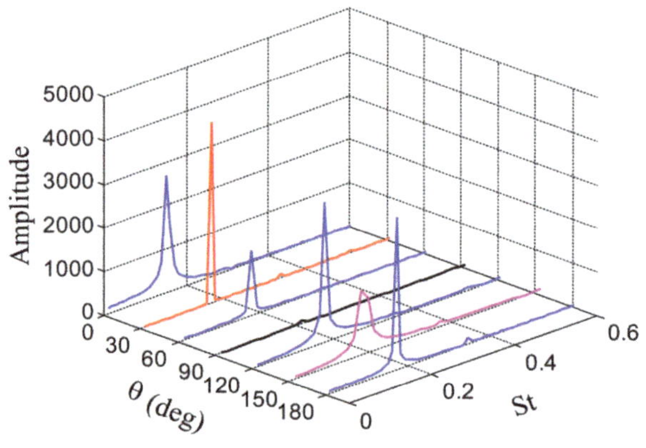

(**b**)

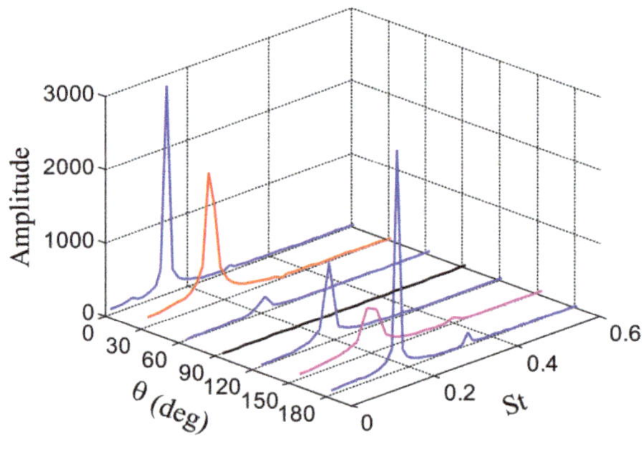

(**c**)

Figure 13. Amplitude spectra of Strouhal number under different wind angles with the change of velocity ratios (co-rotation), (**a**) $k = 0$, (**b**) $k = 2$, (**c**) $k = 4$.

The corresponding Strouhal numbers of the peak values of each curve in Figures 11–13 were calculated and are shown in Figure 14. As can be seen from Figure 14a, the values of the vortex shedding frequency of the square cylinder are close to each other for $k = 2$ and $k = 4$, when the corner circular cylinders rotate inward. When $\theta = 0^{\circ}$, S_t is near 0; the main reason for this is that the alternate shedding vortex has been completely eliminated. When $\theta = 30^{\circ} \sim 90^{\circ}$, the vortex shedding frequency is greater than that at $k = 0$, whereas when $\theta = 120^{\circ} \sim 180^{\circ}$, the vortex shedding frequency is lower than that at $k = 0$. When the circular cylinders rotate outward, the MSBC method cannot eliminate the wake vortex street behind the square cylinder in the entire range of the wind angles. As shown in Figure 14b, when $\theta = 30^{\circ} \sim 90^{\circ}$, the vortex shedding frequency is lower than that at $k = 0$. For the remaining wind angles, the values of the vortex shedding frequency are higher than their corresponding values at $k = 0$. As illustrated in Figure 14c, when $\theta = 60^{\circ}$ and $k = 4$, S_t is equal to 0; this means that the wake vortex street behind the square cylinder has been entirely eliminated. In addition, when $\theta = 90^{\circ}$, $k = 2$, and $k = 4$, the wake vortex street is eliminated, as well, and the MSBC method presents an improved flow control effect.

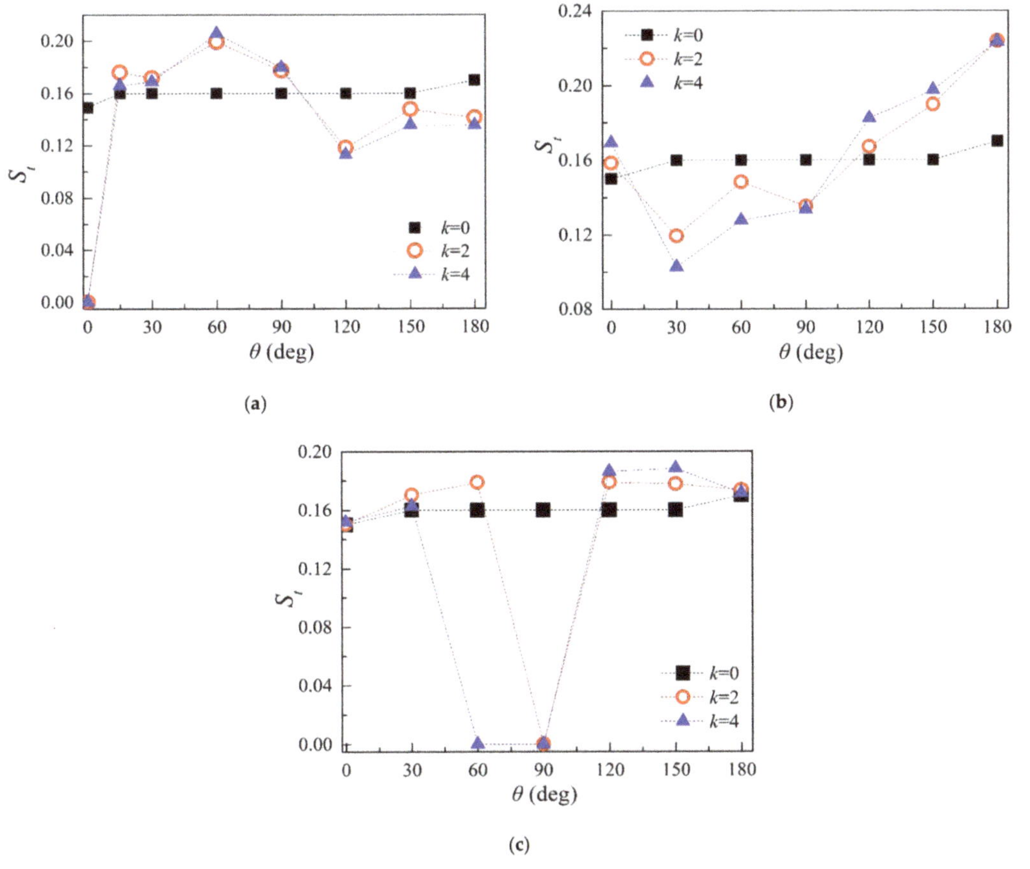

Figure 14. Variation of Strouhal number for the velocity ratios and the wind angles under different rotation modes, (**a**) Inward rotation, (**b**) Outward rotation, (**c**) Clockwise co-rotation.

3.2.2. The Mean Pressure Distribution Characteristics

Figures 15–17 show the distribution characteristics of the mean pressure coefficients around the square cylinder for different rotation modes, different *k* values, and wind angles

θ. Figure 15 shows the distribution of the mean pressure coefficient on the surface of the square cylinder at each wind angle, when the cylinder rotates inward. According to the analysis results in Figure 8, the oscillating wake of the square cylinder is suppressed to a different level when $\theta < 30^{\circ}$. Therefore, the corresponding mean pressure distributions on the surface of the square cylinder at $k = 1 \sim 4$ are illustrated in detail in Figure 15a–c. For the remaining wind angles, only the results for $k = 2$ and $k = 4$ are presented, as shown in Figure 15d–h. When $\theta = 0^{\circ}$, the mean pressure coefficient around the square cylinder is symmetrically distributed. At the windward side of the square cylinder is the flow stagnation zone, resulting in a positive pressure zone on the windward side, and the positive wind pressure reaches its maximum value at point No. 1. The flow separates from the leading edge corner of the square cylinder into the wake region and develops to an alternating shedding vortex street, thus leading to higher negative pressures at points No. 4 and 30. The measuring points No. 9–15 and 19–25 are located around the rotating circular cylinders, and points No. 16–18 are located on the surface of the leeward side between the two circular cylinders. Compared with the mean pressures when $k = 0$, when the small cylinder begins to rotate, the mean pressures of the measuring points No. 9–25 have changed significantly; however, there are small differences among the mean pressures at the remaining measuring points. This suggests that the circular cylinder rotation has a small effect on the upstream flow of the square cylinder. The negative pressure coefficients on the two small circular cylinders increase with the incremental increase in k. Owing to the suction generated by the negative pressure, the shear layers that have separated from the leading edge attach to the upper and lower sides of the square cylinder, and steadily flow into the downstream without vortex formation. The controlled wake has a narrow width and results in drag reduction. When the wind angle gradually increases, the stagnation point moves along the clockwise direction from point No. 1, and the mean pressure distribution becomes asymmetrical. When the wind angle increases from 15° to 60°, the suction from the upper rotating cylinder is larger than that of the lower one. The mean negative pressure distribution of the lower rotating cylinder presents a small difference between $k = 2$ and $k = 0$, whereas the suction of the upper rotating cylinder at $k = 2$ has increased significantly. When $\theta = 90^{\circ}$, the difference of the peak negative pressures between two rotating cylinders has decreased, with a slightly lower negative pressure for the lower cylinder. As shown in Figure 15f–g, the stagnation point moves near the upper rotating cylinder, thus resulting in a reduction of the peak suction on the upper rotating cylinder. When $\theta = 180^{\circ}$, the stagnation zone is located in the region between the two rotating cylinders, and the mean pressure on the surface of the square cylinder presents a symmetrical distribution. The peak values of the negative pressures appear at points No. 9 and 25 in the flow separation zone.

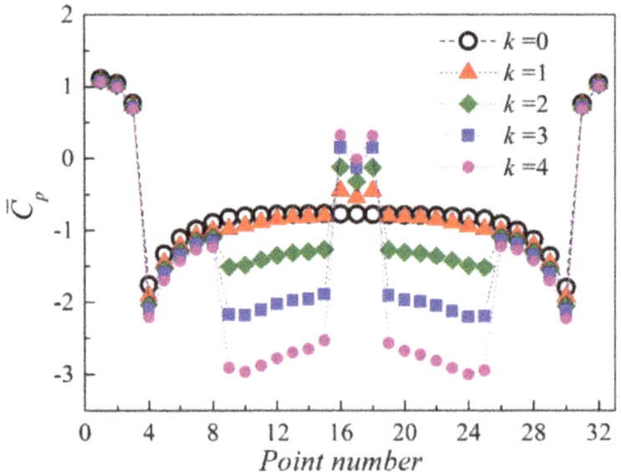

(a)

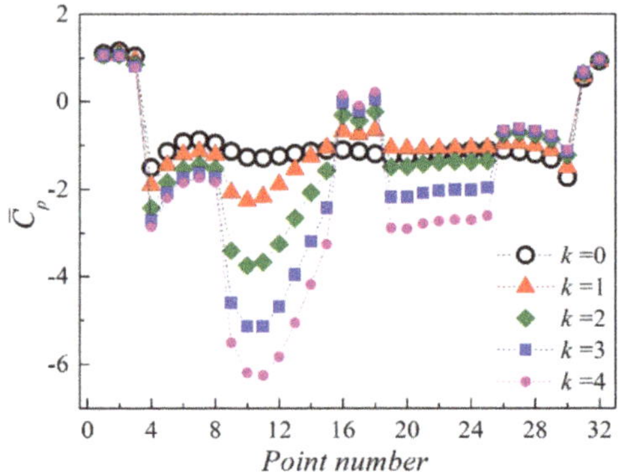

(b)

Figure 15. *Cont.*

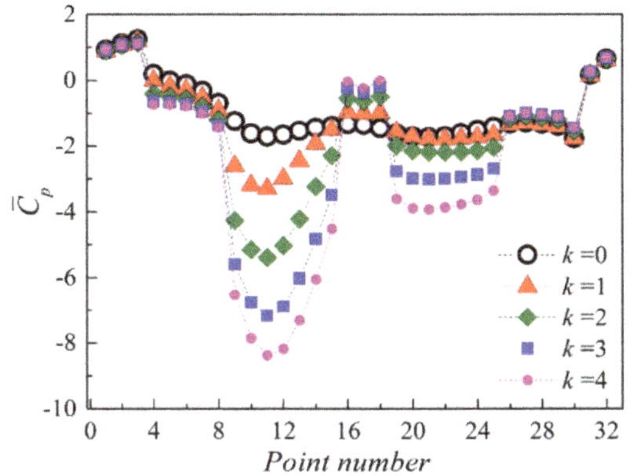

(c)

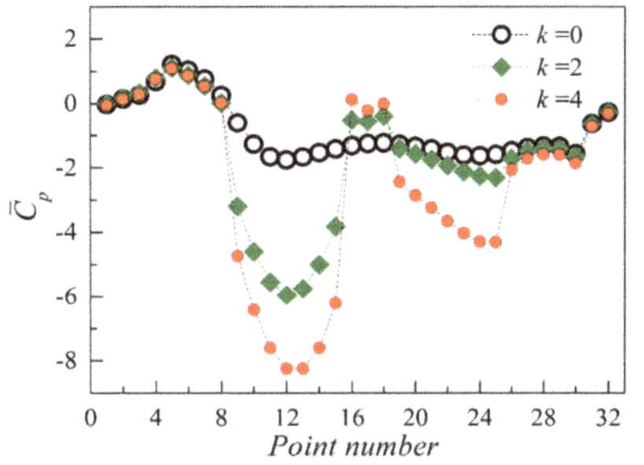

(d)

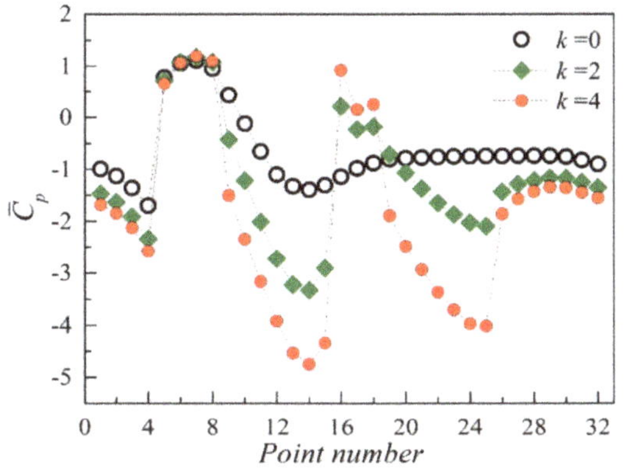

(e)

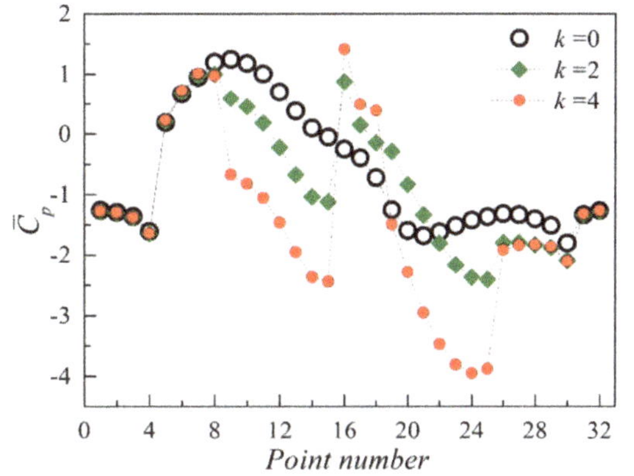

(f)

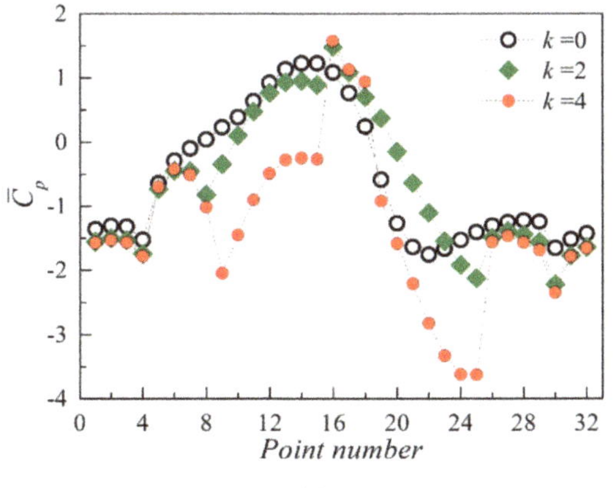

(g)

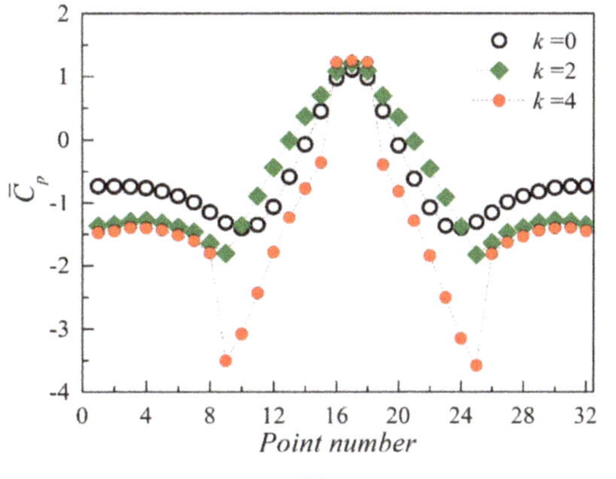

(h)

Figure 15. Mean pressure distribution varies with wind angles and velocity ratios (inward rotation), (**a**) $\theta = 0^\circ$, (**b**) $\theta = 15^\circ$, (**c**) $\theta = 30^\circ$, (**d**) $\theta = 60^\circ$, (**e**) $\theta = 90^\circ$, (**f**) $\theta = 120^\circ$, (**g**) $\theta = 150^\circ$, (**h**) $\theta = 180^\circ$.

Figure 16 shows the mean pressure distribution for different k and θ values when the cylinder rotates outward. When $\theta = 0^\circ$, the negative pressure of measuring points

No. 9–15 and No. 19–25 near the two small circular cylinders obviously increase. The mean pressure coefficient is asymmetrically distributed as the wind angle increases. The peak negative pressures of the lower rotating circular cylinder are higher than those of the upper circular cylinder when $\theta = 30^{\circ} \sim 60^{\circ}$. When $\theta = 90^{\circ}$, the peak negative pressures of the two small rotation circular cylinders are similar; however, they are still asymmetrically distributed. When $\theta = 120^{\circ} \sim 150^{\circ}$, the stagnation point moves to the vicinity of the upper circular cylinder, thus resulting in a significant decrease in the negative pressure, whereas the negative pressure of the lower rotating circular cylinder increases significantly and reaches its maximum value at $\theta = 150^{\circ}$. The stagnation point is located between the two rotating circular cylinders at $\theta = 180^{\circ}$; the peak of the negative pressure appears in measuring points No. 10 and 24, and the mean pressure distribution on the surface of the square cylinder presents a symmetrical state, as well.

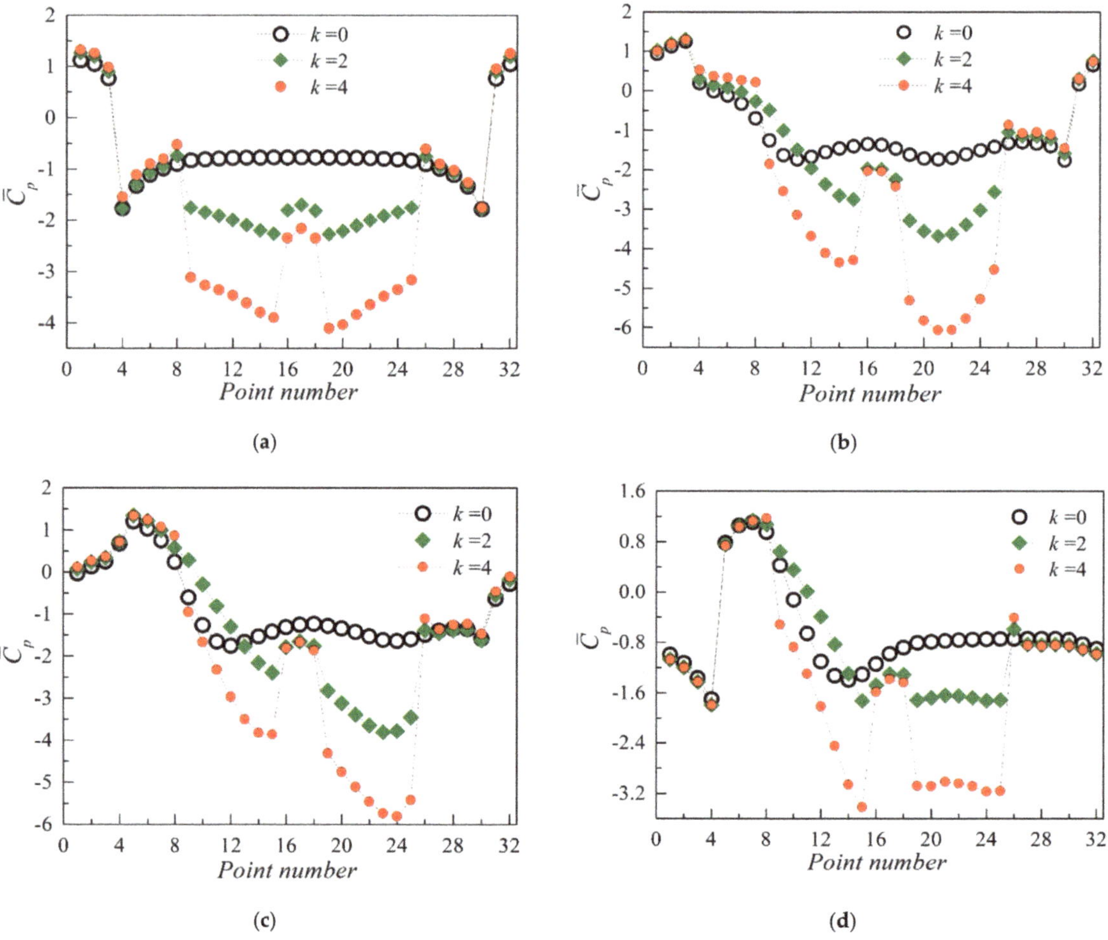

Figure 16. *Cont.*

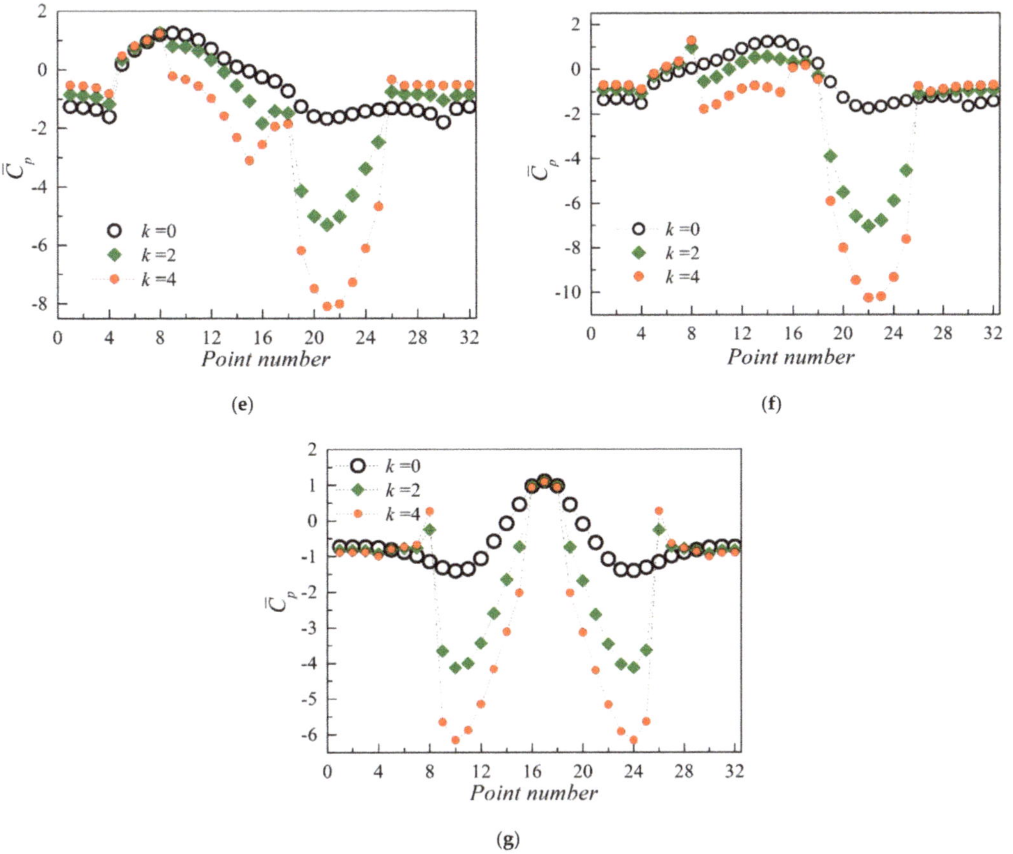

Figure 16. Mean pressure distribution varies with wind angles and velocity ratios (outward rotation), **(a)** $\theta = 0^{\circ}$, **(b)** $\theta = 30^{\circ}$, **(c)** $\theta = 60^{\circ}$, **(d)** $\theta = 90^{\circ}$, **(e)** $\theta = 120^{\circ}$, **(f)** $\theta = 150^{\circ}$, **(g)** $\theta = 180^{\circ}$.

As shown in Figure 17a, the square cylinder position is symmetrical with respect to the incoming flow at $\theta = 0^{\circ}$. However, the two small circular cylinders rotate with the same direction (co-rotation), resulting in an asymmetrical distribution of the mean pressure coefficient on the square cylinder surface and to a similar negative pressure on the measuring points near the small circular cylinders. The negative pressure around the upper rotating circular cylinder is significantly higher than that of the lower rotating circular cylinder at $\theta = 30^{\circ} \sim 60^{\circ}$. The upper circular cylinder is located at the flow separation zone at $\theta = 90^{\circ}$, and the rotating direction is consistent with the flow acceleration direction, which can lead to a higher negative pressure on the upper circular cylinder. The analysis results in Figure 10 show that the wake vortex street of the square cylinder is eliminated in this particular wind angle. When $\theta = 120^{\circ} \sim 150^{\circ}$, the peak of the negative pressure on the surface of the upper cylinder gradually decreases, owing to the stagnation zone. While the lower circular cylinder is in the flow separation zone, the peak of the negative pressure significantly increases. When $\theta = 180^{\circ}$, the rotating direction of the upper small circular cylinder is opposite to the flow direction of the separated shear layer, thus resulting in the decrease in the negative pressure on its surface. However, the rotating direction of the lower cylinder is consistent with the shear layer flow direction, which leads to an increment of negative pressure and to an asymmetrical distribution of the mean pressure.

Therefore, the fluctuation of the lift and drag coefficients increases when $k > 0$, as well, as shown in Figure 10.

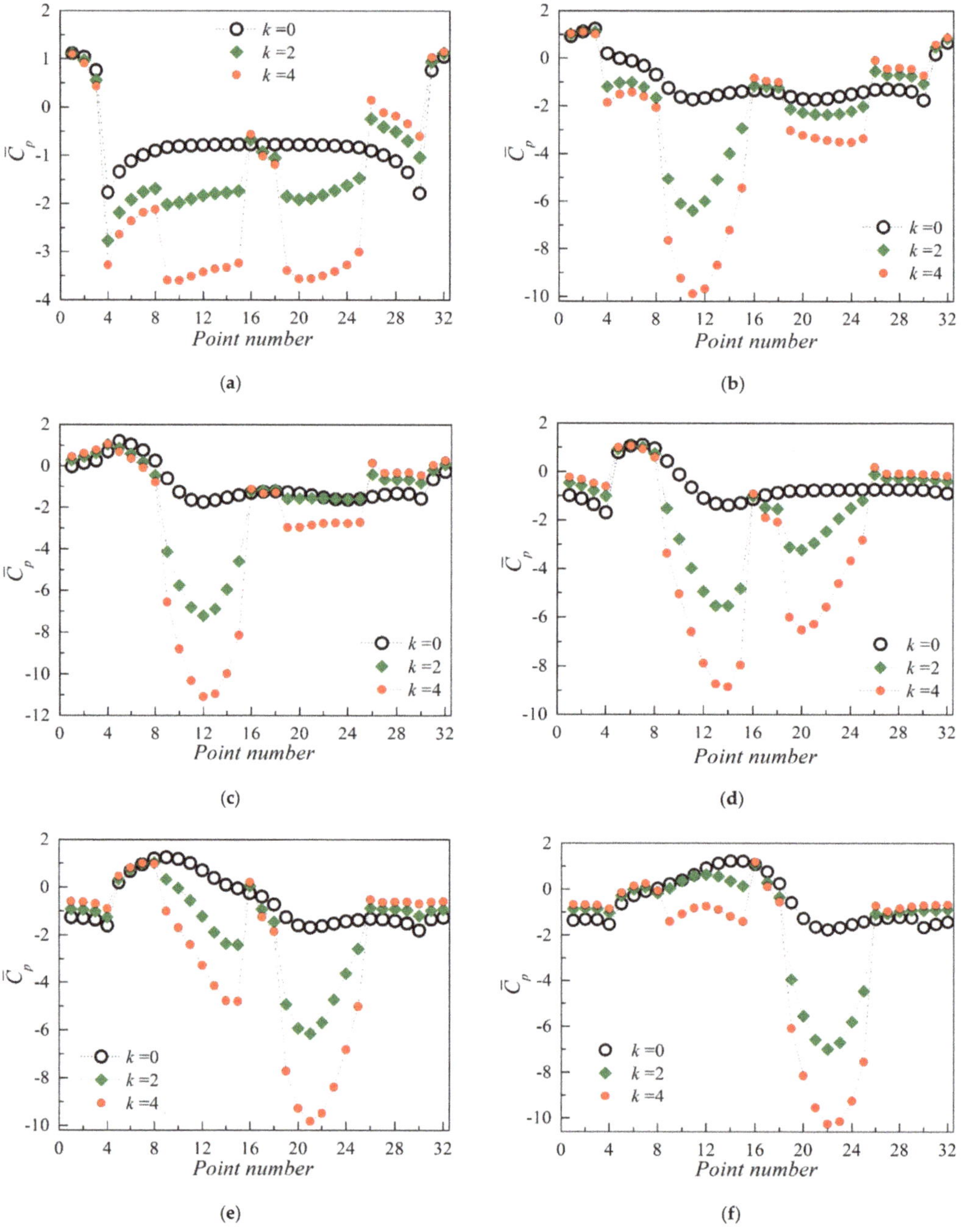

Figure 17. *Cont.*

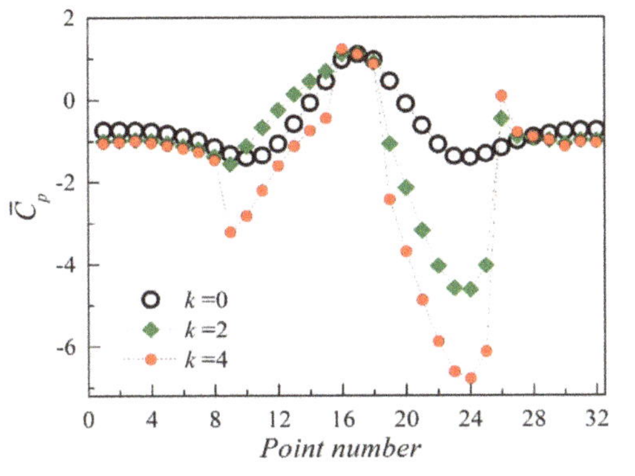

(g)

Figure 17. Mean pressure distribution varies with wind angles and velocity ratios (co-rotation), (**a**) $\theta = 0^\circ$, (**b**) $\theta = 30^\circ$, (**c**) $\theta = 60^\circ$, (**d**) $\theta = 90^\circ$, (**e**) $\theta = 120^\circ$, (**f**) $\theta = 150^\circ$, (**g**) $\theta = 180^\circ$.

3.2.3. Aerodynamic Time History Analysis

This section presents the lift and drag coefficient evolutions of the square cylinder during the entire numerical simulation process, starting from the flow around the fixed square cylinder (simplified as 'FR') up to the MSBC stage. The time histories of the aerodynamic coefficients are compared before and after the rotation of the small circular cylinders. Figures 18 and 19 present the time histories of the lift and drag coefficients of the square cylinder for inward rotation and different k values when the wind angle is 0° and 15°, respectively. At these two wind angles, the MSBC plays an important role in suppressing the oscillating wake. In Figure 18a,b, each time history during the interval of $0 \leq t < 100$ s indicates the lift and drag coefficients of the 'FR' stage. After the small circular cylinders have started to rotate at $t \geq 100$ s, the mean drag coefficient $\overline{C}_d$ decreases from 1.015 to 0.9, and the fluctuation of the lift coefficient decreases at $k = 1$, as well, as shown in Figure 18a; however, the vortex street in the wake of the square cylinder has not been completely eliminated in this case. With the increase in the rotating speed of the small circular cylinders, the input of momentum to the flow field gradually increases, as well. When $k = 2$, the fluctuation of the lift coefficient seems to have been eliminated, and $\overline{C}_d$ further decreases to 0.8. In Figure 18c,d, the MSBC was initialized from the fully developed flow at $t \geq 150$ s. When $k = 3$, the time history of the lift coefficient has the form of an almost straight line, without fluctuations, which indicates that the MSBC method completely eliminates the oscillating wake of the square cylinder. The $\overline{C}_d$ further decreases to 0.71, and the fluctuation of the drag coefficient decreases accordingly. When $k = 4$, $\overline{C}_d$ continues to decrease to 0.65, which indicates that the drag coefficient decreases as k increases; however, the fluctuation of the lift and drag coefficient is larger than that for $k = 3$. This suggests that the case of $k = 2$ can be regarded as the optimal choice of the MSBC method, in terms of the control effect and the external energy consumed.

When $t \geq 200$ s, the square cylinder activates the MSBC from the fully developed flow when the wind angle is 15°, as shown in Figure 19. With the increase in k from 1 to 4, the fluctuation value $C_l{\prime}$ of the lift coefficient decreases significantly, whereas $\overline{C}_l$ increases from 0.29 to 1.54. The drag coefficient $\overline{C}_d$ decreases to 0.87 at $k = 4$, which corresponds to a 54.2% decrease compared to of its value when $k = 0$. It may be observed that the fluctuation of the lift and drag coefficients decreases with the increase in the input of momentum, whereas the mean of the lift coefficient increases, and the mean of the drag coefficient decreases. The MSBC plays an important role in the decrease in the mean of the drag force and in the increment of the mean lift force.

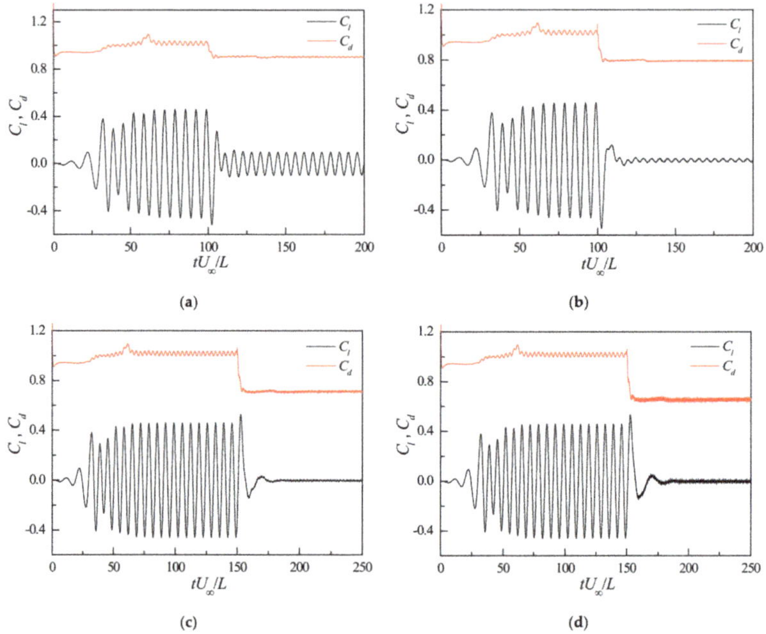

Figure 18. Lift and drag coefficient time histories of the square cylinder with different velocity ratios at $\theta = 0^{\circ}$ (inward rotation), (**a**) $k = 1$, (**b**) $k = 2$, (**c**) $k = 3$, (**d**) $k = 4$.

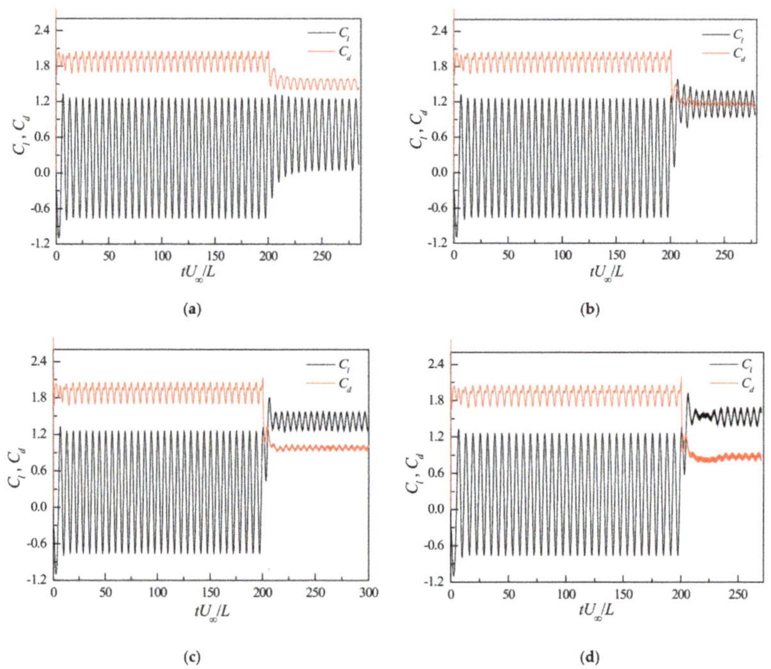

Figure 19. Lift and drag coefficient time histories of the square cylinder with different velocity ratios at $\theta = 15^{\circ}$ (inward rotation), (**a**) $k = 1$, (**b**) $k = 2$, (**c**) $k = 3$, (**d**) $k = 4$.

Figure 20 shows the comparative analysis of time histories of the lift and drag coefficients for inward rotation and $\theta = 0^{\circ}$ when MSBC becomes activated at different times. The simulation is divided into two cases. In the first case, the small circular cylinders are not rotating, and the square cylinder is at the 'FR' stage at $0 \leq t < 100$ s; when the alternating vortex shedding was steady; the MSBC became activated at $t = 100$ s with $k = 1.5$. In the second case, the MSBC becomes activated for $k = 1.5$ to inject the momentum into the flow field from $t = 0$ s. Although the activation time of the MSBC is different, the two cases have the same control effect on the lift and drag forces of the square cylinder. Only a small phase difference can be found in the comparison of the lift coefficient.

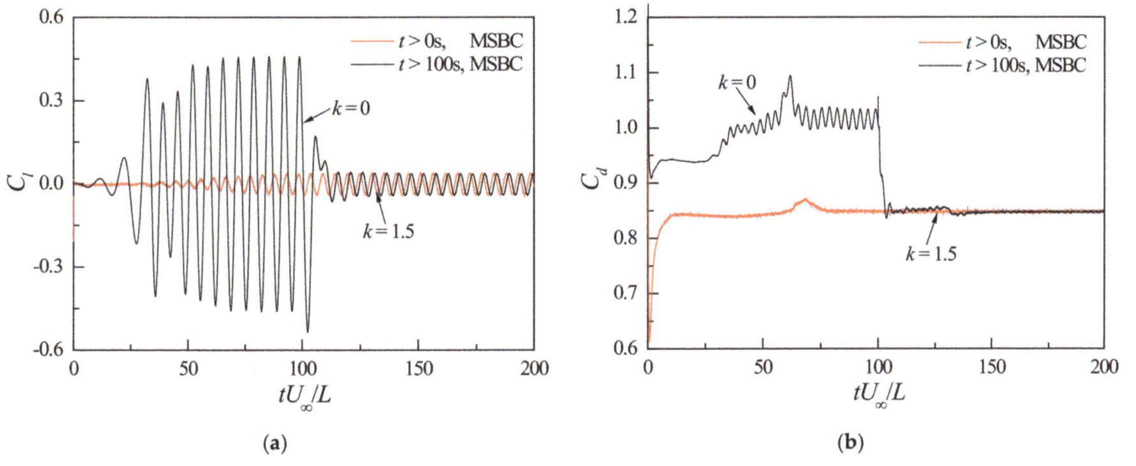

Figure 20. Comparisons of lift and drag coefficient time histories of the square cylinder while starting MSBC at different time under $\theta = 0^{\circ}$ and inward rotation, (**a**) C_l, (**b**) C_d.

Figures 21 and 22 show the lift and drag coefficient evolutions of the square cylinder for outward rotation and different k values when the wind angle is 30° and 120°, respectively. In Figure 21, it may be observed that the fluctuation of the lift coefficient increases with the increase in k when $\theta = 30^{\circ}$ and $t \geq 200$ s; the mean and the fluctuation of the drag coefficient increases with the increase in k, as well. This indicates that the input of momentum that has been generated by the MSBC method enhances the vortex shedding energy in the wake of the square cylinder. Figure 22 shows that the MSBC has an effect of 'drag reduction and lift increment' at $\theta = 120^{\circ}$. The mean and the fluctuation of the drag coefficient decrease with the increase in the value of k after the small circular cylinders start to rotate. Simultaneously, the mean of the lift coefficient increases with the increase in k, and the fluctuation of the lift coefficient decreases with the increase in k, although it still exists when $k = 4$. This shows that the wake vortex shedding of the square cylinder has not been completely eliminated; however, the eddy energy has decreased significantly. Similar conclusions can be drawn from the peak value of the amplitude spectra in Figure 12c.

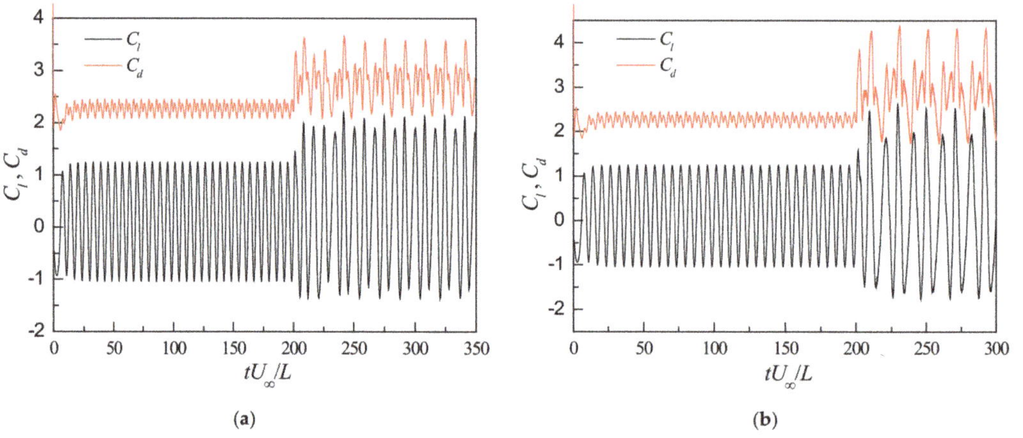

Figure 21. Lift and drag coefficient time histories of the square cylinder with different velocity ratios at $\theta = 30^{\circ}$ (outward rotation), (**a**) $k = 2$, (**b**) $k = 4$.

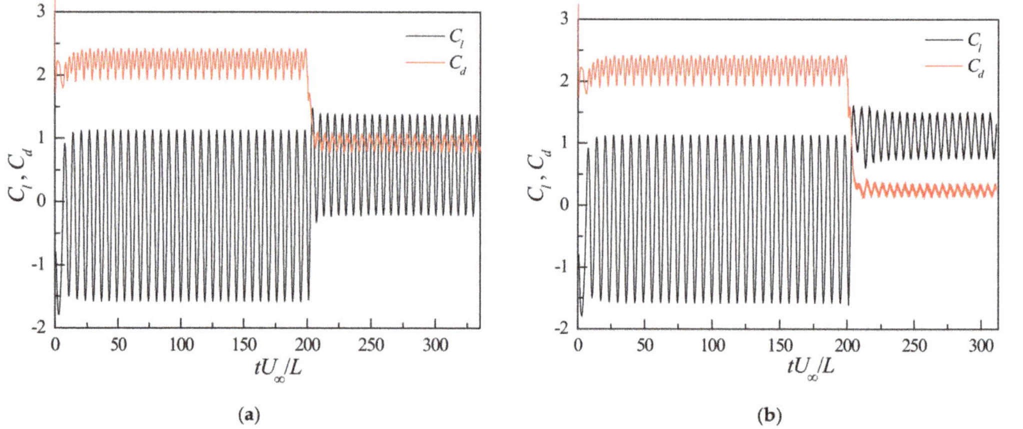

Figure 22. Lift and drag coefficient time histories of the square cylinder with different velocity ratios at $\theta = 120^{\circ}$ (outward rotation), (**a**) $k = 2$, (**b**) $k = 4$.

Figures 23 and 24 show the time histories of the lift and drag coefficients of the square cylinder with clockwise co-rotation and different k values when the wind angle is 60° and 90°, respectively. As shown in Figure 23, the mean and the fluctuation of the drag coefficient significantly decrease; the mean of the lift coefficient increases, and the fluctuation level is still high when $\theta = 60^{\circ}$ and $k = 2$. This indicates that the wake vortex shedding of the square cylinder is still in existence. With the increase in the momentum input, the mean of the drag coefficient continues to decrease, and the mean of the lift coefficient increases when $k = 4$; however, the fluctuation has been significantly reduced. This suggests that the wake vortex street of the square cylinder was suppressed; however, it was not entirely eliminated. As shown in Figure 24, as k further increases, the mean and the fluctuation of the drag coefficient of the square cylinder present a small change; however, the mean of the lift coefficient increases gradually, and the fluctuation of the lift coefficient significantly decreases when $\theta = 90^{\circ}$. This suggests that the oscillating wake of the square cylinder is effectively suppressed. High-frequency parts of the lift and drag coefficients are irrelevant to the vortex frequency, which corresponds to the rotation frequency of the small circular cylinders.

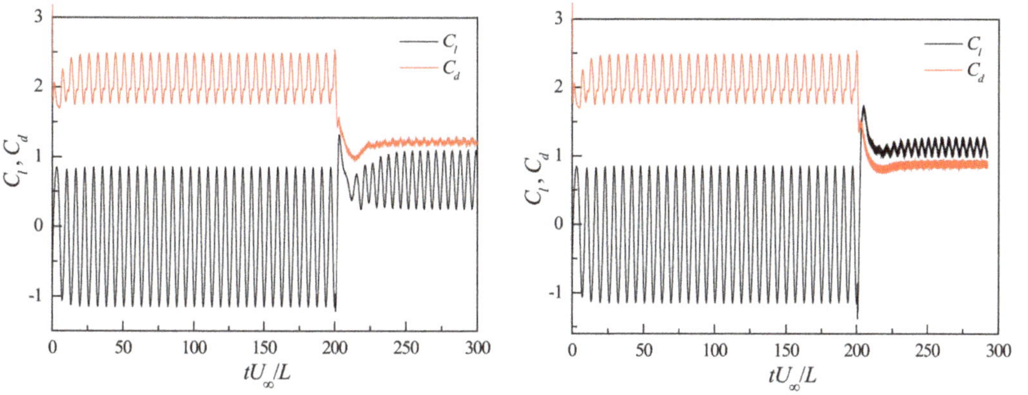

(a) (b)

Figure 23. Lift and drag coefficient time histories of the square cylinder with different velocity ratios at $\theta = 60^\circ$ (co-rotation), (**a**) $k = 2$, (**b**) $k = 4$.

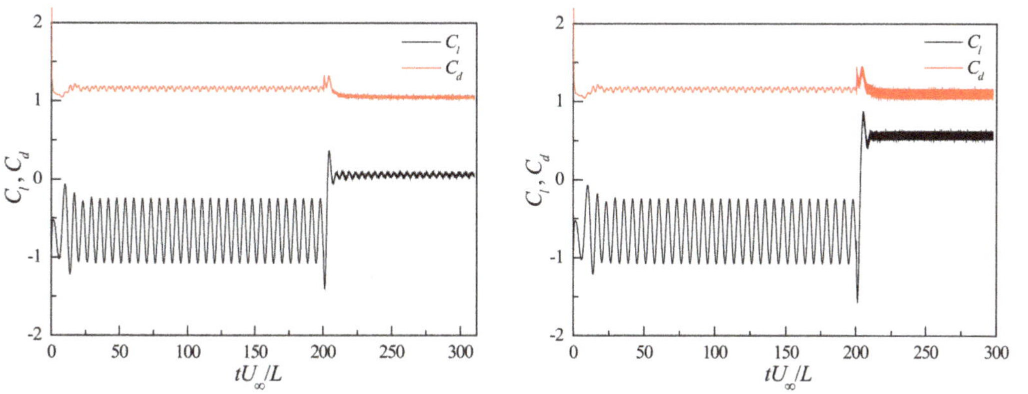

(a) (b)

Figure 24. Lift and drag coefficient time histories of the square cylinder with different velocity ratios at $\theta = 90^\circ$ (co-rotation), (**a**) $k = 2$, (**b**) $k = 4$.

3.2.4. Wake Vortex Structure

The momentum injection of the MSBC method affects the vortex characteristics of the flow field, and the different vortex structures and shedding modes are directly related to the aerodynamic forces of the square cylinder. In order to demonstrate the control effect of the MSBC method on the flow field around the square cylinder, in this study, we present the vortex wake of the square cylinder with representative wind angles for different rotation modes. Figure 25 shows the vorticity contours of the flow field around the square cylinder at different k values for inward rotation and wind angles of 0° and 15°. In Figure 25a, when $k = 0$, the shear layers separate from the leading edge corner of the square cylinder and enter into the wake region; then, they immediately curl and develop into an alternating shedding vortex street, which causes the fluctuation of the lift and drag coefficients. When $k = 1$, the rotating cylinder begins to transfer momentum into the flow field; however, the wake vortex street still exists. Two quasi-stable small vortex structures are formed near the two small circular cylinders at the leeward side of the square cylinder, and they lead to the shear layer separating from the leading edge corner, which rolls up to a slightly farther downstream location. This reduces the lift coefficient by 80%. As the k value increases, the separated shear layer rolls up farther downstream, and the vorticity intensity and scale slightly decrease. Two small symmetric stable vortices on the leeward side of the square cylinder have gradually stretched, thus preventing the interaction between the upper and lower shear layers. When $k = 4$, the large-scale alternate shedding vortex street in the square cylinder wake has been completely eliminated by the small-scale vortices induced by the two rotating cylinders, and the shear layer has not rolled up in the computational domain again. Obviously, for this rotation direction and wind angle, the momentum injected into the flow field by the small cylinders delays the separation of the shear layer and narrows the wake region. It finally eliminates the wake vortex street and forms a symmetrical stable vortex structure in the wake of a steady flow. Figure 25b shows that the vortex near the square cylinder wake is stable and develops into the alternate shedding vortex slightly farther downstream with the increase in k when $\theta = 15^{\circ}$. The fluctuation of the lift coefficient is significantly reduced, as well; however, the vortex street has not been completely eliminated at $k = 4$. The vortex street presents an inclined downward direction, which is the main cause of the increase in the mean lift coefficient. This is consistent with the results shown in Figure 8a.

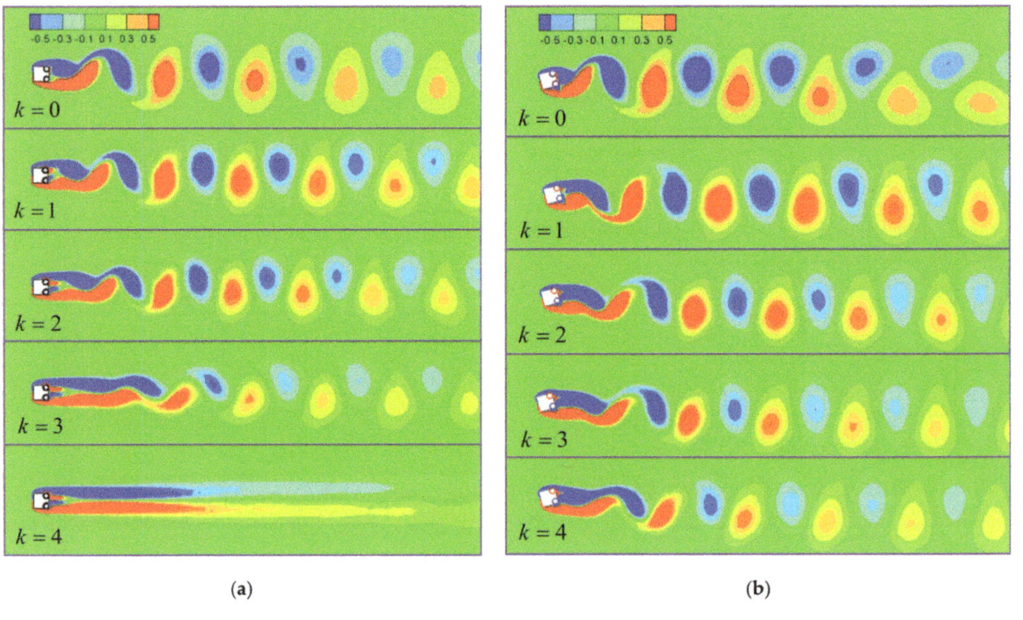

Figure 25. Vorticity contours in the wake of the square cylinder under inward rotation mode, (**a**) $\theta = 0^{\circ}$, (**b**) $\theta = 15^{\circ}$.

For the case of inward rotation mode and $\theta = 0^{\circ}$, the velocity contours and streamlines in the near wake of the square cylinder with different velocity ratios are shown in Figure 26, which can clearly show the close-up view around the region of rotating cylinders. In order to compare the results of different cases more clearly, the upper limit value of the velocity in the legend of each subgraph is consistent, which is the maximum speed in the flow field when $k = 0$.

In Figure 26a, a large-scale vortex is generated in the upper region of the near wake immediately adjacent to the leeward side of the square cylinder, and the lift coefficient of the square cylinder reaches a positive maximum value at this time. The streamline at the downstream of the main vortex is obviously curved, indicating that the Karman vortex street with alternating shedding exists in the wake of the square cylinder. When $k = 1$, it can be seen from Figure 26b that the flow velocity around the small cylinders is increased, and the rotating cylinders started to inject momentum into the wake. At this time, there are still shedding main vortices and obviously curved streamlines in the square cylinder wake. Compared with Figure 26a, the low-velocity region in the wake begins to narrow, the scale of the main vortex is decreased, and the distance between the vortex center and the square cylinder is increased. The influence of the main vortex on the square cylinder begins to weaken, and the fluctuation of the lift coefficient is reduced to a certain extent, which is consistent with the result in Figure 18a. When $k = 2$, 3, as the velocity ratio increases gradually, the velocity around the small cylinders is increased gradually, and the momentum injected into the wake field is also increased gradually, as shown in Figure 26c,d. Meanwhile, two symmetrical small vortices appear in the near wake region of the square cylinder, and the alternating shedding main vortices are pushed farther downstream. The scale and energy of the main vortices are gradually decreased, and the fluctuation of the corresponding lift coefficient is also reduced significantly, as shown in Figure 18b,c.

As shown in Figure 26e, the small circle cylinders rotating with high velocity ratio of $k = 4$ inject more momentum to the wake. The high momentum is mainly concentrated in the near wake range of one time of the side length of the square cylinder and symmetrically distributed on both sides of the centerline of the square cylinder wake. There is a pair of stable and symmetrical small vortices between the high-momentum region and the upper and lower shear layers separated from the corners of the leading edge of the square cylinder. The low-velocity region of the square cylinder wake becomes narrower, the separated shear layers are no longer curled, and the main vortex have entirely disappeared. The streamlines are symmetrically distributed on both sides of the wake centerline and became more and more straight as they move downstream. At this time, the fluctuation of the lift coefficient has disappeared, and the mean value of the drag coefficient is also decreased to its minimum. It can be seen that MSBC achieves the purpose of completely eliminating the alternating shedding vortices by inputting high momentum into the wake of the square cylinder.

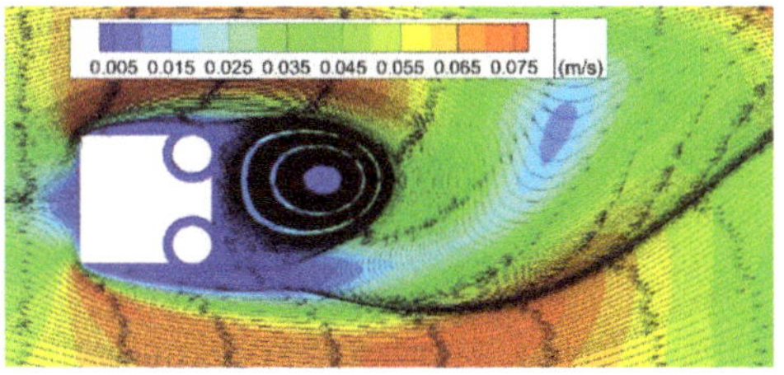

(a)

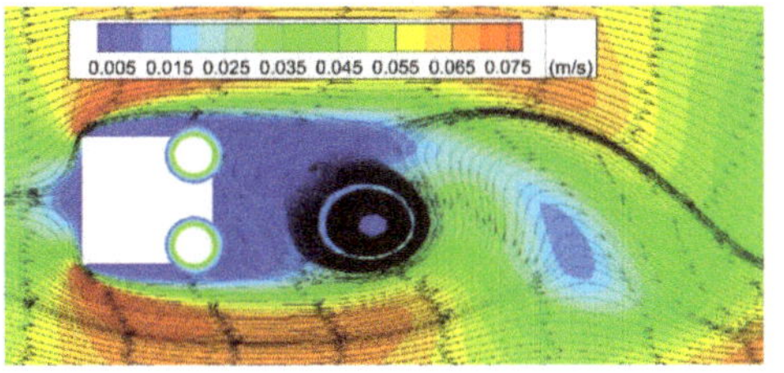

(b)

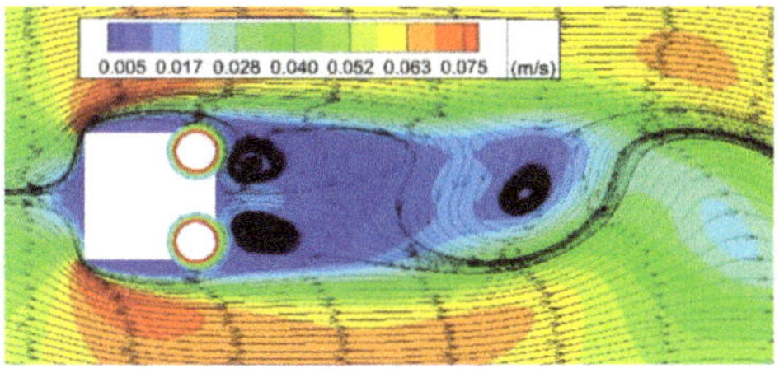

(c)

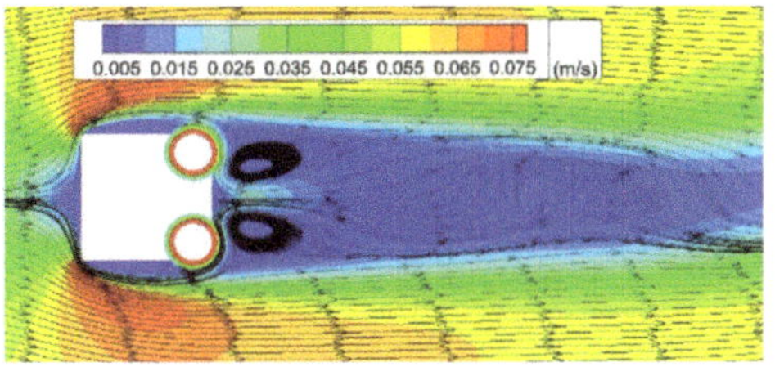

(d)

Figure 26. *Cont.*

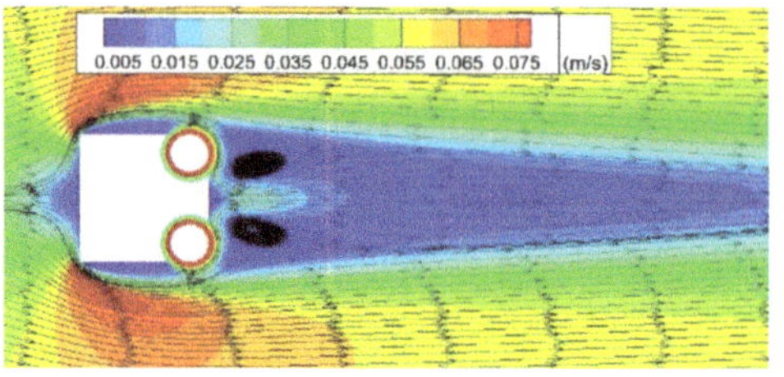

(e)

Figure 26. Velocity contours and streamlines in the near wake of the square cylinder under inward rotation mode and $\theta = 0^{\circ}$, (**a**) $k = 0$, (**b**) $k = 1$, (**c**) $k = 2$, (**d**) $k = 3$, (**e**) $k = 4$.

Figure 27 shows the vorticity contours of the flow field around the square cylinder for different k values, with an outward rotation, and for wind angles of 30° and 120°. The momentum injected by the rotating cylinders significantly influences the wake field at $\theta = 30^{\circ}$. When $k = 2$ and $k = 4$, vortices of different scales appear in the flow field, and the vortex shedding patterns are very complex. As shown in Figure 27a, the vortex shedding pattern is obviously different from the regular vortex shedding model when $k = 0$; the drastic change in the shedding mode and the vortex scale cause a significant increase in the lift and drag coefficients. As shown in Figure 27b, the two small rotation cylinders on the upper side of the square cylinder cannot control the shear layer that has separated from lower side. When $k = 4$, the vortex street still exists, and the entire vortex street presents an inclined downward direction. This leads to a positive mean value of the lift coefficient, which increases with the increase in k.

Figure 28 shows the vorticity contours of the flow field around the square cylinder for different k values, with clockwise co-rotation, and for wind angles of 60° and 90°. When $\theta = 60^{\circ}$, the evolution process of the flow field around the square cylinder for various k values is similar to that of Figure 27b, and the vortex shedding mode presents an effect of 'drag reduction and lift increment' at this moment. The vortex near the square cylinder wake is stable, and the vortex street is formed farther downstream, when $k = 4$. Therefore, the fluctuation of the lift coefficient is significantly reduced, as well. Figure 28b shows that when the small circular cylinders do not rotate, the stable '2S' mode is formed in the square cylinder wake. The shear layer near square cylinder wake is stable and rolls up to a slightly farther downstream location when $k = 2$. At this time, the vortex has not been eliminated; however, the fluctuation value of the lift coefficient has decreased by 90% compared with that when $k = 0$. When $k = 4$, the mean of the lift coefficient is positive, and it increases with the increase in k value because of the overall downward deflection of the steady wake. At this point, the alternating shedding vortex in the wake of the square cylinder has been completely eliminated because the two small circular cylinders are both on the upper side of the square cylinder, and they rotate downstream together; this has a strong suction effect on the upper shear layer. Therefore, the shear layer that separated from lower side of the square cylinder is unable to curl into a vortex, and the wake presents an overall downward deflection, which leads to the increase in the mean of the lift coefficient. This is a newly discovered control mode, that is, when rotating cylinders are arranged on one side of the structure, the control effectiveness can also be achieved. For example, only one side of a long-span bridge deck is used for vehicle traffic, and the other side can be equipped with multiple rotating devices, such as vertical axis wind turbines placed horizontally, which can control the wake of the bridge deck while collecting wind energy.

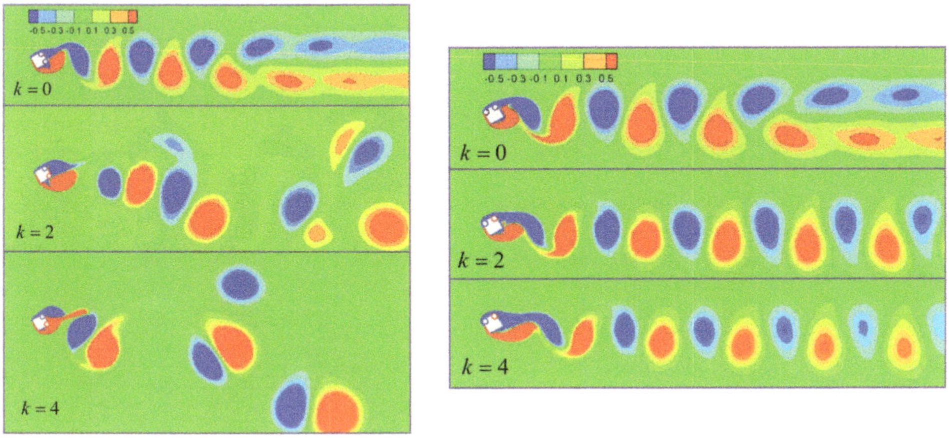

(**a**) (**b**)

Figure 27. Vorticity contours in the wake of the square cylinder under outward rotation mode, (**a**) $\theta = 30^{\circ}$, (**b**) $\theta = 120^{\circ}$.

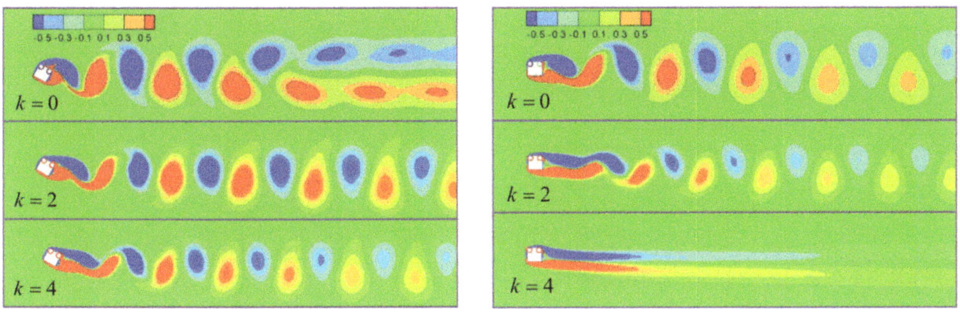

(**a**) (**b**)

Figure 28. Vorticity contours in the wake of the square cylinder under co-rotation mode, (**a**) $\theta = 60^{\circ}$, (**b**) $\theta = 90^{\circ}$.

4. Conclusions

In this study, the computational fluid dynamics (CFD) numerical simulation was employed for the study of the MSBC effect on the oscillating wake of a square cylinder. We conducted a numerical simulation of the entire process, from the flow around a fixed square cylinder to the oscillating wake controlled via the MSBC, and the conclusions obtained from the study are as follows:

For the inward rotation mode and wind angles of $\theta < 60^{\circ}$, the MSBC method plays an important role in suppressing the oscillating wake of the square cylinder; the effectiveness of the control in suppressing the wake was more obvious when the wind angle was smaller. In the other wind direction, the MSBC method was detrimental to the wake stability of the square cylinder. The fluctuation value of the lift coefficient decreased by 98%, and the mean value of the drag coefficient decreased by 36% when $\theta = 0^{\circ}$ and $k = 4$. In this case, the elimination of the oscillation wake of the square cylinder produced the best effect. In terms of control effectiveness and the external energy consumed, the velocity ratio $k = 2$ can be regarded as the optimal choice of the MSBC control method at Re = 200.

For the outward rotation mode and when $\theta = 120^{\circ}$ and 150°, the effect of suppressing oscillating wake of the square cylinder continuously improved with the increase in k. In other wind directions, the momentum input that was generated by the rotating cylinders plays an important role in enhancing the wake of the square cylinder.

For the clockwise co-rotation and when $\theta = 0^{\circ}$ and 180°, the momentum input of the rotating circular cylinders was unfavorable to the wake stability of the square cylinder. In other wind directions, the oscillating wake of the square cylinder was suppressed to different levels. Among them, the effect of eliminating the oscillation wake of the square cylinder was most obvious for wind angles of 60° and 90°.

The high suction region on the surface of the rotating circular cylinders can prevent streamline bending, delay the separation of the shear layer, and form a narrow wake region, which transforms the unsteady flow in the wake of the square cylinder into an approximately steady flow. This leads to the decrease in the mean of the drag coefficient and the fluctuation of the lift coefficient; thus, suppressing the flow-induced vibration of the square cylinder can be achieved.

Author Contributions: Conceptualization, T.S. and F.X.; methodology, T.S. and F.X.; software, X.L.; validation, T.S. and F.X.; formal analysis, F.X.; investigation, T.S. and X.L.; resources, F.X.; data curation, X.L.; writing—original draft preparation, T.S.; writing—review and editing, F.X.; visualization, X.L.; supervision, F.X.; project administration, F.X.; funding acquisition, F.X. All authors have read and agreed to the published version of the manuscript.

Funding: This research was funded by the National Natural Sciences Foundation of China (NSFC) (52078175, 51778199 and U1709207), the Natural Science Foundation of Guangdong Province (2019A15 15012205), the fundamental research funds of Shenzhen Science and Technology plan (JCYJ2019080614 4009332, JCYJ20180306172123896), and the Stability Support Program for colleges and universities in Shenzhen (GXWD20201230155427003-20200823134428001).

Data Availability Statement: Data are contained within this article.

Conflicts of Interest: The authors declare no conflict of interest.

References

1. Choi, K.S. Near-wall structures of a turbulent boundary layer with riblets. *J. Fluid Mech.* **1989**, *208*, 417–458. [CrossRef]
2. Bearman, P.W.; Owen, J.C. Reduction of bluff-body drag and suppression of vortex shedding by the introduction of wavy separation lines. *J. Fluids Struct.* **1998**, *12*, 123–130. [CrossRef]
3. Bearman, P.W.; Brankovic, M. Experimental studies of passive control of vortex-induced vibration. *Eur. J. Mech. B/Fluids* **2004**, *23*, 9–15. [CrossRef]
4. Lee, S.J.; Jang, Y.G. Control of flow around a NACA 0012 airfoil with a micro-riblet film. *J. Fluids Struct.* **2005**, *20*, 659–672. [CrossRef]
5. Laima, S.; Li, H.; Chen, W.; Li, F. Investigation and control of vortex-induced vibration of twin box girders. *J. Fluids Struct.* **2013**, *39*, 205–221. [CrossRef]
6. Chen, W.L.; Li, H.; Hu, H. An experimental study on the unsteady vortices and turbulent flow structures around twin-box-girder bridge deck models with different gap ratios. *J. Wind Eng. Ind. Aerodyn.* **2014**, *132*, 27–36. [CrossRef]
7. Gao, D.L.; Deng, Z.; Yang, W.H.; Chen, W.L. Review of the excitation mechanism and aerodynamic flow control of vortex-induced vibration of the main girder for long-span bridges: A vortex-dynamics approach. *J. Fluids Struct.* **2021**, *105*, 103348. [CrossRef]
8. Li, H.; Chen, W.L.; Xu, F.; Li, F.C.; Ou, J.P. A numerical and experimental hybrid approach for the investigation of aerodynamic forces on stay cables suffering from rain-wind induced vibration. *J. Fluids Struct.* **2010**, *26*, 1195–1215. [CrossRef]
9. Chen, W.L.; Zhang, Q.Q.; Li, H.; Hu, H. An experimental investigation on vortex induced vibration of a flexible inclined cable under a shear flow. *J. Fluids Struct.* **2015**, *54*, 297–311. [CrossRef]
10. Chen, W.L.; Xin, D.B.; Xu, F.; Li, H.; Ou, J.P.; Hu, H. Suppression of vortex-induced vibration of a circular cylinder using suction based flow control. *J. Fluids Struct.* **2013**, *42*, 25–39. [CrossRef]
11. Gao, D.L.; Chen, G.B.; Chen, W.L.; Li, H. Active control of circular cylinder flow with windward suction and leeward blowing. *Exp. Fluids* **2019**, *60*, 26. [CrossRef]
12. Gao, D.L.; Meng, H.; Huang, Y.W.; Chen, G.B.; Chen, W.L. Active flow control of the dynamic wake behind a square cylinder using combined jets at the front and rear stagnation points. *Phys. Fluids* **2021**, *33*, 047101. [CrossRef]
13. Wu, C.J.; Wang, L.; Wu, J.Z. Suppression of the von Karman vortex street behind a circular cylinder by a travelling wave generated by a flexible surface. *J. Fluid Mech.* **2007**, *574*, 365–391. [CrossRef]
14. Xu, F.; Chen, W.L.; Xiao, Y.Q.; Li, H.; Ou, J.P. Numerical study on the suppression of the vortex-induced vibration of an elastically mounted cylinder by a traveling wave wall. *J. Fluids Struct.* **2014**, *44*, 145–165. [CrossRef]

15. Xu, F.; Chen, W.L.; Bai, W.F.; Xiao, Y.Q.; Ou, J.P. Flow control of the wake vortex street of a circular cylinder by using a traveling wave wall at low Reynolds number. *Comput. Fluids* **2017**, *145*, 52–67. [CrossRef]
16. Kubo, C.H.; Chiou, L.C.; Chen, C.C. Wake flow pattern modified by small control cylinders at low Reynolds number. *J. Fluids Struct.* **2007**, *23*, 938–956.
17. Hwang, J.Y.; Yang, K.S. Drag reduction on a circular cylinder using dual detached splitter plates. *J. Wind Eng. Ind. Aerodyn.* **2007**, *95*, 551–564. [CrossRef]
18. Assi, G.R.S.; Bearman, P.W.; Kitney, N. Low drag solutions for suppressing vortex-induced vibration of circular cylinders. *J. Fluids Struct.* **2009**, *25*, 666–675. [CrossRef]
19. Koca, K.; Genç, M.S.; Açıkel, H.H.; Çagdas, M.; Bodur, T.M. Identification of flow phenomena over NACA 4412 wind turbine airfoil at low Reynolds numbers and role of laminar separation bubble on flow evolution. *Energy* **2018**, *144*, 750–764. [CrossRef]
20. Koca, K.; Genç, M.S.; Ozkan, R. Mapping of laminar separation bubble and bubble-induced vibrations over a turbine blade at low Reynolds numbers. *Ocean Eng.* **2021**, *239*, 109867. [CrossRef]
21. Genç, M.S.; Koca, K.; Açikel, H.H. Investigation of pre-stall flow control on wind turbine blade airfoil using roughness element. *Energy* **2019**, *176*, 320–334. [CrossRef]
22. Genç, M.S.; Açikel, H.H.; Koca, K. Effect of partial flexibility over both upper and lower surfaces to flow over wind turbine airfoil. *Energy Convers. Manag.* **2020**, *219*, 113042. [CrossRef]
23. Koca, K.; Genç, M.S.; Bayır, E.; Soguksu, F.K. Experimental study of the wind turbine airfoil with the local flexibility at different locations for more energy output. *Energy* **2022**, *239*, 121887. [CrossRef]
24. Bayramoğlu, N.; Genç, M.S.; Koca, K.; Altunal, A. Electricity production from piezoelectric patches mounted over flexible membrane wing at low Reynolds numbers. *Sigma J. Eng. Nat. Sci.* **2021**, *39*, 70–79.
25. Malekzadeh, S.; Sohankar, A. Reduction of fluid forces and heat transfer on a square cylinder in a laminar flow regime using a control plate. *Int. J. Heat Fluid Flow* **2012**, *34*, 15–27. [CrossRef]
26. Turhal, A.Ö.; Çuhadaroğlu, B. The effects of surface injection through a perforated square cylinder on some aerodynamic parameters. *Exp. Therm. Fluid Sci.* **2010**, *34*, 725–735. [CrossRef]
27. Sohankar, A.; Khodadadi, M.; Rangraz, E. Control of fluid flow and heat transfer around a square cylinder by uniform suction and blowing at low Reynolds numbers. *Comput. Fluids* **2015**, *109*, 155–167. [CrossRef]
28. Chen, W.L.; Gao, D.L.; Li, H.; Hu, H. Passive jet control of flow around a circular cylinder. *Exp. Fluids* **2015**, *56*, 201. [CrossRef]
29. Gao, D.L.; Chen, W.L.; Li, H.; Hu, H. Flow around a circular cylinder with slit. *Exp. Therm. Fluid Sci.* **2017**, *82*, 287–301. [CrossRef]
30. Munshi, S.R.; Modi, V.J.; Yokomizo, T. Aerodynamics and dynamics of rectangular prisms with momentum injection. *J. Wind Eng. Ind. Aerodyn.* **1997**, *11*, 873–892. [CrossRef]
31. Munshi, S.R.; Modi, V.J.; Yokomizo, T. Fluid dynamics of flat plates and rectangular prisms in the presence of moving surface boundary-layer control. *J. Wind Eng. Ind. Aerodyn.* **1999**, *79*, 37–60. [CrossRef]
32. Kubo, Y.; Modi, V.J.; Kotsubo, C.; Hayashida, K.; Kato, K. Suppression of wind-induced vibrations of tall structures through moving surface boundary-layer control. *J. Wind Eng. Ind. Aerodyn.* **1996**, *61*, 181–194. [CrossRef]
33. Kubo, Y.; Yukoku, E.; Modi, V.J.; Yamaguchi, E.; Kato, K.; Kawamura, S.I. Control of flow separation from leading edge of a shallow rectangular cylinder through momentum injection. *J. Wind Eng. Ind. Aerodyn.* **1999**, *83*, 503–514. [CrossRef]
34. Modi, V.J.; Deshpande, V.S. Fluid dynamics of a cubic structure as affected by momentum injection and height. *J. Wind Eng. Ind. Aerodyn.* **2000**, *89*, 445–470. [CrossRef]
35. Patnaik, B.S.V.; Wei, G.W. Controlling wake turbulence. *Phys. Rev. Lett.* **2002**, *88*, 054502. [CrossRef]
36. Muddada, S.; Patnaik, B.S.V. An assessment of turbulence models for the prediction of flow past a circular cylinder with momentum injection. *J. Wind Eng. Ind. Aerodyn.* **2010**, *98*, 575–591. [CrossRef]
37. Mittal, S. Control of flow past bluff bodies using rotating control cylinders. *J. Fluids Struct.* **2001**, *15*, 291–326. [CrossRef]
38. Korkischko, I.; Meneghini, J.R. Suppression of vortex-induced vibration using moving surface boundary-layer control. *J. Fluids Struct.* **2012**, *34*, 259–270. [CrossRef]
39. Sohankar, A.; Norberg, C.; Davidson, L. Low-Reynolds number flow around a square cylinder at incidence: Study of blockage, onset of vortex shedding and outlet boundary condition. *Int. J. Numer. Methods Fluids* **1998**, *26*, 39–56. [CrossRef]
40. Cheng, M.; Whyte, D.S.; Lou, J. Numerical simulation of flow around a square cylinder in uniform-shear flow. *J. Fluids Struct.* **2007**, *23*, 207–226. [CrossRef]
41. Okajima, A. Strouhal number of rectangular cylinders. *J. Fluids Struct.* **1982**, *123*, 379–398. [CrossRef]
42. Jan, Y.J.; Sheu, W.H. A numerical configuration of the dual body vortex flowmeter design. *Comput. Fluids* **2004**, *33*, 1157–1174. [CrossRef]
43. Abograis, A.S.; Alshayji, A.E. Reduction of fluid forces on a square cylinder in a laminar flow using passive control methods. In Proceedings of the COMSOL Conference, Boston, MA, USA, 9–11 November 2013.

applied sciences

MDPI

Article

Experimental Investigation and Validation on Suppressing the Unsteady Aerodynamic Force and Flow Structure of Single Box Girder by Trailing Edge Jets

Guanbin Chen [1,2] **and Wenli Chen** [1,2,*]

[1] Key Lab of Smart Prevention and Mitigation of Civil Engineering Disasters of the Ministry of Industry and Information Technology, Harbin Institute of Technology, Harbin 150090, China; 18b933016@stu.hit.edu.cn
[2] Key Lab of Structures Dynamic Behavior and Control of the Ministry of Education, Harbin Institute of Technology, Harbin 150090, China
* Correspondence: cwl_80@hit.edu.cn

Abstract: In the present investigation, a wind tunnel experiment was performed to evaluate the effectiveness of the trailing edge jets control scheme to mitigate the unsteady aerodynamic force and flow structure of a single box girder (SBG) model. The flow control scheme uses four isolated circular holes for forming the jet flow to modify the periodic vortex shedding behind the SBG model and then alleviate the fluctuation of the aerodynamic force acting on the test model. The Reynolds number is calculated as 2.08×10^4 based on the incoming velocity and the height of the test model. A digital pressure measurement system was utilized to obtain and record the surface pressure that was distributed around the SBG model. The surface pressure results show that the fluctuating amplitude of the aerodynamic forces was attenuated in the controlled case at a specific range of the non-dimensional jet momentum coefficient. The Strouhal number of the controlled case also deviates from that of the original SBG model. Except for the pressure measurement experiment, a high-resolution digital particle image velocimetry system was applied to investigate the detailed flow structure behind the SBG model to uncover the unsteady vortex motion process from the SBG model with and without the trailing edge jets flow control. As the jet flow blows into the wake, the alternating vortex shedding mode is switched into a symmetrical shedding mode and the width of the wake flow is narrowed. The proper orthogonal decomposition was used to identify the energy of the different modes and obtain its corresponding flow structures. Moreover, the linear stability analysis of the flow field behind the SBG model shows that the scheme of trailing edge jets can dramatically suppress the area of unsteady flow.

Keywords: unsteady aerodynamic force; single box girder; Strouhal number; linear stability analysis

Citation: Chen, G.; Chen, W. Experimental Investigation and Validation on Suppressing the Unsteady Aerodynamic Force and Flow Structure of Single Box Girder by Trailing Edge Jets. *Appl. Sci.* **2022**, *12*, 967. https://doi.org/10.3390/app12030967

Academic Editor: Giangiacomo Minak

Received: 25 December 2021
Accepted: 16 January 2022
Published: 18 January 2022

1. Introduction

With the span of bridges increasing, the main girder of the bridge will become more flexible with lower stiffness and damping to resist the dynamic load. The wind load is the most common dynamic force acting on the long-span bridge in the natural environment [1,2]. Airflow flows across the bridge girder and yields the alternating vortex shedding in the wake. When the vortex shedding frequency is closed or equaled to that of the structure, the structure has the potential for the resonance phenomenon to occur, i.e., vortex-induced vibration (VIV). The VIV is the primary issue that is caused by airflow in bridge engineering because it occurs at a low velocity (e.g., the VIV of the Great Belt East Bridge was observed at an incoming flow velocity of 4–12 m/s as presented by Larsen et al. [1] and leads to structural fatigue when undergoing frequent vibration [3,4]. Therefore, it is necessary to present an efficient control scheme to suppress the amplitude of VIV [4–7].

The external shape of the bluff body can influence the performance of the VIV [8]. If the target structure is streamlined, the flow separation cannot occur and the alternating vortex shedding cannot form; it is hard for the VIV to appear in those structures. Therefore, the box girder of the bridge is designed as near streamline for alleviating the VIV. Due to the box girder being not entirely streamlined, the VIV often happens when the airflow velocity attains a specific value, i.e., the wake vortex frequency is equal to that of the box girder at this velocity [2,9]. In addition, the affiliate parts (i.e., railing, maintenance track, and streetlamps) that are set in the box girder also can affect the characteristics of the main girder and the fluctuation of the aerodynamic force. Zhan et al. [10] changed the straight railing along the span-wise direction to the wave railing for forming the three-dimensional disturbance to mitigate the VIV of a long-span bridge. The results indicated that when the wave length of the wavy bar was set as 0.5 or 1 times of the bridge girder height, the vertical and torsional VIV amplitudes were entirely suppressed and both sides of the upper surface were the best installation locations. However, this form of wave railing will occupy more space, impacting usability. The size and location of the horizontal bar of the railing had a dramatic influence on the flow structure of the position near the leading edge and the torsional VIV responses of the main girder as presented by Nagao et al. [11]. Moreover, Xin et al. [12] studied the effect of the inclination angle of the railing on the VIV of the bridge girder. For suppressing the VIV, the control effectiveness of the inclined railing was superior to the railing with the inclination angle of zero.

In addition to using non-structural parts to suppress the wind-induced vibration, many other control strategies, such as the structural and aerodynamic measures, have been presented in recent years [7,13–15]. Larsen et al. [7] designed tuned mass dampers (TMDs) to mitigate the vibration amplitude of the Great Belt East bridge. The results found that the VIV response of the bridge girder with TMDs was at the allowed range using a time-domain simulation. Battista and Pfeil [13] applied the TMDs in the Rio-Niterói Bridge to alleviate the VIV amplitude. They discovered that subsidiary active or tuned vibration absorbers had a remarkable influence on attenuating the dynamic responses on the main girder. Dai et al. [16] studied the parameter determination of the tuned mass damper in alleviating the VIV of the bridge girder. They found that the equivalent damping ratio (EDR) is only determined by the TMD parameters, not the aerodynamic parameters. Moreover, the EDR is closely related to the uncertainty in the stiffness of the bridge according to the result of the uncertainty analysis. The underlying control mechanism of TMD is that changing the vibrational frequency and the structural damping mismatch the frequency of the aerodynamic force. Due to the fluctuation of aerodynamic force or vortex shedding being mainly wind-induced oscillation, El-Gammal et al. [17] used a span-wise sinusoidal perturbation set on the two sides of a plate girder to disturb the span-wise vortex shedding and reduce the dynamic response of VIV. The results indicate that the vibration amplitude was reduced with a higher wave steepness in the span-wise sinusoidal perturbation. Larsen et al. [1] utilized a guide vane that was equipped in the lower sides of the Great Belt East Bridge to alleviate the airflow separation. This investigation indicated that the guide vane could suppress the alternating vortex shedding and mitigate the VIV responses. However, this method possesses slight control effectiveness at a low *Re* because less airflow goes through the guide vane to form the jet flow. The technique of leading-edge and trailing edge holes and the communicating channel formed the self-suction-and-jet (SSJ) flow as designed by Chen et al. [18]. These experimental results showed a reduction of the unsteady surface pressure and aerodynamic force, and the near-wake flow structure was steadier. In addition, the unsteady flow was shifted further downstream by using the control method of SSJ flow. To avoid the strength limit of the suction or jet flow, the control scheme of active suction or jet flow is presented. Zhang et al. [19] adopted active suction holes that were distributed on the bottom surface of the box girder with a specific interval to produce a Mode A secondary unstable wake flow, which can deeply alleviate the fluctuating amplitude of the aerodynamic force that is acting on the bridge girder. Chen et al. [20] designed an active control strategy with a leading-edge suction and trailing edge jet (LSTJ) that were inserted in the box girder to control the unsteady flow structure and aerodynamic force. They found that the best control case alleviated the unsteady lift force by about 37.88% and shrank the unsteady wake flow area by about 69.90%.

Inspired by the investigation result of Chen et al. [20], we only set the jet holes in the trailing edge of a single box girder (SBG) model to control the unsteady surface pressure, aerodynamic force, and unsteady flow structure. The present study is to investigate the control ability of the trailing edge jet for alleviating the unsteady aerodynamic force and the flow field of a single box girder and study what can influence the local separation on the upper and lower surface near to the leading edge of the SBG model. Only the trailing edge jet can reduce the energy consumption compared with the LSTJ as presented by Chen et al. [20]. The analysis methods of proper orthogonal decomposition (POD) and linear stability analysis were used to reveal the underlying control mechanisms of the trailing edge jets. The structure of the present paper is arranged as follows. The experimental setup, test model details with and without the trailing edge jets are given in Section 2. The analysis results of the surface pressure distribution, aerodynamic force, flow structure, POD modes, and linear stability analysis are presented in Section 3. The conclusions are given in the last section.

2. Experimental Setup and Details

The present investigation was performed in a closed-circuit wind tunnel (SMC-WT1) belonging to the Joint Laboratory of Wind Tunnel and Wave Flume, Harbin Institute of Technology, P. R. China. The closed-loop wind tunnel test section, which has a geometric scale of 505 × 505 mm^2 in cross-section and 1050 mm in the airflow direction, possesses optically transparent walls to observe the experimental phenomenon conveniently. The honeycombs, mesh structure, and contraction section were installed upstream of the test channel to produce a high-quality uniform incoming flow entering the square test channel. The incoming flow velocity can be successively regulated at the range of 0–25 m/s. The turbulence intensity level of the free incoming flow was evaluated as less than 0.4%.

2.1. Original Test Model Details

Figure 1 displays the sketch of the SBG model that was used in the current investigation. The Great Belt East Bridge, whose geometric scale of the main girder is 31 m in width and 4.34 m in height, possesses a main span of 1624 m. The SBG model was derived from the Great Belt East Bridge (Denmark) due to this bridge frequently occurring VIV at low airflow speed [1,20]. The SBG model was produced using a high-precision three-dimensional printer with a scale ratio of 1:125. Therefore, the SBG model had a width (B) of 248 mm, a height (H) of 34.72 mm, and a length (L) of 480 mm in the span-wise direction, as shown in Figure 1a,c. The slope surface on the carriageway was considered. The secondary structures on the surface of the bridge girder were neglected in this investigation. The Reynolds number was calculated as $Re = 2.08 \times 10^4$ according to the incoming flow velocity of 9 m/s and the height of the SBG model. It should be declared that despite the Re of the present experiment being much less than that of the real bridge in a natural wind environment, this investigation aimed to study the control effectiveness of the control scheme of the trailing edge jets. There were two iron bars that were used to firmly fix the SBG model in the test channel to prevent displacement and vibration during the experiment. The coordinate system was set as shown in Figure 1a. The positive X and Y are defined as the airflow and upward directions, respectively. The positive Z meets the right-hand rule.

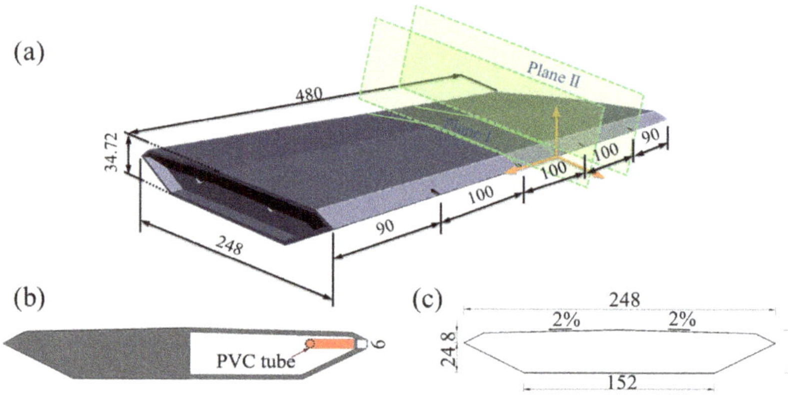

Figure 1. The geometric scale of the present test model (unit: mm). (**a**) The three-dimension view, (**b**) cross-section in plane I, (**c**) cross-section in plane II.

2.2. Test Model with the Trailing Edge Jets

To realize the trailing edge jet flow, four isolated circular jet holes were uniformly spaced in the trailing edge of the SBG model. The space between two neighboring jet-holes was set as 100 mm (~2.88 H), and the diameter of each jet-hole was 6 mm (~0.17H), as shown in Figure 1 a,b. The center of the section of all the jet holes was situated at the cusp of the trailing edge of the SBG model. Polyvinyl chloride (PVC) tubes were used to connect the jet holes with the air accelerators. To allow for real-time control, the velocity of the jet flow of the jet-holes, the mass, and the volumetric flow controllers (FMA-2621A, measuring range: 0–1500 *SLM*) were utilized to set and monitor the airflow rate in the PVC tubes. A total of nine volumetric flow rates (Q_j) were conducted in the experimental investigation, as shown in Table 1. Based on the flow rates and the area of the jet-hole, the average velocity jet out from the jet-hole was obtained. In the present study, the non-dimensional jet momentum coefficient (J_{sj}) was applied to evaluate the strength of the jet flow blowing out from the jet-hole of different controlled cases [18,20–22], which could be defined as Equation (1).

$$J_{sj} = \frac{{U_j}^2 A_h}{{U_0}^2 HL} \tag{1}$$

where U_j and U_0 are the velocity of the jet flow and incoming flow, respectively; A_h is the total area of all the jet holes. The J_{sj} of the SBG model with various jet speeds was calculated and displayed in Table 1.

Table 1. J_{sj} for the SBG model with different jet rates.

Volumetric Flow Rate Q_j (m^3/h)	Jet Velocity U_j (m * s^{-1})	Non-Dimensional Jet Momentum Coefficients J_{sj}
0 (Original main girder model)	0	0
1.80	4.42	0.0016
2.88	7.07	0.0042
3.60	8.84	0.0066
4.32	10.61	0.0094
5.40	13.26	0.0147
7.20	17.68	0.0262
9.00	22.10	0.0409
10.80	26.53	0.0590
12.60	30.95	0.0802

2.3. Measurement Apparatus

The pressure that was distributed on the surface of the SBG model was obtained and recorded by a digital pressure scanner (MPS4264) that possessed 64 pressure measurement ports and a maximum acquiring frequency of 850 Hz. The pressure taps that were distributed around the test model were connected to the pressure scanner by the PVC tube with a length of 0.5 m and a radii of 0.45 mm. Based on the research findings of [23], the distortion effects that were caused by the tubing system that were used for the pressure measurements can be neglectable. All the pressure taps were set in one cross-section located in the middle of the SBG model, which is the same with plane II as given in Figure 1). Both the upper and lower surface of the SBG model had 23 pressure taps, as given in Figure 2. The acquiring frequency and recording time were set as 0.4 kHz and 60 s, respectively, in the present experiment. According to the scale of the SBG model and test channel of SMC-WT1, The blockage ratio was obtained as 6.67%. Therefore, the surface measurement results should be corrected by using the algorithm about the solid blockage and wake blockage presented by [24].

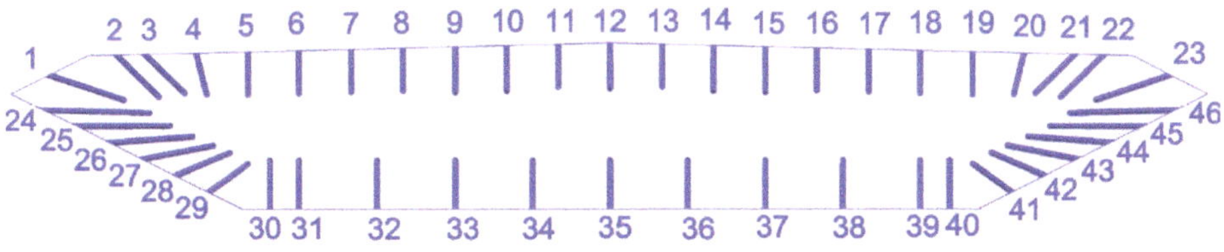

Figure 2. The pressure taps that were distributed around the test model.

To visualize the characteristics of the wake flow structure behind the SBG model with and without the trailing edge jets, a high-resolution particle image velocimetry (PIV) system was used to record the photos of the target flow field. The sketch of the PIV experimental setup in the present measurement is given in Figure 3. The PIV system mainly contains three parts: illuminating, camera, and synchronizer. A high-speed double-pulse Nd: YAG laser (Vlite-Hi 200, Beamtech Optronics Corporation, Beijing, China), which has the highest work frequency of 200 Hz and energy of 100 mJ per pulse, was selected as the illuminating system to light up the target plane. The cylindrical lenses manipulated the laser beam to form a laser sheet with a thickness of about 1.0 mm in the measurement plane. A high-speed CCD camera (PCO: pco. dimax HS4, Kelheim, Germany) was applied to capture and save the photos of the measurement objectives, which possessed 2277-fps at the full-frame of 2000 × 2000 pixels. Moreover, we used a smoke generator that produced oil-droplets with a diameter of about 5 μm that were seeded into the airflow. The oil-droplets was served as the tracer particle whose movement could represent the airflow motion [25].

This investigation measured two vertical target planes, as given in Figure 1a. Planes I and II, the vertical target plane, were situated at the central section of the jet-hole and SBG model, respectively. The position of a CCD camera should be strictly perpendicular to the target plane during the experiment, as displayed in Figure 3. A total of 4000 pairs of photos were acquired at the trigger frequency of 200 Hz in every target plane of different experimental cases. The pixel of photos at the target plane I and II was 2000 × 1600 pixels^2 and 2000 × 1600 pixels^2, respectively. After obtaining the flow field photos, the cross-correlation operation was performed to extract the instantaneous velocity by adopting the interrogation window of 32 × 32 pixels and an efficient overlap rate of 50%. As a result, the spatial resolution of the instantaneous velocity in the measurement planes I and II were calculated as 1.8 × 1.8 mm^2 and 2.2 × 2.2 mm^2, respectively. It should be noted that the measurement uncertainty for the instantaneous velocity of the PIV measurement result was evaluated to be under 5%.

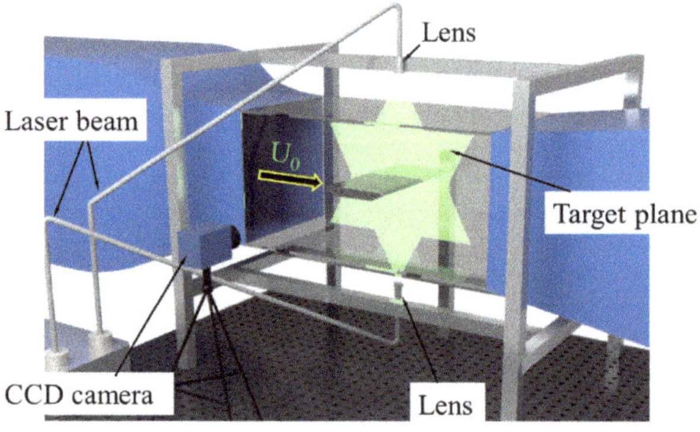

Figure 3. Sketch of the PIV experiment arrangement.

3. Experimental Results and Discussions

3.1. Surface Pressure Distribution

The pressure coefficient ($C_{pi}(t)$) was that the surface pressure results that were acquired by the pressure measurement experiment were nondimensionalized by the dynamic pressure and defined as Equation (2). The mean value of the external surface pressure coefficient (C_p_M) was defined as Equation (3) to illustrate its average features. We used the standard deviation of the external surface pressure coefficient (C_p_R) to describe the fluctuating characters of the pressure distribution, as given in Equation (4).

$$C_{pi}(t) = \frac{2(p_i - p_\infty)}{\rho U_0{}^2} \tag{2}$$

$$C_p\_M = \frac{1}{N}\sum_{i=1}^{N} C_{pi}(t) \tag{3}$$

$$C_p\_R = \sqrt{\frac{1}{N}\sum_{i=1}^{N} \left(C_{pi}(t) - \overline{C_{pi}(t)}\right)^2} \tag{4}$$

where p_i is the static pressure that is acting on the surface SBG model and p_∞ is the static pressure of the free stream. The time histories of p_i and p_∞ are acquired and recorded by the digital pressure scanner. ρ is the air density and N is the length of the total sample.

It should be illustrated that the surface pressure measurement plane was located in the central cross-section (i.e., plane II as given in Figure 1a) was far away from the jet holes. The pressure distribution on other cross-sections may be different. However, the aerodynamic force that was acting on the present cross-section of the controlled case might underestimate the control ability due to the pressure measurement section being the farthest distance from the cross-section with trailing jet hole [26]. The PIV result will confirm that. Figure 4 gives the external surface pressure coefficient that was distributed around the SBG model that was equipped with and without trailing edge jets.

For the case J_{sj} = 0, the mean pressure result of the upper surface on the SBG model shows a dramatic decrease due to an inflection point that existed between the pressure measurement taps 1 and 2. The adverse pressure gradient was observed at the range of X/B from 0.11 to 0.20, as shown in Figure 4a. Therefore, the flow separation occurs in that position, causing the fluctuating pressure to have a maximal value at X/B = 0.11, as shown in Figure 4c. As the X/B was larger than 0.20, the mean pressure distribution of the upper surface had similar values, except for the position of X/B = 0.5 and near the trailing edge. The upper external surface pressure distribution fluctuation shows an enormous value when X/B gets close to 1. The reason for this is that the position nearing the trailing edge will be more influenced by the alternating vortex shedding [20,26]. The adverse pressure gradient also appears in the lower surface of the SBG model at the position of X/B = 0.31–0.41 as shown in Figure 4b, and the peak value of the fluctuating pressure coefficient was obtained at the position of X/B = 0.31, as given in Figure 4d. Therefore, the local flow separation also occurred at the corner point between the pressure measurement taps 29 and 30. The other characteristics of the lower surface pressure are similar to those of the upper surface pressure.

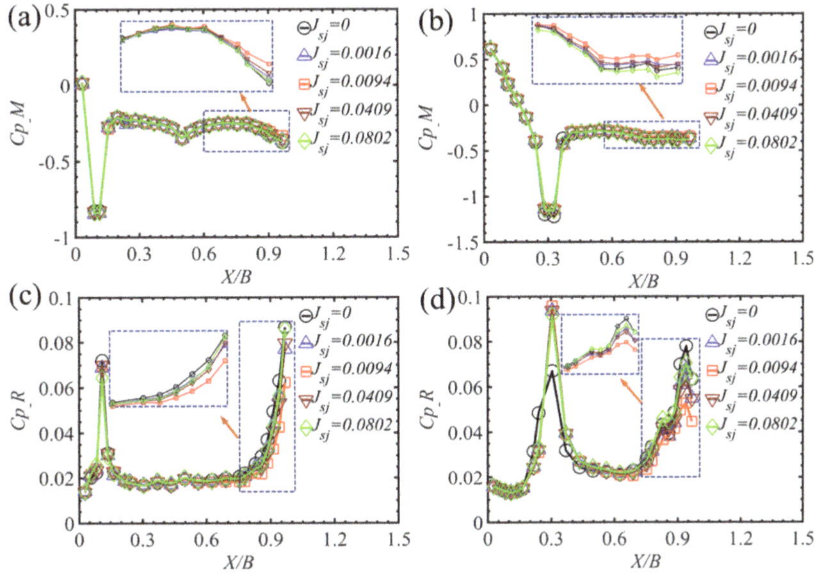

Figure 4. The external surface pressure distributing on the test model with and without the trailing edge jets. The mean of the upper surface (**a**) and lower surface (**b**) pressure distributions, RMS of the upper surface (**c**), and lower surface (**d**) pressure distributions.

When the jet holes were inserted in the SBG model, the external surface pressure was modified compared to that of the case J_{sj} = 0. The mean values of the external surface pressure of the controlled case with different non-dimensional jet momentum coefficients remain close to the uncontrolled case at the position nearing the leading edge of the SBG model. When the pressure taps were located at the trailing edge, the mean value of the external surface pressure was alleviated by the trailing edge jet flow, and the case J_{sj} = 0.0094 exhibited the best control effect. All the controlled cases can slightly attenuate the local flow separation at X/B = 0.11 of the upper surface of the SBG model, as shown in Figure 4c, compared to case J_{sj} = 0. However, the local flow separation was improved in the lower external surface of the controlled cases that caused a more considerable fluctuation of the pressure at X/B = 0.31, as given in Figure 4d. The fluctuation of the surface pressure close to the trailing edge was mitigated on both the upper and bottom surfaces of the SBG model with the trailing edge jets. That is because the jet flow can interrupt the alternating vortex shedding and lead to the wake flow being more steady. According to the pressure measurement results, the case J_{sj} = 0.0094 exhibited the best control ability for alleviating the mean value and fluctuation of the surface pressure. Obtaining better control effectiveness does not need larger non-dimensional jet momentum coefficients. There is an optimal non-dimensional jet momentum coefficient to realize the control effect, which is consistent with the study by Chen et al. [20] and Gao et al. [27].

3.2. Aerodynamic Force

The aerodynamic force can be obtained by integrating the surface pressure distributed around the SBG model. Afterward, the lift coefficient ($C_l(t)$), drag coefficient ($C_d(t)$), and the moment coefficient ($C_m(t)$) are calculated by Equation (5).

$$C_l(t) = \frac{F_l(t)}{\frac{1}{2}\rho U_0{}^2 B},\ C_d(t) = \frac{F_d(t)}{\frac{1}{2}\rho U_0{}^2 B},\ C_m(t) = \frac{F_m(t)}{\frac{1}{2}\rho U_0{}^2 B^2} \tag{5}$$

where $F_l(t)$, $F_d(t)$, and $F_m(t)$ are the time-variant of the lift, drag, and moment force, respectively.

The aerodynamic coefficients that are acting on the SBG model that are equipped with and without control are displayed in Figure 5. The mean value of the drag coefficient of the case where J_{sj} = 0, which was consistent with the experimental results of 0.064 that were presented by Taylor et al. [28], 0.061 by Zhang et al. [9], and the numerical simulating result of 0.060 given by Frandsen [29], was obtained as 0.062. With the increase of the J_{sj}, both the mean value of the drag and lift coefficient shows a decrease at first and then an increase. However, the mean value of the moment coefficient of the controlled case maintained a closing value that was slightly large than the case where J_{sj} = 0, indicating that the different J_{sj} can not influence the moment force that is acting on the SBG model. Note that the main reason for the wind-induced vibration on the main girder is a large fluctuation of the aerodynamic force. The results of the fluctuation of the drag, lift, and moment coefficients that are acting on the SGB model with different J_{sj} are displayed in Figure 5d–f, respectively. The fluctuation of the aerodynamic coefficients were descending at first and then ascending with the increasing of J_{sj}. Therefore, an optimal J_{sj} for alleviating the unsteady aerodynamic force that is being exerted in the SBG model is required. This saturation phenomenon was also found by Chen et al. [20], Gao et al. [27], and Apelt et al. [30]. When the J_{sj} equals 0.0094, the best control ability was possessed in the present investigation. The control effect for the drag, lift, and moment coefficient fluctuation was 16.23%, 36.34%, and 38.74%, respectively. That conclusion agreed well with the results of the surface pressure analysis.

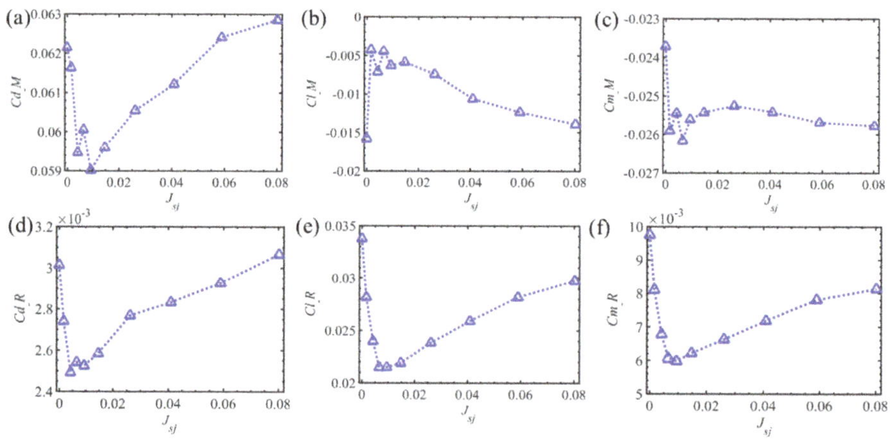

Figure 5. Aerodynamic forces that are acting on the SBG model. The mean value of the drag coefficient (**a**), lift coefficient (**b**), and moment coefficient (**c**). The fluctuation of the drag coefficient (**d**), lift coefficient (**e**), and moment coefficient (**f**).

Figure 6 gives the time series and the power spectral density analysis of the lift coefficient being exerted in the SBG models with and without the trailing edge jets. The lift coefficient when J_{sj} = 0 shows a significant fluctuation during the experiment, as shown in Figure 6a. When the trailing edge jets were conducted in the SBG model with J_{sj} = 0.0094, the fluctuation of the lift coefficient was reduced compared to the case $J_{sj} = 0$. The peak value of the reduced frequency for the case $J_{sj} = 0$ was obtained to be 0.285 based on the fast Fourier transform (FFT) analysis, which was similar to that of 0.28 that was presented by Taylor et al. [28] and Chen et al. [18]. The amplitude of power spectral density analysis of the case $J_{sj} = 0.0094$ was lower than that of the case $J_{sj} = 0$, and the peak value of reduced frequency of the lift force was also changed to 0.279, given in Figure 6b. This indicates that the trailing edge jets that were equipped in the SBG model can modify the frequency and the strength of the vortex shedding behind the test model.

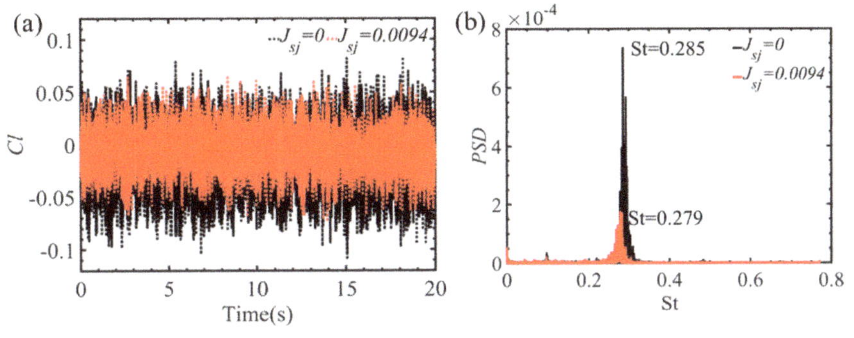

Figure 6. Time histories (**a**) and power spectrum analysis (**b**) of the lift force being exerted in the SBG model of cases $J_{sj} = 0$ and $J_{sj} = 0.0094$.

3.3. PIV Measurement Results

Besides the surface pressure measurement results, the PIV measurement experimental result can assist us in understanding the control mechanism of the trailing edge jets set in the SBG model.

Turbulence kinetic energy (TKE), which serves as a criterion to estimate the unsteady surface pressure and the fluctuating aerodynamic force that is acting on the SBG model [31–33], is calculated as follows:

$$\text{TKE} = \frac{\overline{u'^2} + \overline{v'^2}}{2U_0^2} \tag{6}$$

where u' and v' are the fluctuating velocity components of the streamwise and transverse direction, respectively.

The time-averaged flow fields behind the SBG model in the two target planes with and without the trailing edge jets are given in Figure 7. A cluster of large values of the TKE is concentrated in the wake flow when $J_{sj} = 0$, as shown in Figure 7a. The reason is that the separation flows from the upper and bottom surfaces interact in the wake, leading to the velocity vectors colliding in this region. Therefore, the surface pressure distribution nearing the trailing edge when $J_{sj} = 0$ shows a dramatic fluctuation. As the trailing edge jets is conducted in the SBG model, the distribution area and maximum value of the TKE are reduced. Moreover, the maximum value of the TKE decreases at first and then increases with the enhancement of J_{sj}, as shown in Figure 7a–e, which can explain the changing trend of the fluctuation of the aerodynamic force that is being exerted in the SBG model. Compared to results from the target plane II, both the maximum value and the concentrated area of TKE in the target plane I of the case $J_{sj} = 0.0094$ is reduced as shown in Figure 7c,f. The unsteady aerodynamic force being exerted in the cross-section of plane I is lower than that of plane II, indicating the pressure measurement result that was obtained by plane II would underestimate the control effectiveness of a fluctuating value of surface pressure that is distributed on the SBG model with trailing edge jets. Moreover, the lower TKE value illustrates that the level of turbulence mixing in the wake flow of the controlled cases is lower than the uncontrolled case [34].

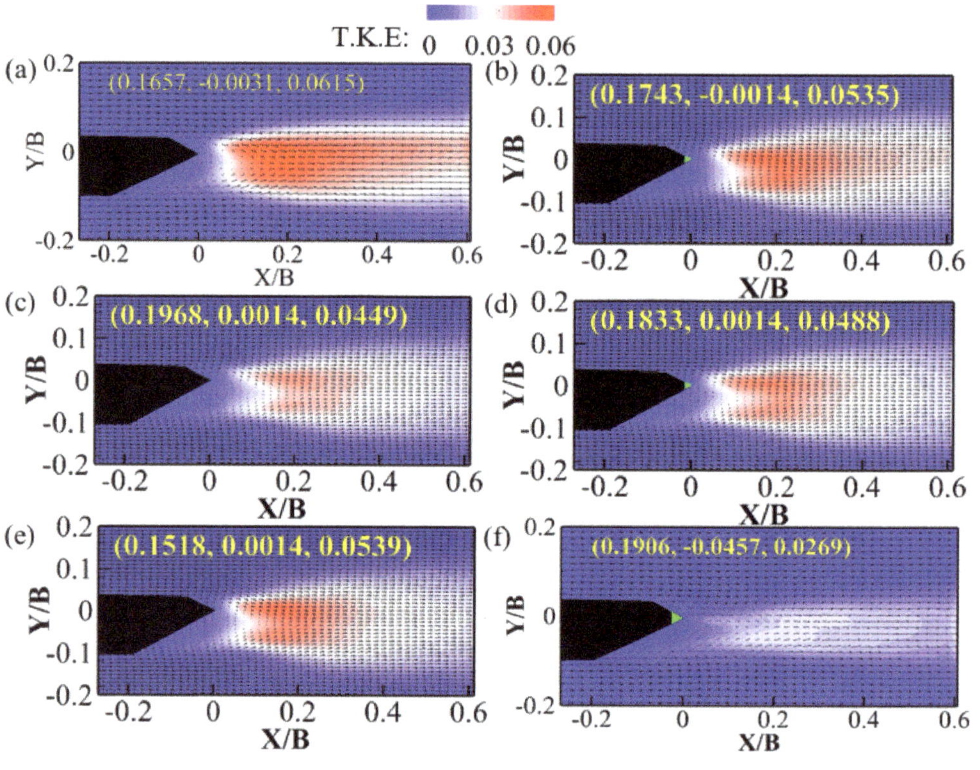

Figure 7. A time-averaged representation of the flow field behind SBG model. (**a**) $J_{sj} = 0$, (**b**) $J_{sj} = 0.0016$, (**c**) $J_{sj} = 0.0094$, (**d**) $J_{sj} = 0.0409$, $J_{sj} = 0.0802$, and (**f**) $J_{sj} = 0.0094$. (**a**–**e**): plane II, (**f**): plane I.

The instantaneous flow field in plane I that is given in Figure 8 illustrates the vortex shedding pattern behind the SBG model. A pair of asymmetric vortices alternating shed from the upper and bottom surface of the SBG model and a Karman vortex street can be obtained in the wake flow, which causes the significant fluctuation of the surface pressure on the trailing edge of the test model without control, as shown in Figure 4. The position where the upper and lower vortices interact leads to the enormous TKE value, which can be seen when comparing Figures 7a and 8a. However, when the trailing edge jets work, a pair of symmetric vortices are formed in the wake flow field and the Karman vortex street is alleviated. The length of the vortex shedding is elongated and shifted further downstream, as given in Figure 8b. In addition, the jet flow interacts with the upper vortex and divides that vortex into some small-scale vortices, illustrating that the unsteady aerodynamic force that is being exerted in the SBG model is mitigated, and then the potential vortex-induced vibration can be prevented with the trailing edge jets [26].

At first, the proper orthogonal decomposition (POD) method was presented in the region of fluid mechanics by Lumley [35]. The snapshot POD that was introduced by Sirovich [36] was utilized in the present investigation. Here, every instantaneous photo that was obtained by the PIV experiment served as a snapshot of the flow structure. The autocovariance matrix was accepted by the product of the matrix of fluctuating part of the two-dimensional velocity and its transpose. The eigenvalues and the eigenvectors of the autocovariance matrix can be obtained. Afterwards, the accumulative energy proportion of each POD mode can be calculated by the eigenvalues of the autocovariance matrix, and the POD modes can be constructed from the basis built up from the snapshots.

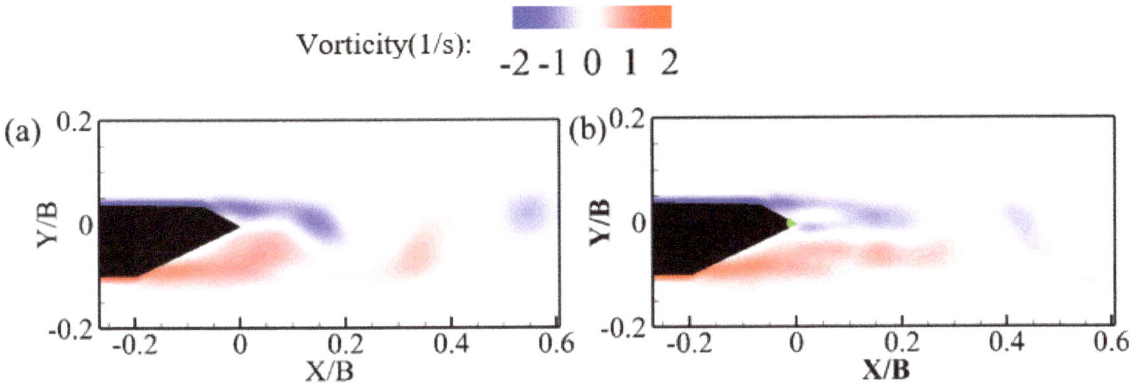

Figure 8. Instantaneous vortex shedding behind SBG model. (**a**) $J_{sj} = 0$, (**b**) $J_{sj} = 0.0094$.

Figure 9 shows the percentage of the accumulative contribution of every POD mode to the total turbulence kinetic energy of the flow field for the different test cases in PIV measurement. It is observed that the first two POD modes possess 64.6% of the total energy for the case $J_{sj} = 0$. The first two POD modes represent the large-scale coherent structure, and the high-order POD modes represent the small-scale coherent structure as given by Feng et al. [37]. Therefore, the large-scale coherent structure plays an important role in the wake flow behind case $J_{sj} = 0$. When the control method is applied in the SBG model, the energy of each POD mode that is exhibited immensely changes. The total energy of the first two POD modes is slightly decreased, as J_{sj} is set as 0.0016, indicating that a little strength of the jet flow cannot significantly influence the vortex. The ratio of the energy of the first and second POD modes to the total kinetic energy for cases $J_{sj} = 0.0016$, $J_{sj} = 0.0094$, $J_{sj} = 0.0409$, and $J_{sj} = 0.0802$ are 62.0%, 33.8%, 53.3%, and 53.2%, respectively. Hence, the total energy of the first two POD modes decreases at first and then increases, as shown in Figure 9 when J_{sj} improves. Moreover, the small-scale coherent structures in the wake flow field behind case $J_{sj} = 0.0094$ occupy most of the energy for the global flow structure, which alleviates the large amplitude of the unsteady aerodynamic force that is being exerted in the SBG model compared to the case $J_{sj} = 0$.

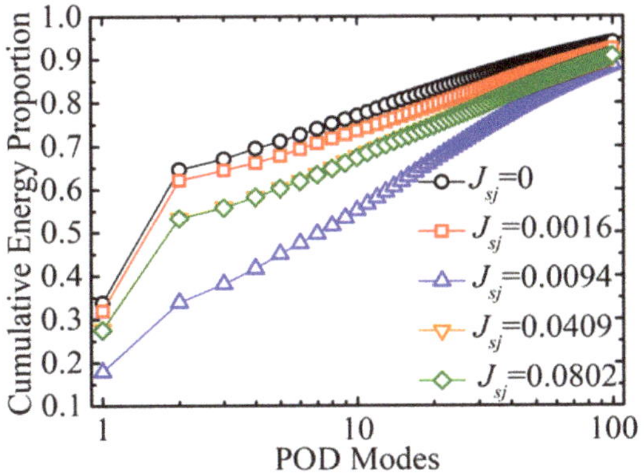

Figure 9. The cumulation energy proportion for the various POD modes.

To identify the characters of the POD modes, the first five POD modes for cases J_{sj} = 0 and J_{sj} = 0.0094 are plotted in Figure 10. For the case where J_{sj} = 0, the first two dominant POD modes are symmetrical about the line Y/B = 0.02 and exhibit alternating vorticity that is distributed in the wake flow. Nevertheless, modes three to five vary from modes one and two: modes three and four are asymmetrical about the line Y/B = 0.02. There are two rows of counter-rotating vorticity concentrations that are displayed in the flow field behind the case J_{sj} = 0 closed to the test model and then changed into the arrowhead structures as the flow field moves downstream, as shown in Figure 10c,d. Moreover, mode five shows some a small-scale vorticity concentration behind case J_{sj} = 0. The antisymmetric vorticity flow fields that are calculated by POD represent the symmetric features of the vortex flow field. Conversely, the symmetrical vortex flow structures stand for an antisymmetric flow field, as Konstantinidis et al. [38] described. In consequence, modes one and two of the case J_{sj} = 0 stand for the large-scale alternating vortex shedding forming the Karman vortex street, while modes three and four denote the small spatial scale vortices with symmetric forms, and mode five represents the irregular distribution of the vorticity structures with small geometric scale. For the SBG model with the trailing edge jets, the flow structures of POD modes show noticeable differences that are in contrast to their counterparts of the uncontrolled case. The vortex concentrations of the case J_{sj} = 0.0094 are dramatically decreased and narrowed laterally and stretched in the flow direction, as shown in the right of Figure 10. This phenomenon of the POD modes is in connection with the elongation of the wake vortex, as shown in Figure 8b. In addition, the wake jet flow disturbs the vorticity. It divides it into two small parts at the nearing wake position as shown in modes one to four, and mode five changes to much smaller vorticity concentrating parts compared with case J_{sj} = 0. Based on the analysis above, the wake flow structure characteristics of the case J_{sj} = 0.0094 yield the unsteady aerodynamic force that is less than that of case J_{sj} = 0, which is consistent with the results in Section 3.2.

To analyze the difference of the velocity profile in the flow field behind the test model with and without the trailing edge jets, the means of the streamwise velocity, streamwise velocity fluctuation, and the transverse velocity were extracted from the PIV measurement plane I. The time-averaged streamwise velocity that was located in different positions of X/B is given in Figure 11a. The velocity profile of the case J_{sj} = 0 is asymmetrical in the location that is close to the SBG model and symmetrical in a further downstream position. The reason is that the geometric shape of the SBG model is asymmetrical along the line of Y/B = 0. The lowest velocity of the case J_{sj} = 0 was less than that of the case J_{sj} = 0.0094 at X/B = 0.03 and 0.12 because the jet flow can alleviate the velocity defect. That indicates the inner region flow field behind case J_{sj} = 0.0094 is more stable than the case J_{sj} = 0 [6]. Conversely, the smallest value of the velocity of the case J_{sj} = 0 was larger than that of case where J_{sj} = 0.0094 at X/B = 0.21 and 0.30. The velocity profile of the case J_{sj} = 0 regained faster than that of the case J_{sj} = 0.0094. Moreover, the width of the wake flow in case J_{sj} = 0 was slightly wider than the case J_{sj} = 0.0094, indicating the large drag force that was acting on the SBG model without control [20].

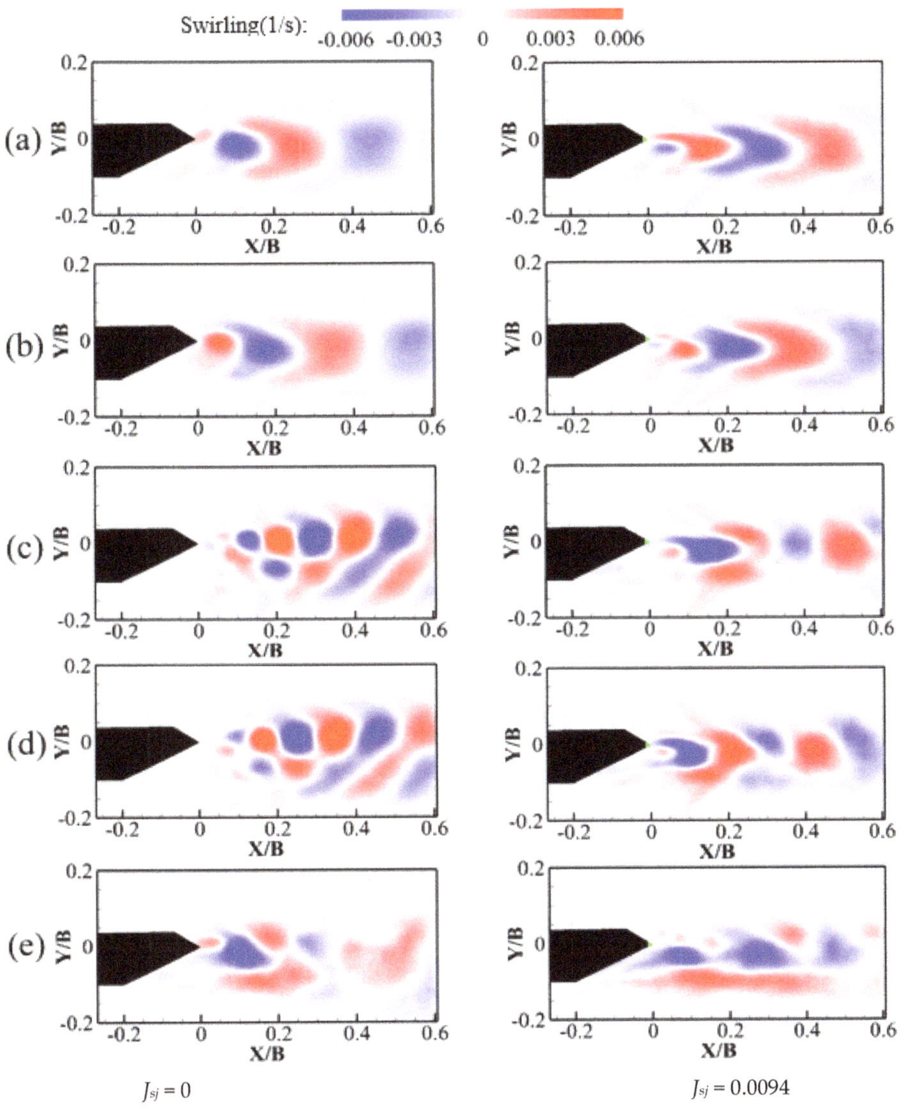

Figure 10. The POD mode of the wake flow in measurement plane I. (**a**) mode 1, (**b**) mode 2, (**c**) mode 3, (**d**) mode 4, and (**e**) mode 5.

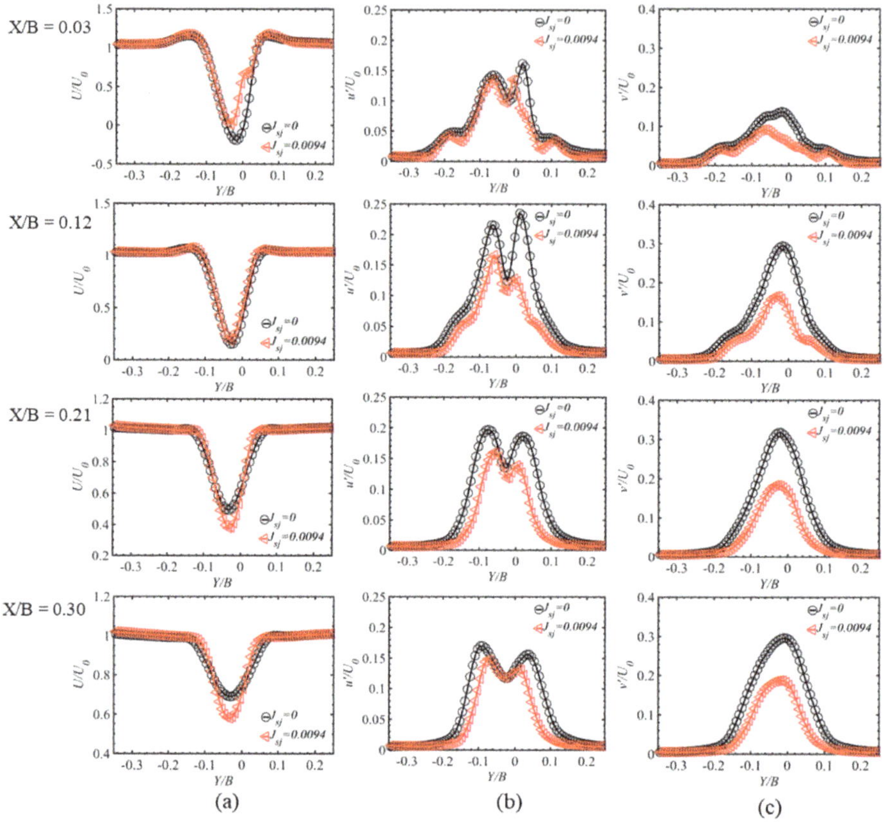

Figure 11. The velocity profile in the different locations at wake flow in measurement plane I. (**a**) Mean of the streamwise velocity, (**b**) fluctuation of the streamwise velocity, (**c**) fluctuation of the transverse velocity.

Figure 11b shows the fluctuation of the streamwise velocity profile at various positions in the wake. For the case J_{sj} = 0, There are two peak values that were observed in the wake flow of the case at X/B = 0.03–0.30, i.e., double-cusp mode. The maximum values appear when the turbulence level grows a dramatically quick ratio, mainly in the shear layers of the separation flow [33]. However, owing to the trailing edge jet flow disturbing the original flow field a the near-wake of X/B = 0.03, this profile for the controlled case exhibited a quadruple cusp mode. The fluctuation of the streamwise velocity profile of the case J_{sj} = 0.0094 presented a double-cusp mode at X/B = 0.12–0.30. It was similar to that of the case J_{sj} = 0, illustrating that the jet flow has a short influence range, which is identical to the investigation by Chen et al. [20]. Moreover, the peak value of the streamwise velocity fluctuation for case J_{sj} = 0.0094 was lower than the case J_{sj} = 0 at all the positions, leading to the unsteady drag force that was being exerted in the SBG model being mitigated.

The profiles of the transverse velocity fluctuation for cases J_{sj} = 0 and J_{sj} = 0.0094 are plotted in Figure 11c. The maximum value appears at approximately Y/B = 0.02, which is the axis of symmetry for the wake flow. The shear layers that are formed from the upper and lower surface are amalgamated and violently interact, causing a significant transverse velocity fluctuation [18,20]. The peak value of the case J_{sj} = 0 shows an increasing trend at the locations of X/B = 0.03 to 0.21 and then a reduction when X/B is more significant than 0.21, which illustrates that the shear layer interaction in the wake flow nearing to the SBG model is weaker than that of downstream. The maximum value of the fluctuation of the transverse velocity for the case J_{sj} = 0.0094 is obviously alleviated at all positions, indicating that the strong interaction of the shear layer decreased. That leads to the unsteady lift force that is acting on the SBG model when the trailing edge jets attenuated, as shown in Figure 5.

The Reynolds shear stress (RSS), which serves as an essential parameter to represent the turbulence in the flow structure, is calculated and given in Figure 12. At the near-wake of X/B = 0.03, the RSS possesses a lower value, illustrating the low turbulence level in this area for the case J_{sj} = 0. The maximum value of RSS is increased at first and then decreased with the improvement of the X/B. The maximum and minimum values of the RSS are obtained at the Y/B = ~−0.01 and 0.06, respectively, because the turbulence level of the shear layer that is separated from the upper and lower surface of the SBG model is strong [33]. This causes a significant fluctuation of the lift force being exerted in the SBG model. As J_{sj} sets as 0.0094, the RSS is mitigated at all the positions in wake flow, and the maximum and minimum values are reduced related to the uncontrolled case, which indicates that the interacting strength of the upper and lower shear layer is mitigated in the wake flow behind SBG model. Therefore, the unsteady lift force is decreased by the trailing edge jets.

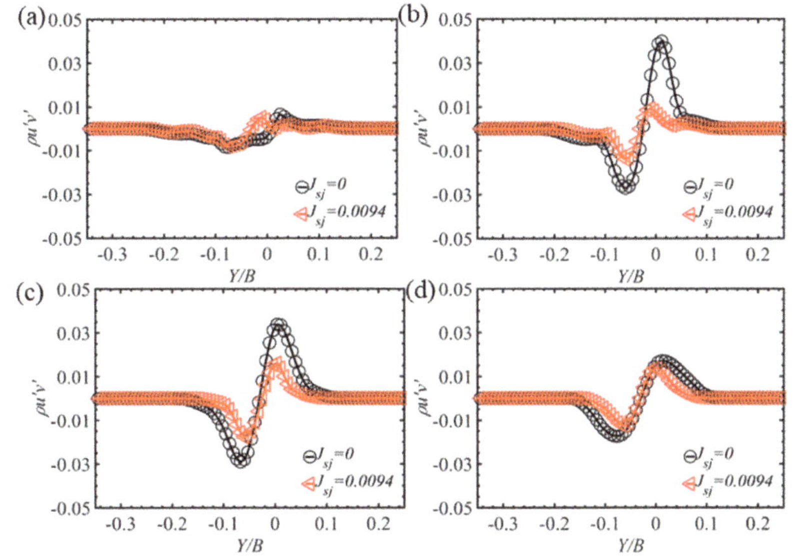

Figure 12. Reynolds stress distributions at different positions behind test model. (**a**) X/B = 0.03, (**b**) X/B = 0.12, (**c**) X/B = 0.21, and (**d**) X/B = 0.30.

3.4. Linear Stability Analysis

Triantafyllou et al. [39] found that the absolute instability of the flow field behind a bluff body is the reason for the vortex appearing in the wake of a stationary bluff body. The time-averaged streamwise velocity profiles, that were extracted from the mean flow field at various locations as shown in Figure 11a, substitute into the inviscid Orr–Sommerfeld equation (OSE). It should be noted that the higher-order small-term occurring in deducing the Orr–Sommerfeld equation and the viscidity of the air is neglected. Solving the OSE obtains the critical value in the complex frequency plane and the stability of the velocity profile by using the mathematical method that was presented by Orszag [40]. The imaginary part of the complex frequency, named as ω_i, determines the stability of the flow structure behind the SBG model. A positive ω_i represents an unstable flow structure because the slight disturbance would evolve over time, leading to a large unstable flow structure [20]. Contrary, the small initial disturbance would vanish with time for a negative ω_i. Figure 13 presents the dispersion relation $\omega = \omega(\alpha)$ maps α_i constant lines on the ω plane that were obtained by solving the OSE using the velocity profile at the position X/B = 0.08 of case J_{sj} = 0. The α is the complex wave number, and the α_i is the imaginary part of α. The imaginary and real parts of the complex frequency at the cusp point are 0.31 and 1.75, respectively. This indicates the velocity profiles behind case J_{sj} = 0 at X/B = 0.08 is absolute instability, causing the TKE to exhibit a considerable value in the wake flow, as shown in Figure 7a. According to the real part of ω, the Strouhal number of the wake flow can be calculated as 0.28, which is close to that which is gained by using the power spectral density analysis for the lift force that is acting on the SBG model as given in Figure 6b.

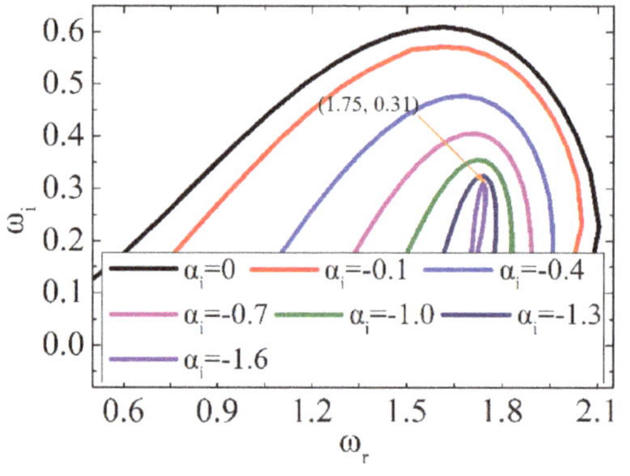

Figure 13. Map of lines α_i= constant in the plane, at X/B = 0.08 for the case J_{sj} = 0.

The ω_i at various locations behind the SBG model with and without trailing edge jets are obtained as given in Figure 14. It can be observed that the ranges of the absolute instability are stretched to the position of X/B = 0.103 for the case J_{sj} = 0. The flow field was slightly stable at the near-wake, and the unstable region was broader for cases J_{sj} = 0.0016, 0.0409, and 0.0802. However, for the case J_{sj} = 0.0094, the unstable area was narrowed to X/B = 0.029, indicating the unstable range decreased about 71.84% compared to that of the case J_{sj} = 0 as shown in the grey region of Figure 14. This phenomenon demonstrates that too strong jet flow isn't needed for controlling the unstable flow field of the SBG model with a trailing edge jet. An appropriate jet-speed causes a steadier flow structure behind the bluff body, which is also uncovered by Gao et al. [41] in investigations of suppressing unsteady aerodynamic force that is being exerted in a circular cylinder. Therefore, a steady flow structure would excite the lower fluctuating amplitude of the aerodynamic force that is acting on the SBG model, as given in Figure 5. Moreover, the stability of the wake flow of the case J_{sj} = 0.0094 is similar to that of the studies of leading-edge suction and trailing-edge jet (LSTJ) that are presented by Chen et al. [20]. The superiority of the present control scheme is that the energy consumption of the case J_{sj} = 0.0094 is half of that of the control method of LSTJ because the present control method only needs the trailing edge jets, not the leading-edge suction.

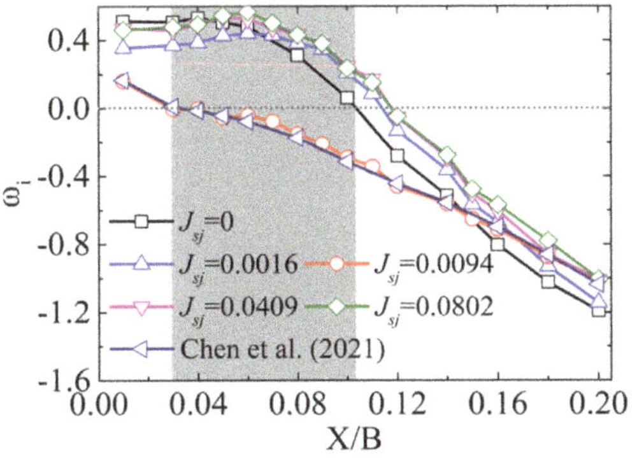

Figure 14. The imaginary part of the critical point at various positions (X/B) for different test cases.

4. Conclusions

This investigation experimentally studies the concept of trailing edge jet control to modify the vortex shedding mode for alleviating the fluctuating amplitude of surface pressure, aerodynamic force, and the unsteady flow structure. The non-dimensional jet momentum coefficient (J_{sj}) serves as a monitor to evaluate the strength of the wake jet flow. The surface pressure and the PIV measurements were conducted to investigate the characteristics of aerodynamic force and flow field of the single box girder (SBG) model with various J_{sj}. Some significant conclusions are given as follows.

The trailing edge jet that flows into the wake flow can obviously mitigate the fluctuating amplitude of the drag, lift, and moment forces. Based on the surface pressure measurement results, the saturation of the control effectiveness for lift force was obtained when the J_{sj} is set as 0.0094. The unsteady drag, lift, and moment forces are estimated to decrease about 16.23%, 36.34%, and 38.74%, respectively, compared to the uncontrolled case. Moreover, the vortex shedding frequency of the case J_{sj} = 0.0094 exhibited a difference.

The PIV measurement results explained the phenomenon that were obtained by the pressure measurement. The TKE distribution behind the SBG model with trailing edge jet were smaller than the case J_{sj} = 0, leading to the more stable wake. The instantaneous wake flow shows that the control method can change the antisymmetric mode into the symmetric mode, which suppresses the alternating vortex shedding and mitigates the unsteady aerodynamic force. Moreover, the unstable wake region of case J_{sj} = 0.0094 was narrowed by about 71.84%, compared to the uncontrolled case.

Author Contributions: Conceptualization, G.C. and W.C.; methodology, W.C.; software, G.C.; validation, W.C., and G.C.; formal analysis, G.C.; investigation, G.C.; resources, W.C.; data curation, G.C.; writing—original draft preparation, G.C.; writing—review and editing, W.C.; visualization, G.C.; supervision, W.C.; project administration, W.C.; funding acquisition, W.C. and G.C. All authors have read and agreed to the published version of the manuscript.

Funding: This research was funded by the National Natural Science Foundation of China grant number 51978222, 51808173, 51722805 and U2106222; The Fundamental Research Funds for the Central Universities grant number HIT.BRETIV 201803.

Institutional Review Board Statement: Not applicable.

Informed Consent Statement: Not applicable.

Data Availability Statement: Not applicable.

Acknowledgments: This research work is supported by the National Natural Science Foundation of China (51978222, 51808173, 51722805 and U2106222) and the Fundamental Research Funds for the Central Universities (HIT.BRETIV 201803). G. B. Chen also acknowledges the financial support from China Scholarship Council (CSC 202106120222).

Conflicts of Interest: The authors declare no conflict of interest.

References

1. Larsen, A.; Esdahl, S.; Andersen, J.E.; Vejrum, T. Storebaelt suspension bridge-vortex shedding excitation and mitigation by guide vanes. *J. Wind Eng. Ind. Aerodyn.* **2000**, *88*, 283–296. [CrossRef]
2. Fujino, Y.; Yoshitaka, Y. Wind-induced vibration and control of Trans-Tokyo Bay Crossing Bridge. *J. Struct. Eng.* **2002**, *128*, 1012–1025. [CrossRef]
3. Yang, Y.; Zhou, R.; Ge, Y.; Zhang, L. Experimental studies on VIV performance and countermeasures for twin-box girder bridges with various slot width ratios. *J. Fluids Struct.* **2016**, *66*, 476–489. [CrossRef]
4. Zhang, H.F.; Xin, D.B.; Zhan, J.; Wang, R.; Zhou, L. Vortex-induced vibration control of a streamline box girder using the wake perturbation of horizontal axis micro-wind turbines. *J. Fluids Struct.* **2022**, *108*, 103444. [CrossRef]
5. Boberg., M.; Feltrin, G.; Martinoli, A. A novel bridge section model endowed with actively controlled flap arrays mitigating wind impact// IEEE International Conference on Robotics and Automation. In Proceedings of the 2015 IEEE International Conference on Robotics and Automation, Seattle, WA, USA, 26–30 May 2015; IEEE: Piscataway, NJ, USA, 2015; pp. 1837–1842.
6. Chen, W.L.; Chen, G.B.; Xu, F.; Huang, Y.W.; Gao, D.L.; Li, H. Suppression of vortex-induced vibration of a circular cylinder by a passive-jet flow control. *J. Wind Eng. Ind. Aerodyn.* **2020**, *199*, 104119. [CrossRef]
7. Larsen, A.; Svensson, E.; Andersen, H. Design aspects of tuned mass dampers for the Great Belt East Bridge approach spans. *J. Wind Eng. Ind. Aerodyn.* **1995**, *54–55*, 413–426. [CrossRef]
8. He, X.H.; Li, H.; Wang, H.F.; Fang, D.X.; Liu, M.T. Effects of geometrical parameters on the aerodynamic characteristics of a streamlined flat box girder. *J. Wind Eng. Ind. Aerod.* **2017**, *170*, 56–67. [CrossRef]
9. Zhang, L.Q.; Chen, G.B.; Chen, W.L.; Gao, D.L. Separation Control on a Bridge Box Girder Using a Bypass Passive Jet Flow. *Appl. Sci.* **2017**, *7*, 501. [CrossRef]
10. Zhan, J.; Xin, D.B.; Ou, J.P.; Liu, Z.W. Experimental study on suppressing vortex-induced vibration of a long-span bridge by installing the wavy railings. *J. Wind Eng. Ind. Aerod.* **2020**, *202*, 104205. [CrossRef]
11. Nagao, F.; Utsunomiya, H.; Yoshioka, E.; Ikeuchi, A.; Kobayashi, H. Effects of handrails on separated shear flow and vortex-induced oscillation. *J. Wind Eng. Ind. Aerodyn* **1997**, *69–71*, 819–827. [CrossRef]
12. Xin, D.B.; Zhan, J.; Zhang, H.F.; Ou, J.P. Control of Vortex-Induced Vibration of a Long-Span Bridge by Inclined Railings. *J. Bridge. Eng.* **2021**, *26*, 04021093. [CrossRef]
13. Battista, R.C.; Pfeil, M.S. Reduction of vortex-induced oscillations of Rio–Niterói bridge by dynamic control devices. *J. Wind Eng. Ind. Aerodyn.* **2000**, *84*, 273–288. [CrossRef]
14. Hu, C.X.; Zhao, L.; Ge, Y.J. Mechanism of suppression of vortex-induced vibrations of a streamlined closed-box girder using additional small-scale components. *J. Wind Eng. Ind. Aerody.* **2019**, *189*, 314–331. [CrossRef]
15. Li, M.; Sun, Y.G.; Jing, H.M.; Li, M.S. Vortex-induced vibration optimization of a wide streamline box girder by wind tunnel test. *J. Civil. Eng.* **2018**, *22*, 5143–5153. [CrossRef]
16. Dai, J.; Xu, Z.D.; Gai, P.P. Parameter determination of the tuned mass damper mitigating the vortex-induced vibration in bridges. *Eng. Struct* **2020**, *221*, 111084. [CrossRef]
17. El-Gammal, M.; King Hangan, H.P. Control of vortex shedding-induced effects in a sectional bridge model by span-wise perturbation method. *J. Wind Eng. Ind. Aerodyn.* **2007**, *95*, 663–678. [CrossRef]
18. Chen, G.B.; Zhang, L.Q.; Chen, W.L.; Gao, D.L.; Yang, W.H.; Li, H. Self-suction-and-jet control in flow regime and unsteady force for a single box girder. *J. Bridge Eng.* **2019**, *24*, 04019072. [CrossRef]
19. Zhang, H.F.; Xin, D.B.; Ou, J.P. Wake control of vortex shedding based on span-wise suction of a bridge section model using Delayed Detached Eddy Simulation. *J. Wind Eng. Ind. Aerodyn.* **2016**, *155*, 100–114. [CrossRef]
20. Chen, G.B.; Chen, W.L.; Gao, D.L.; Yang, Z.F. Active control of flow structure and unsteady aerodynamic force of box girder with leading-edge suction and trailing-edge jet. *Exp. Therm. Fluid Sci.* **2021**, *120*, 110244. [CrossRef]
21. Amitay, M.; Honohan, A.; Trautman, M.; Glezer, A. Modification of the Aerodynamic Characteristics of Bluff Bodies Using Fluidic Actuators. In Proceedings of the 28th Fluid Dynamics Conference, Snowmass Village, CO, USA, 29 June–2 July 1997; AIAA Paper No. 97-2004. p. 2004.
22. Tensi, J.; Boué, I.; Paillé, F.; Dury, G. Modification of the wake behind a circular cylinder by using synthetic jets. *J. Vis.-Jpn.* **2002**, *5*, 37–44. [CrossRef]
23. Irwin, H.P.A.H.; Cooper, K.R.; Girard, R. Correction of distortion effects caused by tubing systems in measurements of fluctuating pressures. *J. Wind Eng. Ind. Aerodyn.* **1979**, *5*, 93–107. [CrossRef]
24. Barlow, B.; Rae, H.; Pope, A. *Low-Speed Wind Tunnel Testing*, 3rd ed.; Wiley: New York, NY, USA, 1999; pp. 330–375.
25. Samimy, M.; Lele, S.K. Motion of particles with inertia in a compressible free shear layer. *Phys. Fluids* **1991**, *3*, 1915–1923. [CrossRef]
26. Chen, W.L.; Gao, D.L.; Yuan, W.Y.; Li, H.; Hu, H. Passive jet control of flow around a circular cylinder. *Exp. Fluids.* **2015**, *56*, 201. [CrossRef]
27. Gao, D.L.; Chen, G.B.; Chen, W.L.; Huang, Y.W.; Li, H. Effects of steady wake-jets on subcritical cylinder flow. *Exp. Therm. Fluid Sci.* **2019**, *102*, 575–588. [CrossRef]

28. Taylor, Z.J.; Gurka, R.; Kopp, G.A. Geometric effects on shedding frequency for bridge sections. In Proceedings of the 11th Americas conference on wind engineering 2009, San Juan, Puerto Rico, 22–26 June 2009.
29. Frandsen, J.B. Comparison of numerical prediction and full-scale measurements of vortex induced oscillations. In Proceedings of the 4th International Colloquium on Bluff Body Aerodynamics and Applications, Ruhu-University of Bochum, Bochum, Germany, 11–14 September 2000.
30. Apelt, C.J.; West, G.S.; Szewczyk, A.A. The effects of wake splitter plates on the flow past a circular cylinder in the range $10^4 \leq$ Re $\leq 5 \times 10^4$. *J. Fluid Mech.* **1993**, *61*, 187–198. [CrossRef]
31. Chen, W.L.; Li, H.; Hu, H. An experimental study on a suction flow control method to reduce the unsteadiness of the wind loads acting on a circular cylinder. *Exp. Fluids* **2014**, *55*, 1707. [CrossRef]
32. Lim, H.C.; Lee, S.J. PIV measurements of near wake behind a U-grooved cylinder. *J. Fluids Struct.* **2003**, *18*, 119–130. [CrossRef]
33. Oruç, V. Passive control of flow structures around a circular cylinder by using screen. *J. Fluids Struct.* **2012**, *33*, 229–242. [CrossRef]
34. Benard, N.; Balcon, N.; Touchard, G.; Moreau, E. Control of diffuser jet flow: Turbulent kinetic energy and jet spreading enhancements assisted by a non-thermal plasma discharge. *Exp. Fluids* **2008**, *45*, 333–355. [CrossRef]
35. Lumley, J.L. The structure of inhomogeneous turbulent flow. In *Atmospheric Turbulence and Radio Wave Propagation*; Yaglom, A.M., Tatarski, V.I., Eds.; Nauka: Moscow, Russia, 1967; pp. 166–178.
36. Sirovich, L. Turbulence and the dynamics of coherent structures. Part I: Coherent structures. *Q. Appl. Maths.* **1987**, *45*, 561–571. [CrossRef]
37. Feng, L.H.; Wang, J.J.; Pan, C. Proper orthogonal decomposition analysis of vortex dynamics of a circular cylinder under synthetic jet control. *Phys. Fluids* **2011**, *23*, 526. [CrossRef]
38. Konstantinidis, E.; Balabani, S.; Yianneskis, M. Bimodal vortex shedding in a perturbed cylinder wake. *Phys. Fluids* **2007**, *19*, 701. [CrossRef]
39. Triantafyllou, G.S.; Triantafyllou, M.S.; Chryssostomidis, C. On the formation of vortex streets behind stationary cylinders. *J. Fluid Mech.* **1986**, *170*, 461–477. [CrossRef]
40. Orszag, S.A. Accurate solution of the Orr-Sommerfeld stability equation. *J. Fluid Mech.* **1971**, *50*, 689–703. [CrossRef]
41. Gao, D.L.; Chen, W.L.; Li, H.; Hu, H. Flow around a circular cylinder with slit. *Exp. Therm.* **2017**, *82*, 287–301. [CrossRef]

applied sciences

MDPI

Article

Aerodynamic Shape Optimization of an Arc-Plate-Shaped Bluff Body via Surrogate Modeling for Wind Energy Harvesting

Tianyi Shi [1], Gang Hu [2] and Lianghao Zou [1,*]

1 Engineering Research Center of Urban Disasters Prevention and Fire Rescue Technology of Hubei Province, School of Civil Engineering, Wuhan University, Wuhan 430072, China; tyshiwhu@163.com
2 School of Civil and Environmental Engineering, Harbin Institute of Technology, Shenzhen 518055, China; hugang@hit.edu.cn
* Correspondence: lhzou@whu.edu.cn

Abstract: Galloping-based piezoelectric wind energy harvesters (WEHs) are being used to supply renewable electricity for self-powered devices. This paper investigates the performance of a galloping-based piezoelectric WEH, with different arc-plate-shaped bluff bodies to improve harvesting efficiency. The Latin hypercube sampling method was employed to design the experiment. After conducting a series of wind tunnel tests, a Kriging surrogate model was then established, with high accuracy. The results show that the wind energy harvester with an arc angle 0.40π and tail length $1.26D$ generated the maximum power. The output power of the proposed WEH was doubled by optimizing the aerodynamic shape of the bluff body. The reasons for the improvement are discussed in detail. The force measurement results indicated that a large value of the transverse force coefficient means a large galloping response of the WEH. The aerodynamic optimization of this study can be applied to improve the performance of galloping-based wind energy harvesters.

Keywords: aerodynamic shape optimization; surrogate model; wind energy harvester; wind tunnel test; galloping

Citation: Shi, T.; Hu, G.; Zou, L. Aerodynamic Shape Optimization of an Arc-Plate-Shaped Bluff Body via Surrogate Modeling for Wind Energy Harvesting. *Appl. Sci.* **2022**, *12*, 3965. https://doi.org/10.3390/app12083965

Academic Editor: Roberto Camussi

Received: 21 February 2022
Accepted: 12 April 2022
Published: 14 April 2022

1. Introduction

Piezoelectric energy harvesters have attracted substantial attention in the last decades. Energy harvesters can generate sustainable power and, hence, are promising as an alternative to batteries, which have a limited life span, while requiring costly maintenance [1,2]. Piezoelectric energy harvesters, therefore, can be an excellent power source for self-powered devices, such as MEMS (microelectromechanical systems) and WSNs (wireless sensor networks). Among the different vibration-based energy resources, wind flow has been suggested as one of the most reliable energy resources, since the wind-induced vibrations of bluff bodies can have considerable amplitudes [3,4]. Consequently, piezoelectric energy harvesters based on wind-induced vibrations of bluff bodies can produce considerable electric energy. Wind-induced vibrations commonly utilized in piezoelectric wind energy harvesters (WEHs) include vortex-induced vibration (VIV) [5,6], galloping [7–10], flutter [11–14], and mixed aerodynamic phenomena [15–17]. Several comprehensive reviews on harnessing energy from wind flow have been published recently [18–24].

Various mathematical methods have been employed to promote harvesting performance [24]. For instance, Abdelkefi et al. [25] developed a coupled model of galloping-based WEH. Their results showed that the electrical load resistance and the Reynolds number play an essential role in determining the level of the harvested power and the onset wind velocity of galloping. Zhao et al. [26] compared the lumped variable model with the distributed variable model. The effects of load resistance, wind exposure area of the bluff body, mass of the bluff body, and length of the piezoelectric sheets on the power output were investigated. Tan et al. [27] solved electromechanical coupled distributed variable equations and theoretically examined the effects of electrical damping and mechanical damping on the performance of the WEH. The parameters were optimized based on mathematical methods for improving the performance of a WEH.

The mathematical models indicate that the geometric shape of the bluff body is one of the critical parameters determining the wind force on the WEH and governing the performance of the WEH [28,29]. Yang et al. [30] compared the effects of bluff bodies with different sections on the performance of a galloping-based WEH. The performance of square, rectangular, triangular, and D-shaped galloping-based WEHs was evaluated through experiments and analytical models. The test results showed that the square section had the lowest onset wind velocity of 2.5 m/s and the highest wind energy power of 8.4 mW. Kluger et al. [31] evaluated the effects of geometric shape and size of the tip body of a transversely galloping-based WEH, experimentally and theoretically. The results showed that the optimal combinations of lateral aerodynamic force coefficients minimized the critical wind velocity for instability and increased the amplitude. Abdelkefi et al. [32] investigated the effects of section geometry, resistance, and wind speed on an energy harvester. The results showed that the section geometry greatly influenced the initial galloping wind velocity. However, the efficiencies of WEHs were limited by the typical bluff body.

Aerodynamic shape optimization is an effective and simple method to improve the efficiency of a WEH. Hu, Tse, and Kwok [7] fitted fins to various corners of a square prism to form the kinetic energy source of a galloping-based piezoelectric energy harvester. They found that attaching fins to the leading corners improved the efficiency of the energy harvester by up to 2.5 times. Hu et al. [33] and Hu et al. [34] investigated the WEH with two small-size rod-shaped attachments on the main circular cylinder. The harvester harnessed the wind energy beyond the defined onset wind velocity and could still work efficiently over the range of lock-in wind velocity. Song et al. [35] examined the effect of a splitter plate placed in the near wake of a circular cylinder on the performance of a piezoelectric WEH. The results showed that the harvester was able to maintain significant amplitude vibrations, even beyond the lock-in wind speed range, and, hence, the range of effective working wind speed was expanded. Alhadidi et al. [36] augmented the rear face of the square prism with Y-shaped fins of different lengths and fork angles. This modification reduced the rise time of the harvester by as much as 75% compared to the finless square prism.

The above aerodynamic shapes were optimized based on the typical cross-section. The bluff bodies were constructed according to research experience and galloping characteristics. A circular arc bluff body has been studied for applications such as a velocity threshold detector and passive stability control of an unmanned flying vehicle. A circular arc bluff body is also capable of producing galloping oscillation. Bot [37] reported and analyzed experimental results on the forces generated by a high-camber thin section with a curved plate, and made measurements of the flow field using particle image velocimetry. Variations of lift were related to changes in the topology of the flow field around the section. These results explained the flow around a high-camber thin section and provided a detailed benchmark database in a simple geometry for model validations. Bot et al. [38] tested arc bodies with different sections and analyzed the force crisis. Souppez et al. [39] tested three geometrically similar circular arcs over a range of Reynolds numbers, from 53,530 to 218,000. The force crisis and boundary layer were analyzed by force measurement and particle image velocimetry, similarly to the circular cylinder. Tucker Harvey et al. [40] investigated a curved arc bluff body WEH experimentally and theoretically. Based on the particle image velocimetry, the separation flow attached to the rear surface of the blade at the initial position, and then the flow pattern enhanced the galloping performance.

A circular arc bluff body is a blade-shaped bluff body with great energy harvesting performance. WEHs with a blade-shaped bluff body can achieve a much higher energy output than those with a square prism. Liu et al. [41] proposed and designed a three-blade bluff body for wind energy harvesting. Both numerical simulations and experiments confirmed that the three-blade structure could achieve a higher energy output than a square prism. Liu et al. [42] presented a fork-shaped bluff body to improve the harvesting efficiency at low wind speed. The results demonstrated that a piezoelectric WEH with a fork-shaped structure could generate much higher output voltages than with the traditional bluff bodies, such as triangular prism and square prism.

From the above literature review, it is noted that the energy harvesting efficiency from the fluid flow is improved through the shape of the bluff body. For the shape of the bluff body, most research work focus on experimental trails, to compare different cross sections and, thus, enhance the energy harvesting. A rare work designed the shape of the bluff body from the perspective of fluid dynamics.

In the present study, an arc-plate-shaped bluff body is designed to improve the performance of a WEH. From a fluid dynamics point of view, the vortex reattachment decreases the lift force of the bluff body with a square cross-section. An arc-plate-shaped bluff body avoids the vortex reattachment and, thus, enhances the vibration. WEHs with arc-plate-shaped bluff bodies were tested in a wind tunnel. The Kriging surrogate model was established, accurately representing the mapping between the output power and the aerodynamic shape variables. The Kriging surrogate model is effective in shape optimization for discovering the change law and determining the optimal solution. The impacts of the shape modification on the efficiency of the WEH were evaluated. The transverse force coefficients of the arc-plate-shaped bluff body were acquired via force measurement tests in a wind tunnel. Based on the transverse force coefficients and quasi-steady theory, the observations in the wind energy harvesting tests were interpreted.

2. Characteristics and Optimization Objectives of a Galloping Bluff Body

2.1. Characteristics of a Galloping Bluff Body

The aerodynamic shape of a galloping bluff body plays a dominant role in the efficiency of converting wind energy into vibration energy. The efficiency of energy harvesting can be improved significantly by optimizing the aerodynamic shape of the bluff body. The bluff body cross-sections of early galloping wind energy harvesters had typical bluff body sections, such as square, D-shape, or triangle [32]. Based on these typical sections, the cross-sections were modified for improving the efficiency of WEHs. Figure 1 shows cross-sections of a few optimized galloping bluff bodies.

In the process of the aerodynamic modification of the galloping bluff bodies, one optimization method is to add fins near the leading edge. The fins form sharp protrusions, resulting in stronger flow separation. A square prism with leading-edge fins [7] is shown in Figure 1a. Circular cylinders with different shape attachments [43,44] are shown in Figure 1b. A circular cylinder with roughness strips [45] is shown in Figure 1c.

Another optimization method is to add a plate near the trailing edge. Adding a plate increases the depth of the bluff body, resulting in increasing cross-wind force. A square prism with a Y-shaped plate near the trailing edge [36] is shown in Figure 1d. A circular cylinder with a splitter plate [35] is shown in Figure 1e.

In summary, the optimized aerodynamic shape of a galloping bluff body has the characteristics of a wide windward surface with sharp protrusions, and overall lengthening with a plate at the tail. As shown in Figure 1f, the blade bluff body has the characteristics of the optimized shape, and, hence, a blade-shaped bluff body has excellent galloping performance [40,46].

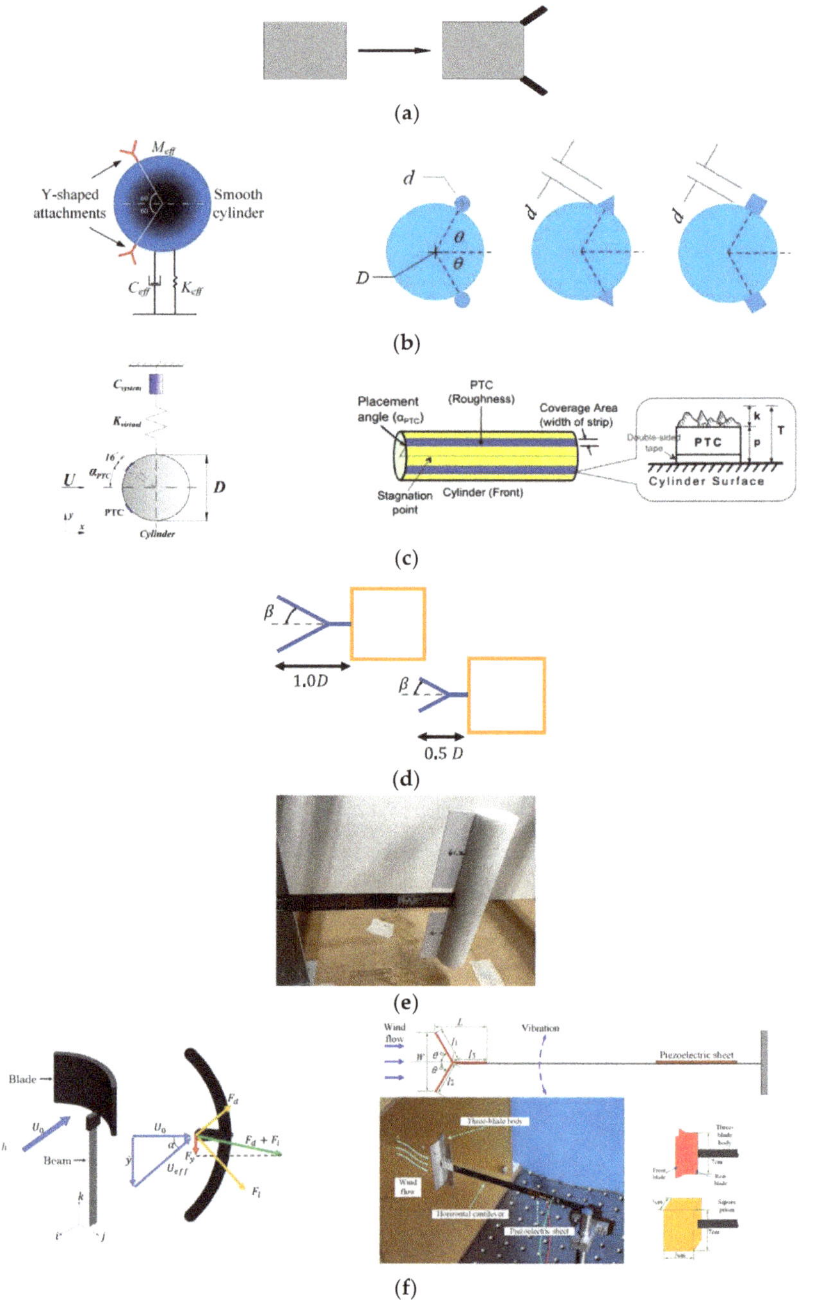

Figure 1. The aerodynamic shapes of optimized galloping bluff bodies, such as (**a**) square prism with leading-edge fins; (**b**) circular cylinder with different shape attachments; (**c**) circular cylinder with roughness strips; (**d**) square prism with Y-shaped fins near the trailing edge; (**e**) circular cylinder with a splitter plate; (**f**) blade bluff bodies.

In order to systematically investigate the performance of the blade bluff body, an arc-plate bluff body, consisting of an arc and a plate at the tail, is proposed, as shown in Figure 2. The windward width D of the different bluff bodies is identical. The geometry of the arc-plate-shaped bluff body is described by two variables, a_1 and a_2, ranging from 0 to 1. The arc angle $\beta = 2\pi a_1$, and the tail plate length $L_t = 2Da_2$. Changing the two variables can create a variety of different shapes. For instance, when the arc angle β is 2π and the plate length L_t is 0, the bluff body is a circular cylinder; when the arc angle β is 2π and the plate length L_t is not 0, the bluff body is a circular cylinder with a plate, following the aerodynamic shape studied by Song, Hu, Tse, Li, and Kwok [35]. When the angle β is $\pi/2$ and the plate length L_t is 0, the bluff body follows the aerodynamic shape studied by Tucker Harvey, Khovanov, and Denissenko [40]. Therefore, the proposed arc-plate bluff body has a good ability to create a variety of aerodynamic shapes using two variables.

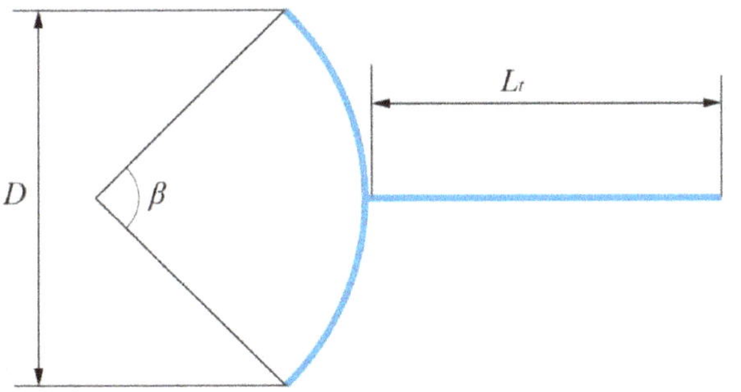

Figure 2. Arc-plate-shaped bluff body.

2.2. Optimization Objectives of a Galloping Bluff Body

The purpose of the shape optimization is to modify the design variables of the shape, to attain the optimal shape with the best potential performance, while satisfying certain constraints. After describing the design variables in terms of a set of n variables collected in the vector of the design variables x, the general optimization problem can be expressed as follows:

$$\begin{array}{ll} \max & y(x) = [f_1(x), f_2(x), \cdots, f_p(x)]^{\mathrm{T}} \\ \text{w.r.t.} & x_l \leq x \leq x_u \\ \text{s.t.} & \begin{cases} h_i(x) = 0 \; i = 1, 2, \cdots, n_h \\ g_j(x) \leq 0 \; j = 1, 2, \cdots, n_g \end{cases} \end{array} \quad (1)$$

where $y(x)$ is the p purpose objective function. x_l and x_u are design variables' upper and lower limits, respectively. $h_i(x)$ and $g_j(x)$ are equality and inequality constraints, respectively. n_h and n_g are the number of constraints, respectively.

The mathematical expression for the aerodynamic shape optimization of the arc-plate bluff body is described as

$$\begin{array}{ll} \text{find} & x = (a_1, a_2) \\ \max & P_e(x) \\ \text{w.r.t.} & x_l \leq x \leq x_u \\ \text{s.t.} & h_1(x) = D \end{array} \quad (2)$$

where $P(x)$is the output energy of the galloping-based WEH with the shape of x.

It is well acknowledged that the output power varies with wind velocity. As a result, the performance of the WEH is assessed by an average output energy P_e weighted by the probability density distribution function of wind speed as

$$P_e = \int_{U_l}^{U_u} P(U) f_p(U) dU \quad (3)$$

where $P(U)$ is the output power for a specific wind velocity U, f_p (U) is the probability density distribution function of wind velocity, and U_U and U_L are the upper and lower limits of the wind speed, respectively. In this paper, the probability density distribution function of wind velocity f_p is assumed as an uniform distribution in the tested wind velocity range, i.e.,

$$f_p(U) = \frac{1}{U_U - U_L}, U_L \leq U \leq U_U \tag{4}$$

3. Surrogate Modeling

3.1. The Kriging Surrogate Modeling Method

Shape optimization is necessary and important in engineering, such as vehicle shapes, airfoil shapes, and building shapes. Surrogate-based optimization has been shown to be a practical approach in aeronautics and astronautics engineering [47,48], civil engineering [49,50], and ocean engineering [51,52]. The surrogate model can replace many expensive objective evaluations with an approximation model. Wang and Shan [53] provided an overview of surrogate models, focusing on solving optimization problems such as global optimization, multi-objective optimization, and probabilistic design optimization. Forrester and Keane [54] discussed details of surrogate modeling methodology, focusing on sampling, surrogate model building, and validation. Westermann and Evins [55] explained the use of surrogate modeling for building design regarding applications in the conceptual design stage, sensitivity and uncertainty analysis, and building design optimization.

Many surrogate models have been adopted in the field of optimization, such as polynomial regression, radial basis functions, neural networks, and support vector regression [56]. In the present work, the widely used Kriging model was adopted [49].

For a stationary random process, the statistical prediction of the unknown function y at an untried x is given by a linear combination of the function values at the original samples and their gradients at observed points, as follows:

$$\hat{y}(\boldsymbol{x}) = \sum_{i=1}^{n} w^{(i)} y^{(i)} \tag{5}$$

where $\hat{y}(x)$ is the predicted value at an untried site $\boldsymbol{x}$, $w^{(i)}$ is the weight coefficient for the function value $y^{(i)}$. To determine the weight coefficient, the output of a deterministic experiment is treated as a realization of a stochastic process, i.e.,

$$Y(\boldsymbol{x}) = \beta_0 + Z(\boldsymbol{x}) \tag{6}$$

where β_0 is an unknown constant mean value and the stationary random processes $Z(x)$ having zero mean and a covariance Cov,

$$\text{Cov}[Z(\boldsymbol{x}), Z(\boldsymbol{x}')] = \sigma^2 R(\boldsymbol{x}, \boldsymbol{x}') \tag{7}$$

where σ^2 is the process variance of $Z(x)$, R is the spatial correlation function, which depends only on the spatial distance between two sites. $\hat{y}(x)$ is treated as random, and we try to minimize its mean squared error (MSE)

$$\text{MSE}[\hat{y}(\boldsymbol{x})] = E\left[\left(\boldsymbol{w}^{\mathrm{T}} \boldsymbol{Y}_S - Y(\boldsymbol{x})\right)^2\right] \tag{8}$$

subject to the unbiasedness constraint

$$E\left[\sum_{i=1}^{n} w^{(i)} Y\left(\boldsymbol{x}^{(i)}\right)\right] = E[Y(\boldsymbol{x})] \tag{9}$$

where $\boldsymbol{Ys}$ is the response of the sample points. Solving this constrained minimization problem, $w^{(i)}$ can be found by solving the following linear equations in matrix form:

$$\begin{bmatrix} \boldsymbol{R} & \boldsymbol{F} \\ \boldsymbol{F}^{\mathrm{T}} & 0 \end{bmatrix} \begin{bmatrix} \boldsymbol{w} \\ \tilde{\mu} \end{bmatrix} = \begin{bmatrix} \boldsymbol{r} \\ 1 \end{bmatrix} \tag{10}$$

where

$$\boldsymbol{F} = \begin{bmatrix} 1 & 1 & \cdots & 1 \end{bmatrix}^{\mathrm{T}} \in \mathbb{R}^n \tag{11}$$

$$\tilde{\mu} = \mu / \left(2\sigma^2\right) \tag{12}$$

$$\boldsymbol{R} = \begin{bmatrix} R\left(\boldsymbol{x}^{(1)}, \boldsymbol{x}^{(1)}\right) & \cdots & R\left(\boldsymbol{x}^{(1)}, \boldsymbol{x}^{(n)}\right) \\ \vdots & & \vdots \\ R\left(\boldsymbol{x}^{(n)}, \boldsymbol{x}^{(1)}\right) & \cdots & R\left(\boldsymbol{x}^{(n)}, \boldsymbol{x}^{(n)}\right) \end{bmatrix} \in \mathbb{R}^{n \times n} \tag{13}$$

$$\boldsymbol{r} = \begin{bmatrix} R\left(\boldsymbol{x}^{(1)}, \boldsymbol{x}\right) \\ \vdots \\ R\left(\boldsymbol{x}^{(n)}, \boldsymbol{x}\right) \end{bmatrix} \in \mathbb{R}^n \tag{14}$$

where $\boldsymbol{R}$ is the correlation matrix representing the correlation between the observed points, and $\boldsymbol{r}$ is the correlation vector representing the correlation between the untried point and the observed points. The kriging predictor can be written in the form

$$\hat{y}(\boldsymbol{x}) = \beta_0 + \boldsymbol{r}^{\mathrm{T}}(\boldsymbol{x})\boldsymbol{R}^{-1}(\boldsymbol{y}_S - \beta_0 \boldsymbol{F}) \tag{15}$$

where $\beta_0 = \left(\boldsymbol{F}^{\mathrm{T}}\boldsymbol{R}^{-1}\boldsymbol{F}\right)^{-1}\boldsymbol{F}^{\mathrm{T}}\boldsymbol{R}^{-1}\boldsymbol{y}_S$.

3.2. Design of Experiments

The purpose of conducting the design of experiments is to generate sampling points for the surrogate model. Since we have no prior knowledge about how the objective output functions vary over the input space, the principle of a sampling plan is to fill the design space with sampling points, as uniformly as possible. For this purpose, a space-filling strategy named Latin hypercube sampling (LHS) for uniform random sampling was employed. The LHS can avoid the over-concentration of the sampled points in a small range and make the sample points evenly distributed in the sampling area. For sampling n_s sample points, n_v design variables are evenly divided into n_{sl} intervals. The total sampling interval is divided into $(n_v)^{n_s}$ subintervals. When each sample point is projected into any dimension, there is only one sample point in each interval.

In the aerodynamic shape optimization of the arc-plate bluff body, 50 sample points were taken for the control variables a_1 and a_2 in the range from 0 to 1. As shown in Figure 3a, 50 samples for the modeling were generated in the two-dimensional input space. The shapes of bluff bodies are shown in Figure 3b, corresponding to all the sample points. Taking shape 2 as an example, a_1 = 0.02 and a_2 = 0.69, the angle of arc β is 0.04π, and the plate length L_t is 0.69D. The corresponding bluff body section is approximately T-shaped. For shape 47, where a_1 = 0.94 and a_2 = 0.10, arc angle β is 1.88π and the plate length L_t is 0.20D. The related bluff body section is roughly circular. For shape 49, where a_1 = 0.98 and a_2 = 0.82, arc angle β is 1.96π and the plate length L_t is 1.64D. The resultant bluff body section follows the aerodynamic shape studied by Song, Hu, Tse, Li, and Kwok [35]. It can be concluded that the arc-plate bluff body controlled by two variables can describe a variety of aerodynamic shapes using LHS.

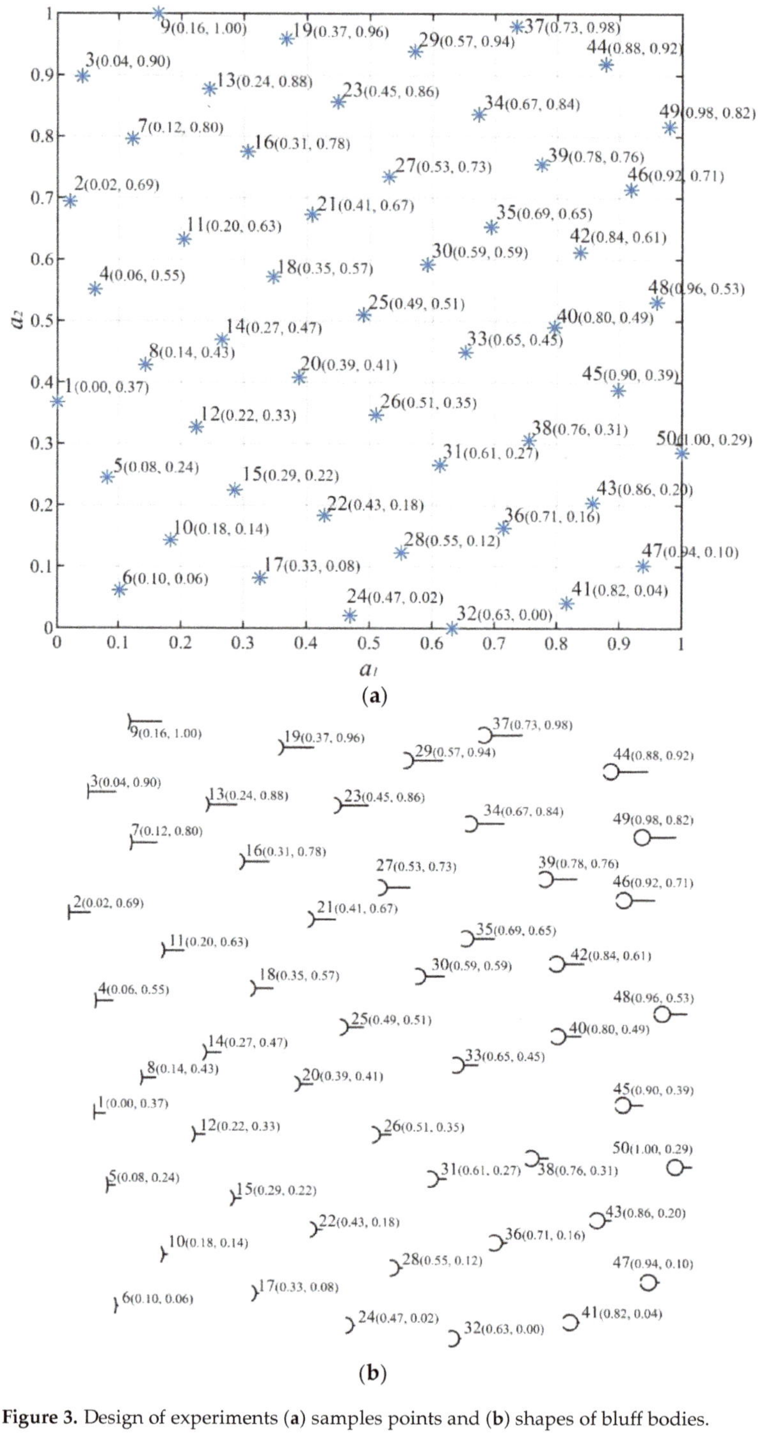

(**a**)

(**b**)

Figure 3. Design of experiments (**a**) samples points and (**b**) shapes of bluff bodies.

4. Wind Tunnel Test

4.1. Piezoelectric Wind Energy Harvesting Tests

The wind tunnel tests in this study were conducted in the open-circuit wind tunnel at Wuhan University, as shown in Figure 4. The wind tunnel has a 40 cm × 40 cm test cross-section and a low-turbulence flow, with a less than 1.5% turbulence intensity. The aluminum cantilever beam's length, width, and thickness were 12, 1.8, and 0.1 cm, respectively. The bluff body was produced with 3D printing technology, with a height of 10 cm and a width D of 5 cm. The piezoelectric layer (MFC-2814P2, Smart Material Corp, Sarasota, FL, USA) was connected to the electrical load resistances R. The voltage V across the load resistance was measured using a DAQ module (NI-9229, National Instruments, Austin, TX, USA). The wind velocity U of the incoming flow was measured using an anemometer (Testo 425, Testo SE & Co., Titisee-Neustadt, Germany).

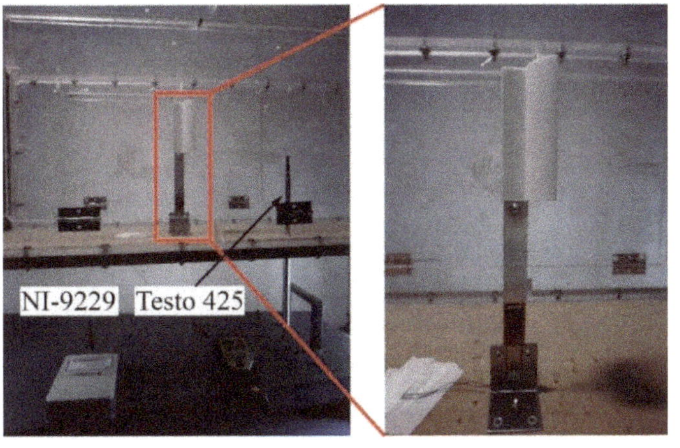

Figure 4. Configuration of piezoelectric wind energy harvesting test.

The natural frequency and damping ratio were 8.65 Hz and 1.2%, which was determined using a free vibration test. The effective mass and stiffness were 27.7 g and 81.8 N/m, which were lumped by a distributed model [57]. The tested wind speeds ranged from 0.5 m/s to 7 m/s, with an interval of 0.5 m/s. Accordingly, the Reynolds number ranged from 1.7×10^3 to 2.4×10^4.

4.2. Force Measurement Tests

A series of force measurement tests were performed on the arc-plate-shaped bluff body in the wind tunnel, to examine the global aerodynamic characteristics, as shown in Figure 5. The bluff body was mounted on a six-dimensional force balance (Gamma, ATI industrial automation, Apex, NC, USA), to measure wind forces on the body. A servo motor determined the rotation angle of the body, simulating different directions of oncoming wind flow.

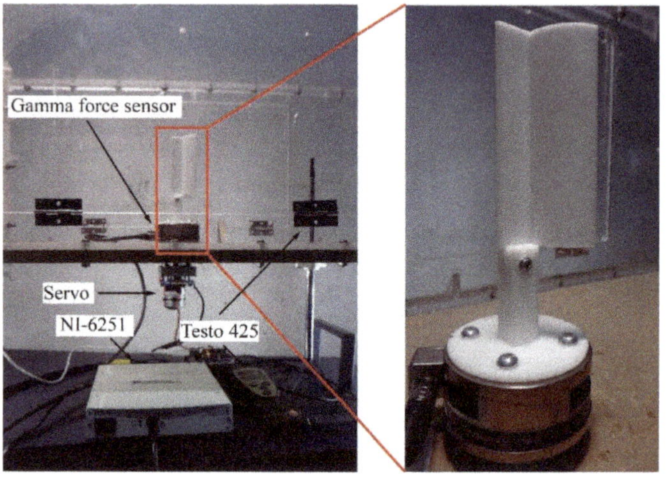

Figure 5. Configuration of force measurement tests.

4.3. Uncertainty Analyses

Experimental results always contain errors, so uncertainty should also be analyzed. The wind velocity U of the incoming flow was measured using an anemometer (Testo 425, Testo SE & Co., Titisee-Neustadt, Germany). The range of velocity measurement was from 0 to 20 m/s. The resolution of the anemometer was 0.01 m/s. The accuracy of the anemometer was ±(0.03 m/s + 5% of measured values). The bluff body was produced with a 3D printer (Form 3 L, Formlabs Corp, Somerville, MA, USA), with a resolution of 0.025 mm. The accuracy of the bluff body's width and length were ±1.2 mm and ±0.2 mm, respectively, because the bluff bodies were thin-walled structures, prone to temperature strain. The force was measured using a six-dimensional force balance (Gamma, ATI industrial automation, Apex, NC, USA). The resolution of the force balance was 0.00625 N. The accuracy of the force balance was ±1%. The resolution of the wind direction was 0.017°. The accuracy of the wind direction was ±0.036°. The resolution of the electric resistance was 1000 Ω, and the accuracy is 1%. The voltage across the load resistance was measured with a DAQ module (NI-9229, National Instruments, Austin, TX, USA). The resolution of the voltage DAQ module was 3.6×10^{-6} V. The accuracy of the voltage DAQ module was ±0.03% of the measured value. The accuracy and resolution of the measuring instruments and equipment are shown in Table 1.

Table 1. The accuracy and resolution of measuring instruments and equipment.

Variable	Resolution	Accuracy
Velocity	0.01 m/s	±(0.03 m/s + 5% of the measured value)
Width of bluff body	0.025 mm	±2.4%
Length of bluff body	0.025 mm	±0.4%
Force	0.00625 N	±1%
Wind direction	0.017°	±0.5%
Resistance	1000 Ω	±1%
Voltage	3.6×10^{-6} V	±0.03%

The average output power of the wind energy harvester was computed using the associated voltage as $P = V_{rms}^2/R$, where V_{rms} is the root mean square of the voltage. The uncertainty was assessed using the Taylor series method (TSM) [58]. The correlated random error terms were assumed to be zero. The standard uncertainty of average output power P is expressed as

$$s_P = \sqrt{4s_V^2 + s_R^2} \tag{16}$$

where S_V and S_R are standard uncertainties of voltage and resistance, respectively. The random uncertainty values for output power P is $\pm 1.0\%$.

The transverse force coefficient C_{Fy} is expressed as

$$C_{Fy} = \frac{F_y}{\frac{1}{2}\rho_a U^2 BL} \tag{17}$$

where ρ_a is the air density, B and L are the width and length of the bluff body, and U is wind velocity.

The errors in the values of the variables on the right-hand side of Equation (17) will cause errors in the experimental result C_{Fy}. The uncertainty was calculated using the Taylor series method (TSM) [58]. It was assumed that only U, B, L, and F_y contributed to the random uncertainty. These correlated random error terms were assumed to be zero. The standard uncertainty of force coefficient is expressed as

$$s_{C_{Fy}} = \sqrt{4s_u^2 + s_{F_y}^2 + s_B^2 + s_L^2} \tag{18}$$

where S_u, S_{Fy}, S_B, and S_L are standard uncertainties of the wind velocity, force, width, and length of the bluff body. The random uncertainty values for C_{Fy} using the TSM are shown in Table 2. The standard uncertainty of force coefficient was caused by the error of the anemometer. The accuracy of the anemometer was $\pm$(0.03 m/s + 5% of measured value), and the force was proportional to the square of the wind velocity.

Table 2. Random uncertainty values for C_{Fy} using the TSM.

U (m/s)	S_{CFy} (%)
1	16.2
2	13.3
3	12.3
4	11.8
5	11.5
6	11.3
7	11.2

5. Results and Discussion

5.1. Surrogate Model

Based on the wind tunnel test results, the Kriging model was constructed for the average output power, as shown in the three-dimensional and two-dimensional cartesian coordinates in Figure 6. The response surface fit well with the sampling points. The variation of the average output power with the aerodynamic shape variables can be observed intuitively. For the two-variable optimization problem, 50 sample points were enough to establish a surrogate model with high accuracy. For instance, Bernardini, Spence, Wei, and Kareem [49] constructed an ordinary Kriging using 90 samples. Ding and Kareem [50] constructed a multi-fidelity Kriging model with 50 samples for a two-variable aerodynamic shape optimization of civil structures.

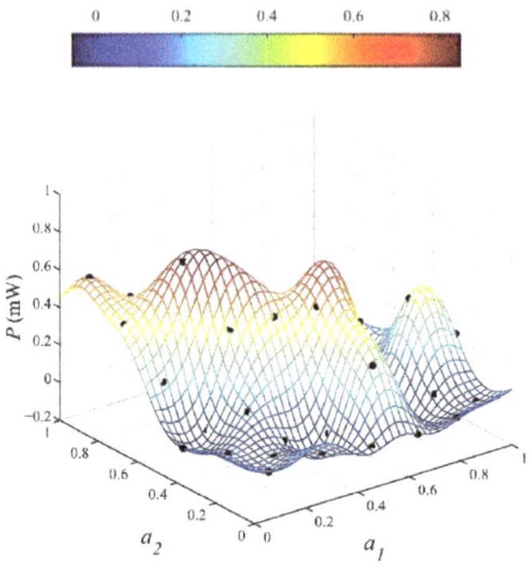

(a)

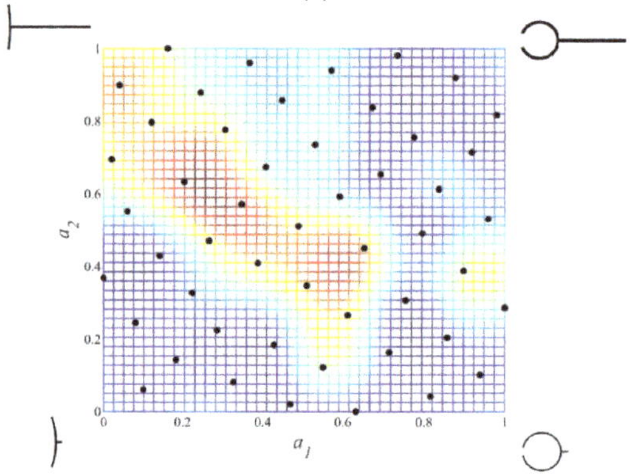

(b)

Figure 6. Kriging model for the average output power in (**a**) three-dimensional and (**b**) two-dimensional cartesian.

According to the experimental results, the average output power of the WEH with shape 11 was maximum for the variables $a_1 = 0.20$ and $a_2 = 0.63$. The arc angle β and tail length L_t were 0.40π and $1.26D$, respectively. The WEH with shape 18 had an excellent performance when $a_1 = 0.35$ and $a_2 = 0.57$. The arc angle β and tail length L_t were 0.70π and $1.14D$, respectively. The surrogate model had a peak region for the average output power P_e, as in the red colored area with an inclination of 45° shown in Figure 6. The WEHs corresponding to the peak region output have abundant power because the bluff body of the WEHs is prone to gallop. For the other areas in Figure 6, the output power declines rapidly and decays to zero. Based on the surrogate model, the impacts of the shape modification on the efficiency of the WEH are further evaluated in the subsequent sections.

5.2. Output Power

In order to investigate effect of the bluff body's shape on the output power, the variation of output power with wind velocity for WEHs with different arc angle bluff bodies is shown in Figure 7. The tail length of the bluff body is almost identical. The WEH starts to work at a wind velocity of 2 m/s. The output power increases with the increase of the wind velocity. The bluff body with shape 18 had the most significant output power. The max output power was 2 mW at a wind velocity of 7 m/s. For the bluff body with a large arc angle, such as shapes 30 and 42, the output power was far lower than that of shape 18, and the max output power was only 0.5 mW at 7 m/s. It can be concluded that the output power tends to decrease with the increase of the arc angle of the bluff body.

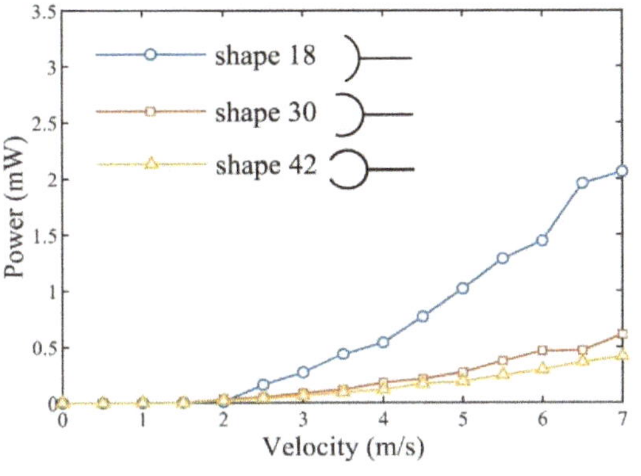

Figure 7. Variation of output power with wind velocity for WEHs with different arc angle bluff bodies.

The variation of output power with wind velocity for WEHs with different tail length bluff bodies is shown in Figure 8. As can be seen, the tail length of shape 18 was between those of shapes 17 and 19. However, the output power of the WEH with shape 18 was the maximum. The WEH with shape 17 almost ceased to work. The output power of the WEH with shape 19 was minimal for all wind velocities and was only 0.5 mW at a wind velocity of 7 m/s. Compared with the WEH with too long and too short a tail length, an optimal tail length can provide the maximal output power.

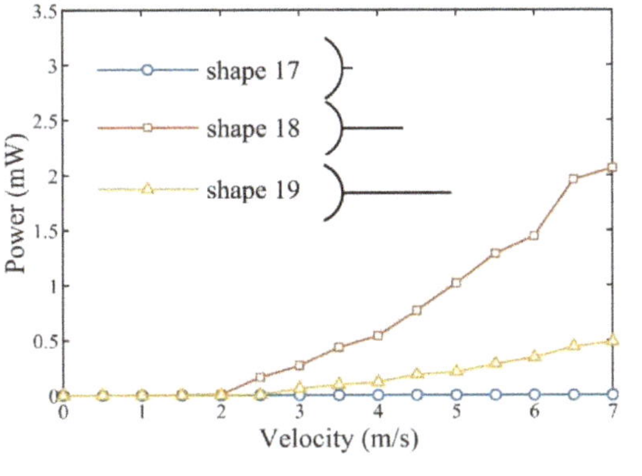

Figure 8. Variation of output power with wind velocity for WEHs with different tail lengths bluff bodies.

The variation of output power with wind velocity for WEHs with prism and arc-plate-shaped bluff bodies is shown in Figure 9. The aerodynamic shapes of the arc-plate-shaped bluff bodies are within the region of the Kriging surrogate model showing the maximum output power. The WEHs with the prism- and arc-plate-shaped bluff bodies have considerable energy output. The critical galloping wind speeds of bluff bodies with different sections were all around 2 m/s. When the wind speed was 7 m/s, the output power of the WEH with shape 11 and the prism bluff body was 2.3 mW and 1.1 mW, respectively. Compared with the WEH with a prism bluff body, the output power of the WEH with shape 11 doubled, indicating the significance of optimization of the aerodynamic shape of the bluff body section. The aerodynamic optimization of this study can, therefore, be directly applied in piezoelectric WEHs, to increase power generation significantly.

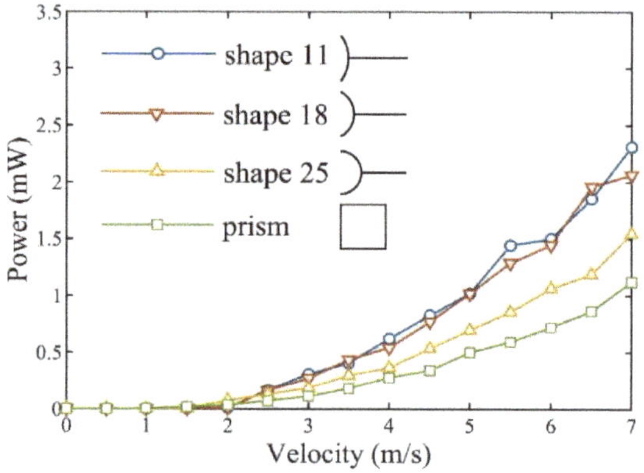

Figure 9. Variation of output power with wind velocity for WEHs with prism and arc-plate-shaped bluff bodies.

5.3. Force Coefficient

According to the quasi-steady theory by Parkinson [59,60], the drag force F_d and lift force F_l at each attack angle α in the process of oscillation are the same as the values measured at the corresponding steady attack angle α. The relative attack angle $\alpha = \arctan(\dot{y}/U)$ and forces were defined, as shown in Figure 9. The transverse force is the resultant of both drag force F_d and lift force F_l

$$F_y = F_L \cos \alpha - F_D \sin \alpha \tag{19}$$

Transverse force F_y was directly measured in the wind tunnel using the force balance. The transverse force coefficient C_{Fy} is expressed as

$$C_{Fy} = \frac{F_y}{\frac{1}{2}\rho_a U_{eff}^2 A} \tag{20}$$

where ρ_a is the air density, A is the projected area of the bluff body, and U_{eff} is effective wind velocity.

The differential equation governing the galloping oscillation is

$$m\ddot{y} + C\dot{y} + Ky = F_y = \frac{1}{2} C_{Fy} \rho_a U_{eff}^2 A \tag{21}$$

where m, C, and K are the mass, damping, and stiffness modified by an electromechanical decoupled model [61]. The net damping ratio of the system is

$$d = C\dot{y} - \frac{1}{2} C_{Fy} \rho_a U_{eff}^2 A \tag{22}$$

If $d > 0$, the system tends toward stability. When $\dot{y}$ and C_{Fy} are in the same direction, as shown in Figure 10, the aerodynamic force is negative. Negative aerodynamic damping is an essential prerequisite for the galloping instability. With the increase of the wind velocity, d decreases gradually and may be negative, and then the system is prone to instability. In terms of the well-known Den Hartog criterion [62,63], the necessary condition for the occurrence of the galloping instability is $\left.\frac{\partial C_{Fy}}{\partial \alpha}\right|_{\alpha=0^\circ} > 0$. In other words, a positive slope of the C_{Fy} versus α curve at $\alpha = 0^\circ$ is an essential prerequisite for galloping instability.

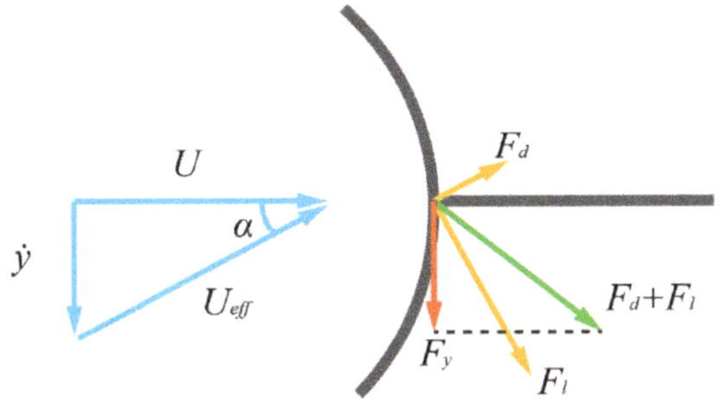

Figure 10. Definitions of wind attack angle and forces in the force measurement tests.

The variation of transverse force coefficient C_{Fy} with wind attack angle α for bluff bodies with different arc angles is shown in Figure 11. The slopes of the transverse force coefficient of all the three bluff bodies are positive when wind attack angle $\alpha = 0°$. As a result, the corresponding WEHs work at high wind velocity, as shown in Figure 7. With the increase of the wind attack angle, the transverse force coefficient of shape 18 is much more significant than those of shapes 30 and 42. This is because the transverse force provides the negative aerodynamic force, and then the output power of the WEH with the shape 18 is much larger than those with the shapes 30 and 42.

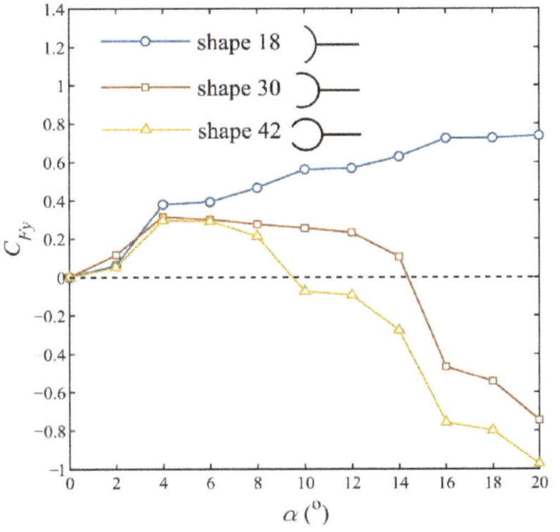

Figure 11. Variation of transverse force coefficient C_{Fy} with wind attack angle α for bluff bodies with different arc angles.

The variation of transverse force coefficient C_{Fy} with wind attack angle α for bluff bodies with different tail lengths is shown in Figure 12. The slope of the transverse force coefficient of shape 17 is positive at $\alpha = 0°$. In other words, shape 17 satisfies the essential prerequisite of galloping. The transverse force coefficient remains positive, but the value of the transverse force coefficient is relatively small, less than 0.1. The WEH with the shape 17 hardly works, because the bluff body cannot produce the aerodynamic damping force. The slope of the transverse force coefficient of shape 19 is positive at $\alpha = 0°$. The bluff body also satisfies the essential prerequisite of galloping. The transverse force coefficient of shape 19 increased at the initial wind attack angle and decreased with the wind attack angle $\alpha > 6°$. Although the WEH with the shape 19 can work, the performance was barely satisfactory.

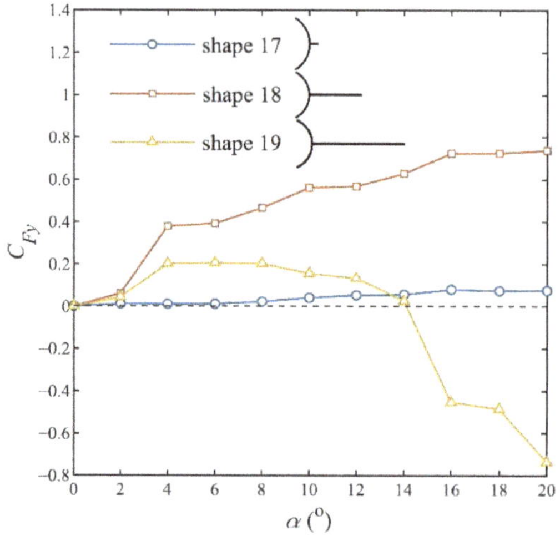

Figure 12. Variation of transverse force coefficient C_{Fy} with wind attack angle α for bluff bodies with different tail lengths.

The variation of transverse force coefficient C_{Fy} with wind attack angle α for bluff bodies with excellent vibration performance is shown in Figure 13, for different wind attack angles α. The selected bluff bodies are within the region of the surrogate model showing peak output energy. For the three bodies shown in Figure 13, the slopes of the transverse force coefficient C_{Fy} are all positive at $\alpha = 0$. The transverse force coefficients of the bluff bodies increase rapidly with α. The transverse force coefficient of shape 25 is less than those of the other two when $\alpha \geq 14°$, and hence the output power of the WEH with the shape 25 bluff body outputs was less than those of the other two.

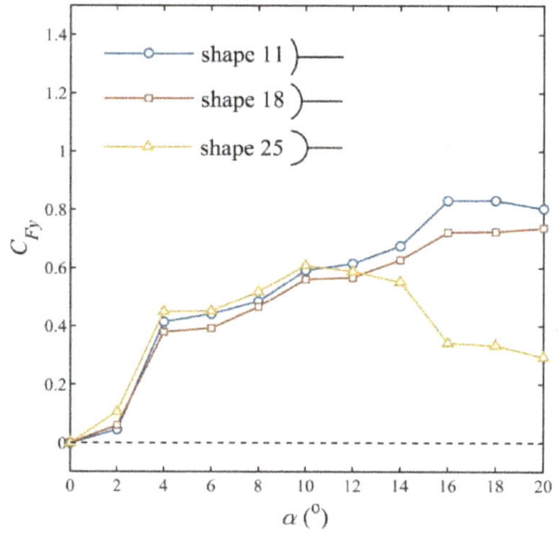

Figure 13. Variation of transverse force coefficient C_{Fy} with wind attack angle α for bluff bodies with excellent vibration performance.

Based on the above discussion, it can be concluded that the performance of WEHs is highly associated with the transverse force coefficient of the bluff bodies. A large value of the transverse force coefficient usually means a large galloping response. Meanwhile, the above results verify the quasi-steady theory and indicate that the static wind tunnel test was accurate and reliable.

5.4. Comparision and Discussion

The performance of a wind energy harvester is mainly affected by the shape and mass of the bluff body, the material characteristics and dimensions of the cantilever beam, the characteristics of the piezoelectric transducer, the conditions of the approaching flow, etc. The harvesting efficiencies of various typical galloping experimental results are summarized in Table 3. The harvesting efficiencies were calculated according to the method from Bernitsas et al. [64]. It can be found that the efficiency of the No.1 harvester was much higher than that of this paper, because the bluff body was made from lightweight materials, with a mass of 1.8 g. The bluff body in the present research was produced by 3D printing technology, with a mass of 25 g. Three-dimensional printing technology was adopted as it replicates the shape more delicately for precise shape optimization. The power of the No.1, 2, and that in this work exceed 1 mW, which is able to power the Internet of Things (IoT) ViPSN platform for a vibration-powered sensing and transmitting system [65]. The efficiencies of No.3, 4, and 5 were much lower than that of this paper, because the characteristics of the piezoelectric transducer were poor.

Table 3. Comparison of various galloping-based wind energy harvesters.

		Bluff Body			Power (mW)	Wind Velocity (m/s)	Efficiency (%)
No.	**Author**	**Shape**	**Width (cm)**	**Height (cm)**			
1	Zhao [66]	Square	2	10	1.25	5	0.77
2	Sirohi [67]	D-shape	3	23.5	1.14	4.7	0.24
3	Hu [33]	Cylinder with rods	4.8	24	0.053	5.5	0.004
4	Hu [7]	Square with fins	2.4	24	0.034	5	0.007
5	Song [35]	Cylinder with plate	4.8	24	0.014	5	0.002
6	Our work	Square	5	10	0.50	5	0.12
7	Our work	Arc-plate	5	10	1.01	5	0.25
8	Our work	Square	5	10	1.12	7	0.10
9	Our work	Arc-plate	5	10	2.31	7	0.21

The results of the harvesters with the square and arc-plate shaped bluff body were compared. The output power of the WEHs with a square and arc-plate-shaped bluff body was 0.50 mW and 1.01 mW at wind velocity of 5 m/s. Meanwhile, the efficiency was improved from 0.12% to 0.25%. For a wind velocity of 7 m/s, the output power was enhanced from 1.12 mW to 2.31 mW, and the efficiency was improved from 0.10% to 0.21%. The aerodynamic shape optimization doubled both the output power and efficiency

For the square cross-section, the vortex reattachment phenomenon often appears near the side faces, as shown in Figure 14a. This phenomenon causes up-side and down-side pressure fluctuations and, thus, reduces the total lift force of the bluff body [68]. Fortunately, the arc-plate-shaped bluff body avoids the vortex reattachment and, thus, guarantees the coordination of vortex shedding, as shown in Figure 14b. Therefore, the lift force of the arc-plate-shaped cross-section is larger than that of the square cross-section.

The above comparison demonstrates the superiority of the aerodynamic shape optimization of an arc-plate-shaped bluff body. This finding has great potential to be applied in energy harvesters based on galloping. In the past few years, significant efforts have been made to develop various types of piezoelectric energy harvesters based on wind-induced vibrations as the power source for wireless sensors. The work of this study can be directly applied to this type of piezoelectric wind energy harvester, to increase the power generation significantly.

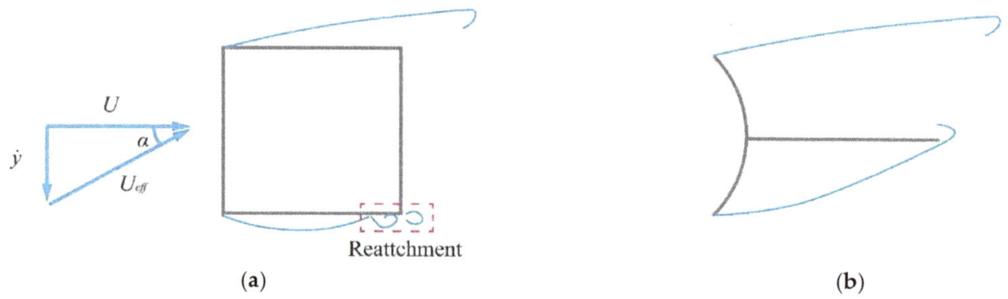

Figure 14. The schematic of flow separation for (**a**) square cross-section and (**b**) arc-plate-shaped cross-section.

Bernitsas's team [64] invented a device converting ocean/river current's hydrokinetic energy to electricity using a circular bluff body. Different-shaped bluff bodies were also investigated for energy harvesting, such as prism-, triangular-, and pentagon-shaped bluff bodies [69,70]. Although this converter was driven by ocean/river currents, the findings achieved with wind flow reported in the current study are potentially applicable.

6. Conclusions

In this paper, the aerodynamic shape of a arc-plate bluff body was optimized, based on the output energy of a galloping-based WEH. The Latin hypercube sampling method was employed for the design of the experiment. According to the wind tunnel test results, a Kriging surrogate model with a high accuracy was established. The variation of transverse force coefficients of the bluff body with wind attack angle was examined in detail. Experiments were employed to prove the superiority of the arc-plate-shaped bluff body for energy harvesting. The following conclusions were obtained.

(a) Through the Latin hypercube sampling method, the arc-plate shape determined by the two variables can describe a variety of aerodynamic shapes. The Kriging surrogate model established by the wind tunnel test showed a high degree of fit with the sampling points, which laid the foundation for the subsequent shape modification.

(b) The output power of the WEH was significantly increased by optimizing the aerodynamic shape. The WEH with the arc angle 0.40π and tail length $1.26D$ generated the maximum power.

(c) The output power of the WEH was related to the transverse force coefficient. A large transverse force coefficient usually means a large output power of a galloping-based WEH.

(d) Optimizing the aerodynamic shape of the bluff body section doubled the output power of the WEH, compared with the WEH with a prism bluff body.

Author Contributions: Conceptualization, G.H.; methodology, formal analysis, and investigation, T.S.; writing—original draft preparation, T.S.; writing—review and editing, L.Z.; supervision, L.Z.; project administration, L.Z.; funding acquisition, L.Z. All authors have read and agreed to the published version of the manuscript.

Funding: This study was funded by the National Natural Science Foundation of China, under project No. 51478369.

Institutional Review Board Statement: Not applicable.

Informed Consent Statement: Not applicable.

Data Availability Statement: Data sharing is not applicable.

Conflicts of Interest: The authors declare no conflict of interest.

References

1. Shaikh, F.K.; Zeadally, S. Energy harvesting in wireless sensor networks: A comprehensive review. *Renew. Sustain. Energy Rev.* **2016**, *55*, 1041–1054. [CrossRef]
2. Alaei, E.; Afrasiab, H.; Dardel, M. Analytical and numerical fluid–structure interaction study of a microscale piezoelectric wind energy harvester. *Wind Energy* **2020**, *23*, 1444–1460. [CrossRef]
3. Larsen, A. A generalized model for assessment of vortex-induced vibrations of flexible structures. *J. Wind Eng. Ind. Aerodyn.* **1995**, *57*, 281–294. [CrossRef]
4. Arunachalam, S.; Lakshmanan, N. Across-wind response of tall circular chimneys to vortex shedding. *J. Wind Eng. Ind. Aerodyn.* **2015**, *145*, 187–195. [CrossRef]
5. Abdelkefi, A.; Hajj, M.R.; Nayfeh, A.H. Phenomena and modeling of piezoelectric energy harvesting from freely oscillating cylinders. *Nonlinear Dyn.* **2012**, *70*, 1377–1388. [CrossRef]
6. Dai, H.; Abdelkefi, A.; Wang, L. Theoretical modeling and nonlinear analysis of piezoelectric energy harvesting from vortex-induced vibrations. *J. Intell. Mater. Syst. Struct.* **2014**, *25*, 1861–1874. [CrossRef]
7. Hu, G.; Tse, K.T.; Kwok, K.C.S. Enhanced performance of wind energy harvester by aerodynamic treatment of a square prism. *Appl. Phys. Lett.* **2016**, *108*, 123901. [CrossRef]
8. Hu, G.; Wang, J.; Su, Z.; Li, G.; Peng, H.; Kwok, K.C.S. Performance evaluation of twin piezoelectric wind energy harvesters under mutual interference. *Appl. Phys. Lett.* **2019**, *115*, 073901. [CrossRef]
9. Javed, U.; Abdelkefi, A. Role of the galloping force and moment of inertia of inclined square cylinders on the performance of hybrid galloping energy harvesters. *Appl. Energy* **2018**, *231*, 259–276. [CrossRef]
10. Zhao, L.; Yang, Y. An impact-based broadband aeroelastic energy harvester for concurrent wind and base vibration energy harvesting. *Appl. Energy* **2018**, *212*, 233–243. [CrossRef]
11. Abdelkefi, A.; Nayfeh, A.H.; Hajj, M.R. Enhancement of power harvesting from piezoaeroelastic systems. *Nonlinear Dyn.* **2012**, *68*, 531–541. [CrossRef]
12. Orrego, S.; Shoele, K.; Ruas, A.; Doran, K.; Caggiano, B.; Mittal, R.; Kang, S.H. Harvesting ambient wind energy with an inverted piezoelectric flag. *Appl. Energy* **2017**, *194*, 212–222. [CrossRef]
13. McCarthy, J.M.; Watkins, S.; Deivasigamani, A.; John, S.J.; Coman, F. An investigation of fluttering piezoelectric energy harvesters in off-axis and turbulent flows. *J. Wind Eng. Ind. Aerodyn.* **2015**, *136*, 101–113. [CrossRef]
14. Doaré, O.; Michelin, S. Piezoelectric coupling in energy-harvesting fluttering flexible plates: Linear stability analysis and conversion efficiency. *J. Fluids Struct.* **2011**, *27*, 1357–1375. [CrossRef]
15. Petrini, F.; Gkoumas, K. Piezoelectric energy harvesting from vortex shedding and galloping induced vibrations inside HVAC ducts. *Energy Build.* **2018**, *158*, 371–383. [CrossRef]
16. Shan, X.; Tian, H.; Cao, H.; Xie, T. Enhancing Performance of a Piezoelectric Energy Harvester System for Concurrent Flutter and Vortex-Induced Vibration. *Energies* **2020**, *13*, 3101. [CrossRef]
17. Yang, K.; Su, K.; Wang, J.; Wang, J.; Yin, K.; Litak, G. Piezoelectric wind energy harvesting subjected to the conjunction of vortex-induced vibration and galloping: Comprehensive parametric study and optimization. *Smart Mater. Struct.* **2020**, *29*, 075035. [CrossRef]
18. Abdelkefi, A. Aeroelastic energy harvesting: A review. *Int. J. Eng. Sci.* **2016**, *100*, 112–135. [CrossRef]
19. Li, D.; Wu, Y.; Da Ronch, A.; Xiang, J. Energy harvesting by means of flow-induced vibrations on aerospace vehicles. *Prog. Aeosp. Sci.* **2016**, *86*, 28–62. [CrossRef]
20. Rostami, A.B.; Armandei, M. Renewable energy harvesting by vortex-induced motions: Review and benchmarking of technologies. *Renew. Sustain. Energy Rev.* **2017**, *70*, 193–214. [CrossRef]
21. Wei, C.; Jing, X. A comprehensive review on vibration energy harvesting: Modelling and realization. *Renew. Sustain. Energy Rev.* **2017**, *74*, 1–18. [CrossRef]
22. Zou, H.; Zhao, L.; Gao, Q.; Zuo, L.; Liu, F.; Tan, T.; Wei, K.; Zhang, W. Mechanical modulations for enhancing energy harvesting: Principles, methods and applications. *Appl. Energy* **2019**, *255*, 113871. [CrossRef]
23. Watson, S.; Moro, A.; Reis, V.; Baniotopoulos, C.; Barth, S.; Bartoli, G.; Bauer, F.; Boelman, E.; Bosse, D.; Cherubini, A.; et al. Future emerging technologies in the wind power sector: A European perspective. *Renew. Sustain. Energy Rev.* **2019**, *113*, 109270. [CrossRef]
24. Wang, J.; Geng, L.; Ding, L.; Zhu, H.; Yurchenko, D. The state-of-the-art review on energy harvesting from flow-induced vibrations. *Appl. Energy* **2020**, *267*, 114902. [CrossRef]
25. Abdelkefi, A.; Hajj, M.R.; Nayfeh, A.H. Power harvesting from transverse galloping of square cylinder. *Nonlinear Dyn.* **2012**, *70*, 1355–1363. [CrossRef]
26. Zhao, L.; Tang, L.; Yang, Y. Comparison of modeling methods and parametric study for a piezoelectric wind energy harvester. *Smart Mater. Struct.* **2013**, *22*, 125003. [CrossRef]
27. Tan, T.; Yan, Z.; Lei, H. Optimization and performance comparison for galloping-based piezoelectric energy harvesters with alternating-current and direct-current interface circuits. *Smart Mater. Struct.* **2017**, *26*, 075007. [CrossRef]
28. Bibo, A.; Daqaq, M.F. An analytical framework for the design and comparative analysis of galloping energy harvesters under quasi-steady aerodynamics. *Smart Mater. Struct.* **2015**, *24*, 094006. [CrossRef]

29. Bibo, A.; Daqaq, M.F. On the optimal performance and universal design curves of galloping energy harvesters. *Appl. Phys. Lett.* **2014**, *104*, 023901. [CrossRef]
30. Yang, Y.; Zhao, L.; Tang, L. Comparative study of tip cross-sections for efficient galloping energy harvesting. *Appl. Phys. Lett.* **2013**, *102*, 064105. [CrossRef]
31. Kluger, J.M.; Moon, F.C.; Rand, R.H. Shape optimization of a blunt body Vibro-wind galloping oscillator. *J. Fluids Struct.* **2013**, *40*, 185–200. [CrossRef]
32. Abdelkefi, A.; Yan, Z.; Hajj, M.R. Performance analysis of galloping-based piezoaeroelastic energy harvesters with different cross-section geometries. *J. Intell. Mater. Syst. Struct.* **2014**, *25*, 246–256. [CrossRef]
33. Hu, G.; Tse, K.T.; Kwok, K.C.S.; Song, J.; Lyu, Y. Aerodynamic modification to a circular cylinder to enhance the piezoelectric wind energy harvesting. *Appl. Phys. Lett.* **2016**, *109*, 193902. [CrossRef]
34. Hu, G.; Liu, F.; Li, L.; Li, C.; Xiao, Y.; Kwok, K.C.S. Wind energy harvesting performance of tandem circular cylinders with triangular protrusions. *J. Fluids Struct.* **2019**, *91*, 102780. [CrossRef]
35. Song, J.; Hu, G.; Tse, K.T.; Li, S.W.; Kwok, K.C.S. Performance of a circular cylinder piezoelectric wind energy harvester fitted with a splitter plate. *Appl. Phys. Lett.* **2017**, *111*, 223903. [CrossRef]
36. Alhadidi, A.H.; Alhussein, H.; Daqaq, M.F. Improving the sensitivity of galloping energy harvesters to flow fluctuations. *Appl. Phys. Lett.* **2020**, *116*, 263902. [CrossRef]
37. Bot, P. Force Variations Related to Flow Pattern Changes Around a High-Camber Thin Wing. *AIAA J.* **2019**, *58*, 1906–1912. [CrossRef]
38. Bot, P.; Rabaud, M.; Thomas, G.; Lombardi, A.; Lebret, C. Sharp Transition in the Lift Force of a Fluid Flowing Past Nonsymmetrical Obstacles: Evidence for a Lift Crisis in the Drag Crisis Regime. *Phys. Rev. Lett.* **2016**, *117*, 234501. [CrossRef]
39. Souppez, J.-B.R.G.; Bot, P.; Viola, I.M. Turbulent flow around circular arcs. *Phys. Fluids* **2022**, *34*, 015121. [CrossRef]
40. Harvey, S.T.; Khovanov, I.A.; Denissenko, P. A galloping energy harvester with flow attachment. *Appl. Phys. Lett.* **2019**, *114*, 104103. [CrossRef]
41. Liu, F.-R.; Zou, H.-X.; Zhang, W.-M.; Peng, Z.-K.; Meng, G. Y-type three-blade bluff body for wind energy harvesting. *Appl. Phys. Lett.* **2018**, *112*, 233903. [CrossRef]
42. Liu, F.-R.; Zhang, W.-M.; Peng, Z.-K.; Meng, G. Fork-shaped bluff body for enhancing the performance of galloping-based wind energy harvester. *Energy* **2019**, *183*, 92–105. [CrossRef]
43. Hu, G.; Tse, K.T.; Wei, M.; Naseer, R.; Abdelkefi, A.; Kwok, K.C.S. Experimental investigation on the efficiency of circular cylinder-based wind energy harvester with different rod-shaped attachments. *Appl. Energy* **2018**, *226*, 682–689. [CrossRef]
44. Wang, J.; Zhou, S.; Zhang, Z.; Yurchenko, D. High-performance piezoelectric wind energy harvester with Y-shaped attachments. *Energy Convers. Manag.* **2019**, *181*, 645–652. [CrossRef]
45. Ding, L.; Zhang, L.; Bernitsas, M.M.; Chang, C.-C. Numerical simulation and experimental validation for energy harvesting of single-cylinder VIVACE converter with passive turbulence control. *Renew. Energy* **2016**, *85*, 1246–1259. [CrossRef]
46. Zhou, Z.; Qin, W.; Zhu, P.; Shang, S. Scavenging wind energy by a Y-shaped bi-stable energy harvester with curved wings. *Energy* **2018**, *153*, 400–412. [CrossRef]
47. Zhang, Y.; Han, Z.-H.; Shi, L.; Song, W.-P. Multi-round Surrogate-based Optimization for Benchmark Aerodynamic Design Problems. In Proceedings of the 54th AIAA Aerospace Sciences Meeting, San Diego, CA, USA, 4–8 January 2016; AIAA SciTech Forum. American Institute of Aeronautics and Astronautics: Reston, VA, USA, 2016.
48. Zhang, X.; Xie, F.; Ji, T.; Zhu, Z.; Zheng, Y. Multi-fidelity deep neural network surrogate model for aerodynamic shape optimization. *Comput. Methods Appl. Mech. Eng.* **2021**, *373*, 113485. [CrossRef]
49. Bernardini, E.; Spence, S.M.J.; Wei, D.; Kareem, A. Aerodynamic shape optimization of civil structures: A CFD-enabled Kriging-based approach. *J. Wind Eng. Ind. Aerodyn.* **2015**, *144*, 154–164. [CrossRef]
50. Ding, F.; Kareem, A. A multi-fidelity shape optimization via surrogate modeling for civil structures. *J. Wind Eng. Ind. Aerodyn.* **2018**, *178*, 49–56. [CrossRef]
51. Kuhn, A.M.; Fennel, K. Evaluating ecosystem model complexity for the northwest North Atlantic through surrogate-based optimization. *Ocean Model.* **2019**, *142*, 101437. [CrossRef]
52. Liu, X.; Zhao, W.; Wan, D. Linear reduced order method for design-space dimensionality reduction and flow-field learning in hull form optimization. *Ocean Eng.* **2021**, *237*, 109680. [CrossRef]
53. Wang, G.G.; Shan, S. Review of Metamodeling Techniques in Support of Engineering Design Optimization. *J. Mech. Des.* **2006**, *129*, 370–380. [CrossRef]
54. Forrester, A.I.J.; Keane, A.J. Recent advances in surrogate-based optimization. *Prog. Aeosp. Sci.* **2009**, *45*, 50–79. [CrossRef]
55. Westermann, P.; Evins, R. Surrogate modelling for sustainable building design—A review. *Energy Build.* **2019**, *198*, 170–186. [CrossRef]
56. Queipo, N.V.; Haftka, R.T.; Shyy, W.; Goel, T.; Vaidyanathan, R.; Tucker, P.K. Surrogate-based analysis and optimization. *Prog. Aeosp. Sci.* **2005**, *41*, 1–28. [CrossRef]
57. Zhao, L.; Yang, Y. Enhanced aeroelastic energy harvesting with a beam stiffener. *Smart Mater. Struct.* **2015**, *24*, 032001. [CrossRef]
58. Coleman, H.W.; Steele, W.G. *Experimentation, Validation, and Uncertainty Analysis for Engineers*, 3rd ed.; John Wiley & Sons: Hoboken, NJ, USA, 2009.
59. Parkinson, G.V.; Brooks, N. On the aeroelastic instability of bluff cylinders. *J. Appl. Mech.* **1961**, *28*, 252–258. [CrossRef]

60. Parkinson, G.V.; Smith, J.D. A square prism as an aeroelastic non-linear oscillator. *Q. J. Mech. Appl. Math.* **1964**, *17*, 225–239. [CrossRef]
61. Tan, T.; Yan, Z.; Hajj, M. Electromechanical decoupled model for cantilever-beam piezoelectric energy harvesters. *Appl. Phys. Lett.* **2016**, *109*, 101908. [CrossRef]
62. Den Hartog, J.P. *Mechanical Vibrations*; McGraw-Hill Book Company: New York, NY, USA, 1956.
63. Païdoussis, M.P.; Price, S.J.; de Langre, E. *Fluid-Structure Interactions: Cross-Flow-Induced Instabilities*; Cambridge University Press: Cambridge, UK, 2010.
64. Bernitsas, M.M.; Raghavan, K.; Ben-Simon, Y.; Garcia, E.M.H. VIVACE (Vortex Induced Vibration Aquatic Clean Energy): A New Concept in Generation of Clean and Renewable Energy from Fluid Flow. *J. Offshore Mech. Arct. Eng.* **2008**, *130*, 041101. [CrossRef]
65. Li, X.; Teng, L.; Tang, H.; Chen, J.; Wang, H.; Liu, Y.; Fu, M.; Liang, J. ViPSN: A Vibration-Powered IoT Platform. *IEEE Internet Things J.* **2020**, *8*, 1728–1739. [CrossRef]
66. Zhao, L.; Tang, L.; Yang, Y. Synchronized charge extraction in galloping piezoelectric energy harvesting. *J. Intell. Mater. Syst. Struct.* **2016**, *27*, 453–468. [CrossRef]
67. Li, S.; Yuan, J.; Lipson, H. Ambient wind energy harvesting using cross-flow fluttering. *J. Appl. Phys.* **2011**, *109*, 026104. [CrossRef]
68. Zhao, D.; Hu, X.; Tan, T.; Yan, Z.; Zhang, W. Piezoelectric galloping energy harvesting enhanced by topological equivalent aerodynamic design. *Energy Convers. Manag.* **2020**, *222*, 113260. [CrossRef]
69. Zhang, B.; Mao, Z.; Song, B.; Ding, W.; Tian, W. Numerical investigation on effect of damping-ratio and mass-ratio on energy harnessing of a square cylinder in FIM. *Energy* **2018**, *144*, 218–231. [CrossRef]
70. Zhang, B.; Song, B.; Mao, Z.; Tian, W.; Li, B. Numerical investigation on VIV energy harvesting of bluff bodies with different cross sections in tandem arrangement. *Energy* **2017**, *133*, 723–736. [CrossRef]

applied sciences

MDPI

Article

Buffeting Response Prediction of Long-Span Bridges Based on Different Wind Tunnel Test Techniques

Yi Su [1,2], Jin Di [1,2,*], Shaopeng Li [1,2,*], Bin Jian [3] and Jun Liu [4]

[1] Key Laboratory of New Technology for Construction of Cities in Mountain Area, Chongqing University, Chongqing 400450, China; windsu888@cqu.edu.cn
[2] College of Civil Engineering, Chongqing University, Chongqing 400450, China
[3] Research Centre for Wind Engineering, Southwest Jiaotong University, Chengdu 610031, China; janebim@126.com
[4] School of Civil Engineering and Geomatics, Southwest Petroleum University, Chengdu 610500, China; liujun_sichuan@126.com
* Correspondence: dijin@cqu.edu.cn (J.D.); lishaopeng0314@163.com (S.L.)

Abstract: The traditional method for calculating the buffeting response of long-span bridges follows the strip assumption, and is carried out by identifying aerodynamic parameters through sectional model force or pressure measurement wind tunnel tests. However, there has been no report on predicting the buffeting response based on the sectional model vibration test. In recent years, the author has proposed a method, based on the integrated transfer function, for predicting the buffeting response of long-span bridges through theoretical and full-bridge tests. This provided an idea for predicting the buffeting response based on the sectional model vibration test. Unfortunately, the effectiveness and accuracy of this method have not been proven or demonstrated through effective tests. To solve this problem, a long-span suspension bridge was taken as a background. Parameters such as aerodynamic admittance were identified through a sectional model force measurement test and the integrated transfer functions were identified through a sectional model vibration test. A taut strip model test was also conducted. Furthermore, the buffeting response prediction results based on three kinds of wind tunnel test techniques were compared. The results showed that if the strip assumption was established, the results of the three methods aligned well, and that selecting a reasonable model aspect ratio for the test could effectively reduce the influence of the 3D effect; moreover, identifying the integrated transfer function by the sectional model vibration test could effectively predict the long-span bridge buffeting response. Furthermore, when the strip assumption failed, the results of the traditional calculation method using 3D aerodynamic admittance became smaller. A larger result would be obtained by neglecting the influence of aerodynamic admittance.

Keywords: long-span bridge; buffeting response; wind tunnel test; sectional model; aerodynamic admittance; integrated transfer function

Citation: Su, Y.; Di, J.; Li, S.; Jian, B.; Liu, J. Buffeting Response Prediction of Long-Span Bridges Based on Different Wind Tunnel Test Techniques. *Appl. Sci.* **2022**, *12*, 3171. https://doi.org/10.3390/app12063171

Academic Editors: Wenli Chen, Zifeng Yang, Gang Hu, Haiquan Jing and Junlei Wang

Received: 21 February 2022
Accepted: 16 March 2022
Published: 20 March 2022

1. Introduction

Among the many vibration forms of bridge structures, buffeting is a random forced vibration generated by the structure under the action of natural wind fluctuation components. It is one of the main areas of research content for the wind-induced vibration of long-span bridges. Low wind speed causes buffeting, and continuous buffeting may cause local fatigue failure of the bridge structure, affecting the bridge's lifespan. In addition, excessive buffeting response of a bridge structure under strong winds negatively affects the safety of construction personnel, equipment, and the comfort of drivers and pedestrians during the operation of the structure. Its concept was proposed in 1930 and originated from the vibration of aircraft in turbulent flow [1]. In the mid-1950s, Scruton first used the concept of buffeting to describe the forced vibration caused by wake flow when he was studying the dynamic response of the Runcorn–Widnes bridge [2]. The theoretical analysis of the buffeting problem of long-span bridges began when Davenport [3,4] applied the statistical analysis method of buffeting in the aviation field to bridge structures. In the 1960s, based on the theory of aerodynamics, Davenport [3,4] took the lead in proposing the concept of aerodynamic admittance, defined the joint acceptance function, and considered the time and space distribution of the aerodynamic force on the structure. Furthermore, the spanwise correlation of the fluctuating wind was used to describe the spanwise correlation of the buffeting force, and a relatively complete buffeting analysis method for long-span bridges and other slender linear structures was established. In the 1970s, Scanlan et al. [5–8] extended their research results from the flutter theory to the analysis of the buffeting response, and believed that the self-excited force in the flutter analysis would affect the buffeting response of the bridge structure. They expressed the self-excited force caused by structural motion in the form of flutter derivatives, used aerodynamic stiffness and aerodynamic damping to modify Davenport's theory, and then introduced the mechanical admittance function to reflect the influence of structural motion on the system transfer function. Since then, Davenport's buffeting theory, in combination with Scanlan's theory, has become the basis for the theoretical buffeting analysis and calculation method for long-span bridge structures. In the following decades, many scholars [9–15] did fruitful work regarding the buffeting response prediction for long-span bridges under turbulent flow.

So far, the buffeting response has mainly been obtained by two methods: wind tunnel tests and theoretical calculation. The latter is usually based on the principle of aerodynamics, establishes a mathematical model of the relevant wind load, and then applies the structural dynamics method to solve the wind-induced response of the structure. The current theoretical method is based on the pioneering work achieved by Davenport, Scanlan, etc., and has been improved in many subsequent studies. However, due to the characteristics of atmospheric turbulence and the complexity and diversity of bridge section forms, a perfect analytical model cannot be established for the prediction of the buffeting response of bridges. It is difficult to calculate the aerodynamic force and wind-induced responses of bridge structures through purely theoretical analysis. The wind tunnel test is an indispensable and important method, and is still mainly used to identify parameters such as aerodynamic admittance and flutter derivatives.

At present, wind tunnel test techniques [16–18] for long-span bridge buffeting response prediction are divided into three categories: full-bridge aeroelastic model wind tunnel tests, taut strip model wind tunnel tests and sectional model wind tunnel tests. Among them, the full-bridge aeroelastic model wind tunnel test is the most direct test method for measuring the buffeting response of long-span bridges. Based on accurately simulating the bridge model, it directly measures the dynamic response of the bridge to the wind in a simulated natural wind field, and then converts the response of the model measured in the test to the real bridge through a scale ratio. This method is a complete-scale simulation of a real bridge, which can naturally simulate the bridge's three-dimensional aeroelastic effect and dynamic mode shape. It obtains the buffeting response of the bridge more intuitively, and avoids the identification of the aerodynamic parameters of the bridge section. However, the full-bridge test features the drawbacks of high cost, lengthy duration, and difficult model design and production. With the continuous increase of current bridge spans, full-bridge tests of long-span bridges require larger test wind tunnel sizes. In contrast, the sectional model mainly simulates the mid-span section of the main span of the bridge according to the geometric similarity principle, and has a large scale ratio. It has the advantages of low cost, less difficulty in design and production, short preparation time, etc. In addition, the shape of the sectional model can easily be changed, and the aerodynamic shape of the bridge can be optimized in time according to the test results, so it is more widely used. For the prediction of the buffeting response of long-span bridges, the sectional model test is an important auxiliary means of conducting the theoretical calculation method. The theoretical calculation of the buffeting response of long-span bridges has always used sectional model tests to identify the unsteady aerodynamic parameters of the bridge sections. The buffeting response of the bridge is then obtained through the buffeting analysis theory. However, because the sectional model wind tunnel test cannot simulate turbulent flow fields with large turbulence integral scales, the three-dimensional effect has affected the prediction results of the buffeting response for many years. In recent years, many scholars have proposed empirical correction methods for this problem, but they have not solved the problem with respect to the theoretical foundation or experimental technology [19]. The taut strip model, an aeroelastic model between the full-bridge aeroelastic model and the sectional model, has often been used in previous studies to compare and analyze the difference between the test results of the sectional model and the full-bridge aeroelastic model. The taut strip model test is used to measure the buffeting response of the bridge girder under the condition that the additional aerodynamic interference caused by the structural auxiliary facilities (such as suspenders, cables, towers, etc.) is eliminated. The method considers the first several vibration modes of the bridge, and can consider both the two-dimensional effect of the sectional model and the three-dimensional vibration effect of the bridge structure.

For a long time, sectional model tests only used force and pressure measurements to identify the aerodynamic admittance to assist with the theoretical calculations [20]. Due to the influence of factors such as the applicability of the strip assumption [21,22], there is no case wherein the buffeting response of long-span bridges is predicted by sectional model vibration tests. However, the force measurement test is difficult in the early leveling, correlation measurement, etc., and the accuracy is difficult to guarantee; the pressure measurement test has limitations such as the inability to measure the truss girder and accurately measure the overall force of the bridge in the completed state. In addition, Yan et al. [23] identified the aerodynamic admittance of the bridge section in the free vibration state. The results show that the aerodynamic admittance is related to the vibration state of the main girder, so it is necessary to consider the effect of the actual bridge vibration on the aerodynamic admittance [24,25]. This makes the prediction results biased by the traditional calculation method based on the rigid sectional model pressure or force measurement test to identify aerodynamic admittance on account of the ignorance of the vibration of the structure. Fortunately, the vibration test can effectively avoid the limitations of force and pressure tests. At the same time, it can truly reflect the structural motion state, and comprehensively consider the effects of self-excited force terms such as aerodynamic damping and aerodynamic stiffness. Recently, Su et al. [26] proposed the concept of the integrated transfer function based on the research of Li et al. [27–29] on the influence of the structural aspect ratio on the three-dimensional effect of turbulence, and gave a long-span bridge buffeting response prediction method based on the integrated transfer function. The method points out that the aspect ratio of the test model will have a significant impact on the prediction accuracy of the buffeting response, and can correct the deviation of the prediction result of the buffeting response caused by the inaccurate simulation of the wind field parameters in the wind tunnel. However, even though Su et al. [26] proved the feasibility of the integrated transfer function in the prediction of the buffeting response of long-span bridges through rigorous theoretical derivation and verified the independence of the integrated transfer function from the wind field characteristics through full-bridge aeroelastic model wind tunnel tests, they only expected the results to provide ideas for the buffeting response prediction of long-span bridges based on the identification of integrated transfer functions, and the accuracy and effectiveness of the method have not been directly verified by sectional model vibration tests. In addition, they considered that the sectional model test would also eliminate the additional aerodynamic interference of the structural ancillary facilities. At the same time, compared with the full-bridge test, the taut strip model test has the characteristics of low cost and short duration, and can also consider the three-dimensional vibration of the structure. The taut strip model has been selected to replace the full-bridge aeroelastic model to perform the buffeting response test, and the results are compared with the prediction results based on the sectional model test in this paper. For this reason, the paper uses a long-span bridge with a basic section as an example, takes the turbulent characteristics simulated in the taut strip model test as the target wind field, and relies on three different test techniques to predict the buffeting response of the structure. These techniques are as follows:

1. Identifying the aerodynamic parameters, such as aerodynamic admittance, through sectional model force measurement tests, and then calculating the response according to the traditional buffeting response calculation method;

2. Identifying the integrated transfer function of the structure through the sectional model vibration test, then predicting the buffeting response by using the method proposed by Su et al. [26];

3. Measuring the buffeting response of the structure through the taut strip model test.

Each buffeting response prediction result was then compared to investigate the long-span bridge buffeting response prediction methods based on the above three different test techniques and verify the accuracy and effectiveness of the prediction method based on the sectional model vibration test to identify the integrated transfer function.

2. Theoretical Analysis

At present, in the more commonly used analysis process, although the higher-order modes and the coupling effects between modes have a certain degree of influence on the structural buffeting response, the fundamental frequency still dominates [30]. Therefore, the coupling between modes in different directions is usually ignored without affecting the calculation accuracy. This paper will analyze the problem without considering the modal coupling effects. In addition, since the theoretical principles of the vertical, lateral and torsional buffeting responses are consistent, to simplify the derivation and demonstration process, this paper takes the vertical buffeting response as an example to study.

2.1. Traditional Buffeting Response Calculation

As mentioned above, most of the current bridge structure buffeting analysis uses Davenport's correction through the introduction of the aerodynamic admittance function into Scanlan's quasi-steady aerodynamic force expression. Although much work has been completed by many scholars to continuously improve the theories, these theoretical approaches are based on the pioneering work of Davenport and Scanlan.

Because the studies on buffeting by Su et al. [26] and Li et al. [27–29] used the wavenumber domain, in order to compare the traditional buffeting response calculation method with the following method, the buffeting response in the frequency domain is converted to the wavenumber domain for expression according to the relationship between the wavenumber and the frequency $k = 2\pi n/U$. Based on the random vibration theory and the buffeting analysis principle, the power spectrum of the bridge buffeting displacement response can be expressed as:

$$S_{h_i}(y,k_1) = \frac{\varphi_{h_i}^2(y)}{M_{h_i}^2}(2\rho U b C_L)^2|\chi(k_1)|^2|H_{h_i}(k_1)|^2|J_{h_i}(k_1)|^2 S_u(k_1) \quad (1)$$

where $S_{h_i}(y,k_1)$ is the power spectral density of the vertical buffeting displacement response at the axial coordinate y of the bridge; $\varphi_{h_i}(y)$ is the i-th mode shape of the vertical motion; M_{h_i} is the vertical generalized mass; ρ is the air density; U is the mean wind velocity; $b = B/2$ is the half-width of the bridge, where B is the width; C_L is the lift coefficient of the bridge section; $k_1 = 2\pi n/U$ is the longitudinal wavenumber, where n is the frequency; $\chi(k_1)$ is the one-wavenumber aerodynamic admittance function; $H_{h_i}(k_1)$ is the vertical one-wavenumber frequency response function; $J_{h_i}(k_1)$ is the vertical one-wavenumber joint acceptance function of the structure; $S_u(k_1)$ is the longitudinal one-wavenumber fluctuating wind velocity spectrum.

As mentioned above, the traditional method for calculating the buffeting response is to identify the aerodynamic admittance and the aerodynamic derivative in the frequency response function of the bridge section through sectional model tests, and then obtain the power spectrum of the structural buffeting displacement response based on Equation (1).

Among them, the aerodynamic admittance function is an important transfer function for calculating the buffeting response, which is identified by the sectional model pressure or force measurement test. The RMS (root mean square) of the buffeting displacement can be obtained by integrating the buffeting response power spectrum:

$$\sigma_{h_i}^2(y) = \int_0^{+\infty} S_{h_i}(y, k_1)\mathrm{d}k_1 \tag{2}$$

Finally, according to the SRSS (Square Root of Sum Square) method, the total buffeting response without considering the coupling effect between modes can be obtained:

$$\mathrm{SRSS}(h_i) = \sqrt{\sigma_{h_i,1}^2 + \sigma_{h_i,2}^2 + \cdots + \sigma_{h_i,m}^2} \tag{3}$$

where m is the number of structural modes.

2.2. Calculation of Buffeting Response Based on Sectional Model Vibration Test

Traditional buffeting response calculation and analysis has always been based on the strip assumption; it divides the bridge into several strip units with independent aerodynamic properties in the spanwise direction, so as to obtain the aerodynamic response of the whole bridge through the integration along the spanwise direction [31]. From another perspective, it is assumed that the spanwise correlation of the buffeting force acting on the structure is equal to the correlation of the fluctuating wind. Based on the one-wavenumber aerodynamic admittance identified through methods such as wind tunnel tests to calculate the structural buffeting response, the calculation and analysis of the buffeting force and the buffeting response can be greatly simplified. However, many scholars [32–37] found that when the spanwise wavelength of the incoming flow is not much larger than the structure width, the three-dimensional effect of turbulence cannot be ignored, and the spanwise correlation of the buffeting force is much greater than that of the wind. This will lead to the failure of the strip assumption, which means that there will be a certain error in the calculation of the buffeting response of the bridge when using the traditional one-wavenumber aerodynamic admittance. This indicates that the influence of the spanwise wavenumber k_2 should not be ignored when calculating the buffeting response of the structure. In turbulent analysis, compared with the autocorrelation of different turbulent fluctuation components, the influence of the cross-correlation is small and can be ignored [38–40]. To simplify the theoretical elaboration, according to the two-wavenumber buffeting analysis considering the three-dimensional effect of turbulence, and under the premise of ignoring the correlation between the horizontal fluctuating wind velocity and the vertical fluctuating wind velocity, the contribution of the longitudinal and vertical fluctuating wind velocities to the lift is equated to an equivalent aerodynamic admittance that considers the contribution of the fluctuating wind velocity in both directions to the lift. The two-wavenumber buffeting lift spectrum of a long-span bridge is:

$$S_{L_i}(k_1,k_2) = (\rho U b)^2 |\chi(k_1,k_2)|^2 \left[4C_L^2 \left|J_{u_{h_i}}(k_1,k_2)\right|^2 S_u(k_1) + (C_L' + C_D)^2 \left|J_{w_{h_i}}(k_1,k_2)\right|^2 S_w(k_1)\right] \tag{4}$$

where $S_{L_i}(k_1,k_2)$ is the unsteady buffeting lift spectrum corresponding to the i-th mode of the structure; $\chi(k_1,k_2)$ is the equivalent two-wavenumber aerodynamic admittance; C_D is the drag coefficient of the bridge section; C_L' represents the slope of the lift coefficient to the wind attack angle; $J_{u_{h_i}}(k_1,k_2)$ and $J_{w_{h_i}}(k_1,k_2)$ represent the two-wavenumber joint acceptance function corresponding to the longitudinal and vertical fluctuating wind velocities, respectively; $S_w(k_1)$ is the vertical one-wavenumber fluctuating wind velocity spectrum.

According to recent theoretical and experimental studies [26–29] of the influence of the aspect ratio on the accuracy of the strip assumption and the three-dimensional effect of turbulence, for long-span bridges and other slender linear structures with large aspect ratios, the strip assumption is accurate enough. The influence of the spanwise wavenumber on the aerodynamic admittance can be ignored in the calculation and analysis, and Equation (4) can be simplified as:

$$S_{L_i}(k_1,k_2) = (\rho U b)^2 |\chi(k_1)|^2 \left[4C_L^2 \left|J_{u_{h_i}}(k_1,k_2)\right|^2 S_u(k_1) + (C_L' + C_D)^2 \left|J_{w_{h_i}}(k_1,k_2)\right|^2 S_w(k_1)\right] \quad (5)$$

The one-wavenumber buffeting lift spectrum commonly used in the buffeting response calculation is easily obtained by integrating the above formula against k_2. At this time, it is consistent with the one-wavenumber buffeting lift spectrum in the process of the traditional calculation method in the previous section. The difference lies in the fact that the traditional buffeting analysis theory does not consider the influence of the three-dimensional effect of turbulence, which many scholars believe will cause a large error in the calculation of the buffeting response in the traditional buffeting analysis theory. The theoretical analysis mentioned earlier in this section pointed out that even if the influence of the turbulent three-dimensional effect is considered, as long as the turbulence integral scale is not much smaller than the structure width, the influence of the turbulent three-dimensional effect can also be ignored for long-span bridges and other linear slender structures with large aspect ratios [26–29]. The one-wavenumber aerodynamic admittance can be used instead of the two-wavenumber aerodynamic admittance, and the influence of the spanwise wavenumber k_2 can be ignored. That is, the traditional buffeting analysis theory is considered to be accurate at this time. At present, the two-wavenumber buffeting analysis considering the three-dimensional effect of turbulence returns to the traditional buffeting theory analysis process, and the power spectral density function of the buffeting displacement response can also be expressed by Equation (1). In the formula, structural parameters such as geometric dimensioning, mode shape function, equivalent mass, natural frequency and damping ratio are only related to the inherent characteristics of the structure and do not change with the change of turbulence. The one-wavenumber aerodynamic admittance function only contains longitudinal wavenumbers, which reflects the aerodynamic transfer relationship between the 2D fluctuating velocity and the buffeting force, and can be expressed as a function of the dimensionless reduced frequency. It is only related to the bridge cross-section geometry and longitudinal wavenumber and has nothing to do with the characteristics of the turbulent flow field. In addition, considering the significance of buffeting research, and the fact that the study of the buffeting response is limited to ranges where the wind velocity is relatively small, the buffeting force is the main fluctuating load at this time, and the aerodynamic self-excited force accounts for a small proportion. Buffeting will not cause catastrophic consequences such as wind-induced flutter instability. It mainly affects comfort, safety and the fatigue damage of components in the use stage of the structure. Therefore, the study of buffeting under high wind velocity will, to some extent, be meaningless. For example, long-span bridges will be stopped when the wind velocity is high, limiting the passage of vehicles and pedestrians. Based on the above factors, Su et al. [26] proposed the concept of the integrated transfer function, which is the product of the one-wavenumber aerodynamic admittance and the one-wavenumber frequency response function. It is only determined by the structural characteristic parameters and has nothing to do with the turbulent flow characteristics:

$$|T_{h_i}(k_1)|^2 = |\chi(k_1)|^2 |H_{h_i}(k_1)|^2 \quad (6)$$

At this time, the buffeting response spectrum of the structure can be written as:

$$S_{h_i}(y,k_1) = \frac{\varphi_{h_i}^2(y)}{M_{h_i}^2}(2\rho U b C_L)^2 |T(k_1)|^2 |J_{h_i}(k_1)|^2 S_u(k_1) \quad (7)$$

Based on Equation (7), a sectional model with a reasonable aspect ratio can be selected for vibration testing to identify the integrated transfer function. Since the function has nothing to do with the turbulent flow characteristics, it can be used to predict the buffeting response of bridge structures under any wind field by using Equations (1)–(3).

Two points need to be explained:

(1) Since the identification of the integrated transfer function is based on the measured value of the buffeting response at a certain point of the structure, when it is applied to the prediction of the buffeting response, the function identified at a point (such as the mid-span position of a long-span bridge) can only be used to predict the buffeting response of the actual structure at the same position;

(2) Due to the inconsistency between the mode shape functions of different orders, the integrated transfer functions corresponding to the mode shapes of different orders are also different. In other words, there is a fixed integrated transfer function corresponding to the different order mode shapes of the structure. Therefore, when using the integrated transfer function to predict the buffeting response of the actual structure, it is necessary to calculate the response of the different modes corresponding to the real bridge through the integrated transfer function of different orders, after which the SRSS method can be used to calculate the total response of the structure.

2.3. Buffeting Response Prediction Based on Taut Strip Model Test

The taut strip model buffeting response test is similar to the full-bridge aeroelastic model test. They both measure the 3D buffeting response of the 3D model under the action of 3D turbulence, but the former eliminates the additional aerodynamic interference of the structural ancillary facilities. The buffeting response spectrum $S_h(y, k_1)$ of the structure under the simulated turbulent flow field is measured through the taut strip wind tunnel test, and the RMS of the buffeting response can then be calculated by Equation (8):

$$\sigma_h^2(y) = \int_0^{+\infty} S_h(y, k_1) \mathrm{d}k_1 \tag{8}$$

3. Test Preparation

In the comparative study of different buffeting response calculation methods in this paper, taking a long-span bridge as the engineering background and taking the vertical buffeting response at the mid-span as an example, three test methods were used to predict and compare the buffeting response of the structure. It should be noted that considering the influence of the model aspect ratio on the three-dimensional effect [26–29], it is necessary to use a small scale ratio to make the model with a large aspect ratio in the sectional model test to minimize the influence of the three-dimensional effect of turbulence. This will lead to difficulty in the simulation of structures such as railings and maintenance channels in the sectional model test, so the 100% construction state of the bridge is selected as the research object.

3.1. Structural Overview

A long-span steel box girder suspension bridge was taken as an example to study in this paper. The main span of the bridge is 1020 m. The stiffening beam is an integral steel box girder with a width of 30.5 m and a height of 3.0 m. The tower height is 177.5 m, and the basic wind velocity is 25.0 m/s. The overall layout of the bridge structure is shown in Figure 1, and the standard cross-section of the main girder is shown in Figure 2.

3.2. Analysis of Structural Dynamic Characteristics

Structural dynamic characteristics constitute the premise of structural dynamic response analysis. Through the analysis of structural dynamic characteristics, the frequency distribution and mode shape characteristics of the structure can be understood, and parameters can be provided for the design of the wind tunnel test model. In this section, the ANSYS software is used to analyze and calculate the dynamic characteristics of the bridge.

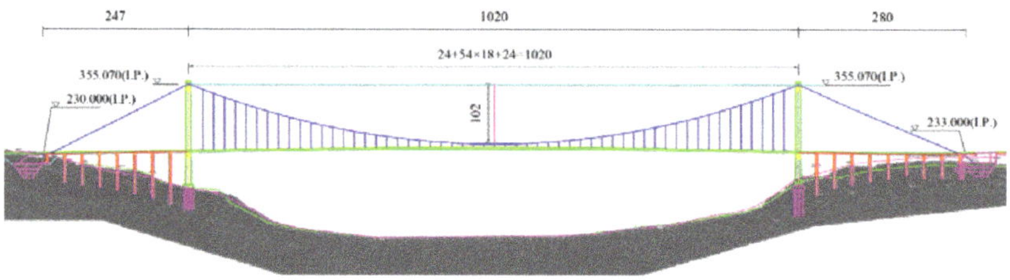

Figure 1. General layout drawing of the bridge (unit: m).

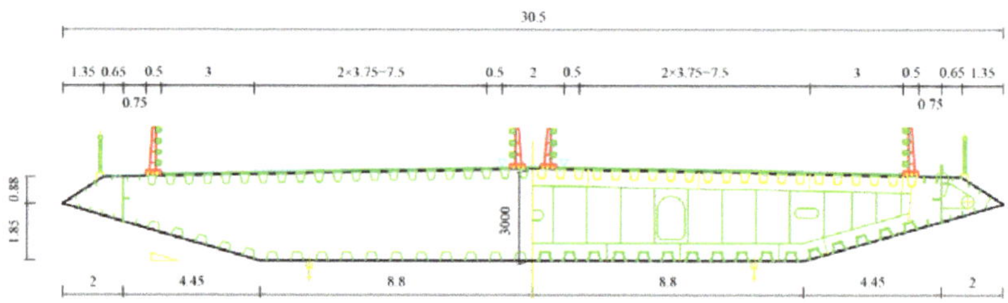

Figure 2. Standard cross-section of main beam (unit: m).

The calculation model of the bridge deck adopted the traditional "fishbone beam" model, as shown in Figure 3. The main girder and tower were discretized into three-dimensional beam elements (Beam4) according to the actual spatial position, the cables were discretized into three-dimensional cable elements (Link10), and the piers and junction piers were also discretized into three-dimensional beam elements (Beam4). The section properties and material properties of each element were assigned for the calculation. The mass and mass moment of inertia of the main beam were simulated by mass point element (Mass21).

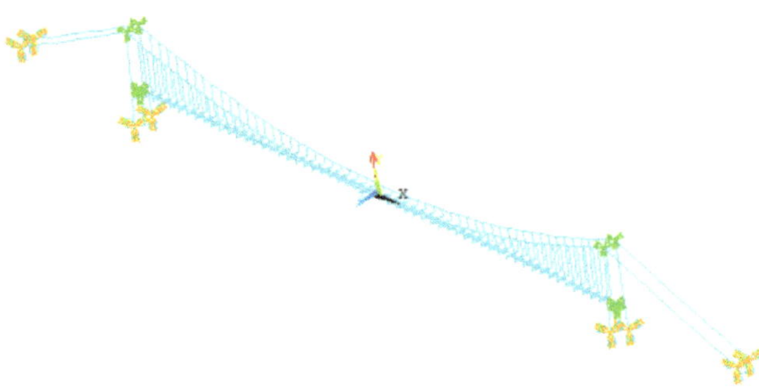

Figure 3. Bridge finite element model diagram.

The boundary constraints in the model were set as follows:

(1) The bottom of the cable tower was embedded with the top surface of the cushion cap, that is, the degrees of freedom in six directions were restricted;

(2) The main cable was fixed at the anchorage; the top connection of the main tower was a master–slave relationship;

(3) The connection between the main beam and the main tower: The connection between the rotational degrees of freedom of the main beam in the transverse, vertical and around the bridge axis and the middle of the lower beam under the main tower was a master–slave relationship. The remaining three degrees of freedom were relaxed.

Using ANSYS software, the subspace iteration method was used to calculate the first 30-order frequencies and mode shapes of the bridge, which included three vertical modes: $n_1 = 0.165$ Hz, $n_2 = 0.230$ Hz and $n_3 = 0.343$ Hz.

3.3. Test Equipment

The sectional model test was conducted in the XNJD-1 wind tunnel at Southwest Jiaotong University. The size of the test section was 16 m (length) × 2.4 m (width) × 2.0 m (height), and the wind velocity could be adjusted within a range of 1.0 to 45.0 m/s. In the vacant state of the wind tunnel, the turbulence intensity of the incoming flow was less than 0.5%, and the average airflow deflection angles in both the longitudinal and vertical directions were less than 0.5°. The taut strip model test was conducted in the XNJD-3 wind tunnel of Southwest Jiaotong University. The size of the test section was 36 m (length) × 22.5 m (width) × 4.5 m (height), and the wind velocity could be adjusted within a range of 1.0 to 16.0 m/s. The turbulence intensity of the incoming flow in the vacant state of the wind tunnel was less than 1%, and the average airflow deflection angles in both the longitudinal and vertical directions were less than 1°.

The turbulent wind field characteristics were collected using a TFI Cobra three-dimensional fluctuating anemometer (Cobra Probe), which is a four-hole pressure probe measuring instrument that can measure the real-time dynamic three-component velocity (longitudinal, horizontal and vertical), pitch angle and yaw angle. Its linear frequency measurement range is 0 Hz (uniform flow) to 2000 Hz, and its measurable wind velocity range is 2 m/s to 100 m/s. Its allowable error for measured wind velocity is ±0.1 m/s, and its measurable wind direction angle range is ±45°. The buffeting force measurement was conducted using a high-frequency dynamic six-axis force ATI Gamma balance, whose horizontal and vertical force ranges were 65 N and 200 N, respectively, with respective corresponding accuracies of 1/80 N and 1/40 N. The displacement response measurement used a non-contact laser displacement measurement sensor with a range of 200 mm and a static test accuracy of 40 μm.

3.4. Turbulence Field Simulated

The taut strip model test was conducted in a large-scale wind tunnel, and passive simulation devices such as spires, baffles and rough elements were used to simulate the turbulent flow field, as shown in Figure 4.

Figure 4. Spire turbulence field installed in XNJD-3 wind tunnel.

Because the turbulence integral scale simulated in the spire turbulence field is usually larger than that in the grid turbulence field, the test identification of the aerodynamic admittance and integrated transfer function of the structure were conducted in the simulated spire turbulence field, which can increase the ratio of the turbulence integral scale to the model width in order to minimize the influence of the turbulent three-dimensional effect. It can also simulate the shear layer characteristics, which is more convenient for analysis and comparison with the taut strip model test results. Therefore, the sectional model test was also conducted in a spire turbulence field, as shown in Figure 5.

Figure 5. Spire turbulence field installed in XNJD-1 wind tunnel.

To study the buffeting response under the turbulence field, it is necessary to first measure the flow field quality to ensure the accuracy of the study. It should be noted that, according to the buffeting theory of long-span bridges, it is not difficult to find that the fluctuating components of the turbulence field along the transverse wind direction do not contribute to the calculation of the buffeting response, so the transverse turbulent fluctuating characteristics were not repeated here. Among many turbulent fluctuating wind velocity spectrum models, the von Kármán spectrum [41], which is widely used in aviation and structural wind engineering, is more suitable for describing atmospheric turbulence higher above the ground and the turbulent characteristics simulated in a wind tunnel [42,43]. To ensure the quality of the simulated turbulence field in the test, it is usually fitted to the von Kármán spectrum. Figures 6 and 7 show the longitudinal and vertical fluctuating wind velocity spectrum of the above turbulence fields, and the von Kármán spectrum was used to fit the results. The results show that the simulated fluctuating wind velocity spectra fit well with the von Kármán spectrum, and met the test requirements.

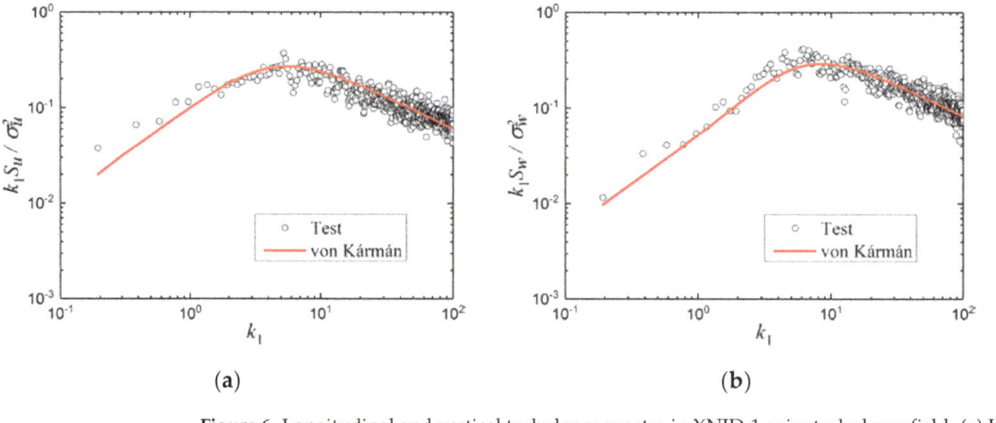

Figure 6. Longitudinal and vertical turbulence spectra in XNJD-1 spire turbulence field. (**a**) Longitudinal; (**b**) vertical.

In addition, Table 1 gives the other two important parameters of the turbulence field: turbulence intensity and turbulence integral scale. In Table 1, I and L represent the turbulence intensity and turbulence integral scale, respectively, and the subscripts u and w represent the longitudinal and vertical directions of the turbulence field, respectively.

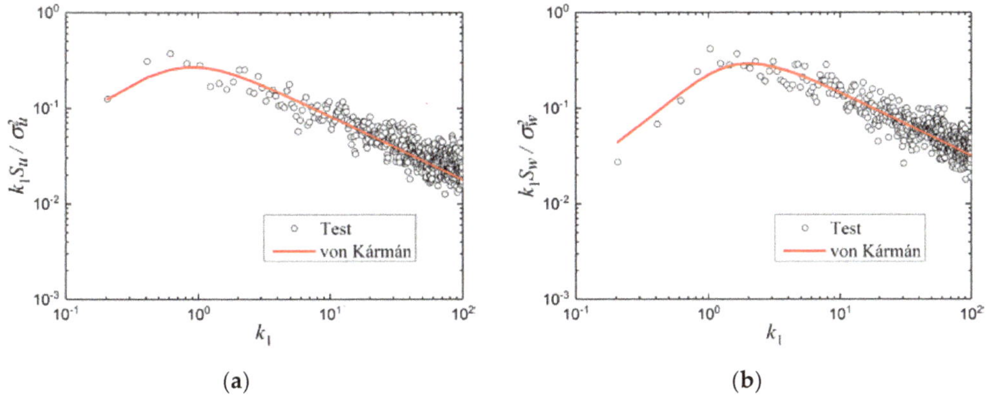

Figure 7. Longitudinal and vertical turbulence spectra in XNJD-3 spire turbulence field. (**a**) Longitudinal; (**b**) vertical.

Table 1. Turbulence characteristics parameters.

Turbulence Field	Turbulence Intensity		Turbulence Integral Scale	
	I_u	I_w	L_u	L_w
XNJD-1	14.2	12.3	0.153	0.092
XNJD-3	15.7	11.3	1.187	0.650

4. Test Arrangement

It should be noted that it is important to keep the dimensionless quantities consistent when using scaled models for wind tunnel testing. In the related tests in this paper, on the premise of ensuring that the bridge models were similar to the actual structure geometry, the sizes of the models that could effectively ensure the test operability were selected according to the sizes of the wind tunnels, and the geometric scale ratios were determined. Subsequently, based on the conversion relationship of geometric ratio, frequency ratio, and wind speed ratio, the model frequencies and test wind speeds were determined. The wind tunnel tests conducted at this time were considered to truly reflect the vibration performance of the bridge under the actual wind field.

4.1. Three-Component Force Test

Through the static sectional model test of the stiffened beam, the three-component force coefficients were measured at different attack angles, and the results were provided for the subsequent prediction and analysis of the buffeting response of the structure. The sectional model of the stiffening beam, which was made of high-quality wood, adopted a geometric scale ratio of 1:50. The model was 2.095 m long, 0.61 m wide and 0.061 m high.

Stiffening beams were made of high-quality wood, and auxiliary facilities such as railings and maintenance tracks were carved from plastic plates as a whole according to the size of the drawings. The test was conducted in the XNJD-1 wind tunnel. The test section was equipped with a side wall support and a force balance system specially used for the static three-component force test of the bridge sectional model, and the attitude angle of the model (the angle of attack of the incoming flow relative to the model) was controlled by the computer. A three-component strain balance was used to measure the static three-component force. The sectional model ends were mounted directly on the three-component force test balance. In order to ensure the two-dimensional flow, end plates were set at both ends of the stiffening girder model to avoid the flow field being disturbed. The model installed in the wind tunnel is shown in Figure 8.

Figure 8. Three-component force test of sectional model.

4.2. Aerodynamic Derivative Identification Test

The aerodynamic derivative identification test model was the same as the three-component force coefficient test model in the previous section. It was suspended on the bracket by eight tension springs to form a two-degree-of-freedom vibration system that could move vertically and rotate around the model axis. End plates were installed at both ends of the model to reduce the end effects and ensure the bidimensionality of the flow. The test support was placed outside the wind tunnel wall to avoid disturbing the turbulence field. The test model is shown in Figure 9.

Figure 9. Aerodynamic derivative identification test.

4.3. Aerodynamic Admittance Identification Test

The traditional buffeting response calculation adopts the 2D one-wavenumber aerodynamic admittance combined with the correlation of the fluctuating wind. According to the research of Larose et al. on the influence of the three-dimensional effect, the commonly used aerodynamic admittance identified through sectional model tests is the generalized 3D one-wavenumber aerodynamic admittance. It is necessary to use the correlation of the buffeting force to calculate the buffeting response of long-span bridges. In recent years, based on studies by Su et al. [26] and Li et al. [27–29], it was found that, in addition to the ratio of the turbulence integral scale to the structure width, the aspect ratio is also an important parameter that controls the calculation accuracy of the buffeting response of long-span bridges. Even if the integral scale is close to or even slightly smaller than the structure width, as long as the structural aspect ratio is sufficient, the influence of the three-dimensional effect is very small, the strip assumption is valid, and the calculation of the buffeting response will also have high accuracy. At this time, the correlation of the fluctuating wind can be used instead of that of the buffeting force. Therefore, to minimize the influence of the three-dimensional effect, in the test identification of the aerodynamic admittance in this section and the integrated transfer function in the next section, under the premise that the test was conducted in the spire turbulence field with a large integral scale, a sectional model with a relatively large aspect ratio was used. The model was 1.35 m long and 0.15 m wide.

The identification of the 2D aerodynamic admittance was carried out in a spire-generated turbulent flow using the high-frequency force-balance measurement technique. The spires were 0.198 m and 0.160 m in width at the base and top, respectively, and 2.0 m in height. The distance between the spires was 0.8 m. The characteristics of turbulent wind field were measured using the TFI Cobra Probe. The three section models were supported on a steel platform approximately 4.5 m downstream of the spires, as shown in Figure 10. The test model was divided into three parts, namely the force model, two end plates (0.4 m × 0.3 m) and their connected pseudo-models (0.1 m in length). The gap between the pseudo-models and the force model was 1 mm. In addition, the balance and connections were covered by a fairing to prevent the influence of the incoming flow. The force measurement model was horizontally mounted at a 0-attack angle on the high-frequency dynamic balance.

Figure 10. Aerodynamic admittance function identification test.

4.4. Integrated Transfer Function Identification Test

According to the structural scale ratio, the test wind velocity was taken as 2.2 m/s. Considering the influence of the scale ratio, simulating the high-order modes of the sectional model is very difficult. In this paper, only the first 30 modes of the 100% construction state of the structure were considered, including three vertical modes with respective frequencies of 0.165 Hz, 0.23 Hz and 0.343 Hz. The corresponding frequencies of the test model were 2.948 Hz, 4.01 Hz and 6.231 Hz, respectively. The model was 1.35 m long and 0.15 m wide. In the test, two laser displacement sensors were symmetrically arranged on both sides of the width direction of the model, and synchronous measurement was performed. The mean value of the buffeting response measured by the two sensors was the vertical buffeting response value of the structure. The integrated transfer function identification test is shown in Figure 11.

Figure 11. Integrated transfer function identification test.

4.5. Taut Strip Model Test

Considering the size of the XNJD-3 wind tunnel and the span of the bridge, the scale ratio of the taut strip model was set to 1:100. The model was 10.2 m long and 0.305 m wide, and it was divided into 20 sections of equal length. Steel strands with a diameter of 2 mm were selected for the bracing wires, and the two ends of the steel strands connecting the models were anchored by screws to the end plates made of 10 mm thick steel. According to the similarity principle, the wind velocity ratio was 1:10 and the frequency ratio was 10:1. The test wind velocity was taken as 2.5 m/s. The three-order vertical frequencies of the model were 1.648 Hz, 2.304 Hz and 3.426 Hz, respectively. The taut strip model installed in the wind tunnel is shown in Figure 12.

Figure 12. Taut strip model test conducted in XNJD-3 wind tunnel.

5. Test Results

5.1. Three-Component Force Coefficient

The static three-component force test was conducted in uniform flow. The test wind attack angles were $\alpha = -12^\circ \sim +12^\circ$ and $\Delta\alpha = 1^\circ$. The static three-component force coefficient in the body–axis coordinate system is usually used for theoretical calculation and practical application. Figure 13 shows the relationship curve of the three-component force coefficient with the wind attack angle.

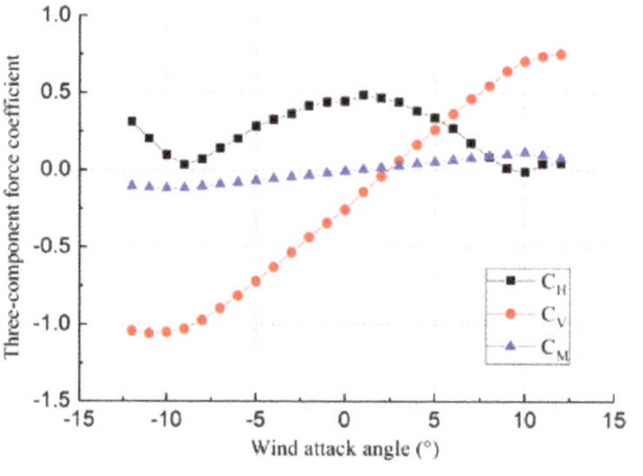

Figure 13. Three-component force coefficient curve of main beam in construction state.

5.2. Aerodynamic Derivative

As mentioned above, the paper only takes the vertical buffeting response of the structure as an example, and it is only related to H_1^* in the aerodynamic derivatives. The result of H_1^* identified in the tests is shown in Figure 14.

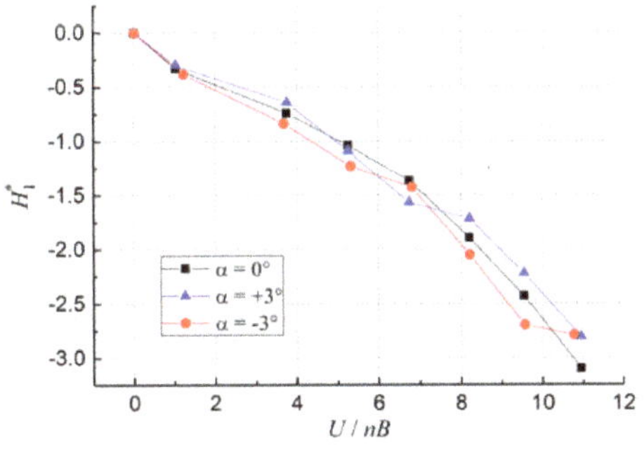

Figure 14. Aerodynamic derivative H_1^* identification result.

5.3. Aerodynamic Admittance

The 2D aerodynamic admittance can be obtained through the 3D aerodynamic admittance identified by the test and the 3D effect influencing factor [44,45]:

$$\left|\chi_L^{2D}(k_1)\right|^2 = \frac{1}{g_L} \cdot \left|\chi_L^{3D}(k_1)\right|^2 \tag{9}$$

where $\chi_L^{2D}(k_1)$ is the 2D one-wavenumber aerodynamic admittance; $\chi_L^{3D}(k_1)$ is the 3D one-wavenumber aerodynamic admittance; g_L is the 3D effect influencing factor, and:

$$g_L = \frac{\int_{-\infty}^{+\infty} \sin c^2(k_2\delta)\left[4C_L^2 S_u(k_1,k_2) + (C_L' + C_D)^2 S_w(k_1,k_2)\right] dk_2}{\left[4C_L^2 S_u(k_1) + (C_L' + C_D)^2 S_w(k_1)\right]} \tag{10}$$

where δ is the structural aspect ratio.

According to the above formulas, the 3D and 2D aerodynamic admittances of the structure were identified in the test, and the following general expression of the 2D aerodynamic admittance was used to fit the test results of the 2D aerodynamic admittance:

$$\left|\chi_L^{2D}(k_1)\right|^2 = \frac{1}{1 + \alpha k_1^{\beta}} \tag{11}$$

The two fitting parameters obtained were $\alpha = 6.584$ and $\beta = 1.444$. In addition, the 3D aerodynamic admittance was inconsistent with the general expression of the above 2D aerodynamic admittance, so a fitting tool was used to fit it for application to the subsequent comparison of the buffeting response results. Figure 15 shows the identification results and fitting curves of the aerodynamic admittance.

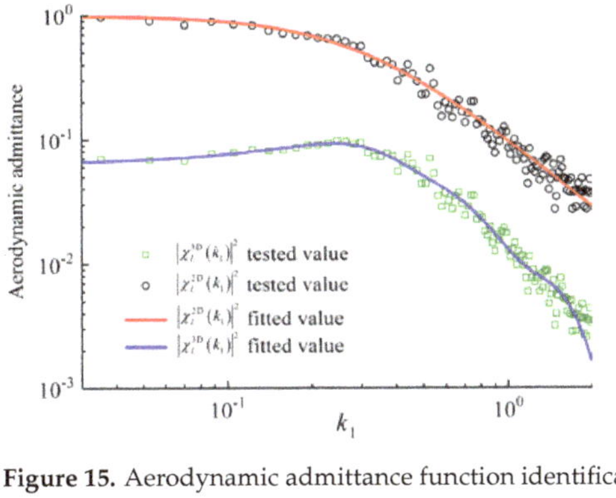

Figure 15. Aerodynamic admittance function identification result.

5.4. Integrated Transfer Function

The measured buffeting response results of the three modes were combined, and the results are shown in Figure 16. V-S-1, V-S-2 and V-S-3 represent the first- to third-order vertical mode shapes of the structure, respectively.

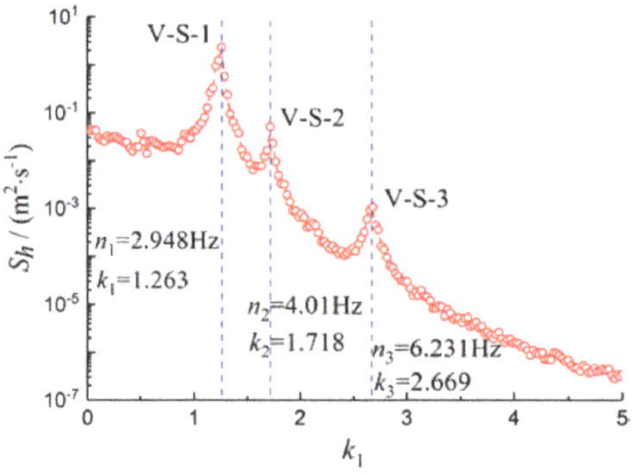

Figure 16. Vertical buffeting response of structure in XNJD-1 spire turbulence field.

As shown in the analysis results of the turbulence field characteristics simulated in the wind tunnel, the simulated turbulent fluctuating wind velocity spectrum conforms to the von Kármán spectral model. By substituting the turbulence field characteristic parameters into the von Kármán spectral model and combining them with the measured buffeting response of the structure, the integrated transfer function corresponding to each mode of the streamlined box girder can be calculated according to Equation (7). In addition, in order to clarify the displayed results, and based on the qualitative findings of Yang [45], Li [46], etc., who indicate that the results are less affected by the 3D effect at high-frequency positions while greatly affected at low-frequency positions, the test results of the integrated transfer function in this section are displayed in double logarithmic coordinates, as shown in Figure 17.

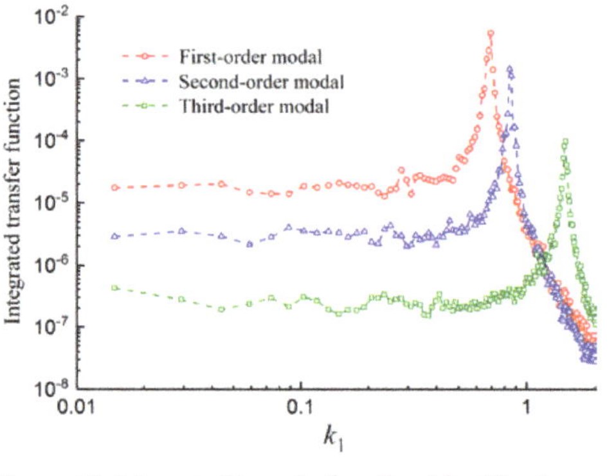

Figure 17. Integrated transfer function identification results.

5.5. Taut Strip Model Test

In the taut strip model test, two laser displacement sensors were symmetrically arranged on both sides of the width direction of the model and synchronous measurement was performed. The mean value of the buffeting response measured by the two sensors was the vertical buffeting response value of the structure. The measured buffeting response results of the three modes were combined, and the results are shown in Figure 18.

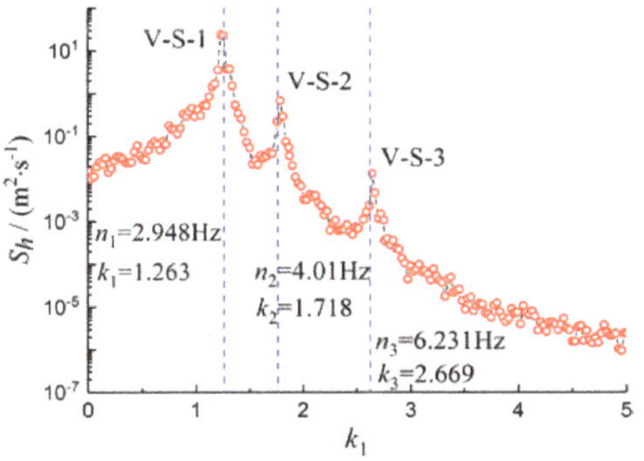

Figure 18. Buffeting response measured in taut strip model test.

6. Buffeting Response Prediction and Analysis

Relying on the aerodynamic admittance and aerodynamic derivative identified by the above sectional model test, the buffeting response result of the structure can be calculated according to Equations (1)–(3). Relying on the integrated transfer function identified by the sectional model vibration test, the prediction result of the buffeting response can be calculated according to Equation (7) and Equations (2) and (3). The buffeting response test result of the taut strip model can be calculated using Equation (8). A comparison of the three sets of results is shown in Figure 19. In addition, because the acquisition of the 2D aerodynamic admittance requires the identification of the 3D aerodynamic admittance, the results predicted using the 3D aerodynamic admittance and ignoring the influence of the aerodynamic admittance, based on the identification results of each aerodynamic admittance shown in Figure 12, are also displayed for comparison.

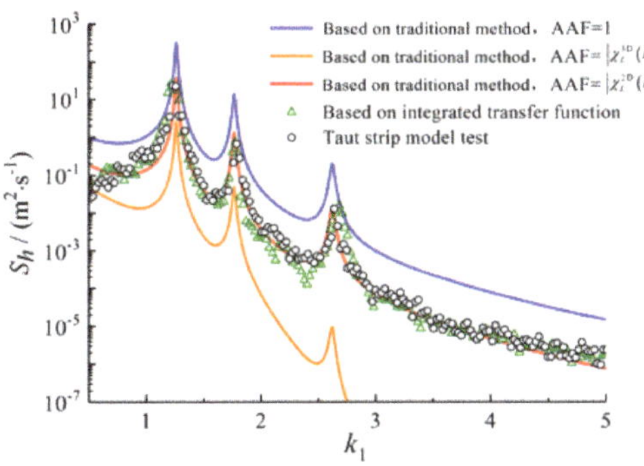

Figure 19. Buffeting response prediction results.

In Figure 18, the black and green scatter plots are the results of the taut strip model test and the prediction results based on the integrated transfer function, respectively. The solid curves with different colors are based on the traditional buffeting response calculation method, using the 2D and 3D aerodynamic admittances, respectively, and without considering the influence of the aerodynamic admittance. To facilitate this distinction and make the graph line clearer, the calculations based on the traditional method are based on the fitting formula of the aerodynamic admittance identification result.

As shown in the figure, there are three obvious main frequency peaks, which align well with the vertical natural frequency of the large-span bridge structure in the buffeting response prediction results obtained through different methods. The traditional buffeting response calculation method ignores the influence of the three-dimensional effect. Using the traditional sectional model test technique, which identifies the aerodynamic admittance without considering the influence of the model aspect ratio, the traditional method uses the correlation of the fluctuating wind instead of the correlation of the buffeting force, which will cause a certain deviation in the calculation results. However, based on the research of Li et al. [27–29] regarding the influence of the aspect ratio on the accuracy of the strip assumption, it was found that when the integral scale is not much larger than the structure width, if the structural aspect ratio is sufficient, the three-dimensional effect can be ignored, and the strip assumption is accurate enough. At this time, the correlation of the fluctuating wind can be used to replace the correlation of the buffeting force; that is, the traditional buffeting response calculation method has high accuracy. Su et al. [26] then proposed the concept of the integrated transfer function and used it as the basis for a method for predicting the buffeting response of long-span bridges. The taut strip model test is carried out under the premise of accurately simulating the dynamic characteristics of the bridge. As shown in Figure 18, the buffeting response prediction results based on the identification of the 3D aerodynamic admittance and the integrated transfer function of the model with a large aspect ratio fit well with the taut strip model test results. We have reason to propose a reasonable sectional model test technique. That is, if the turbulence integral scale is not much larger than the structure width, the aspect ratio can be increased to make the traditional buffeting response calculation method and the method based on the integrated transfer function highly accurate. On the other hand, these two calculation methods are derived strictly based on Davenport's theory, so the two calculation results fit well with the theoretical expectations. In addition, this result also verifies from the side that Davenport and Scanlan's analysis methods based on the strip assumption apply to long-span bridges, and reveals the problems existing in the current identification methods for the aerodynamic parameters required in theory. A solution is also given, which is to select a reasonable aspect ratio to make the model.

On the other hand, the results show that the prediction results of the integrated transfer function identified based on the sectional model vibration test have high accuracy when using the effective test technique. This paper verifies the accuracy and applicability of the integrated transfer function proposed by Su et al. [26] in the buffeting response prediction of long-span bridges through effective tests for the first time. This paper rarely realizes the prediction of the buffeting response of long-span bridges through sectional model vibration tests, which broadens the application scope of sectional model tests to a certain extent, and improves the long-span bridge buffeting response prediction efficiency. It is more conducive to the consideration of the buffeting performance of the bridge during the aerodynamic selection in the preliminary design stage of the structure.

In addition, the calculation results using the 3D aerodynamic admittance are obviously small and become smaller and smaller as the wavenumber increases. This shows that when using the 3D aerodynamic admittance directly identified by the test, there is a large error in calculating the buffeting response using the correlation of the fluctuating wind. It is again demonstrated that when the influence of the three-dimensional effect is considered, the correlation of the fluctuating wind should not be used instead of the correlation of the buffeting force. The research of Larose et al. [32–37] on the three-dimensional effect is also verified from the opposite side. When the correlation of the fluctuating wind is used instead of the correlation of the force to calculate the buffeting response, a small buffeting response will be obtained, leading to an unsafe result.

However, ignoring the influence of the aerodynamic admittance, that is, in the case of AAF = 1, the buffeting response calculation result is higher. At this time, this means that the unsteady characteristics of the buffeting force are not considered, and the buffeting force acting on the main beam of the bridge can be calculated according to the quasi-steady theory. Obviously, for a bridge in the atmospheric boundary layer, the buffeting force acting on the bridge girder has strong unsteady characteristics, and ignoring the aerodynamic admittance will result in a more conservative analysis result.

It should be noted that this paper takes the streamlined box girder section as an example to compare various buffeting response prediction methods, and obtains the expected results. Although overall positive in terms of scientific significance, in view of the increasingly complex cross-section forms of current long-span bridges, the conclusions of this paper still need further in-depth research to verify applicability for other bridge cross-section forms.

7. Conclusions

Taking a long-span suspension bridge as an example, this paper predicts the buffeting response of the bridge under the design wind velocity in the simulated large-scale turbulence field through different calculation methods based on different sectional model test techniques. The research results show the following:

(1) The feasibility of the method for predicting the buffeting response of long-span bridges based on the integrated transfer function has been verified by an effective wind tunnel test for the first time. Through reasonable test methods, the integrated transfer function can be identified through the sectional model vibration test to predict the buffeting response of long-span bridges with high accuracy;

(2) The problems existing in the current test methods for identifying the theoretically required aerodynamic parameters are pointed out, and an effective solution is proposed. That is, if the turbulence integral scale is not much larger than the structure width, increasing the model's aspect ratio can effectively reduce the influence of the three-dimensional effect. This provides an effective technical means of predicting the buffeting response of long-span bridges in future sectional model tests;

(3) When the 3D aerodynamic admittance is used for the calculation of the buffeting response, if the correlation of the fluctuating wind is used instead of that of the buffeting force, a small buffeting response and an unsafe result will be obtained. Neglecting the effect of aerodynamic admittance will overestimate the buffeting response of long-span bridges and obtain a more conservative analysis result. The unsteady characteristics of the buffeting force should be considered when calculating the buffeting response of long-span bridges.

In conclusion, by adopting the reasonable test techniques (selecting a reasonable model aspect ratio for the test to reduce the influence of the 3D effect), the buffeting response prediction results obtained through the three kinds of wind tunnel tests aligned well with the expected results. Even so, there were still slight deviations in the results, which were mainly caused by reasons such as the coupling between modes being ignored in the sectional model vibration test, the influence of turbulence on self-excited forces and the motion state of structure being ignored in the rigid sectional model force test, the Reynolds number effect, etc. Considering that the wind tunnel test is still an indispensable method in the buffeting response prediction for bridges, more effective test techniques are required in subsequent research to further improve the accuracy of the prediction methods.

Author Contributions: Conceptualization, Y.S. and J.D.; methodology, Y.S. and J.D.; experiments, Y.S. and J.L.; validation, Y.S. and S.L.; investigation, Y.S., S.L. and B.J.; data curation, Y.S. and J.L.; writing—original draft, Y.S.; writing—review and editing, Y.S. and B.J.; supervision, J.D.; funding acquisition, J.D. All authors have read and agreed to the published version of the manuscript.

Funding: This work was financially supported by the National Key R&D Program of China (Grant No. 2021YFF0501004) and the National Natural Science Foundation of China (52008357).

Institutional Review Board Statement: Not applicable.

Informed Consent Statement: Not applicable.

Data Availability Statement: Data are contained within this article.

Acknowledgments: The authors thank the reviewers for their great help on the article during its review progress.

Conflicts of Interest: The authors declare no conflict of interest.

References

1. Fung, Y.C. Fluctuating lift and drag acting on a cylinder in a flow at supercritical Reynolds numbers. *J. Aerosp. Sci.* **1960**, *27*, 801–814. [CrossRef]
2. Scruton, C. Aerodynamic buffeting on bridges. *Engineer* **1955**, *199*, 654–667.
3. Davenport, A.G. The response of slender, line-like structures to a gusty wind. *ICE Proc.* **1962**, *23*, 389–408. [CrossRef]
4. Davenport, A.G. Buffeting of a suspension bridge by storm winds. *J. Struct. Div.* **1962**, *88*, 233–270. [CrossRef]
5. Scanlan, R.H.; Tomko, J. Airfoil and bridge deck flutter derivatives. *J. Eng. Mech. Div.* **1971**, *97*, 1717–1737. [CrossRef]
6. Scanlan, R.H.; Gade, R.H. Motion of suspended bridge spans under gusty wind. *J. Struct. Div.* **1977**, *103*, 1867–1883. [CrossRef]
7. Scanlan, R.H. The action of flexible bridges under wind, I: Flutter theory. *J. Sound Vib.* **1978**, *60*, 187–199. [CrossRef]
8. Scanlan, R.H. The action of flexible bridges under wind, II: Buffeting theory. *J. Sound Vib.* **1978**, *60*, 201–211. [CrossRef]
9. Lin, Y.K.; Yang, J.N. Multimode bridge response to wind excitation. *J. Struct. Mech.* **1983**, *109*, 586–603. [CrossRef]
10. Kiviluoma, R. Coupled-mode buffeting and flutter analysis of bridges. *Comput. Struct.* **1999**, *70*, 219–228. [CrossRef]
11. Jain, A.; Jones, N.P.; Scanlan, R.H. Coupled flutter and buffeting analysis of long-span bridges. *J. Struct. Eng.* **1996**, *122*, 716–725. [CrossRef]
12. Xu, Z.; Wang, H.; Zhang, H.; Zhao, K.; Zhu, Q. Non-stationary turbulent wind field simulation of long-span bridges using the updated non-negative matrix factorization-based spectral representation method. *Appl. Sci.* **2019**, *9*, 5506. [CrossRef]
13. Kim, S.; Jung, H.; Kong, M.J.; Li, D.K.; An, Y.K. In-situ data-driven buffeting response analysis of a cable-stayed bridge. *Sensors* **2019**, *19*, 3048. [CrossRef]
14. Yan, L.; Ren, L.; He, X.; Lu, S.; Guo, H.; Wu, T. Strong wind characteristics and buffeting response of a cable-stayed bridge under construction. *Sensors* **2020**, *20*, 1228. [CrossRef]
15. Domaneschi, M.; Martinelli, L. Refined optimal passive control of buffeting-induced wind loading of a suspension bridge. *Wind Struct.* **2014**, *18*, 1–20. [CrossRef]
16. Chen, W.; Li, H.; Hu, H. An experimental study on the unsteady vortices and turbulent flow structures around twin-box-girder bridge deck models with different gap ratios. *J. Wind Eng. Ind. Aerodyn.* **2014**, *132*, 27–36. [CrossRef]
17. Li, H.; Chen, W.; Xu, F.; Li, F.; Ou, J. A numerical and experimental hybrid approach for the investigation of aerodynamic forces on stay cables suffering from rain-wind induced vibration. *J. Fluids Struct.* **2010**, *26*, 1195–1215. [CrossRef]
18. Chen, W.; Zhang, Q.; Li, H.; Hu, H. An experimental investigation on vortex induced vibration of a flexible inclined cable under a shear flow. *J. Fluids Struct.* **2015**, *54*, 297–311. [CrossRef]
19. Tao, T.; Wang, H.; Yao, C.; He, X. Parametric sensitivity analysis on the buffeting control of a long-span triple-tower suspension bridge with MTMD. *Appl. Sci.* **2017**, *7*, 395. [CrossRef]
20. Liu, H.; Lei, J.; Zhu, L. Identification and application of the aerodynamic admittance functions of a double-deck truss girder. *Appl. Sci.* **2019**, *9*, 1818. [CrossRef]
21. Gao, D.; Chen, W.; Li, H.; Hu, H. Flow around a circular cylinder with slit. *Exp. Therm. Fluid Sci.* **2017**, *82*, 287–301. [CrossRef]
22. Chen, W.; Gao, D.; Li, H.; Hu, H. Passive jet control of flow around a circular cylinder. *Exp. Fluids.* **2015**, *56*, 201. [CrossRef]
23. Yan, L.; Zhu, L.; He, X.; Flay, R.G.J. Experimental determination of aerodynamic admittance functions of a bridge deck considering oscillation effect. *J. Wind Eng. Ind. Aerodyn.* **2019**, *190*, 83–97. [CrossRef]
24. Chen, W.; Xin, D.; Xu, F.; Li, H.; Ou, J.; Hu, H. Suppression of vortex-induced vibration of a circular cylinder using suction-based flow control. *J. Fluids Struct.* **2013**, *42*, 25–39. [CrossRef]
25. Laima, S.; Li, H.; Chen, W.; Li, F. Investigation and control of vortex-induced vibration of twin box girders. *J. Fluids Struct.* **2013**, *39*, 205–221. [CrossRef]
26. Su, Y.; Li, M. Integrated transfer function for buffeting response evaluation of long-span bridges. *J. Wind Eng. Ind. Aerodyn.* **2019**, *189*, 231–242. [CrossRef]
27. Li, M.; Yang, Y.; Li, M.; Liao, H. Direct measurement of the Sears function in turbulent flow. *J. Fluid Mech.* **2018**, *847*, 768–785. [CrossRef]
28. Li, M.; Li, M.; Yang, Y. A statistical approach to the identification of the two-dimensional aerodynamic admittance of streamlined bridge decks. *J. Fluids Struct.* **2018**, *83*, 372–385. [CrossRef]
29. Li, M.; Li, M.; Zhong, Y.; Luo, N. Buffeting response evaluation of long-span bridges with emphasis on three-dimensional effects of gusty winds. *J. Sound Vib.* **2019**, *439*, 156–172. [CrossRef]
30. Chen, X.; Kareem, A. Equivalent static wind loads for buffeting response of bridges. *J. Struct. Eng.* **2001**, *127*, 1467–1475. [CrossRef]
31. Fung, Y.C. *An Introduction to the Theory of Aeroelasticity*; John Wiley & Sons: New York, NY, USA, 1955.
32. Larose, G.L.; Livesey, F.M. Performance of streamlined bridge decks in relation to the aerodynamics of a flat plate. *J. Wind Eng. Ind. Aerodyn.* **1997**, *69*, 851–860. [CrossRef]
33. Larose, G.L.; Mann, J. Gust loading on streamlined bridge decks. *J. Fluids Struct.* **1998**, *12*, 511–536. [CrossRef]
34. Hjorth-Hansen, E.; Jakobsen, A.; Strømmen, E. Wind buffeting of a rectangular box girder bridge. *J. Wind Eng. Ind. Aerodyn.* **1992**, *42*, 1215–1226. [CrossRef]
35. Jakobsen, J.B. Span-wise structure of lift and overturning moment on a motionless bridge girder. *J. Wind Eng. Ind. Aerodyn.* **1997**, *69*, 795–805. [CrossRef]

36. Kimura, K.; Fujino, Y.; Nakato, S.; Tamura, H. Characteristics of buffeting forces on flat cylinders. *J. Wind Eng. Ind. Aerodyn.* **1997**, *69*, 365–374. [CrossRef]
37. Ma, C.; Wang, J.; Li, Q.; Liao, H. 3D aerodynamic admittances of streamlined box bridge decks. *Eng. Struct.* **2019**, *179*, 321–331. [CrossRef]
38. Ma, C. 3D Aerodynamic Admittance Research of Streamlined Box Bridge Decks. Ph.D. Thesis, Southwest Jiaotong University, Chengdu, China, 2007. (In Chinese)
39. Li, S.; Li, M.; Larose, G.L. Aerodynamic admittance of streamlined bridge decks. *J. Fluids Struct.* **2018**, *78*, 1–23. [CrossRef]
40. Su, Y.; Li, M.; Yang, Y.; Mann, J.; Liao, H.; Li, X. Experimental investigation of turbulent fluctuation characteristics observed at a moving point under crossflows. *J. Wind Eng. Ind. Aerodyn.* **2020**, *197*, 104079. [CrossRef]
41. von Kármán, T.; Howarth, L. On the statistical theory of isotropic turbulence. *Proc. R. Soc. Lond. Ser. A-Math. Phys. Sci.* **1938**, *164*, 192–215. [CrossRef]
42. Simiu, E.; Scanlan, R.H. *Wind Effects on Structures*; John Wiley & Sons: New York, NY, USA, 1996.
43. Xie, J.; Hunter, M.; Irwin, P. Experimental and analytical approaches in wind engineering studies for bridges. In Proceedings of the Budapest: Responding to Tomorrow's Challenges in Structural Engineering, Budapest, Hungary, 13–15 September 2006.
44. Li, M.; Li, M.; Su, Y. Experimental determination of the two-dimensional aerodynamic admittance of typical bridge decks. *J. Wind Eng. Ind. Aerodyn.* **2019**, *193*, 103975. [CrossRef]
45. Yang, Y. Aerodynamic Admittances of Airfoil and Rectangular Cylinder. Ph.D. Thesis, Southwest Jiaotong University, Chengdu, China, 2019. (In Chinese)
46. Li, M. Buffeting Response Analysis of Long-Span Bridges with Emphasis on the Three-dimensional Effects of Turbulence and the Study on the Equivalent Static Wind Loads. Ph.D. Thesis, Southwest Jiaotong University, Chengdu, China, 2019. (In Chinese)

applied sciences

MDPI

Article

Effects of the Configuration of Trailing Edge on the Flutter of an Elongated Bluff Body

Jie Feng [1,2,3], Buchen Wu [1,2,4,5] and Shujin Laima [1,2,*]

1 Key Lab of Smart Prevention and Mitigation for Civil Infrastructures, Ministry of Industry and Information Technology, Harbin 150090, China; 20b954002@stu.hit.edu.cn (J.F.); buchenwu@u.nus.edu (B.W.)
2 Key Lab of Structures Dynamic Behavior and Control of the Ministry of Education, Harbin Institute of Technology, Harbin 150090, China
3 School of Civil Engineering and Environment, Harbin Institute of Technology (Shenzhen), Shenzhen 518073, China
4 Department of Mechanical Engineering, National University of Singapore, 10 Kent Ridge Crescent, Singapore 119260, Singapore
5 Guangdong Provincial Key Laboratory of Turbulence Research and Applications, Department of Mechanics and Aerospace Engineering, Southern University of Science and Technology, Shenzhen 518055, China
* Correspondence: laimashujin@hit.edu.cn

Abstract: Wind-tunnel experiments are performed to investigate the effects of trailing-edge reattachment on the flutter behaviors of spring-suspended trailing-edge-changeable section models. Different Trailing edges (TE) were fixed at the back of a body to adjust reattachment of the vortex. A laser-displacement system was used to acquire the vibration signals. The relationship between flutter characteristics and TEs that affects the wake mode was analyzed. The results show that the motion of the wake vortex has a certain correlation with the flutter stability of the bridge deck. Limit cycle flutter (LCF) occurs to a section model with a 30° TE, whose amplitude gradually increases as the wind speed increases, and the vibration develops into a hard flutter when the wind speed is 12.43 m/s. A section model with 180 TE reaches a hard flutter when the wind speed is 15.31 m/s, without the stage of LCF. As the TE becomes more and more blunt, the critical wind speed, U_s, gradually increases, meaning the flutter stability gradually increases. The results reveal that LCF may still occur to the bridge section with a streamlined front edge, and, in some cases, it also may have a range of wind speeds in which LCF occurs.

Keywords: trailing-edge reattachment; trailing edge; trailing-edge-changeable streamlined section mode; limit cycle flutter; hard flutter; flutter stability

Citation: Feng, J.; Wu, B.; Laima, S. Effects of the Configuration of Trailing Edge on the Flutter of an Elongated Bluff Body. *Appl. Sci.* **2021**, *11*, 10818. https://doi.org/10.3390/app112210818

Academic Editors: Wenli Chen, Zifeng Yang, Gang Hu, Haiquan Jing and Junlei Wang

Received: 22 October 2021
Accepted: 12 November 2021
Published: 16 November 2021

1. Introduction

Flutter is a phenomenon of aerodynamic instability caused by the interaction of fluid and the vibration of structure, which was first found on thin wings. The research of wind-induced vibration of bridge decks began in 1940. Tacoma Narrows Bridge of Washington State, USA, which had just been completed and opened to traffic for 4 months, experienced severe torsional vibration under a low-speed wind of 19 m/s. After more than an hour of vibration, the bridge collapsed. Since then, the wind-induced vibration of the bridge decks has been highly reported, and a large number of research results have been obtained after 80 years of continuous in-depth research.

Bleich in 1948 [1] regarded the deck of the truss stiffening beam, commonly used in suspension bridges as a flat plate, and ignored the air force on the hollow truss. It was proposed that the Theodorsen's unsteady aerodynamic theory [2] can be used to predict the vibration of bridge. However, the critical flutter wind speed, calculated based on Theodorsen's unsteady aerodynamic theory, is much greater than the wind speed when the Tacoma Narrows Bridge collapsed. Kloppel et al. [3] also used Theodorsen's function to solve the critical flutter wind speed of the bridge, but for a specific actual bridge section.

A reduction factor obtained by wind tunnel tests was introduced to modify the critical flutter wind speed. Scanlan et al. [4] believed that it is difficult to obtain an aerodynamic expression similar to Theodorsen's function for bridges with bluff-body section. Therefore, it is recommended to use aerodynamic derivatives to establish a small-amplitude, unsteady aerodynamic model of bridges in which the self-excited aerodynamic force of the main girder section of bridges is approximated by a linear function of the two-dimensional section's motion state vector composed of flutter derivatives, and the determination of the aerodynamic derivatives requires a special wind tunnel test on segment model. The vertical and torsional self-excited aerodynamic force are obtained:

$$L = \frac{1}{2}\rho U^2 B\left[KH_1^*\frac{\dot{h}}{U} + KH_2^*\frac{B\dot{\alpha}}{U} + K^2H_3^*\alpha + K^2H_4^*\frac{h}{B}\right] \quad (1)$$

$$M = \frac{1}{2}\rho U^2 B\left[KA_1^*\frac{\dot{h}}{U} + KA_2^*\frac{B\dot{\alpha}}{U} + K^2A_3^*\alpha + K^2A_4^*\frac{h}{B}\right] \quad (2)$$

where L and M are the self-excited aerodynamic forces in the vertical and torsional directions of the main girder of the bridge, respectively; H_i^*, A_i^*, and $i = 1, 2, 3, 4$, are the flutter derivative of the main girder; ρ is the density of air; B is the width of the bridge; U is the average wind speed; K is the converted frequency, $K = \omega B/U$, where ω is the circular frequency; h and α are the displacements in the vertical and torsional directions respectively; and $\dot{h}$ and $\dot{\alpha}$ are the speeds in the vertical and torsional directions respectively. Since all the motion parameters that have an influence on the self-excited aerodynamic force are considered, this method has strong universality. Therefore, Scanlan's flutter theory has been widely used in the design of the flutter stability of bridge structures and has become a classic theory of bridge flutter. In the research of the mechanism of aerodynamic coupling of flutter, Matsumoto et al. [5–7] proposed the use of step-by-step analysis to analyze bridge flutter. This method divides the analysis of flutter into torsional branch and vertical branch. In each branch, the coupling effects of torsion and vertical motion are considered separately. Not only can the change in system's damping and frequency with wind speed be obtained, but it can also reflect the role of an aerodynamic derivative in the change of vibration characteristics and the flutter derivatives' impact on a system's damping. Based on Scanlan's linear flutter theory, Bartoli et al. [8] proposed an approximate method to calculate the critical wind speed and frequency of flutter by the analysis of a large number of dynamics and aerodynamics data, which only requires three flutter derivatives (H_1^*, A_2^*, and A_3^*) or two flutter derivatives (H_1^* and A_2^*) to perform the calculation, and greatly simplifies the approximate calculation of critical wind speed and frequency of flutter. It is suitable for bridges with the section whose bending-torsion coupling is only affected by the shape of a structural mode or a higher mode, but it cannot completely replace more accurate analysis methods. Lee et al. [9] started by simplifying the coefficients in the third and fourth degree polynomials of ω, derived from the singular condition of the frequency matrix. Under the quasi-steady assumption, they proposed an approximate formula for calculating the flutter of the aeroelastic system. The critical wind speed of flutter of long-span bridges can be estimated well within the range of bending-torsion coupling, and the calculation result fits well with the results of the wind tunnel test. The rapid and approximate calculation of bridges' flutter stability helps to better understand the mechanism of flutter and represents the first step towards a simple engineering tool, which is also of great significance for speeding up the design process of bridges.

Researchers found in wind tunnel tests that when the model has a large wind attack angle, or the bridge's deck cross section is a blunt-body section, there will be no obvious divergent flutter, but large-amplitude and limiting-amplitude vibration, called soft flutter, which is also known as limit cycle flutter (LCF) may occur. Correspondingly, the divergent flutter is called hard flutter [10–12]. Ying et al. [13] used a fluid-structure-coupling model to numerically simulate the aeroelastic response of the bridge deck and verified it with the theoretical solution of the flutter response of a thin plate. It is proposed that the increase in the wind attack angle makes the bridge deck more prone to LCF; the stable amplitude of LCF has nothing to do with the initial excitation, and the increasing the damping of structure can significantly increase the critical wind speed of LCF of the bridge, but it has little effect on the vibration frequency. Gao et al. [14] analyzed the nonlinear aeroelastic behavior of a typical passivated bridge section (that is, the bridge deck with double-sided guardrails opened) through a series of LCF tests of the Spring-Suspended Sectional Model (SSSM). A model of self-excited nonlinear force suitable for large-amplitude vibration is proposed, and its forecast results are compared with experimental results. The identified aerodynamic parameters further indicate that the mechanism of LCF is due to the negative aerodynamic damping provided by the linear term as the energy source to drive the increase in vibration amplitude and the nonlinear positive aerodynamic damping provided by the third-order angular velocity. In the tested wind speed range, the observed stable LCF amplitude increased linearly with the decrease in wind speed. Zhang et al. in 2017 [15] researched the flutter characteristics of streamlined bridge decks by using a comprehensive wind tunnel test and introduced a nonlinear mathematical model to simulate the aeroelastic behavior of nonlinear torsional flutter. It was found that nonlinear aerodynamics are the main source of nonlinear vibration. Based on the instantaneous amplitude and instantaneous frequency calculated by the normalized Hilbert transform, a system identification method was developed to extract linear and nonlinear aerodynamic parameters at the same time. On this basis, Zhang et al., in 2019, [16] simulated an LCF process of the bridge based on computational fluid dynamics (CFD) and analyzed the energy input in this process. The results show that only the first-order component of the self-excited force contributes significantly to the energy input of LCF of the bridge deck, while the influence of the higher-order force component is negligible, which means considering only the first-order component of the self-excited nonlinear force can well predict the displacement response of LCF of the bridge deck.

The self-excited aerodynamic force is sensitive to the aerodynamic shape, which means a small change in the aerodynamic shape may cause a huge change in aerodynamic force. Researchers have conducted in-depth research on the influence of aerodynamic shape on the flutter performance of bridges. Larsen [17] researched in detail the influence of aerodynamic components—such as railings and wind screens—on the critical wind speed of flutter by a series of wind-tunnel tests of the Great Belt Bridge in Denmark, including tests of a section model, a taut strip model and a full bridge model. The result provides reliable advice for the design and maintenance of the Great Belt Bridge. Bruno et al. [18] investigated the influence of deck equipment of bridges—such as middle divider, edge safety barrier, and guardrail—on the aerodynamic performance of bridges. Using numerical simulations with and without barriers, such members' noticeable effect on the analysis bridge aerodynamics is demonstrated. The simulation results are compared with each other and then compared with wind-tunnel-test results, showing that bridge deck equipment cannot be ignored in bridge aerodynamic analysis because the existence of barriers increases the overall degree of bluffness of the section. Wilde et al. [19] used control flaps attached to the edge of the bridge deck to aerodynamically control the wind-induced instability of long-span bridges. It is proposed that the control system whose flap rotation center is located at the edge of the deck is the most effective, which can provide satisfactory aerodynamic damping and has sufficient stability. Lin et al. [20] investigated the influence of cross-section geometry and oncoming turbulence on the flutter of cable-supported bridges through wind-tunnel tests on multiple section models, under different flow conditions and subsequent flutter analysis. The results show that the width-to-depth ratio of the bridge deck plays an important role in the aerodynamic performance of the bridge. Increasing the width-to-depth ratio and the flatness of the bridge deck can improve the flutter stability of the bridge deck. In addition, the greater the intensity of turbulence, the greater the flutter stability of the bridge deck, which means the turbulence tends to enhance the flutter stability of the bridge. Based on the nonlinear aerodynamic stability analysis, Zhang et al. [21] conducted parametric studies on the Runyang bridge over the Yangtze River to investigate the effects of some design parameters on the aerodynamic. It was proposed that depth of the deck has an important effect on the aerodynamic stability of the bridge.

Based on Scanlan's flutter theory, it is difficult to obtain the aerodynamic characteristics of a specific flow pattern and the flutter behavior caused by it, which play an important role in the design of bridge sections with higher flutter stability. Based on the basic theories of fluid mechanics and fluid-structure interaction, this paper carried out wind-tunnel experiments to investigate the effect of wake flow on the flutter stability of bridge decks.

2. Experiment Setup

The tests were performed in a blow-down wind tunnel at the Joint Laboratory of Wind Tunnel and Wave Flume, Harbin Institute of Technology, China, and the panoramic view of the wind tunnel is shown in Figure 1. The test section is 1150 mm long and has a cross-sectional area of 450 mm × 450 mm. The wind speed in the test section can be adjusted within the range of 1–30 m/s, and the turbulence of the incoming flow is lower than 0.6%, which meets the test requirements.

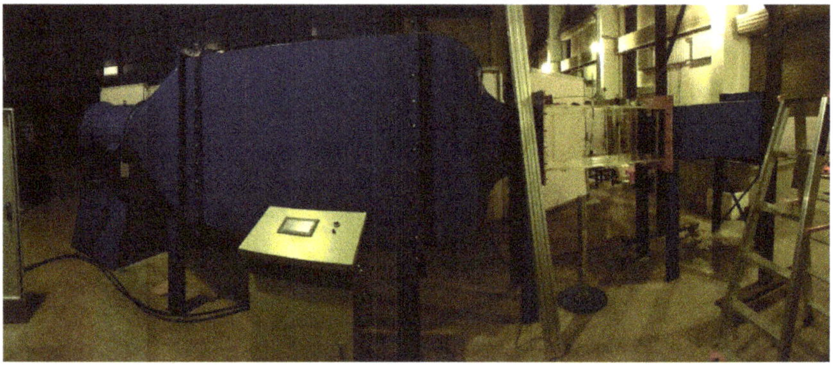

Figure 1. Wind tunnel.

2.1. Model Details

As is shown in Figure 2, the overall aspect ratio of the model is 1:10, the height H is 20 mm, the width B is 200 mm, and the axial length is 430 mm. In order to study the effect of wake flow on the flutter stability of the bridge, the streamlined front edge (FE), whose section is an isosceles triangle with an apex angle of 30°, is used to reduce the separation of flow at the FE of the model. For the FE, rounded corners are used in the transition section to minimize the separation here, which is illustrated in Figure 2c. For the trailing edge (TE), a different apex angle, β, is configured to construct TEs with different degrees of streamline to obtain different wake flow. In this paper, five cases of apex angle of the TE β, from 30° to 180° (β = 30°, 45°, 60°, 90°, 180°), were studied to investigate the effects of wake flow on the flutter stability of bridge decks, using β to quantify the effects of wake flow. The details of the section are shown in Figure 2b. The test model consists of three parts: one FE, five TEs, and a foundation deck. Since the mass of the FE and TE is not negligible relative to the mass of the entire model, simply designing the TE with different shapes will cause the model of bridge deck to be seriously off-center, which may cause large errors in the test results. In order to reduce this error as much as possible, masses are added at appropriate locations inside the TE, so that the mass of each TE and moment of inertia of each TE relative to the center axis of the bridge deck are nearly equal to that of the FE (±3% theoretically). The perspectives of the FE and 5 TEs are shown in Figure 2d. Each part is made of resin material through 3D printing technology. Unfortunately, due to the limitations of 3D printing technology, the error mentioned above may be magnified.

The overall rigidity of the foundation bridge deck is guaranteed by a model shell with a thickness of at least 3 mm and 3 steel bars fixed inside, and the overall rigidity of the TE and FE is guaranteed by a model shell with a thickness of at least 2 mm and the internal mass block.

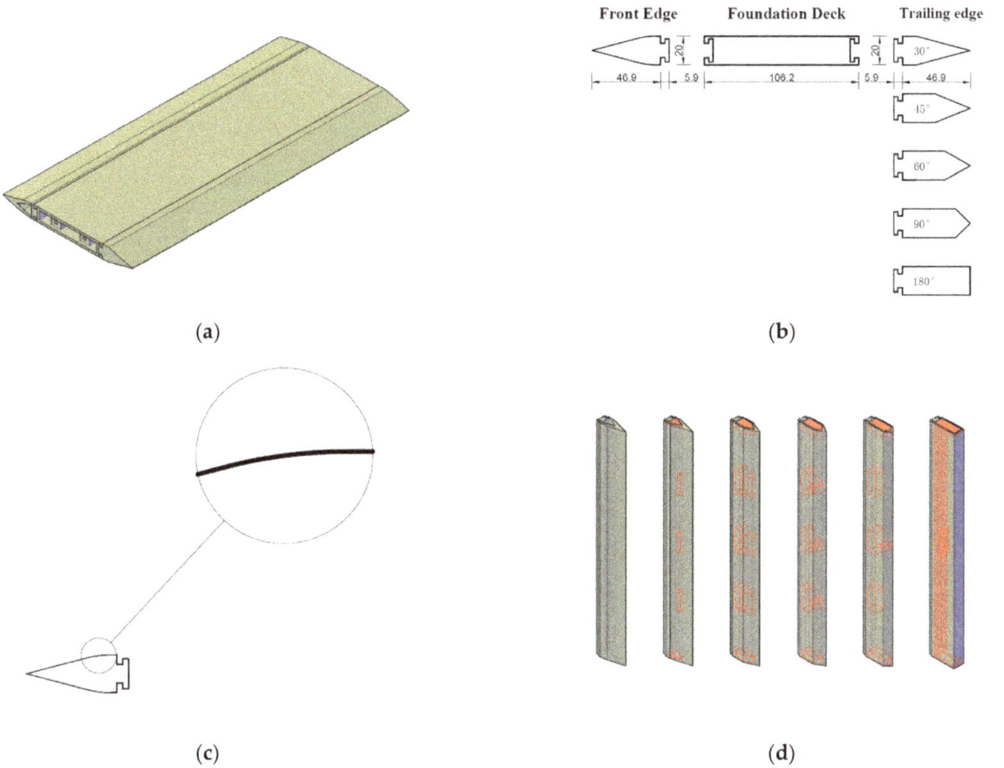

Figure 2. Details of the test model. (**a**) Overview test model (case 30c-30). (**b**) Cross section. (**c**) Details about FE. (**d**) Details about the masses added inside TE.

A test model was installed in the test section of the wind tunnel through a spring suspension system. As shown in Figure 3, the test model is elastically supported by eight springs of equal stiffness. The tests of free vibration were performed on the five cases to obtain their natural frequency. The vertical and torsional natural frequency between each case, ω_v and ω_t, are basically equal, 2.6 Hz and 5.0 Hz, respectively. f_v and f_t, representing the vertical and torsional damping ratio, respectively, are both less than 0.2%, and the details of the model's dynamic characteristics are shown in Table 1. The blockage ratio is 4.25%.

2.2. Measurement Details

As mentioned above, five cases of β were studied to explore the effects of wake flow on the flutter characteristics of bridge decks. The model was installed in the test section of the wind tunnel through the spring suspension system, the wind speed, U_{in}, was gradually increased to the model's critical wind speed of flutter, and the laser-displacement system was used to collect the displacement response of the model. The laser-displacement system includes two IL-100 sensors produced by Keyence, USB-6288 produced by National

Instruments, and a personal computer. The sampling frequency for all tests is 1000 Hz. The torsional and vertical responses are calculated as follows:

$$\begin{cases} x = (x_1 + x_2)/2 \\ \theta = 180 \times (x_1 - x_2)/\pi d \end{cases} \quad (3)$$

where x_1 and x_2 are the displacement signals collected by the laser-displacement system; d is the distance between two measure points; x is the vertical displacement of the deck model; and θ is the torsional displacement of the deck model in radians.

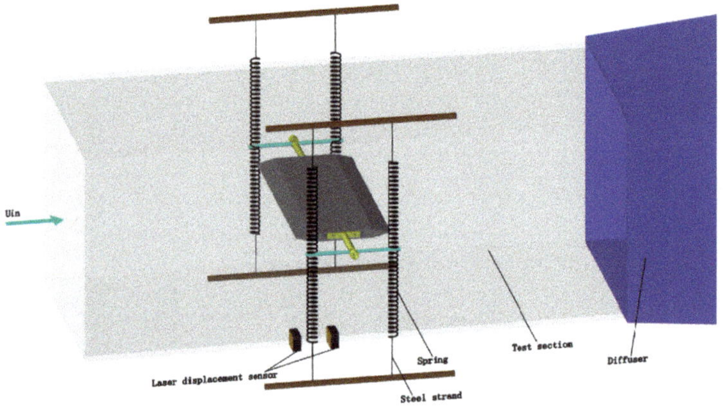

Figure 3. The setup of the vibration system. The steel strand is tightened by bolts. The model of bridge deck is placed in the center of the test section.

Table 1. The dynamic characteristics of model.

Serial Number	Fundamental Frequency (Hz)		Damping Ratio (%)	
	Vertical	Torsional	Vertical	Torsional
30c-30	2.640	5.066	0.193	0.141
30c-45	2.63	5.028	0.125	0.092
30c-60	2.625	5.020	0.104	0.145
30c-90	2.640	5.035	0.157	0.170
30c-180	2.623	5.005	0.176	0.126

3. Results and Discussions

3.1. Dynamic Response of Flutter with Different Trailing Configurations

3.1.1. 30c-30 Case

Figure 4a,b, respectively, show the vertical and torsional vibration of deck model in case 30c-30, where U_r is reduced velocity, $U_r = U_{in}/\omega_t H$. The blue vertical dashed line is used to distinguish several parts in the figure, each part is the continuous displacement response of the model within 10s under the corresponding U_r.

When $U_r < 61.25$, the deck does not vibrate. As $U_r > 70.00$, there is an obvious single-frequency vibration in the torsional degree of freedom of the bridge deck, but it is still a multi-frequency and low-amplitude vibration in the vertical degree of freedom. Until $U_r = 105.00$, the vibration of the model is a torsional single-frequency LCF. When $U_r = 113.75$, a large and stable vibration appears in the vertical degree of freedom of the bridge deck, and its vibration frequency is equal to that of the torsional degree of freedom, which shows vertical and torsional vibrations are completely coupled. As $U_r = 113.75 \sim 122.50$, the vibration of the deck is a bending-torsional coupling LCF. The bridge deck in the 30c-30 case has a wide range of wind speeds in which LCF occurs. When

$U_r = 70.00 \sim 122.50$, LCF occurs. Among them, pure torsional single-frequency vibration occurs at lower wind speeds, and when U_r is higher, it develops into bending-torsional coupling vibration.

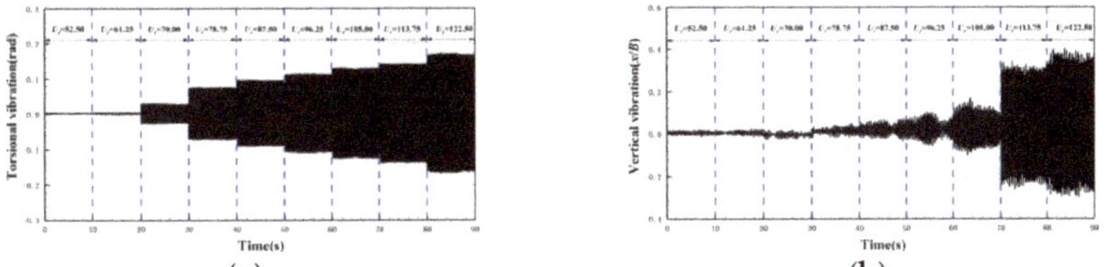

(a) (b)

Figure 4. Time history of displacement of the bridge deck in case 30c-30. (**a**) Torsional degree of freedom; and (**b**) vertical degree of freedom.

When $U_r = 122.50$, stable and high-amplitude LCF occurs to the bridge deck, and as U_r is increased to 124.25, the LCF develop into hard flutter. Figure 5 shows this process. The amplitudes of the vertical and torsional degrees of freedom gradually increase to divergence.

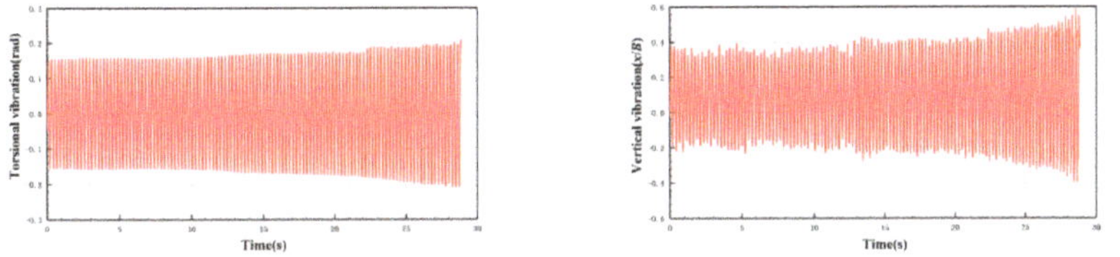

(a) (b)

Figure 5. Time history of vibration divergence in 30c-30 case. (**a**) Torsional degree of freedom; and (**b**) vertical degree of freedom.

3.1.2. 30c-45 Case

As Figure 6 shows, in the case of 30c-45, when $U_r < 61.25$, the torsional and vertical vibration of the bridge deck are both multi-frequency and low-amplitude vibration. As U_r is increased to 87.50, the bridge begins to undergo torsional vibration, while the vibration is low-amplitude in the vertical degree of freedom and the effect of static wind is relatively large. When U_r is in the range of 87.50~96.25, there is stable single-frequency vibration on the torsional freedom of the bridge deck, while there are some low frequency components in the vertical degrees of freedom. In terms of vertical degrees of freedom, the effect of static wind takes a larger proportion. The coupling degree of vertical and torsional vibration is low.

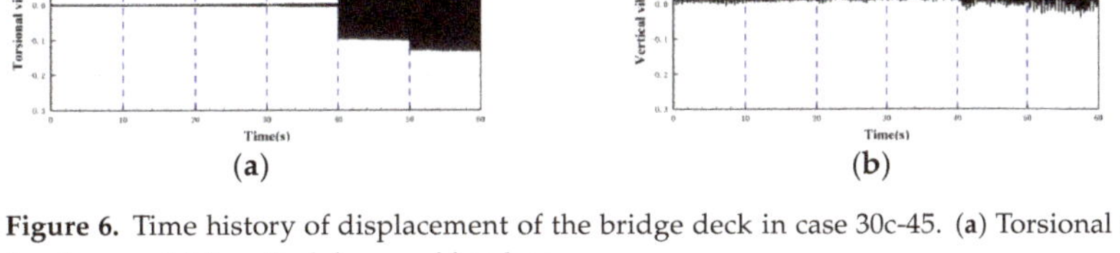

(a) (b)

Figure 6. Time history of displacement of the bridge deck in case 30c-45. (**a**) Torsional degree of freedom; and (**b**) vertical degree of freedom.

Figure 7 shows the process of that vibration of bridge deck develops from LCF to hard flutter when U_r is increased from 96.25 to 97.13. There is no stable vibration in the vertical degree of freedom, which means the bridge deck's vibration is pure torsional vibration during the process of LCF to hard flutter.

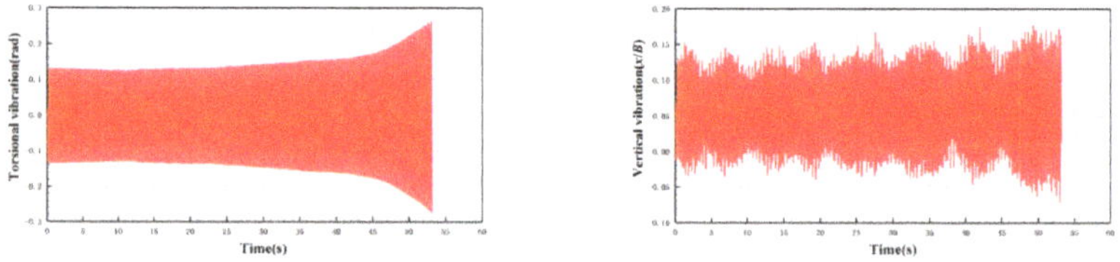

(**a**) (**b**)

Figure 7. Time history of vibration divergence in 30c-45 case. (**a**) Torsional degree of freedom; and (**b**) vertical degree of freedom.

3.1.3. 30c-60 Case

In the 30c-60 case, as Figure 8 shows, when $U_r < 113.75$, there is multi-frequency and low-amplitude vibration in both vertical and torsional degree of freedom of the bridge deck. As $U_r = 122.50$, there is stable large-amplitude vibration in both vertical and torsional degree of freedom of the bridge deck, showing bending-torsional coupling LCF occurs.

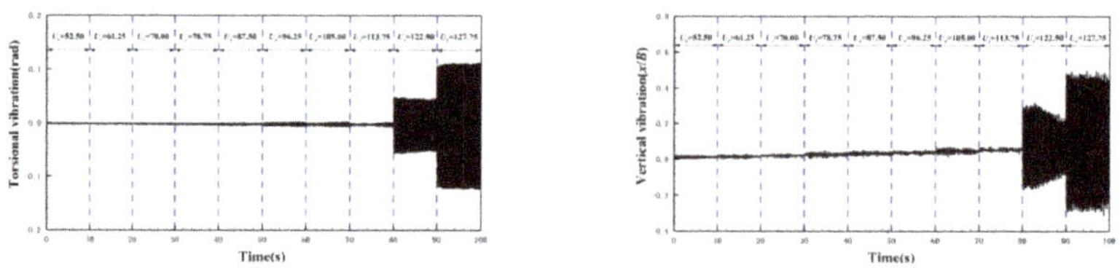

(**a**) (**b**)

Figure 8. Time history of displacement of the bridge deck in case 30c-60. (**a**) Torsional degree of freedom; and (**b**) vertical degree of freedom.

As Figure 9 shows, when the U_r is increased from 127.75 to 128.63, the vibration is developed from stable LCF to hard flutter. In this process, both vertical degree of freedom and torsional degree of freedom have large amplitudes, and both have trend of divergence.

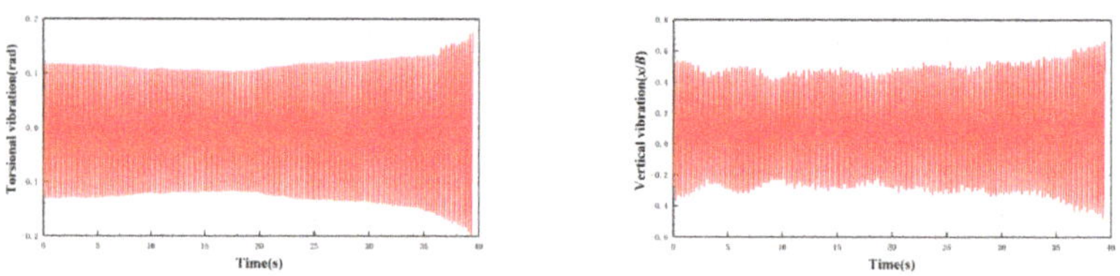

(**a**) (**b**)

Figure 9. Time history of vibration divergence in 30c-60 case. (**a**) Torsional degree of freedom; and (**b**) vertical degree of freedom.

3.1.4. 30c-90 Case

As for the 30c-90 case, when $U_r < 131.25$, there is no obvious vibration observed on the bridge deck, which can be seen from Figure 10. When U_r is increased to 140.00, large-amplitude and bending-torsion coupling LCF occurs to the bridge deck. The amplitude is greatly increased compared to the case of $U_r < 131.25$. The vibration frequency of the vertical degrees of freedom and the torsional degrees of freedom are equal, indicating that the vibrations of the two degrees of freedom are completely coupled.

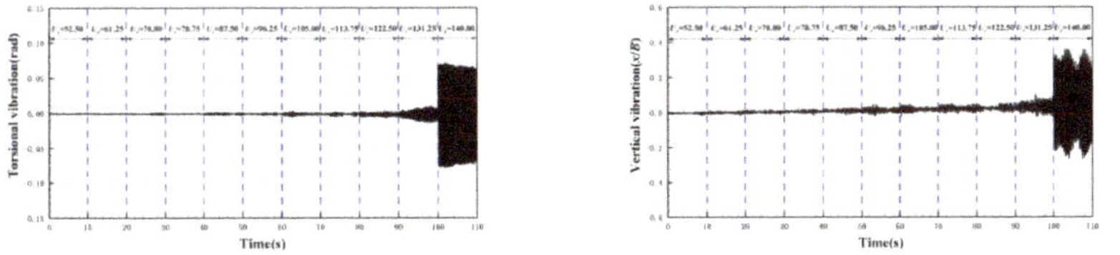

(a) (b)

Figure 10. Time history of displacement of the bridge deck in case 30c-90. (**a**) Torsional degree of freedom; and (**b**) vertical degree of freedom.

As U_r increased from 140.00 to 140.88, the amplitude of the bridge deck gradually diverged, which is shown in Figure 11. In this process, both vertical degree of freedom and torsional degree of freedom have a trend of divergence.

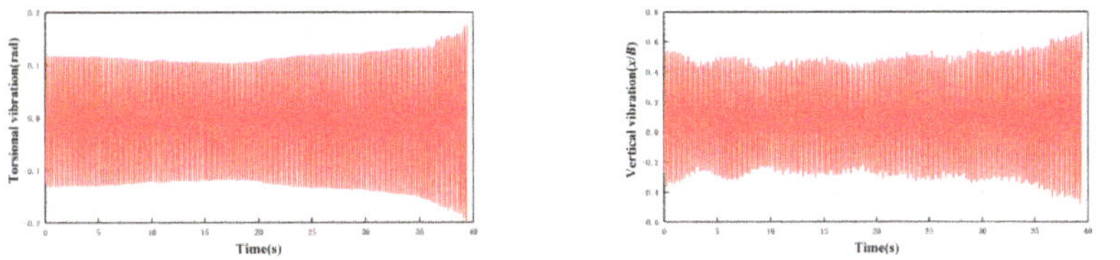

(a) (b)

Figure 11. Time history of vibration divergence in 30c-90 case. (**a**) Torsional degree of freedom; and (**b**) vertical degree of freedom.

3.1.5. 30c-180 Case

In the 30c-180 case, as is shown in Figure 12, when $U_r < 148.75$, there is no obvious vibration occurring in the model. As Figure 13 shows, when U_r is increased from 148.75 to 153.13, hard flutter occurs, instead of LCF first. Both vertical degrees of freedom and torsional degrees of freedom have obvious trends of divergence, and their vibration frequencies are equal.

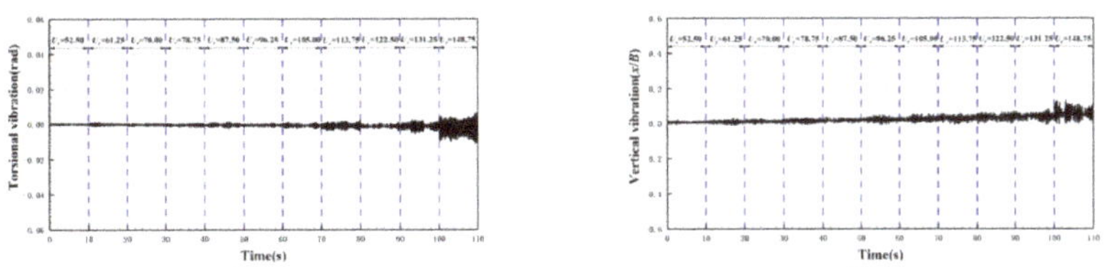

(a) (b)

Figure 12. Time history of displacement of the bridge deck in case 30c-180. (**a**) Torsional degree of freedom; and (**b**) vertical degree of freedom.

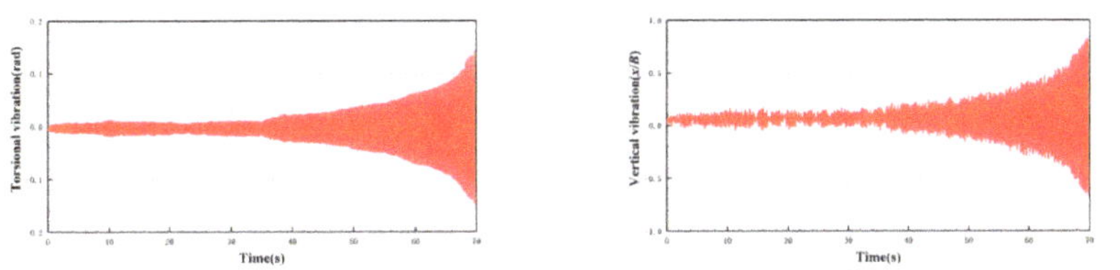

(a) (b)

Figure 13. Time history of vibration divergence in 30c-180 case. (**a**) Torsional degree of freedom; and (**b**) vertical degree of freedom.

3.2. Frequency Domain Analysis

Figure 14 shows the vibration frequency of the model at different wind speeds in each case. It should be noted that the vibration frequency mentioned above is the frequency corresponding to the vibration mode with the highest amplitude.

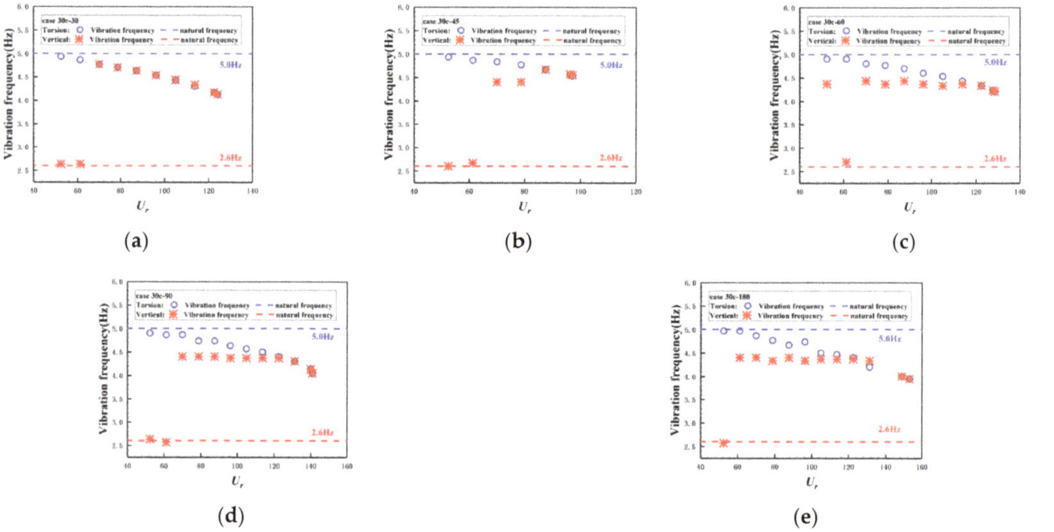

Figure 14. Vibration frequency of each case. (**a**) 30c-30 case; (**b**) 30c-45 case; (**c**) 30c-60 case; (**d**) 30c-90 case; and (**e**) 30c-180 case.

As Figure 14 shows, as the wind speed increases, the vibration frequency of the vertical degree of freedom shows an upward trend, and the vibration frequency of the torsional degree of freedom shows a downward trend. It is most interesting that the torsional vibration frequency has continuous change regularities, however, the vertical vibration frequency exists as a mutation phenomenon and is attracted by the torsional vibration frequency. As the wind speed continues to increase, the main frequency of vertical vibration is equal to the torsional frequency, indicating that the vertical vibration and torsional vibration begin to couple. Since then, as the wind speed continues to increase, the vibration frequency of the model, which is the common frequency of vertical and torsional vibration, continues to decrease.

3.3. Wavelet and Hilbert Analysis

The flutter characteristics of the model are analyzed by the wavelet transform and Hilbert transform [22,23]. Since the 30c-30 case has a wide range of LCF wind speeds, and hard flutter directly takes place in the 30c-180 case when U_r is increased from 148.75 to 153.13, the two cases are analyzed below.

The wavelet transform is performed on the vibration time history. Figure 15 shows the frequency domain distribution of the vibration of the model in case 30c-30 at each moment when $U_r = 122.5$. It can be seen that the frequency of vibration is mainly concentrated in the range of 4.0 to 5.0. The distribution of torsional vibration in the frequency domain is very stable, while that of vertical vibration changes more obviously with time. Figure 16 shows the distribution in the frequency domain of the vibration of the model in case 30c-180 at each moment after U_r is increased from 148.75 to 153.13. As time passes, the frequency band of vibration in both the torsional and vertical degrees of freedom gradually widens and the amplitude gradually increases. In addition, during the period when the model vibrates obviously, the frequency band of torsional vibration is wider than that of vertical vibration.

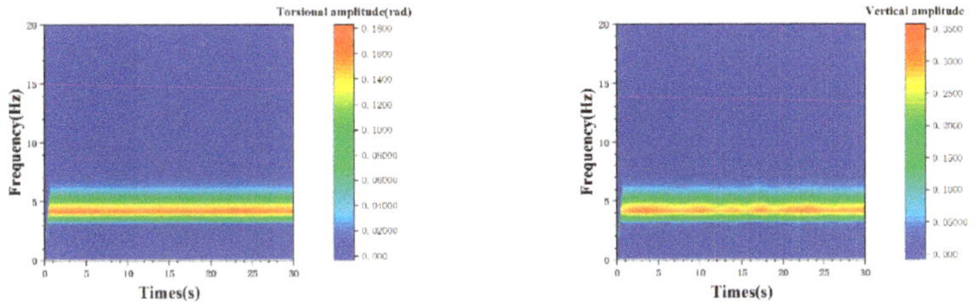

(a) (b)

Figure 15. Distribution of vibration in case 30c-30 when U_r is 122.5. (**a**) Torsional degree of freedom; and (**b**) vertical degree of freedom.

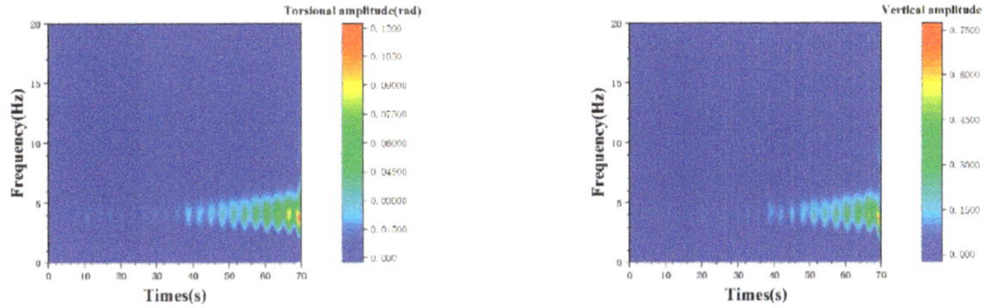

(a) (b)

Figure 16. Distribution of vibration in case 30c-30 when U_r is increased from 148.75 to 153.13. (**a**) Torsional degree of freedom; and (**b**) vertical degree of freedom.

The Hilbert transform is used to obtain the phase trajectory in phase plane. The plane composed of the vibration displacement as the abscissa and the vibration velocity as the ordinate is called the phase plane, and the curve formed by the points corresponding to the system motion state on the plane is called the phase trajectory. Figure 17 shows the phase trajectory of case 30c-30 when U_r is 113.75 and 122.50. There are relatively stable rings in the vertical and torsional degrees of freedom, and the higher the wind speed, the larger the ring. The torsional vibration ring is thinner than that of vertical vibration, indicating that the torsional vibration is more stable. Figure 18 shows the phase trajectory of the model in different time periods in case 30c-180 when U_r is increased from 148.75 to 153.13, where the time origin is the moment when the measurement starts. It can be seen that the phase trajectory continuously expands outwards on the counterclockwise direction. With the passage of time, the width of the phase trajectory within the same length of time gradually increases, indicating that the divergence speed of vibration gradually increases.

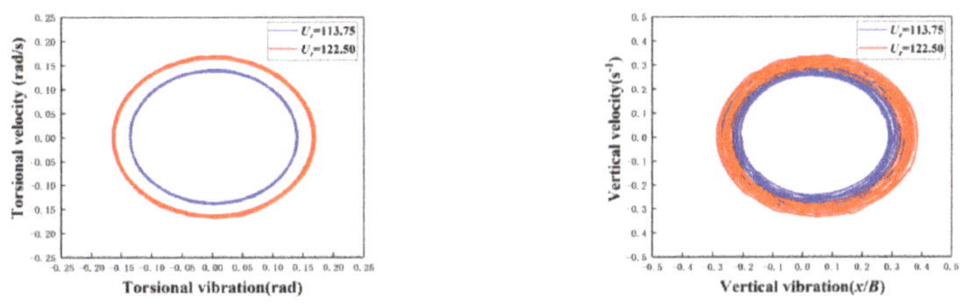

(a) (b)

Figure 17. Phase trajectory of case 30c-30 under different wind speeds. (**a**) Torsional degree of freedom; and (**b**) vertical degree of freedom.

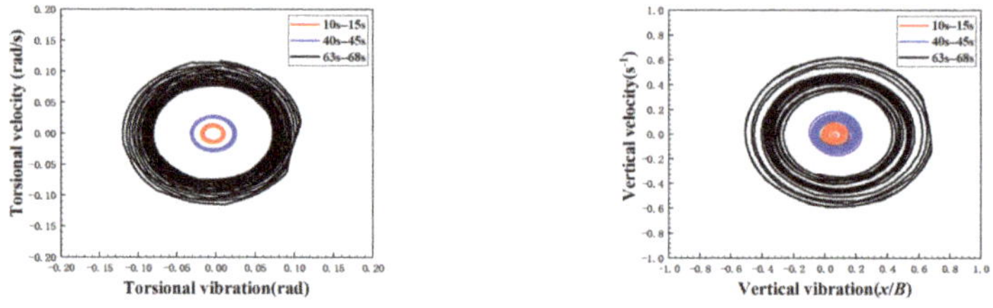

(**a**) (**b**)

Figure 18. Phase trajectory of case 30c-180 in different time periods when U_r is increased from 148.75 to 153.13. (**a**) Torsional degree of freedom; and (**b**) vertical degree of freedom.

3.4. Phase Lag between Torsional and Vertcial Vibration

In order to investigate the coupling characteristics of the vertical and torsional degrees of freedom when the bridge deck vibrates, the amplitude ratio of torsional vibration to vertical vibration, and the phase difference φ_d between the torsional degrees of freedom φ_t, and the vertical degrees of freedom φ_v when bending-torsional coupling flutter occurs are analyzed.

The amplitude ratio of torsional vibration to vertical vibration is calculated as follows:

$$\gamma = \frac{\mathrm{RMS}(\theta \times B/2)}{\mathrm{RMS}(y)} \tag{4}$$

Figure 19 shows the amplitude ratio of torsional vibration to vertical vibration at various apex angles of the trailing edge. It can be found that the torsional vibration dominates the bending-torsional coupling system in case 30c-45, where the amplitude ratio of torsional vibration to vertical vibration is 5.96, and then amplitude of vertical vibration gradually increase as the trailing edge becomes blunter. At case 30c-180, the amplitude ratio of torsional vibration to vertical vibration is 0.65.

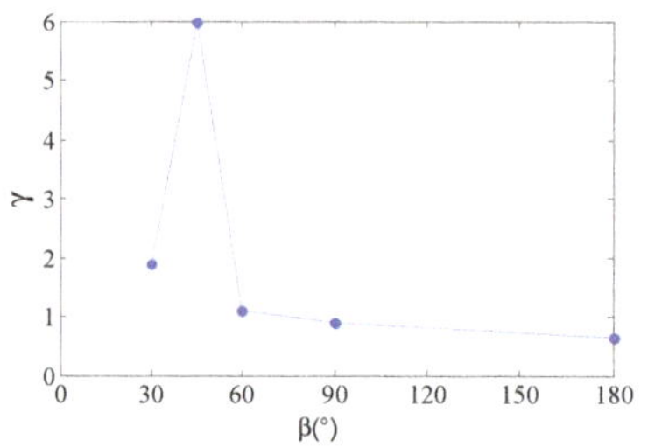

Figure 19. The amplitude ratio of torsional vibration to vertical vibration at various apex angles of the trailing edge.

The phase difference between torsional vibration and vertical vibration is calculated as follows:

$$\varphi_d(f) = \left\{ \begin{array}{cc} \arctan\frac{\mathrm{imag}\{S_{\theta y}(f)\}}{\mathrm{real}\{S_{\theta y}(f)\}} & \text{if} \quad \mathrm{real}(S_{\theta y}) \geq 0 \\ \arctan\frac{\mathrm{imag}\{S_{\theta y}(f)\}}{\mathrm{real}\{S_{\theta y}(f)\}} + \pi & \text{if} \quad \mathrm{real}(S_{\theta y}) < 0 \end{array} \right\} \tag{5}$$

where φ_d is phase difference, $S_{\theta y}$ is the cross-power spectrum, and f is the coupling vibration frequency.

Figure 20 shows the phase difference between torsional vibration and vertical vibration at various apex angles of the trailing edge. For the 30c-30, 30c-45, 30c-60, 30c-90, and 30c-180 cases, the vertical vibration lags behind the torsional vibration by −3.8, 33.5, −7.8,

−12.6, and −17.2 degrees, respectively. Except in the case 30c-45, the phase gradually decreases (gradually moves away from 0 degrees) as the trailing edge becomes blunter.

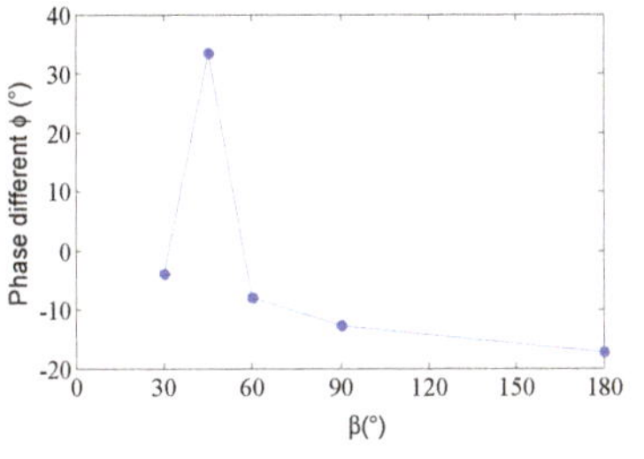

Figure 20. The phase difference between torsional vibration and vertical vibration at various apex angles of the trailing edge.

3.5. Discussion of Effect of TE

Figures 21 and 22 show the variation of deck vibration amplitude with wind speed. In order to clearly reflect the vibration characteristics of the bridge deck in each case, the amplitude of the non-divergent vibration is calculated using the average amplitude within 30 s, and the amplitude of the hard flutter is expressed by the maximum amplitude during the vibration process. The flutter stability of the bridge deck increases with the increase in the β except in case 30c-45. Similar to the torsional amplitude, except in case 30c-45, the critical flutter wind speed on the vertical degree of freedom increases gradually with the increase in the angle of the TE β.

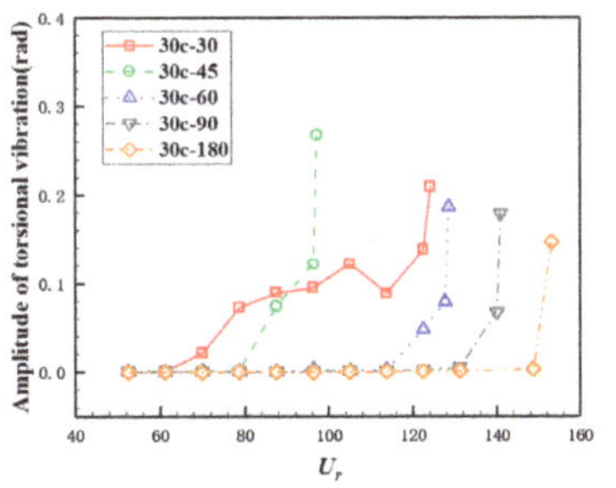

Figure 21. Torsional amplitude changes with U_r.

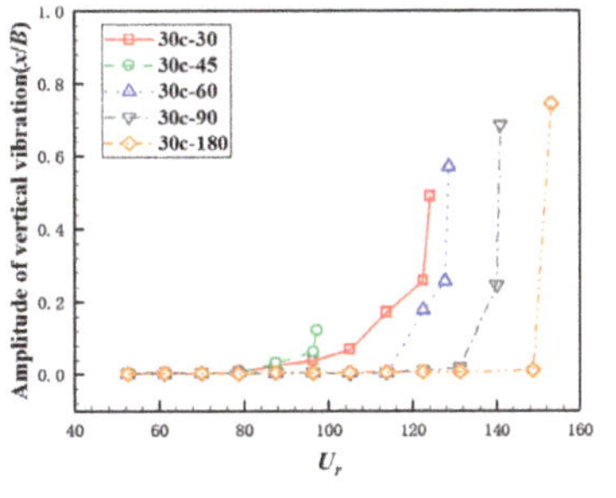

Figure 22. Vertical amplitude changes with U_r.

Next, this paper compares and analyzes the critical wind speed when the bridge deck starts to vibrate U_s, the LCF wind speed range V_L, and the critical wind speed of the hard flutter vibration U_f. Figure 23 shows curves of U_s, V_L, and U_f followed by the change in β. Among them, U_s gradually increases with the gradual increase in β while the curve of V_L and U_f changing with β drops at $\beta = 45°$, and then gradually increase as β increases.

It shows that when $\beta > 45°$, the increase in β has a significant inhibitory effect on flutter, which means the more blunt the TE of the bridge deck, the stronger the flutter stability. In this paper, the value of $\cot(\beta)$ is used to describe the streamline performance of the TE. Figure 24 shows the change of streamline performance with β. It can be seen that the absolute value of the slope of the curve gradually decreases with the increase in the β. That means the decreasing speed of streamline performance of the TE gradually decreases, which is basically consistent with curves in Figure 23 when $\beta \neq 45°$. The streamline performance of the TE has obvious correlation with the flutter characteristics and flutter stability of the bridge deck, while the streamline performance of the TE has a direct impact on the vortex in the wake region. The poorer the streamline performance of the TE is, the more complicated the distribution and movement of the vortex structure in the wake area will be, which means the distribution and motion law of the vortex in the wake area have an important influence on the flutter characteristics and flutter stability of the bridge deck.

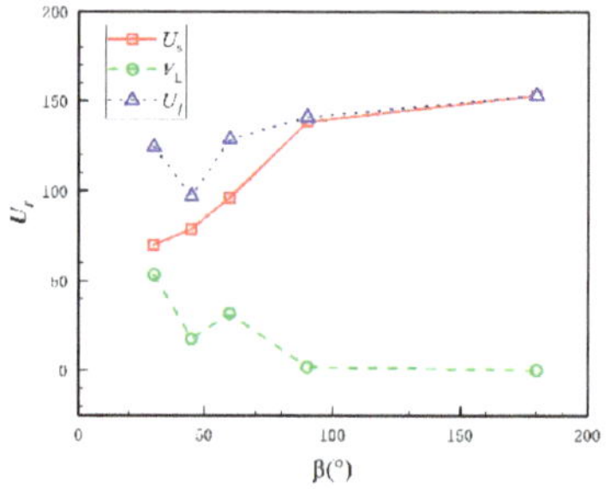

Figure 23. The flutter characteristics of each case.

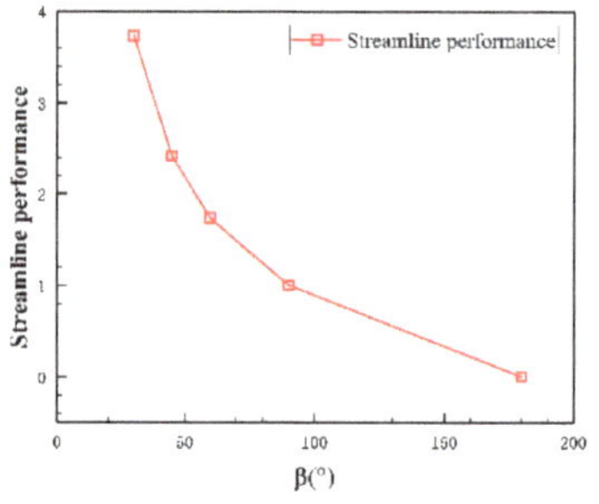

Figure 24. The relationship between streamline performance of the TE and β.

As for 30c-45 ($\beta = 45°$), the flutter characteristics of the model are quite different from other conditions. First, during the process of fluttering, there is basically no vibration in the vertical degree of freedom, which indicates until hard flutter develops, the vertical and torsional vibrations of the model are still not completely coupled. Secondly, when $\beta \geq 30°$, as β increases, $U_f(\beta \neq 45°)$ will gradually increase and $V_L(\beta \neq 45°)$ will gradually decrease, while $U_f(\beta = 45°) < U_f(\beta = 30°)$ and $V_L(\beta = 45°) < V_L(\beta = 60°)$. The specific reasons need to be further experimentally studied.

4. Conclusions

In this paper, by changing the aerodynamic shape of the TE of the model, the flow pattern of the model wake is controlled, and the displacement response of the model under different wind speeds and different TEs is collected by laser displacement sensors, and then the effect of different wake flow patterns on flutter characteristics and the flutter stability of the bridge deck is studied. The vibration response of the segment model under different TEs and different incoming wind speeds is compared and analyzed, after which the flutter characteristics of the segment model with different TEs are analyzed, and the following conclusions are drawn:

(1) Limit cycle flutter may still occur in the bridge section with a streamlined front edge and, in some cases, it also has a wider limit cycle flutter wind speed range.
(2) When the bending-torsional coupling flutter is fully developed, the torsional degree of freedom slightly lags behind the vertical degree of freedom.
(3) As the trailing edge becomes more and more blunt, the critical wind speed U_s gradually increases, and the flutter stability gradually increases. The wake vortex motion has a certain correlation with the flutter stability of the bridge deck.

Author Contributions: Conceptualization, S.L.; methodology, J.F. and S.L.; experiments, B.W., J.F. and S.L.; software, J.F.; validation, J.F.; formal analysis, J.F. and S.L.; investigation, J.F., B.W. and S.L.; resources, S.L.; data curation, J.F. and S.L.; writing—original draft preparation, J.F.; writing—review & editing, J.F. and S.L.; supervision, S.L.; project administration, S.L.; funding acquisition, S.L. All authors have read and agreed to the published version of the manuscript.

Funding: This study is financially supported by the NSFC under Grant No. 51878230, Natural Science Foundation of Heilongjiang Province under Grant No. YQ2021E033, Postdoctoral scientific research development fund of Heilongjiang Province under Grant No. LBH-Q20021 and supported by Heilongjiang Touyan Team and Fundamental Research Funds for the Central Universities.

Institutional Review Board Statement: Not applicable.

Informed Consent Statement: Not applicable.

Data Availability Statement: Data are contained within this article.

Conflicts of Interest: The authors declare no conflict of interest.

References

1. Bleich, F. Dynamic instability of truss-stiffened suspension bridges under wind action. *Trans. Am. Soc. Civ. Eng.* **1948**, *114*, 1177–1222. [CrossRef]
2. Theodorsen, T. *General Theory of Aerodynamic Instability and the Mechanism of Flutter*; NACA Report No. 496; National Aeronautics and Space Administration: Washington, DC, USA, 1935; pp. 291–311.
3. Kloppel, K.; Thiele, F. Modellversuche im Windkanal zur Bemessung von Brucken gegen die Gefahr winderregter Schwingungen. *Stahlbau* **1967**, *32*, 353–365.
4. Scanlan, R.H.; Tomko, J.J. Airfoil and Bridge Deck Flutter Derivatives. *J. Eng. Mech. Div.* **1971**, *97*, 1717–1737. [CrossRef]
5. Matsumoto, M.; Kobayashi, Y.; Shirato, H. The influence of aerodynamic derivatives on flutter. *J. Wind Eng. Ind. Aerodyn.* **1996**, *60*, 227–239. [CrossRef]
6. Matsumoto, M.; Daito, Y.; Yoshizumi, F.; Ichikawa, Y. Torsional flutter of bluff bodies. *J. Wind Eng. Ind. Aerodyn.* **1997**, *69*, 871–882. [CrossRef]
7. Matsumoto, M.; Matsumiya, H.; Fujiwara, S.; Ito, Y. New consideration on flutter properties based on step-by-step analysis. *J. Wind Eng. Ind. Aerodyn.* **2010**, *98*, 429–437. [CrossRef]
8. Bartoli, G.; Mannini, C. A simplified approach to bridge deck flutter. *J. Wind Eng. Ind. Aerodyn.* **2008**, *96*, 229–256. [CrossRef]
9. Lee, H.; Vu, T.; Yoo, S.; Lee, H.Y. A Simplified Evaluation in Critical Frequency and Wind Speed to Bridge Deck Flutter. *Procedia Eng.* **2011**, *14*, 1784–1790. [CrossRef]
10. Schmid, P.J. Application of the dynamic mode decomposition to experimental data. *Exp. Fluids* **2011**, *50*, 1123–1130. [CrossRef]
11. Wu, T.; Kareem, A.; Ge, Y. Linear and nonlinear aeroelastic analysis frameworks for cable-supported bridges. *Nonlinear Dyn.* **2013**, *74*, 487–516. [CrossRef]
12. Amandolese, X.; Michelin, S.; Choquel, M. Low speed flutter and limit cycle oscillations of a two-degree-of-freedom flat plate in a wind tunnel. *J. Fluids Struct.* **2013**, *43*, 244–255. [CrossRef]
13. Ying, X.; Xu, F.; Zhang, M.; Zhang, Z. Numerical explorations of the limit cycle flutter characteristics of a bridge deck. *J. Wind Eng. Ind. Aerodyns.* **2017**, *169*, 30–38. [CrossRef]
14. Gao, G.; Zhu, L.; Han, W.; Li, J. Nonlinear post-flutter behavior and self-excited force model of a twin-side-girder bridge deck. *J. Wind Eng. Ind. Aerodyn.* **2018**, *177*, 227–241. [CrossRef]
15. Zhang, M.; Xu, F.; Ying, X. Experimental Investigations on the Nonlinear Torsional Flutter of a Bridge Deck. *J. Bridge Eng.* **2017**, *22*, 04017048. [CrossRef]
16. Zhang, M.; Xu, F.; Zhang, Z.; Ying, X. Energy budget analysis and engineering modeling of post-flutter limit cycle oscillation of a bridge deck. *J. Wind Eng. Ind. Aerodyn.* **2019**, *188*, 410–420. [CrossRef]

17. Larsen, A. Aerodynamic aspects of the final design of the 1624 m suspension bridge across the great belt. *J. Wind Eng. Ind. Aerodyn.* **1993**, *48*, 261–285. [CrossRef]
18. Bruno, L.; Mancini, G. Importance of Deck Details in Bridge Aerodynamics. *Struct. Eng. Int.* **2002**, *12*, 289–294. [CrossRef]
19. Wilde, K.; Omenzetter, P.; Fujino, Y. Suppression of bridge flutter by active deck-flaps control system. *J. Eng. Mechan.* **2001**, *127*, 80–89. [CrossRef]
20. Lin, Y.Y.; Cheng, C.M.; Wu, J.C.; Lan, T.L.; Wu, K.T. Effects of deck shape and oncoming turbulence on bridge aerodynamics. *J. Appl. Sci. Eng.* **2005**, *8*, 43–56.
21. Zhang, X.; Sun, B. Parametric study on the aerodynamic stability of a long-span suspension bridge. *J. Wind Eng. Ind. Aerodyn.* **2004**, *92*, 431–439. [CrossRef]
22. Mariotti, A. Axisymmetric bodies with fixed and free separation: Base-pressure and near-wake fluctuations. *J. Wind Eng. Ind. Aerodyn.* **2018**, *176*, 21–31. [CrossRef]
23. Rocchio, B.; Mariotti, A.; Salvetti, M.V. Flow around a 5:1 rectangular cylinder: Effects of upstream-edge rounding. *J. Wind Eng. Ind. Aerodyns.* **2020**, *204*, 104237. [CrossRef]

applied sciences

MDPI

Article

Effect of Topography Truncation on Experimental Simulation of Flow over Complex Terrain

Zhen Wang [1], Yunfeng Zou [1,2,*], Peng Yue [1], Xuhui He [1,2], Lulu Liu [1] and Xiaoyu Luo [3]

[1] School of Civil Engineering, Central South University, Changsha 410075, China; wangzhen@csu.edu.cn (Z.W.); yuepeng2016@csu.edu.cn (P.Y.); xuhuihe@csu.edu.cn (X.H.); liululu2020@csu.edu.cn (L.L.)

[2] Hunan Provincial Key Laboratory for Disaster Prevention and Mitigation of Rail Transit Engineering Structures, Changsha 410075, China

[3] Electric Power Science Research Institute of Guangdong Power Grid, Guangzhou 510080, China; lxy86@163.com

* Correspondence: yunfengzou@csu.edu.cn; Tel.: +86-138-7499-7131

Featured Application: This article analyzes the effect of topography truncation on wind characteristics of flow over complex terrain in a wind tunnel simulation.

Abstract: Wind tunnel tests are a commonly used method for studying wind characteristics of complex terrain; but truncation of the terrain model is usually unavoidable and affects the accuracy of the test results. For this reason, the effects of truncated and original terrain models on the simulation of wind characteristics for complex terrain were investigated by considering both nontruncated and truncated models, with the truncated model considering the applicability of two types of transition sections. The results show that the effect of topographic truncation on profiles of mean velocity and turbulence intensity is different for regions and that inclination angle profiles are extremely sensitive to the changing topographic features upwind. In those cases, the spectra of streamwise velocity were overestimated in the low-frequency range but underestimated in the high-frequency range due to topographic truncation. At the same time, the less negative value of the slope of the spectra was found at the inertial subrange. Furthermore, the normalized bandwidth was also influenced by topographic truncation, which was narrowed in windward and leeward regions and broadened in the valley region. We should note that the performance of the transition sections used in this study was quite limited and even resulted in inaccuracies in the simulation.

Keywords: wind characteristics; wind tunnel testing; complex terrain; model truncation; transition section

Citation: Wang, Z.; Zou, Y.; Yue, P.; He, X.; Liu, L.; Luo, X. Effect of Topography Truncation on Experimental Simulation of Flow over Complex Terrain. *Appl. Sci.* **2022**, *12*, 2477. https://doi.org/10.3390/app12052477

Academic Editors: Wenli Chen, Zifeng Yang, Gang Hu, Haiquan Jing, Junlei Wang and Mohamed Benbouzid

Received: 14 December 2021
Accepted: 23 February 2022
Published: 27 February 2022

1. Introduction

Flow over complex terrain remains a formidable issue that plays an important role in many practical applications, including wind power micrositing, pollution dispersion as well as wind loading on bridges and transmission towers. Flow separation and reattachment regions induced by topographies such as ridges, cliffs and escarpments not only lead to misconceptions about flow characteristics but also cause difficulties in wind tunnel testing and numerical simulation. Flow characteristics in a region close to a topographic feature and to what extent the flow is influenced by the terrain geometry are not exactly specific [1]. An in-depth understanding of complex turbulent flows is essential for greatly improving simulation accuracy, especially for numerical simulations including the large eddy simulation (LES) and the unsteady Reynolds-averaged Navier-Stokes (URANS). Significant effort has been devoted in recent years to gaining a better understanding of flow characteristics in complex terrains with wind tunnel testing [2–6]. Furthermore, numerical simulation can be compared to validate results [7]. Kilpatrick et al. [8] investigated the effect of inflow conditions on the characteristics of flow over Bolund Hill. Bolund Hill is a well-studied test case for wind power misting and numerical simulations, as its steep geometry poses a challenge for numerical models to resolve [9]. Lange et al. [10] studied the mean flow behavior and turbulence intensity when flowing over Bolund Hill under a round or sharp escarpment leading edge; wind energy production was also analyzed. To further study the unsteady flow behavior around Bolund Hill, a 1/115 scale model was employed in the wind tunnel experiment by Yeow et al. [11]. Several field measurements had also been carried out to verify the experimental and numerical models [12–14]. In addition, early studies such as Askervein Hill [15] and Kettles Hill [16] are classic cases for studying the characteristics of flow over complex terrain.

The above-mentioned studies primarily focused on terrains such as regular independent hills, slopes and steep escarpments to reveal the mechanism of the flow over different terrains. There was little attention devoted to real complex terrains where the topography and geometry are more intricate. Due to the limitations of the wind tunnel test and of the numerical computational domain, complex terrain models should be truncated. The bottom of a terrain model is determined by the lowest elevation in the real terrain. Because the elevation differences of real topography such as deep-cutting gorges and valleys are often extreme, the edges of the model are higher than the bottom side of the wind tunnel and computational domain. This phenomenon, named "artificial cliff" [17], leads to flow separation and baffling flow at the leading edge, changing inflow conditions and inaccuracies in the test. Several methods have been proposed to address the "artificial cliff"[18–20]. For example, a transition section, which was used to connect the floor of the wind tunnel or numerical computational domain and the terrain model edge, was proposed to avert flow separation. There are various forms of transition sections, such as ramps, bicubic transition curves, and Witoszynski transition curves. Jubayer and Hangan [21] implemented 3.2 mm thick masonite sheets at the model front and sides to smoothen the height transition from the wind tunnel floor to the model edge. Maurizi et al. [22] used a ramp transition section with a slope of 5.7° to simulate the wind characteristics of a mountainous terrain by numerical simulation. Although the form of the ramp transition section is simple and easy to use, its applicability still needs to be verified. For topography with large elevation differences, the length of the ramp transition section may be too long to be suitable for a wind tunnel simulation because of the limitation of the test section. Moreover, the slope of the ramp transition section should be small, otherwise flow separation and baffling flow will still occur. Accordingly, the ramp transition section is not applicable for highly undulating topography. In this case, Hu et al. [23] proposed a theoretical curve based on the potential flow theory around the cylinder and then conducted a performance comparison between the theoretical curve and ramp transition. For application convenience, Hu et al. [24] simplified the expression of the theoretical curve based on the original. Huang et al. [17] proposed a comprehensive method to evaluate the performance of several curve transitions including the Hu et al. [24] improved theoretical curve. Chen et al. [25] considered different weights of speed-up factors, wind attack angle, and turbulence intensity by combining a weighted solution based on the correlation degree to ameliorate the evaluation method for determining the optimal form of the transition section.

Although transition sections are widely implemented in experiments and numerical simulations, the feasibility is still difficult to explain, because neither wind tunnel nor numerical simulations can avoid "artificial cliff". Additionally, comprehensively detailed field measurement data of flow over complex terrain, which is time-consuming and highly expensive, are also difficult to obtain. Hence, analyzing the effect of the transition section on flow characteristics is not possible. Nevertheless, based on previous studies on wind tunnel simulations of intricate topographic flows, the details of terrain models play an important role [10,26,27]. In the study of Lange et al. [10], the variation in the escarpment leading edge, which had round or sharp geometry, generated completely different flow behavior on the rear. The difference in annual energy production estimated based on the two scenarios differed from 20% to 51%. Furthermore, the structural dynamic response and aerodynamic characteristics are highly sensitive to changes in the flow characteristics [28–31]. Accordingly, a systematic study is needed to evaluate the effect of topographic truncation on flow over complex terrain.

Based on the above reasons, a wind tunnel simulation of flow over intricate terrain was conducted to obtain wind characteristics and analyze the effect of topographic truncation. The terrain model was scaled from an isolated island, which is the optimum simulation object to avoid the "artificial cliff" because the transition between island topography and sea level is smooth. Flow over complex terrain was investigated by assessing how changes to the forms of transition sections, including the original topography, affect flow behavior at different measurement locations. The flow parameters investigated included mean velocity, inclination angle, turbulence intensity, and velocity spectra.

2. Experimental Method

2.1. Terrain Scale Model

The topography investigated herein is an island located in the coastal region of South China. A view of this region is shown in Figure 1 along with the measurement locations for this study. For the wind tunnel test, wind speeds were measured at a total of five locations on the island, which are potential wind power sites. The center region of the island has the highest elevation with a mountain running northwest to southeast. There is a valley to the east of the mountain with another mountain stretching from the northeast to the southwest to the east of the valley. Overall, the topography of the island can be characterized as high-west and low-east. According to the topographic features of the island and the incoming flow direction, measurement locations were divided into three regions: windward region, leeward region, and valley region. As shown in Figure 1, location A1 was situated in the windward region, which is the windward slope of the center mountain, while locations A2 and A3 were situated at the leeward slope. The valley region representing the valley terrain at the island center included measurement locations of A4 and A5.

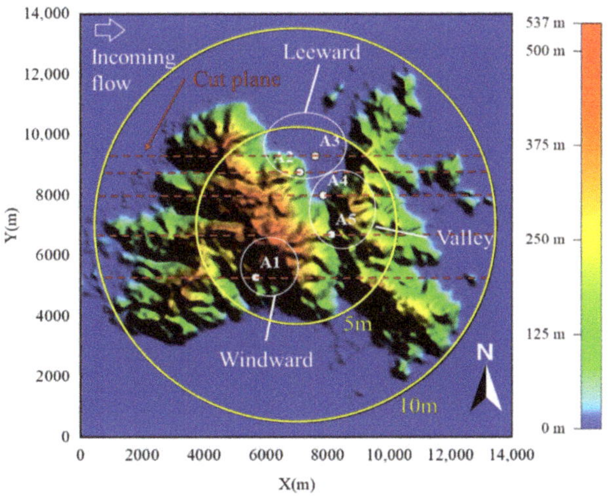

Figure 1. Topography of island and measurement locations.

For the wind tunnel testing, a 1:1300 scale terrain model was made of polyethylene foam and overlapped layer by layer according to the terrain contour line. The scale ratio was much larger than the ratio of 1:5000 recommended by Bowen [32] to avoid the error of turbulence viscous dissipation. The diameter of the test terrain model was 10 m for the full island model and 5 m for the truncation model with a maximum height of 0.41 m. The covered area of the terrain model was different according to the model diameter. The large one covered an area of approximately 132.7 km^2, and the small one covered an area approximately one-fourth the size of the larger one. The terrain of the full island is shown in Figure 1, and two circles indicate the areas covered by the terrain model with diameters of 10 m and 5 m. The blockage ratio was approximately 3.27% for the terrain model with a diameter of 10 m, which is lower than the maximum allowable blockage ratio of 5% recommended by Holmes [33].

2.2. Transition Sections

The transition section connecting the wind tunnel floor or the bottom of the numerical computational domain was proposed to handle the "artificial cliff" generated by topographic truncation, as shown in Figure 2a. This idea was inspired by the contraction curve of the wind tunnel, which can achieve good flow quality and avoid flow separation.

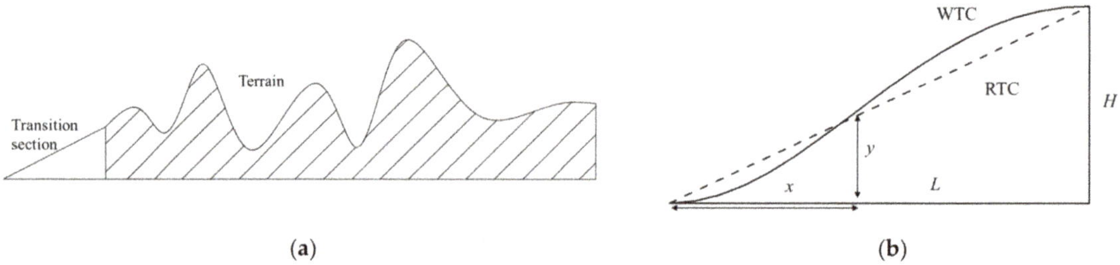

Figure 2. Transition section: (**a**) sketch of transition section; (**b**) sketch of transition curve.

Four cases conducted in this study are shown in Figure 3: original topography (OT), without transition section (WTS), Witozinsky transition curve (WTC), and ramp transition curve (RTC). According to previous studies [17,23–25], the performance of the transition curve was evaluated by comparing the characteristics of the flow after the transition section with incoming flow, and the less differences observed, the better. However, the studies of Kilpatrick et al. [1] and Lange et al. [10] highlighted the significant effect of the leading edge geometry of the terrain model on the flow behavior over the Bolund Hill. The implementation of a transition section to smooth the altitude difference between the model edge and the wind tunnel floor may change the leading-edge geometry, but the degree to which the flow is influenced is not completely understood. Furthermore, when transition sections were applied in the complex terrain, flow characteristics observed at measurement locations were affected not only by transition sections but also by complex terrain. Hence, it is not appropriate to evaluate the performance of transition sections by comparing flow characteristics with incoming flow. For this purpose, the test case with original topography was carried out as criterion instead of incoming flow, and the case without transition sections and the one with two forms of transition sections were also conducted in this study for comparison. Figure 2b shows a sketch of these two transition curves. The first one is the WTC, which is widely applied in low-speed wind tunnels. The formula of WTC can be described as follows:

$$y = H\left\{1 - \frac{\left[1-(x/L)^2\right]^2}{\left[1+\alpha(x/L)^2\right]^3}\right\} \tag{1}$$

where $\alpha = 50$, H is the altitude difference between the model edge and the wind tunnel floor, and L is the length of the transition curve at certain points corresponding to H to maintain the equivalent slope $H/L = 0.5774$. The other is RTC with the same slope but a straight line for the transition curve.

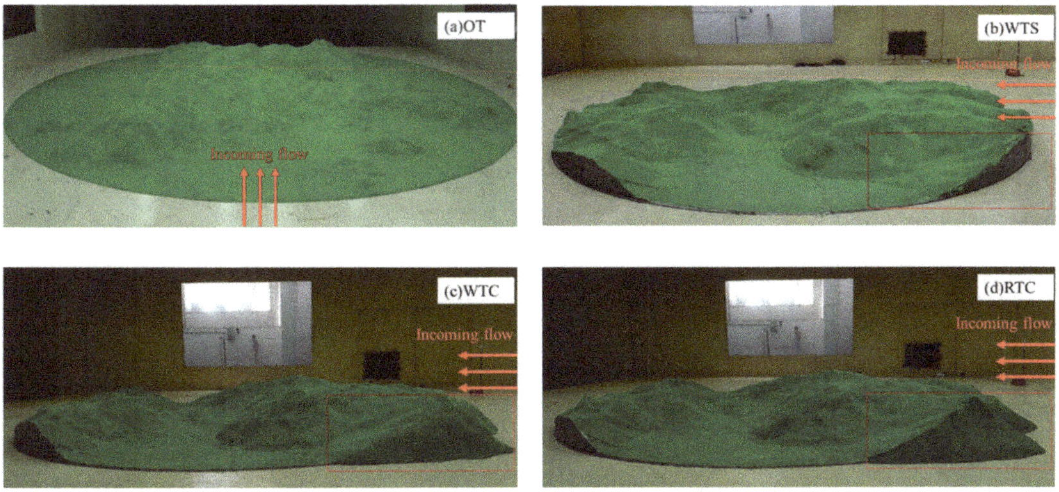

Figure 3. The test model at wind tunnel: (**a**) OT; (**b**) WTS; (**c**) WTC; (**d**) RTC.

2.3. Experimential Set-Up

The testing facility was the wind tunnel laboratory of Central South University, which is a closed-circuit atmospheric boundary layer wind tunnel with two test sections. The test was carried out in the low-speed section with a width of 12 m and a height of 3.5 m. The test wind speed was set to 10 m/s with a uniform profile. The incoming flow came from the west to the island, which is the most critical wind direction based on the local meteorological statistics. Furthermore, the altitude difference caused by topographic truncation is most obvious on the west side due to the high-west and low-east topography of the island. The Reynolds numbers based on model height h can be calculated as Re = U h/ν, where U is the streamwise velocity and $\nu = 1.47 \times 10^{-5}$ m^2/s is the kinematic viscosity of air at 20 °C. The value of Reynolds numbers in this study was approximately 3.0×10^5, which is in the range of 1.7×10^5–4.6×10^5 studied by Kilpatrick et al. [1]. According to their research, the flow behavior is generally less affected by the Reynolds number. For the other nondimensional numbers determining the flow similarity between the wind tunnel and the real island, the Richardson number and the Eckert number similarities were satisfied due to the consideration of the neutrally stratified boundary layer, and the Prandtl number similarity was satisfied by the air working fluid. Depending on the early study of flow over complex terrain [34], the Rossby similarity was relaxed because of the practical difficulties of modeling the Coriolis force in the wind tunnel. The Froude similarity was also relaxed, because no thermal stratification was reproduced in the wind tunnel.

A turbulence flow instrument cobra probe with a high-frequency up-to 10 kHz and measurement accuracy within ±0.5 m/s was used to measure the velocity at different locations on the terrain model. In this study, the data were sampled at 2 kHz. A minilaser range finder with measurement accuracy within ±1 mm was used to measure the distance above the ground.

3. Mean wind Characteristics

3.1. Mean Velocity

Profiles of streamwise velocity U, which were normalized by the mean velocity at the top height of each location U_r to reduce the influence of variation in incoming flow velocity, are shown in Figure 4 for four test cases with uniform inflow conditions but different forms of the transition section. For the convenience of analysis and interpretation, Figure 5 provides east-west cutting planes of measurement locations.

Figure 4 shows the normalized velocity profiles of location A1, where obvious differences in the four test cases are found near the ground. The influenced area is approximately 450 m above the ground in true magnitude, in which the profiles of WTC and RTC are higher than the other two. Because A1 sits at the windward of the center mountain near the edge of the terrain model, the speed-up effects generated by transition sections control this area. For OT and WTS, the mean velocity profiles are similar under a height of approximately 400 m. However, it should be noted that the flow behaviors in these two cases are completely different. In case OT, there is a side mountain range running northeast to southwest at the front of location A1. As shown in Figure 5, the mountain peak lies ahead of location A1 with a height of approximately 400 m, leading to a blockage effect and flow separation. Hence, the mean velocity profile in the OT case shows a smaller speed at the part below approximately 400 m. In addition, for the case WTS, the front mountain peak was truncated and created an "artificial cliff" with a height of 287 m. Owing to the larger height difference between the terrain model edge and the wind tunnel floor, separation bubbles occurred, and flow separation controlled this region. However, this does not mean that the WTS shows better performance than the WTC and RTC. It is just a special case with a mountain peak at the front of location A1. Moreover, both the WTC and RTC could accelerate the flow near the ground. When the transition section is implemented, the speed-up effect should be considered, which may considerably affect the experimental accuracy, especially for the locations close to the terrain model leading edge.

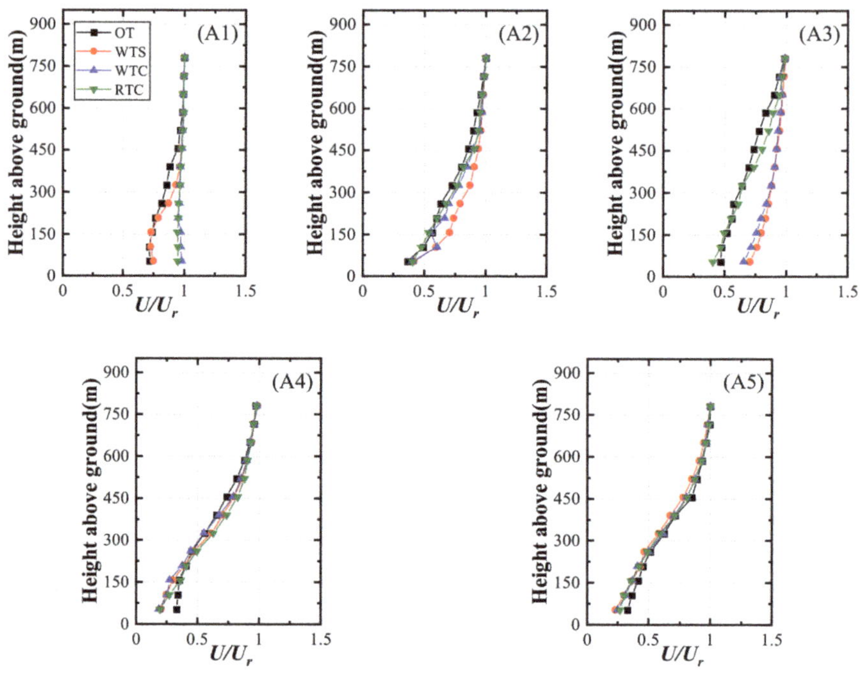

Figure 4. Profiles of mean velocity at measurement locations: (A1) location A1; (A2) location A2; (A3) location A3; (A4) location A4; (A5) location A5.

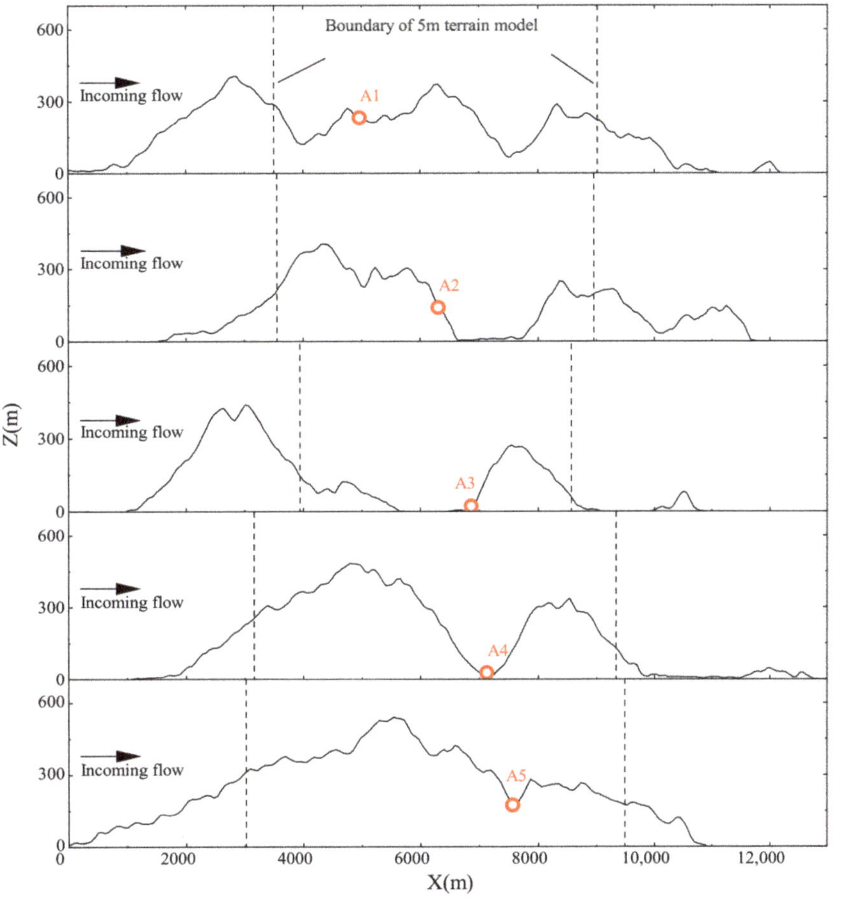

Figure 5. East-West cutting planes of measurement locations: (A1) location A1; (A2) location A2; (A3) location A3; (A4) location A4; (A5) location A5.

The leeward region, which includes locations A2 and A3, is on the downhill slope to the leeward of the peak. These locations are located north of the island with the location A2 at the slope and location A3 at the foot of the mountain. As shown in Figure 4, the velocity profiles of location A2 in the four cases are similar, but WTS is slightly larger at heights below approximately 600 m. From the cutting plane of location A2 in Figure 5, it shows that the truncated part of topography is the windward slope of the mountain, and the geometry is similar to the transition section of WTC and RTC. However, the case of WTS with a cliff at the edge of the model may result in flow separation. Thus, differences in velocity profiles were observed. For location A3, the results can almost be divided into two parts according to profiles. One is OT and RTC. The other one is WTS and WTC. As shown in Figure 5, there is a mountain peak in the front of location A3. When the terrain model is truncated, the mountain peak has also been cut down, which may lead to disparate flow behavior in the downwind region. Flow separation generated by the mountain peak controls the wind velocity at location A3 to a great extent. However, the ramp transition section can make up this difference instead of the mountain peak to a certain extent by generating flow separation. Coincidentally, Conan [35] studied the wind characteristics in complex terrain and successfully obtained analogous results by replacing the mountain peak with a ramp of a specific slope. Therefore, these results indicate that the flow behavior in front of the target locations should be completely simulated to improve the accuracy in wind tunnel simulation. The velocity profiles at the valley region, including the measurement locations of A4, A5, and A6, at the bottom of the valley, show virtually no distinction because these locations are considerably affected by recirculation flow in the enveloping separation bubble. The cutting planes of these three locations are shown in Figure 5. From the cutting planes, when the terrain model was truncated, most topographic features at the front of the measurement locations were retained, and only the ridge ahead of location A6 was truncated. Thus, few differences can be found in the velocity profiles.

The velocity profiles at the valley region, including the measurement locations of A4 and A5 at the bottom of the valley, show virtually no distinction because these locations are considerably affected by recirculation flow in the enveloping separation bubble. The cutting planes of these two locations are shown in Figure 5. From the cutting planes, when the terrain model has been truncated, most topography features at the front of measurement locations are retained, particularly the mountain ridge. Thus, there are few differences in the velocity profiles.

According to the above analysis, the velocity profiles of measurement locations at windward and leeward regions are deeply influenced by topographic truncation. Implementation with transition sections could not solve the problem, but accelerated the flow after it and brought inaccuracy to the location near the model edge. The measurement locations in the valley region are slightly affected by truncated terrain because the upwind flow behaviors are little changed. This means that the truncating the terrain causes a change in topographic characteristics, leading to different flow behaviors upstream, which influence the velocity characteristics downstream. Hence, repeating upwind flow behavior is essential for simulating velocity characteristics.

3.2. Inclination Angle

The inclination angle means the vertical flow velocity is normalized by the horizontal velocity, which is the most intractable issue in the application of the transition section. Because of transition sections with nearly slope geometry, flow is accelerated through the transition section. This could not be avoided due to the altitude difference between the model edge and the wind tunnel floor, whichever type of transition curve was used. Additionally, this issue can only be minimized by keeping the distance between measurement locations and the model edge long enough to reduce the adverse impact of the transition section.

Figure 6 shows profiles of the inclination angle at the measurement locations. The profiles are distinctly different from OT at almost every location, particularly in the region near the ground. For location A1 in the windward region, the inclination angle profiles corresponding to the terrain model with a transition section show analogous shapes to OT compared to the completely different profile in the terrain model without a transition section. The flow separation that occurred at the leading edge of the terrain model in the case of WTS greatly changed the profile and resulted in a negative value of inclination angle at the lower part.

At A2 and A3 in the leeward region, the profiles of the inclination angle in cases with transition sections are quite different from OT with a negative value. Although the profiles of WTS are not the same as those of OT, they are numerically closer, which indicates that when better velocity profile results are obtained, the additional inclination angle induced by the transition section cannot be ignored.

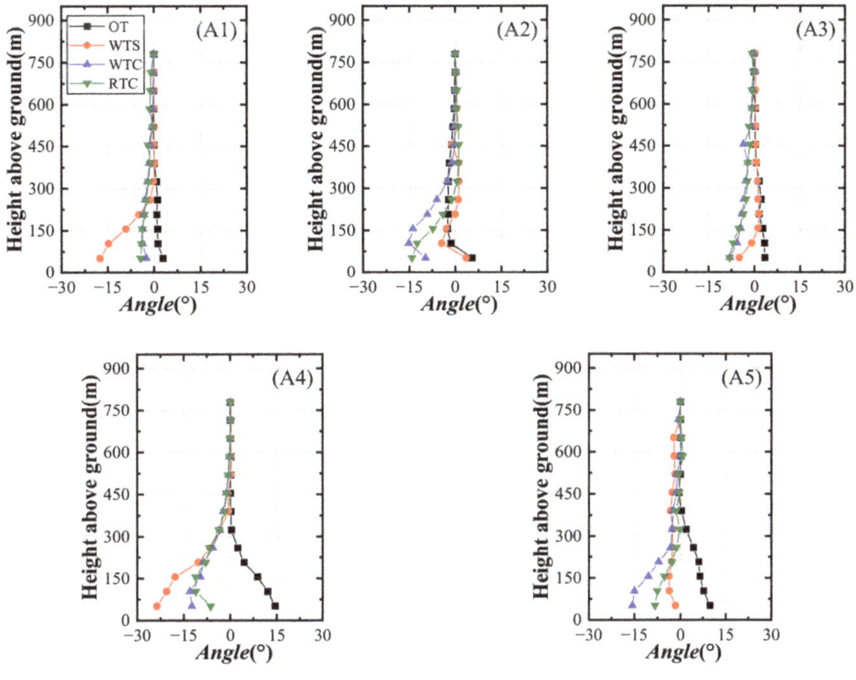

Figure 6. Profiles of inclination angle at measurement locations: (A1) location A1; (A2) location A2; (A3) location A3; (A4) location A4; (A5) location A5.

In the valley region, there are two measurement locations: A4 and A5. Due to the influence of flow separation in the previous analysis of mean velocity, the profiles are almost unchanged in different cases. However, it is entirely diverse in profiles of inclination angle. The inclination angle profiles in the cases of WTS, WTC, and RTC are almost opposite that of OT at heights below approximately 450 m for locations A4 and A5. Positive values have been observed in the region near the ground in the case of OT, while negative values have been found in the other cases. As shown in Figure 6, there is only a single peak ahead of locations A4 and A5. According to previous studies, flow separation off a peak or the edge of a bluff body creates a detached shear layer whose character depends strongly on the separation location [36]. For the single peak topography, separation occurs over crests and on the windward face, while the upwind topography of multiple peaks may advance the separation location. Thus, when the terrain model has been truncated, the single peak topography may be changed to multiple peaks by "artificial cliff", and locations of the valley region show opposite inclination angles with OT. For this reason, the values of the inclination angle at locations A4 and A5 become negative in the cases with the truncated terrain model. Additionally, the differences in the cases of WTS, WTC, and RTC indicate that the separation location is determined not only by the slope of the peak but also by its geometry.

Profiles of inclination angle show more sensitivity to upwind topography than mean velocity. Although the same results of mean velocity were obtained, the opposing results were observed in inclination angle such as at locations A4 and A5. Implementing transition sections in the simulation reduces the adverse effect of flow separation at the leading edge of the terrain model to a certain extent, but there is still a discrepancy compared with the original terrain.

4. Turbulence Characteristics

4.1. Turbulence Intensity

Turbulence intensity is given by the standard deviation of streamwise wind velocity normalized by mean velocity. The turbulence intensity profiles of the measurement locations, shown in Figure 7, intensely change in the region near the ground. Location A1 in the windward region shows higher turbulence intensities at the lower part of the profile for WTS, which indicates that flow separation induced by topographic truncation enhanced turbulence production in the region near the model edge to a greater extent, leading to inaccurate experimental results. Combined with the analysis of the mean velocity in Section 3.1, the study noted that flow separation in the case of WTS generated a similar profile in mean velocity but a different profile in turbulence intensity with OT. This means that flow separation induced by terrain truncation at the region near the terrain model edge is a difficult issue in turbulence simulation. Moreover, the lower level of turbulence intensity in cases with transition sections indicates that implementing transition sections in the experiment can effectively avoid flow separation, but there is no avoidance of the speed-up effects.

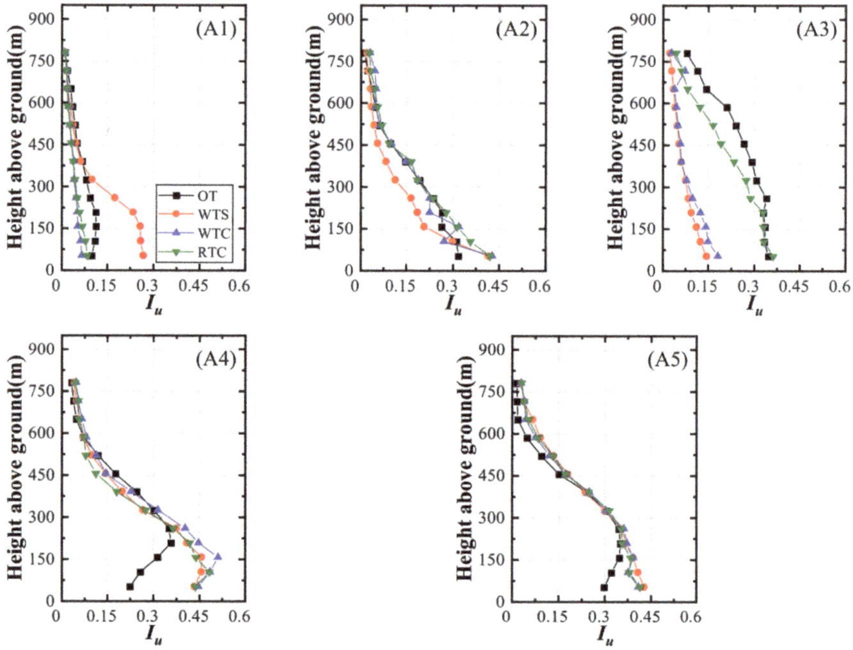

Figure 7. Profiles of turbulence intensity at measurement locations: (A1) location A1; (A2) location A2; (A3) location A3; (A4) location A4; (A5) location A5.

In the leeward region, for locations A2 and A3, there is at least one case with transition sections consistent with OT. Compared to WTS with lower turbulence intensity, both WTC and RTC obtained similar results with OT at location A2. As the cutting plane of location A2 is shown in Figure 5, the artificial cliff of WTS induced upwind flow separation, which was surmounted in the WTC and RTC cases. For location A3, RTC is the only case that obtained results approaching OT, which may be attributed to the fact that the ramp transition section took the place of the mountain peak in the front with producing flow separation. Coincidentally, Mcauliffe and Larose [26] used ramps to generate flow separation instead of the original topography with a mountain peak, and they also obtained acceptable results.

For the valley region, although the mean velocity profiles of the two measurement locations showed similar trends, the turbulence profiles were quite different at the elevations in the lowest 300 m above the ground. Compared with the case of OT, the other three show higher turbulence levels with decreasing height, and the largest discrepancy is found at the lower height, which is almost twice that of the OT. It seems that the increase in turbulence level caused by topographic truncation is inevitable regardless of implementation with transition sections when measurement locations are in the valley region.

4.2. Velocity Spectra

The spectra of streamwise velocity, normalized by $nS_u(n)/\sigma^2$, where S_u is the streamwise velocity spectra, n is the frequency, and σ is the standard deviation of velocity, are shown in Figure 8. The abscissa axis is the normalized frequency, expressed as $f = nZ/U(Z)$, where Z is the height above ground and $U(Z)$ is the mean velocity at height Z. The height of the spectrum shown in Figure 9 is 104 m, which is the hub height of most wind turbines.

At location A1 in the windward region, the spectra of cases with transition sections are higher than those of the WTS case, which may be attributed to flow separation suppression at the leading edge of the terrain model. Thus, large-scale vortices representing the energy of the low-frequency range are preserved. As shown in Figure 5, the side mountain range at the front of location A1 in the case of OT generated flow separation and broke up large-scale eddies into small-scale eddies, leading to lower spectra in the low-frequency range, similar to the case of WTS.

For the leeward region, spectra of four cases show little differences at location A2, but at location A3, the spectra of OT are lower than the other three in the low-frequency range. Despite the similarity in profiles of mean velocity and turbulence intensity obtained between OT and RTC, the discrepancy remains, owing to the truncation of the mountain peak ahead of location A3 in the RTC case.

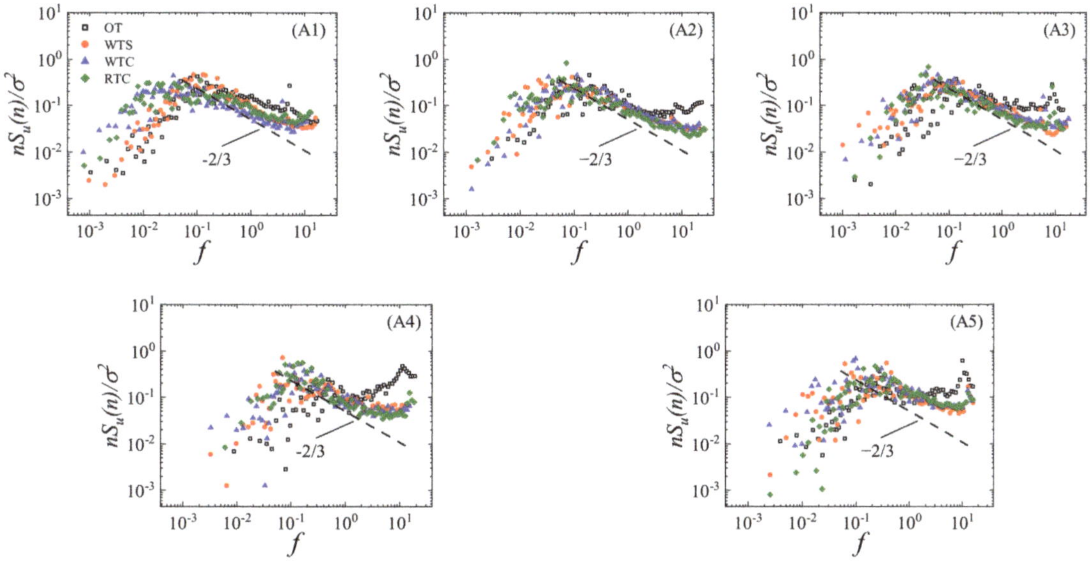

Figure 8. Normalized streamwise spectra for measurement locations at a height of 104 m: (A1) location A1; (A2) location A2; (A3) location A3; (A4) location A4; (A5) location A5.

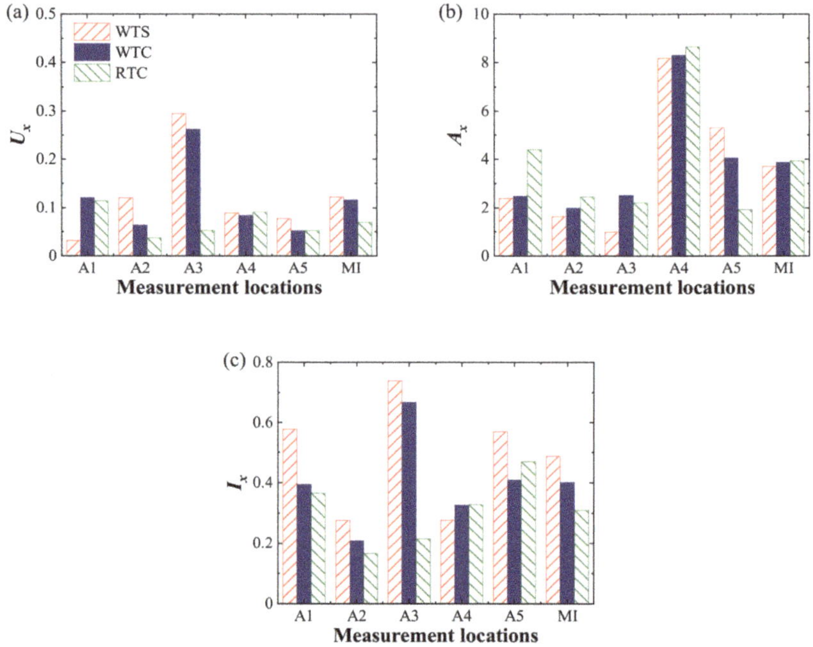

Figure 9. Special indexes at measurement locations: (**a**) mean velocity; (**b**) inclination angle; (**c**) turbulence intensity.

According to spectra of measurement locations in the valley region, one of the distinct differences is the shift of the spectra peak toward the high-frequency range and a reduction in the low-frequency range in all cases. Strong flow separation in the valley region may account for this difference, and another peak at high-frequency might support this explanation, which is associated with a vortex-shedding process that was also measured in the wake region of an escarpment by Alvaro et al. [37]. In the case of OT, owing to the strong perturbation from complete terrain, spectra at high frequency are higher than those in the other three cases.

As indicated in Figure 8, the inertial subrange of all cases corresponds roughly with the asymptotic relation of the $-2/3$ slope derived by Kaimal et al. [38] in windward and leeward regions, while the slopes of spectra for measurement locations in the valley region are mismatched with less negative values, which is more significant in the case of OT. This is similar to the research of Mcauliffe and Larose [26] in which the slope of spectra is less negative in the flow separation region. In addition, experiments with the truncated terrain model may underestimate spectra in the high-frequency range due to the incompleteness of topography ahead, and the slope of spectra is also dramatically influenced with a less negative value.

5. Quantitative Analyses on Effect of Topographic Truncation

5.1. Metrics on Profiles Differences

Based on the previous analysis, topographic truncation had significant effects on mean wind characteristics and turbulence characteristics when simulating flow over an island in the wind tunnel. Hence, to quantify the differences in mean velocity, inclination angle, and turbulence intensity, the special indicators proposed by Huang et al. [17] were used, which are defined as:

$$U_x = \sum_{i=1}^{N} \frac{|U_i - U_{i,OT}|}{U_{i,OT}} \cdot \frac{\Delta h_i}{H_w} \tag{2}$$

$$A_x = \sum_{i=1}^{N} \frac{|A_i - A_{i,OT}|}{|A_{i,OT}|} \cdot \frac{\Delta h_i}{H_w} \tag{3}$$

$$I_x = \sum_{i=1}^{N} \frac{|I_i - I_{i,OT}|}{I_{i,OT}} \cdot \frac{\Delta h_i}{H_w} \tag{4}$$

where N is the total number of measurement points along height at each measurement location, U_i is the normalized streamwise mean velocity at i-th point above the ground, $U_{i,OT}$ is the normalized mean velocity at i-th point above the ground in the case of OT, Δh_i is the height difference between i-th point and (i-1)-th point, H_w is the total height at the measurement location, A_i is the inclination angle at i-th point above the ground, $A_{i,OT}$ is the inclination angle at i-th point above the ground in the case of OT, I_i is the turbulence intensity at i-th point above the ground, and $I_{i,OT}$ is the turbulence intensity at i-th point above the ground in the case of OT. It should be noted that in contrast to previous work [17], the absolute value is applied to Equations (2), (3), and (4) to quantify the overall differences. We also define a mean indicator (*MI*) to evaluate the topographic truncation effect in different cases, which can be calculated as:

$$MI = \sum_{j=1}^{M} a_x \cdot \frac{1}{M} \tag{5}$$

where M is the total number of measurement locations and a_x means special indicators: U_x, A_x, and I_x.

Special indicators of the mean velocity, inclination angle, and turbulence intensity at each measurement location are plotted in Figure 9. The special indicator of the inclination angle shows the largest magnitude compared with the other two, while the special indicators of the mean velocity and turbulence intensity show similar magnitudes. This supports the fact that the inclination angle is likely the characteristic most influenced by topographic truncation. Hence, structures sensitive to the inclination angle should be emphasized when obtaining wind characteristics by wind tunnel simulation. As illustrated in Figure 9, the MIs of the cases with transition sections are much lower than that of the WTC in terms of the mean velocity and inclination angle. However, in terms of the inclination angle, the MIs of the cases with transition sections are slightly higher than that of the WTC. From the point of view of mean velocity and turbulence intensity, installing the transition section at the edge of the terrain model can improve the accuracy of a simulation to some extent, but only to a very limited extent. In contrast, installing transition sections at the edges of the terrain model could not improve the accuracy of the inclination angle in the wind tunnel simulation but brought additional errors in the simulation.

5.2. Metrics on Spectra Shifts

To quantify the shift of energy for spectra in frequency, three normalized frequencies [37]: f_l, f_m, and f_u, which are the cut-off frequencies of the different energy intervals, can be calculated from Equations (6) and (7):

$$\int_0^{n_c} n s_u(n)/\sigma^2 d[\ln(n)] - a_c = 0 \tag{6}$$

$$f_c = n_c Z/U(Z) \tag{7}$$

where n is the frequency, $nS_u(n)/\sigma^2$ is nonnegative and integrates to one in $(-\infty, +\infty)$, and $c = l, m, u$. The value of a_c varies with c, and $a_l = 0.05$, $a_m = 0.5$, and $a_u = 0.95$. Hence, the bandwidth between f_l and f_u, which means 90% energy interval bandwidth, is calculated from Equation (8):

$$\Delta f = \log_{10}(f_u/f_l) \tag{8}$$

As illustrated in Figure 10, $f_{m,OT}$ and Δf_{OT} mean median of normalized frequency and bandwidth of normalized frequency in the case of OT, respectively. As seen in Figure 10a, the normalized median of all cases and measurement locations shows values lower than 1. This indicates that the spectra obtained from the truncation terrain model show a higher value than that from OT, which also supports our analysis in Sect. 4.2. In addition, it can be concluded that energy shifts to a low-frequency range when using the truncation terrain model to obtain spectra.

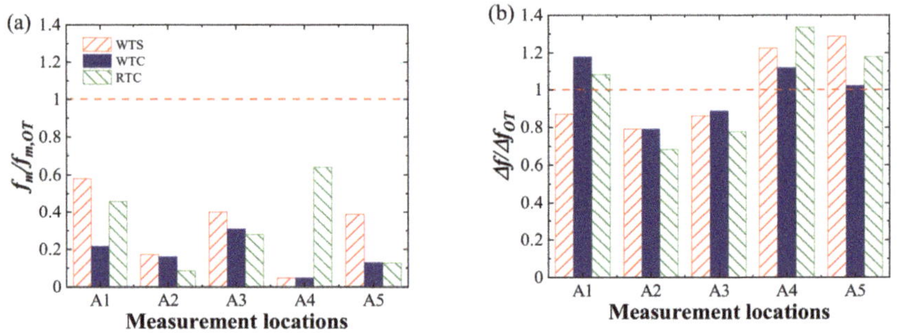

Figure 10. Metrics on spectra shifts at measurement locations: (**a**) median of the normalized frequency; (**b**) the bandwidth of the normalized frequency.

As displayed in Figure 10b, a decrease in this bandwidth is found at the measurement locations of the leeward region (A2 and A3) where the normalized bandwidth is lower than 1. However, the trend of the measurement locations at the valley region (A4 and A5) is markedly different with normalized bandwidth higher than 1. In addition, the situation for the measurement location at the windward region (A1) can be divided into two parts according to whether the transition section is applied. The value of the normalized frequency bandwidth in the case with transition section is higher than 1, while the value is lower than 1 in the case of WTS. This shows that topographic truncation brought inaccuracy into spectral bandwidth, but its effect is not certain and seems to be related to the character of the terrain where the measurement is located. Additionally, implementing transition sections at the edge of the terrain model greatly influenced the normalized frequency bandwidth, which may be attributed to the change in topography ahead leading to different flow behaviors.

6. Conclusions

A wind tunnel test for simulating flow over complex terrain was conducted in the wind tunnel laboratory of Central South University to investigate the effect of topographic truncation on wind characteristics. The analysis was carried out in two perspectives: mean wind characteristics, including mean velocity and inclination angle, and turbulence characteristics, including turbulence intensity and velocity spectra, which are fundamental parameters for flow characteristic description and are also essential for engineering applications. From the wind tunnel test, conclusions can be summarized as follows:

(1) Experimental results show that the effect of topographic truncation on profiles of mean velocity and turbulence intensity is different for different regions. A greater impact of the truncation was found in windward and leeward regions. The truncation of the terrain leads to a change in topographic features, causing a change in flow behavior upwind, and is the main reason for this difference. Accurate simulation of flow behavior upwind is crucial for repeating mean velocity and turbulence intensity profiles at target locations.

(2) By comparing the mean velocity and turbulence intensity, profiles of the inclination angle are more sensitive to topographic changes upstream, or they are more sensitive to changes in upstream flow behavior.

(3) Overestimation of streamwise velocity spectra was found in cases with the truncated terrain model in the low-frequency range but underestimated in the high-frequency range. Meanwhile, the slope of the spectra is influenced by a less negative value at the inertial subrange. Moreover, the normalized bandwidth representing the 90% energy interval is influenced by the topographic truncation, but the effect relates to the measurement locations. The bandwidth of windward and leeward regions narrows down, while the bandwidth of the valley region broadens.

(4) Transition sections used in this study have only limited effectiveness. Transition sections can only curb flow separation at the edge of the terrain model, but the change in flow behavior caused by the absence of topographic features due to topographic truncation cannot be resolved. In addition, the implementation of transition sections at the model edge may introduce additional errors into the experiment.

Author Contributions: Conceptualization, Z.W. and Y.Z.; methodology, X.H. and Y.Z.; software, P.Y.; validation, Z.W., Y.Z and L.L.; formal analysis, Z.W.; investigation, Z.W.; resources, Y.Z.; data curation, P.Y.; writing—original draft preparation, Z.W.; writing—review and editing, Y.Z.; visualization, P.Y. and L.L.; supervision, X.H.; project administration, Y.Z.; funding acquisition, X.L. All authors have read and agreed to the published version of the manuscript.

Funding: This research was funded by the National Natural Science Foundation of China, grant numbers 52078504, U1934209 and 51925808, and the Science and Technology Innovation Program of Hunan Province, grant number 2021RC3016.

Institutional Review Board Statement: Not applicable.

Informed Consent Statement: Not applicable.

Data Availability Statement: Data are contained within this article.

Conflicts of Interest: The authors declare no conflict of interest.

Nomenclature

LES	Large Eddy Simulation
URANS	Unsteady Reynolds-averaged Navier-Stokes
OT	Original topography
WTS	Without transition section
WTC	Witozinsky transition curve
RTC	Ramp transition curve
α	A constant coefficient and equal to 50
H	Altitude difference between model edge and wind tunnel floor
L	Length of transition curve at the certain points corresponding to H
Re	Reynolds number
h	Model height
ν	Kinematic viscosity
U	Streamwise velocity
U_r	Mean velocity at the top height of each location
I_u	Turbulence intensity
S_u	Streamwise velocity spectra
n	Frequency
σ	Standard deviation of velocity
f	Normalized frequency
Z	Height above ground
U(Z)	Mean velocity at height Z
U_x	Special indicator for mean velocity profiles
A_x	Special indicator for inclination angle profiles
I_x	Special indicator for turbulence intensity profiles
N	Total number of measurement points alone height at each measurement location

U_i	Normalized streamwise mean velocity at i-th point above the ground
$U_{i,OT}$	Normalized mean velocity at i-th point above the ground in case OT
Δh_i	Height difference between i-th point and (i-1)-th point
H_w	Total height at the measurement location
A_i	Inclination angle at i-th point above the ground
$A_{i,OT}$	Inclination angle at i-th point above the ground in case OT
I_i	Turbulence intensity at i-th point above the ground
$I_{i,OT}$	Turbulence intensity at i-th point above the ground in case OT
MI	Mean indicators
f_l	Cutoff frequency of 0.05 energy interval
f_m	Cutoff frequency of 0.5 energy interval
f_u	Cutoff frequency of 0.95 energy interval
Δf	90% energy interval bandwidth

References

1. Kilpatrick, R.J.; Hangan, H.; Siddiqui, K.; Lange, J.; Mann, J. Turbulent Flow Characterization Near the Edge of a Steep Escarpment. *J. Wind Eng. Ind. Aerodyn.* **2021**, *212*, 104605. [CrossRef]
2. Ren, H.; Wu, Y. Turbulent Boundary Layers Over Smooth and Rough Forward-facing Steps. *Phys. Fluids* **2011**, *23*, 45102. [CrossRef]
3. Wu, Y.; Ren, H. On the Impacts of Coarse-scale Models of Realistic Roughness on a Forward-facing Step Turbulent Flow. Int. *J. Heat Fluid Flow* **2013**, *40*, 15–31. [CrossRef]
4. Conan, B.; van Beeck, J.; Aubrun, S. Sand Erosion Technique Applied to Wind Resource Assessment. *J. Wind Eng. Ind. Aerodyn.* **2012**, *104*, 322–329. [CrossRef]
5. Jiao, A.; Shen, Y.; Wang, Z.; Chen, T.; Tao, H.; Xu, Z.; Fan, C. Experimental Study on the Effect of Canyon Cross Wind Yaw Angle on Airflow and Flame Characteristics in a Tunnel. *J. Wind Eng. Ind. Aerodyn.* **2021**, *213*, 104616. [CrossRef]
6. Nanos, E.M.; Yilmazlar, K.; Zanotti, A.; Croce, A.; Bottasso, C.L. Wind Tunnel Testing of a Wind Turbine in Complex Terrain. *J. Phys. Conf. Ser.* **2020**, *1618*, 032041. [CrossRef]
7. Ayotte, K.W.; Hughes, D.E. Observations of Boundary-layer Wind-tunnel Flow Over Isolated Ridges of Varying Steepness and Roughness. *Bound.-Layer Meteorol.* **2004**, *112*, 525–556. [CrossRef]
8. Kilpatrick, R.; Hangan, H.; Siddiqui, K.; Parvu, D.; Lange, J.; Mann, J.; Berg, J. Effect of Reynolds Number and Inflow Parameters on Mean and Turbulent Flow Over Complex Topography. *Wind Energ. Sci.* **2016**, *1*, 237–254. [CrossRef]
9. Berg, J.; Mann, J.; Bechmann, A.; Courtney, M.S.; Jørgensen, H.E. The Bolund Experiment, Part I: Flow Over a Steep, Three-dimensional Hill. *Bound.-Layer Meteorol.* **2011**, *141*, 219. [CrossRef]
10. Lange, J.; Mann, J.; Berg, J.; Parvu, D.; Kilpatrick, R.; Costache, A.; Jubayer, C.; Siddiqui, K.; Hangan, H. For Wind Turbines in Complex Terrain, the Devil Is in the Detail. *Environ. Res. Lett.* **2017**, *12*, 94020. [CrossRef]
11. Yeow, T.; Cuerva-tejero, A.; Perez-alvarez, J. Reproducing the Bolund Experiment in Wind Tunnel. *Wind Energy* **2015**, *18*, 153–169. [CrossRef]
12. Lange, J.; Mann, J.; Angelou, N.; Berg, J.; Sjöholm, M.; Mikkelsen, T. Variations of the Wake Height Over the Bolund Escarpment Measured by a Scanning Lidar. *Bound.-Layer Meteorol.* **2016**, *159*, 147–159. [CrossRef]
13. Liao, H.; Jing, H.; Ma, C.; Tao, Q.; Li, Z. Field Measurement Study on Turbulence Field by Wind Tower and Windcube Lidar in Mountain Valley. *J. Wind Eng. Ind. Aerodyn.* **2020**, *197*, 104090. [CrossRef]
14. Jing, H.; Liao, H.; Ma, C.; Tao, Q.; Jiang, J. Field Measurement Study of Wind Characteristics at Different Measuring Positions in a Mountainous Valley. *Exp. Therm. Fluid Sci.* **2020**, *112*, 109991. [CrossRef]
15. Taylor, P.A.; Teunissen, H.W. The Askervein Hill Project: Overview and Background Data. *Bound.-Layer Meteorol.* **1987**, *39*, 15–39. [CrossRef]
16. Salmon, J.R.; Teunissen, H.W.; Mickle, R.E.; Taylor, P.A. The Kettles Hill Project: Field Observations, Wind-tunnel Simulations and Numerical Model Predictions for Flow Over a Low Hill. *Bound.-Layer Meteorol.* **1988**, *43*, 309–343. [CrossRef]
17. Huang, G.; Cheng, X.; Peng, L.; Li, M. Aerodynamic Shape of Transition Curve for Truncated Mountainous Terrain Model in Wind Field Simulation. *J. Wind Eng. Ind. Aerodyn.* **2018**, *178*, 80–90. [CrossRef]
18. Han, Y.; Shen, L.; Xu, G.; Cai, C.S.; Zhang, J. Multiscale Simulation of Wind Field on a Long Span Bridge Site in Mountainous Area. *J. Wind Eng. Ind. Aerodyn.* **2018**, *177*, 260–274. [CrossRef]
19. Zhang, M.; Zhang, J.; Li, Y.; Yu, J.; Wu, L. Wind Characteristics in the High-altitude Difference at Bridge Site by Wind Tunnel Tests. *Wind Struct.* **2020**, *30*, 548–557.
20. Zhang, M.; Yu, J.; Zhang, J.; Wu, L.; Li, Y. Study on the Wind-field Characteristics Over a Bridge Site Due to the Shielding Effects of Mountains in a Deep Gorge Via Numerical Simulation. *Adv. Struct. Eng.* **2019**, *22*, 3055–3065.
21. Jubayer, C.M.; Hangan, H. A Hybrid Approach for Evaluating Wind Flow Over a Complex Terrain. *J. Wind Eng. Ind. Aerodyn.* **2018**, *175*, 65–76. [CrossRef]
22. Maurizi, A.; Palma, J.; Castro, F.A. Numerical Simulation of the Atmospheric Flow in a Mountainous Region of the North of Portugal. *J. Wind Eng. Ind. Aerodyn.* **1998**, *74–76*, 219–228. [CrossRef]

23. Hu, P.; Li, Y.; Huang, G.; Kang, R.; Liao, H. The Appropriate Shape of the Boundary Transition Section for a Mountain-gorge Terrain Model in a Wind Tunnel Test. *Wind Struct.* **2015**, *20*, 15–36. [CrossRef]
24. Hu, P.; Han, Y.; Xu, G.; Li, Y.; Xue, F. Numerical Simulation of Wind Fields at the Bridge Site in Mountain-gorge Terrain Considering an Updated Curved Boundary Transition Section. *J. Aerosp. Eng.* **2018**, *31*, 4018008. [CrossRef]
25. Chen, X.; Liu, Z.; Wang, X.; Chen, Z.; Xiao, H.; Zhou, J. Experimental and Numerical Investigation of Wind Characteristics Over Mountainous Valley Bridge Site Considering Improved Boundary Transition Sections. *Appl. Sci.* **2020**, *10*, 751. [CrossRef]
26. Mcauliffe, B.R.; Larose, G.L. Reynolds-number and Surface-modeling Sensitivities for Experimental Simulation of Flow Over Complex Topography. *J. Wind Eng. Ind. Aerodyn.* **2012**, *104–106*, 603–613. [CrossRef]
27. Røkenes, K.; Krogstad, P. Wind Tunnel Simulation of Terrain Effects on Wind Farm Siting. *Wind Energy* **2009**, *12*, 391–410. [CrossRef]
28. Gao, D.L.; Deng, Z.; Yang, W.H.; Chen, W.L. Review of the excitation mechanism and aerodynamic flow control of vortex-induced vibration of the main girder for long-span bridges: A vortex-dynamics approach. *J. Fluids Struct.* **2021**, *105*, 103348. [CrossRef]
29. Chen, W.L.; Zhang, Q.Q.; Li, H.; Hu, H. An experimental investigation on vortex induced vibration of a flexible inclined cable under a shear flow. *J. Fluids Struct.* **2015**, *54*, 297–311. [CrossRef]
30. Li, H.; Chen, W.L.; Xu, F.; Li, F.C.; Ou, J.P. A numerical and experimental hybrid approach for the investigation of aerodynamic forces on stay cables suffering from rain-wind induced vibration. *J. Fluids Struct.* **2010**, *26*, 1195–1215. [CrossRef]
31. He, X.H.; Zuo, T.H.; Zou, Y.F.; Yan, L.; Tang, L.B. Experimental study on aerodynamic characteristics of a high-speed train on viaducts in turbulent crosswinds. *J. Cent. South Univ.* **2020**, *27*, 2465–2478. [CrossRef]
32. Bowen, A. Modelling of Strong Wind Flows Over Complex Terrain at Small Geometric Scales. *J. Wind Eng. Ind. Aerodyn.* **2003**, *91*, 1859–1871. [CrossRef]
33. Holmes, J.D. *Wind Loading of Structures*, 3rd ed.; CRC Press: Los Angeles, CA, USA, 2015; pp. 187–188.
34. Cermak, J.E. Physical Modelling of Flow and Dispersion Over Complex Terrain. In *Boundary Layer Structure*, 1st ed.; Hadassah, K., Nathan, D., Eds.; Springer: Dordrecht, The Netherlands, 1984; pp. 261–292.
35. Conan, B. Wind Resource Assessment in Complex Terrain by Wind Tunnel Modelling. Ph.D. Thesis, Orléans University, Orléans, France, 2012.
36. Derickson, R.; Peterka, J. Development of a Powerful Hybrid Tool for Evaluating Wind Power in Complex Terrain: Atmospheric Numerical Models and Wind Tunnels. In Proceedings of the 23rd Asme Wind Energy Symposium: American Institute of Aeronautics and Astronautics, Reno, NV, USA, 5–8 January 2004.
37. Cuerva-tejero, A.; Avila-sanchez, S.; Gallego-castillo, C.; Lopez-garcia, O.; Perez-alvarez, J.; Yeow, T.S. Measurement of spectra over the Bolund hill in wind tunnel. *Wind Energy* **2018**, *21*, 87–99. [CrossRef]
38. Kaimal, J.C.; Wyngaard, J.C.; Izumi, Y.; Coté, O.R. Spectral characteristics of surface-layer turbulence. *Q. J. R. Meteorol. Soc.* **1972**, *98*, 563–589. [CrossRef]

applied sciences

MDPI

Article

Aerodynamics of a Train and Flat Closed-Box Bridge System with Train Model Mounted on the Upstream Track

Hui Wang [1,2,3], Huan Li [1,2,3,*] and Xuhui He [1,2,3]

[1] School of Civil Engineering, Central South University, Changsha 410075, China; wanghui111@csu.edu.cn (H.W.); xuhuihe@csu.edu.cn (X.H.)
[2] National Engineering Laboratory for High Speed Railway Construction, Changsha 410075, China
[3] Joint International Research Laboratory of Key Technology for Rail Traffic Safety, Changsha 410075, China
[*] Correspondence: lihuan2016@csu.edu.cn

Abstract: The aerodynamic features of a train and flat closed-box bridge system may be highly sensitive to train-bridge aero interactions. For the generally utilized railway bridge-deck with two tracks (the upstream and downstream ones), the aero interactions above are occupied-track-dependent. The present paper thus aims to reveal the aero interactions stated above via a series of wind tunnel tests. The results showed that the aero interactions of the present train-bridge system display four typical behaviors, namely, the underbody flow restraining effect, bridge deck shielding effect, flow transition promoting effect, and the flow separation intensifying effect. The above four aero interactions result in obvious reductions in the aerodynamic forces of the train in wind angle of attack α of [$-4°$, $12°$] and in the static stall angle of the bridge-deck, and leads to sensible increases in the absolute values of the bridge aerodynamic forces in α of [$-4°$, $12°$]. Upon comparing the results with the same train and bridge system but with the train model mounted on the downstream track, the quasi-Reynolds number effect was non-detectable when the train model was moved to the upstream track. Thus, no drag crisis and other saltatory aerodynamic behaviors were observed in the present study in α of [$0°$, $12°$].

Keywords: wind engineering; wind tunnel test; wind-train-bridge system; flow visualization

Citation: Wang, H.; Li, H.; He, X. Aerodynamics of a Train and Flat Closed-Box Bridge System with Train Model Mounted on the Upstream Track. *Appl. Sci.* **2022**, *12*, 276. https://doi.org/10.3390/app12010276

Academic Editors: Wenli Chen, Zifeng Yang, Gang Hu, Haiquan Jing and Junlei Wang

Received: 25 November 2021
Accepted: 25 December 2021
Published: 28 December 2021

1. Introduction

Currently, railway vehicles need to run on or through long-span bridges, i.e., Hutong Yangtze river railway bridge, Jinshajiang railway bridge, and Strait of PingTan Bridge [1]. The serviceability of trains running on such long-span railway bridges is of major concern around the world [2–4]. In order to predict, control, and improve the serviceability of the train-bridge system in crosswinds, their aerodynamic properties and underlying flow physics should be fully investigated in advance.

From the train point of view, modern vehicles are more sensitive to crosswinds due to their higher operating speed and lighter weight [2,5]. To guarantee the operating comfortability and safety of a train running on scenarios of different types (flat ground, embankment, viaduct, and long-span bridge) in crosswinds, two kinds of risk analysis methods have been put forward, i.e., the mathematically rigorous approach developed by many previous researchers [4,6–10] and the characteristic wind curve (CSC) as documented in specification [11]. However, no matter which method is utilized, the wind loads acting on the train are the basic input parameters.

From the infrastructure scenario point of view, the long-span railway bridge is generally the most critical site in a segment [12]. Relative to other scenarios, the long-span railway bridge is characterized by smaller stiffness, lower damping, lighter weight, and severer wind environment. Moreover, the dynamic responses of the train and bridge system induced by crosswinds are chiefly resisted by the bridge. Given all the above characteristics, the crosswind aerodynamic properties of the railway bridge scenario should be taken into consideration in the risk analysis of the train and bridge system. Within the range of the CSC, the bridge deck may suffer from buffeting and vortex-induced vibration [13–15], which should have a strong negative effect on the serviceability of the train in crosswinds. To assess the dynamic responses of the long-span railway bridge induced by crosswinds, the aerodynamic properties of the bridge deck need to be measured, analyzed, and classified.

From the train-bridge system point of view, the aerodynamic and dynamic interactions between the train and bridge are very complex [16–18]. On one hand, the presence of the bridge deck should significantly impact the approaching and underbody flows of the train. For example, the leading-edge separated flow of the bridge-deck may lead to an overspeed effect on the approaching flow of the train, similar to Baker's hypothesis [19]. Meanwhile, the flow separation-reattachment of the bridge-deck would be noticeably altered by the presence of the train. On the other hand, the dynamic responses of the train-bridge system are obviously intensified in crosswinds [4,6–10]. The above aerodynamic and dynamic interactions were mainly carried out on train-viaduct and train-truss bridge systems [4,18,20–29]. Associated studies on the train and flat closed-box bridge system are limited. Currently, the flat closed-box deck has been widely utilized in long-span railway bridges in China since 2014. For such a wind-sensitive bridge deck, systematic investigations on the aerodynamics of the train and flat closed-box bridge system are urgently required.

Generally, the relative spatial position of two objects (i.e., circular cylinders, square and rectangular prisms) plays a determining role in their aerodynamic forces, flow structures, pressure distributions, velocity fields, and near wakes [30–32]. When it comes to the train-infrastructure system, the change of the relative spatial position of the train and infrastructure is mainly influenced by the variations of the occupied track the train is running on. In the past decades, several studies have already been carried out to discuss the influences of the aforementioned occupied track variations on the aerodynamic properties of trains [12,22,26,33]. However, the influences of the occupied track variation on the crosswind aerodynamic characteristics and their underlying flow mechanisms of the train and bridge system are not yet fully understood. Thus, the present research carried out a series of wind tunnel tests to study the crosswind aerodynamic characteristics of a train and flat closed-box bridge system with the train model mounted on the upstream track, as well as the underlying flow physics. Compared to the case with the train model mounted on the downstream rack [34], the influences caused by the above occupied track variation are revealed.

Pressure field measurements and flow visualization results of a train and flat closed-box bridge are reported in the present study, which may be a meaningful complement to the aerodynamic interactions of the train-bridge system. The remainder of this paper is organized as follows. Section 2 depicts the detailed information of the wind tunnel test setups, including the wind tunnels, the testing models, the testing instruments, and the arrangements of the testing points. Sections 3 and 4 discuss the aerodynamic interactions and properties of the train and flat closed-box bridge system, respectively. The results were then compared to the results of the same train-bridge system but with the train model mounted on the downstream track, and the influences caused by the occupied track variation were analyzed and classified. In Section 5, two core findings are summarized.

2. Configurations of Wind Tunnel Setup

2.1. Wind Tunnels and Models

Two wind tunnels belonging to Central South University were utilized in these two experiment as shown in Figure 1a,b. The width and height of the test sections of these two wind tunnels are 450 mm × 450 mm and 3000 mm × 3000 mm, respectively. For clear representation, the smaller wind tunnel is named as WT-I, whereas the larger one is named as WT-II. Without any turbulence generators, the flow in the WT-I and WT-II with turbulence intensities smaller than 0.6% could be considered as uniform based on the current specification [11]. Please see [34] for more detailed geometric parameters of the above two wind tunnels.

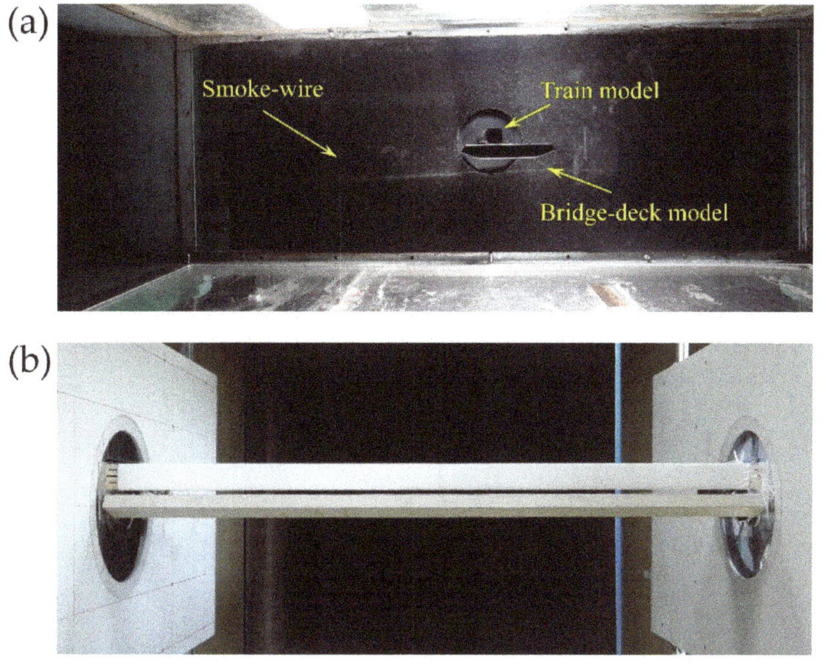

Figure 1. Mode installations: (**a**) Models for flow visualization, and (**b**) Models for pressure field and flow profile measurements.

Two groups of wind tunnel test models were made to match the WT-I and WT-II, which are simplified from a practical engineering model in China as shown in Figure 1. On one hand, as depicted in Figure 1a, the first group models scaled by 1:200 were used for smoke-wire visualization in WT-I. The widths, heights, and lengths of the train and bridge-deck models are $b \times d \times l$ = 16.3 mm × 18.5 mm × 450.0 mm and $B \times D \times L$ = 98.0 mm × 14.0 mm × 450.0 mm, respectively, resulting in a maximum blockage ratio of 6.2%. In this flow visualization experiment, the models were made by 3D printing technique. The wind angle of attack α was set in [−12°, 12°] with an interval of 2°. The incoming flow velocity is U_∞ = 2 m/s, and the corresponding Reynolds number (*Re*) based on U_∞ and d is of 2.5 × 10^3. On the other hand, the second group of models scaled by 1:40 were used for pressure field and flow profile measurements as shown in Figure 1b in WT-II. The widths, heights, and lengths of the train and bridge-deck models are $b \times d \times l$ = 81.6 mm × 92.3 mm × 2000.0 mm and $B \times D \times L$ = 490.0 mm × 70.0 mm × 2000.0 mm, respectively, resulting in a maximum blockage ratio of 4.6%. For brevity, only the geometrics of this group of models are depicted in Figure 2. In the pressure field and profile measurement experiment, the train model was still made by 3D printing technique, whereas the bridge-deck model was made of wood with an aluminum spine embedded inside. The test range of α was also set in [−12°, 12°] with an interval of 2°, while the incoming flow velocity is U_∞ = 20 m/s corresponding to Re = 1.25 × 10^5 (based on U_∞ and d). Please note that, only the train's shoulders are rounded without any sharp corners, thus all the Reynolds numbers in this paper are calculated based on d.

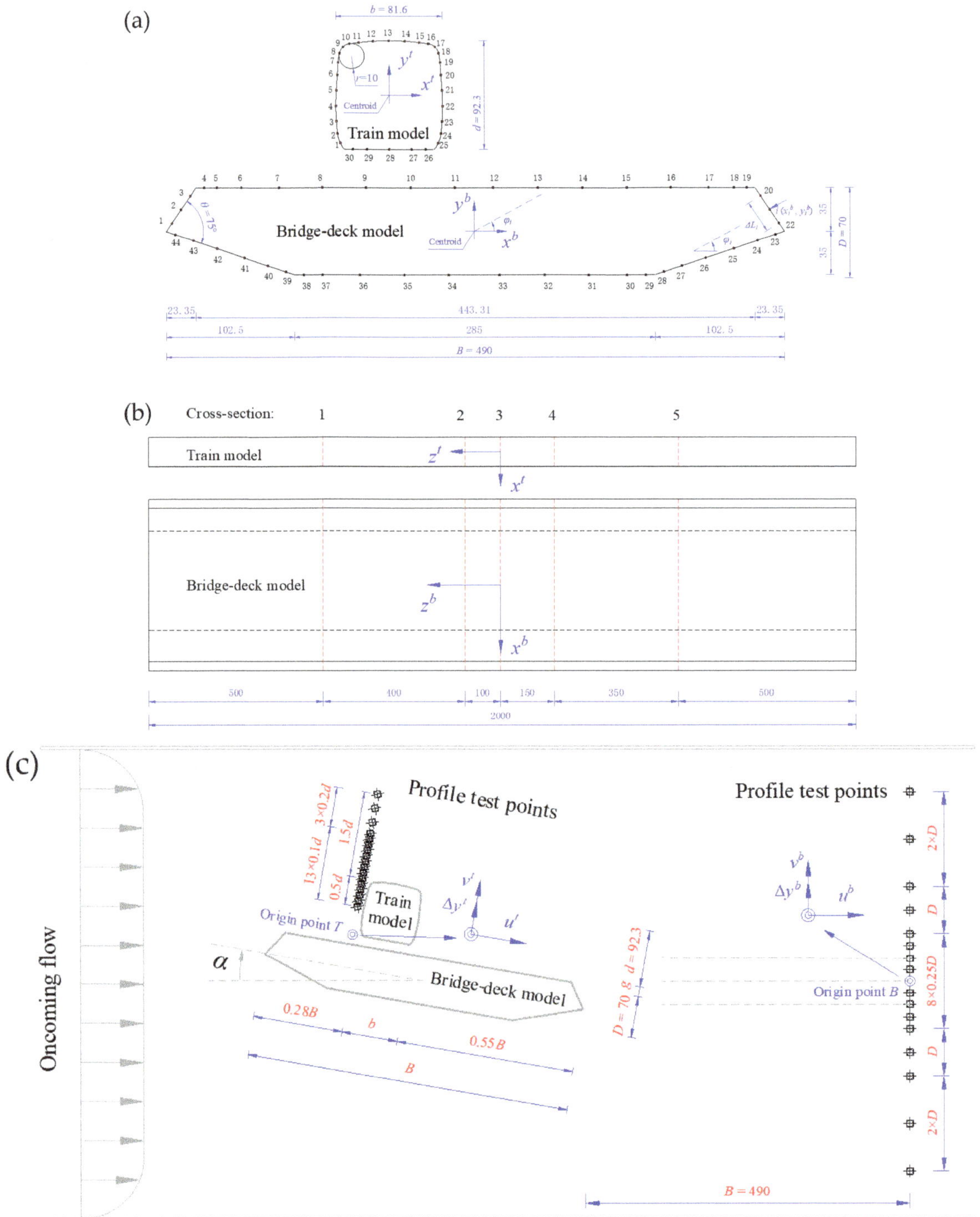

Figure 2. Model geometric parameters: (**a**) pressure test points in the lateral direction; (**b**) pressure test sections in the longitudinal direction; and (**c**) flow profile test points. (Unit: mm).

As described in Figure 2, x^t–y^t and x^b–y^b denote the geometric coordinate systems of the train and bridge-deck models, respectively. The coordinate origins are fixed at the centroids of the cross sections of the test models.

2.2. Testing Points and Instruments

The pressure fields around the test models were measured by pressure taps pre-installed on their surfaces as shown in Figure 2a,b. In the streamwise direction, there are 30 and 44 taps distributed around the train and bridge-deck models, respectively. In the spanwise direction, there are five pressure test cross-sections. In total, six scanner (Scanivalve, ZOC33/64PxX2) modes were used to record the pressure signals of all the above 370 pressure taps. The sampling frequency and duration of all six scanner modes were modulated as 625 Hz and 32 s, respectively. Note that, the Strouhal number of the present model is smaller than 0.18 corresponding to a vortex shedding mode less than 40 Hz, thus the time delay 1/625 s between any two taps caused by the scanner is neglected. The pressure tap and scanner were connected by PVC tubing system with a length of 500 mm. As demonstrated by [17], the influence caused by this PVC tubing system is less than 8%, which is also neglected in this research.

A Cobra probe was utilized to monitor the flow profiles of the present models, which was mounted on a 3D traverse system with a displacement resolution of 0.02 mm as demonstrated in Figure 2c. The sampling frequency and duration of the Cobra probe were modulated as 2000 Hz and 30 s, respectively.

As depicted in Figure 2c, the flow profile coordinate system is defined as u (the streamlined component) and v (the vertical component). For the train approaching flow, the coordinate origin A is fixed at the bottom of the train model. For the bridge-deck near wake, the coordinate origin B is fixed on the nose-tail line of the bridge-deck. $\Delta y^t/\Delta y^b$ are the vertical coordinates of the test points.

3. Aerodynamic Interactions

This section depicts the instantaneous flow structures around the present models, the approaching flow profiles of the train model, and the pressure fields and near wake profiles of the bridge-deck model. Based on the above experimental results, the train-bridge aero interactions in crosswinds were analyzed and compared to the study of [34] in order to reveal the effects of the occupied track variation (with train model mounted on the upstream or downstream track) on the aero interactions of the train and bridge system.

3.1. Train Model

3.1.1. Underbody Flow Restraining Effect

Figures 3 and 4 depict the instantaneous flow structures around the test models in the train-only, bridge-only, and train-bridge (with the train model mounted on the upstream track) cases. As shown in the left columns of Figures 3 and 4, the underbody flow of the train-only model generally transits from full separation to intermittent reattachment with an increase in α in $[-12°, 12°]$. Interacting with the roof separated flow, a typical Karman vortex street appears in the near wake of the train model similar to that of a circular cylinder. However, as demonstrated in the right columns of Figures 3 and 4, the underbody flow separation of the train model is noticeably decaying with the appearance of the bridge-deck model, as was evidenced by only a small piece of smoke-wire going through the gap between the train and bridge models. Even through the roof flow separation, reattachment is still detectable; no Karman vortex street can be observed in the near wake of the train model. Obviously, the chief reason for the above weakened underbody flow separation is the presence of the bridge deck model. Based on the above-mentioned discussion, the bridge deck has a strong restraining effect on the underbody flow separation of the train model.

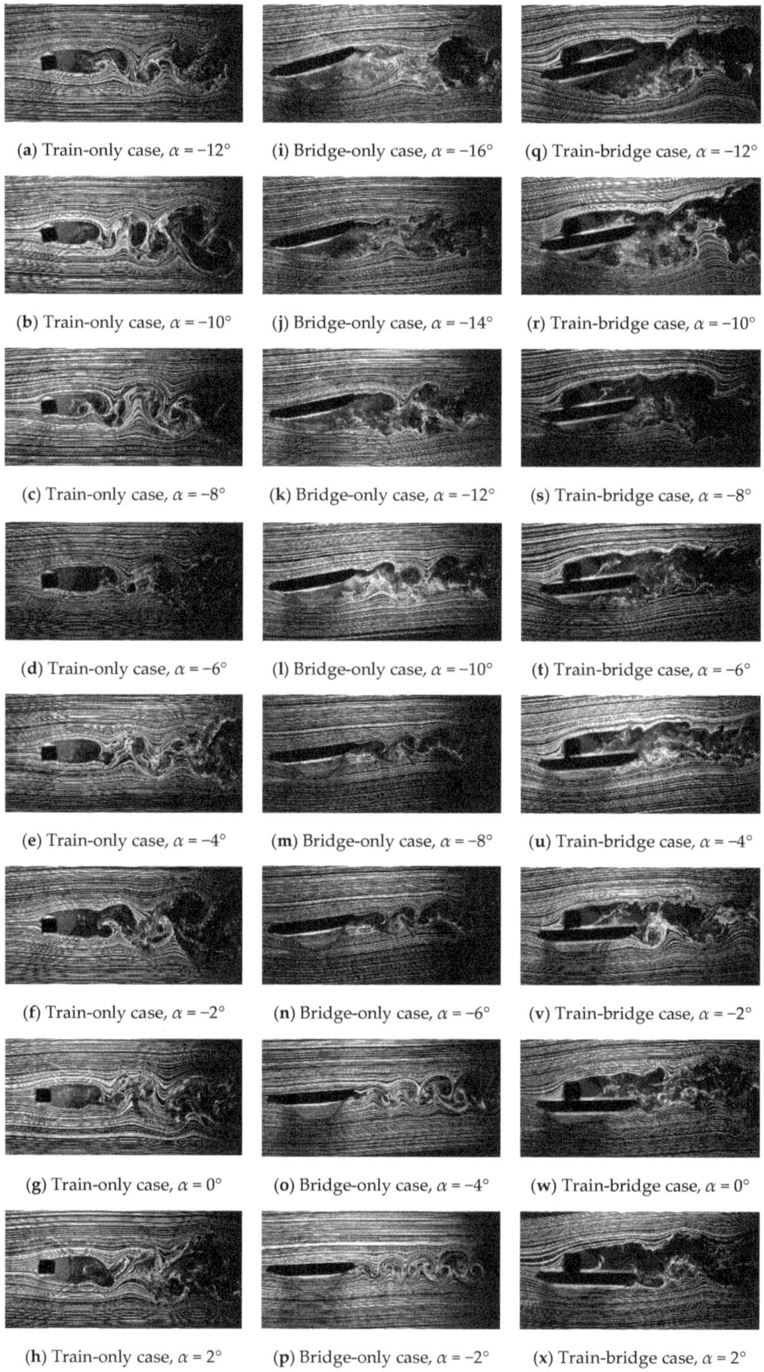

(**a**) Train-only case, $\alpha = -12°$ (**i**) Bridge-only case, $\alpha = -16°$ (**q**) Train-bridge case, $\alpha = -12°$

(**b**) Train-only case, $\alpha = -10°$ (**j**) Bridge-only case, $\alpha = -14°$ (**r**) Train-bridge case, $\alpha = -10°$

(**c**) Train-only case, $\alpha = -8°$ (**k**) Bridge-only case, $\alpha = -12°$ (**s**) Train-bridge case, $\alpha = -8°$

(**d**) Train-only case, $\alpha = -6°$ (**l**) Bridge-only case, $\alpha = -10°$ (**t**) Train-bridge case, $\alpha = -6°$

(**e**) Train-only case, $\alpha = -4°$ (**m**) Bridge-only case, $\alpha = -8°$ (**u**) Train-bridge case, $\alpha = -4°$

(**f**) Train-only case, $\alpha = -2°$ (**n**) Bridge-only case, $\alpha = -6°$ (**v**) Train-bridge case, $\alpha = -2°$

(**g**) Train-only case, $\alpha = 0°$ (**o**) Bridge-only case, $\alpha = -4°$ (**w**) Train-bridge case, $\alpha = 0°$

(**h**) Train-only case, $\alpha = 2°$ (**p**) Bridge-only case, $\alpha = -2°$ (**x**) Train-bridge case, $\alpha = 2°$

Figure 3. Instantaneous flow patterns: (**a–h**) Train-only case at $\alpha = -12°{\sim}2°$, (**i–p**) Bridge-only case at $\alpha = -12°{\sim}2°$, and (**q–x**) Train-bridge case at $\alpha = -16°{\sim}{-2°}$.

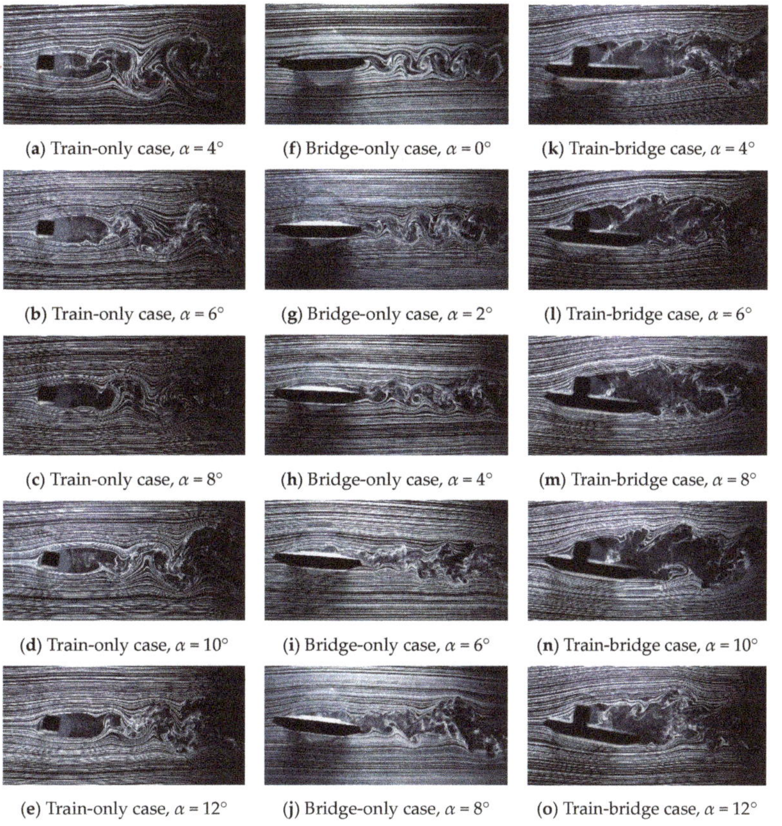

Figure 4. Instantaneous flow patterns: (**a–e**) Train-only case at $\alpha = 4°{\sim}12°$, (**f–j**) Bridge-only case at $\alpha = 4°{\sim}12°$, and (**k–o**) Train-bridge case at $\alpha = 0°{\sim}8°$.

In comparison with the study of [34], the underbody flow separation of the train model in the present study is still detectable, as evidenced by a small piece of smoke-wire, suggesting the underbody flow restraining effect is a smoother dependence in the occupied track variation.

3.1.2. Bridge Deck Shielding Effect

In the bridge-only case, the shear layer thickness of the leading-edge separated flow over the bridge-deck upper surface is limited in $\alpha = [-16°, 2°]$ as evidenced by the middle columns of Figures 3i–o and 4f,g. With an increase in $\alpha = [4°, 8°]$, the above shear layer thickness increases noticeably as shown in Figure 4h–j. In the train-bridge case, the above leading-edge separated flow reattaches on the upstream surface of the train model, and then bifurcates into two parts, as shown in the right columns of Figures 3 and 4. With an increase in $\alpha = [-12°, 12°]$, the stagnation point (bifurcation) of the leading-edge shear layer is always located at the 1/4 height of the train model, which is larger than the train-bridge gap, suggesting that this leading-edge shear layer may have a shielding effect on the train model.

To quantitatively identify the above-mentioned bridge deck shielding effect, the approaching flow profiles of the train model were monitored by a Cobra probe as shown in Figure 5. Due to the $\pm 45°$ conical measurement area of the Cobra probe, the flow profiles in the range of $0 \le \Delta y^t \le 0.5$ could not be effectively tested and thus are not shown here. In Figure 5a, the streamlined component of the normalized mean velocity u/U_∞ is always smaller than 1.0 in Δy^t = [0.5, 1.5], evidencing the above achieved shielding effect in all the test ranges of α = [−12°, 12°]. However, in the range of α = [−12°, −6°], this bridge-deck shielding effect was limited, as is evidenced by the $u/U_\infty \approx 1.0$ flow profile. With an increase in α = [−4°, 12°], this bridge deck shielding effect is remarkably enhanced as is evidenced by the noticeably decaying u/U_∞ profile and obviously intensifying turbulence intensities I_u & I_v profiles in Δy^t = [0.5, 1.0], respectively.

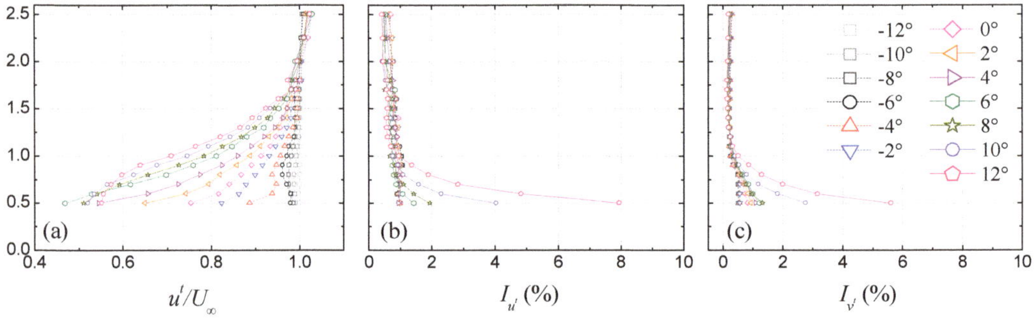

Figure 5. Profiles of flow approaching the train model at various angles of attack: (**a**) Normalized mean velocity in streamwise direction; (**b**) Turbulence intensity in streamwise direction; and (**c**) Turbulence intensity in vertical direction.

When compared to the study of [34], two striking differences can be summarized here. Firstly, no speed-up shear layer can be detected in the approaching flow of the train model mounted on the upstream track, which leads to the disappearance of the quasi-Reynolds number effect [34]. Then, the space upstream from the train is compressed when the train model is mounted on the upstream track, which leads to limited room for the development of the leading-edge shear layer of the bridge-deck model. Thus, the stagnation point of the train approaching flow is almost α-independent. Otherwise, the location of this stagnation point shall be a function of α as discussed in [34].

3.2. Bridge-Deck Model

3.2.1. Flow Transition Promoting Effect

In Figures 3 and 4, the instantaneous flow structures of the bridge-only model in α = [−16°, 12°] are depicted in the middle columns. Based on previous studies [31,35,36], the lower half bridge-deck is characterized by leading-edge vortex shedding (LEVS) in α = [−16°, −12°], impinging leading-edge vortices (ILEV) in α = [−10°, −8°], and trailing-edge vortex shedding (TEVS) in α = [−6°, 12°]. However, the corresponding α ranges of the above three typical flow patterns (i.e., LEVS, ILEV, and TEVS) in the train-bridge case are α = [−12°, −8°], α = [−6°, −4°], and α = [−2°, 12°] as shown in the right columns of Figures 3 and 4, respectively. In other words, the critical $|\alpha|$ of the above three typical flow patterns transiting from one to another on the lower half bridge-deck is reduced by about 4° with the presence of the train model. Moreover, Figure 6 can quantitively re-confirm the above flow transition promoting phenomenon. By comparison, the pressure distribution in x_p = 7.3~14.1 at α_i in the train-bridge case is in line with that at $\alpha_i - 4°$ in the bridge-only case. That is to say, the critical $|\alpha|$ of the flow transition of the lower half bridge-deck model is reduced by about 4° with the presence of the train model.

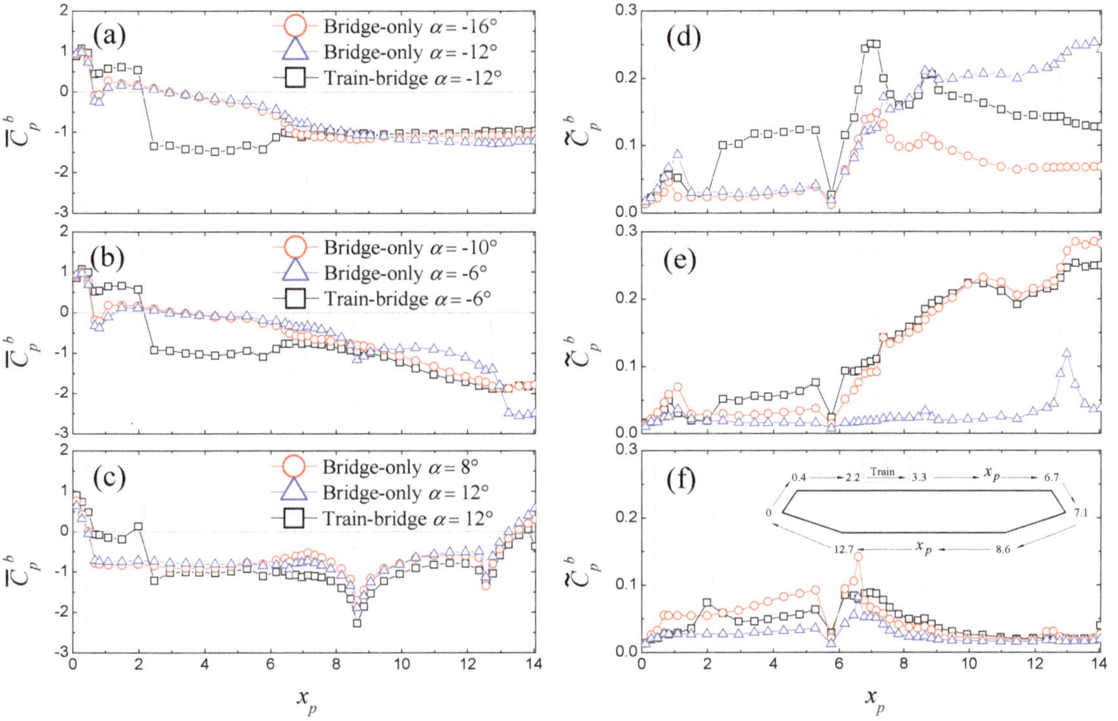

Figure 6. Pressure distributions around the bridge-deck model at various angles of attack: (**a**–**c**) Mean pressure coefficients; (**d**–**f**) Fluctuating pressure coefficients.

Upon comparison with the study of [34], the flow behaviors around the lower half bridge-deck model are almost the same as when the train model is mounted on the upstream and downstream tracks, suggesting the flow transition promoting effect is insensitive to the occupied track variation.

3.2.2. Flow Separation Intensifying Effect

As shown in the middle columns of Figures 3 and 4, the dominated flow patterns of the upper half bridge-deck model in the bridge-only case are still LEVS, ILEV, and TEVS, similar to those of the lower half bridge-deck. However, with the presence of the train model, the upper half bridge-deck model is always controlled by flow separation from the upper half of the train model, as described in the right columns of Figures 3 and 4. Generally, the flow separated from the train roof overshoots the entire upper half bridge-deck model resulting in a wider near wake than that of the corresponding bridge-only case, which can also be evidenced by a selected comparison of the wake profiles of the bridge-deck model as shown in Figure 7.

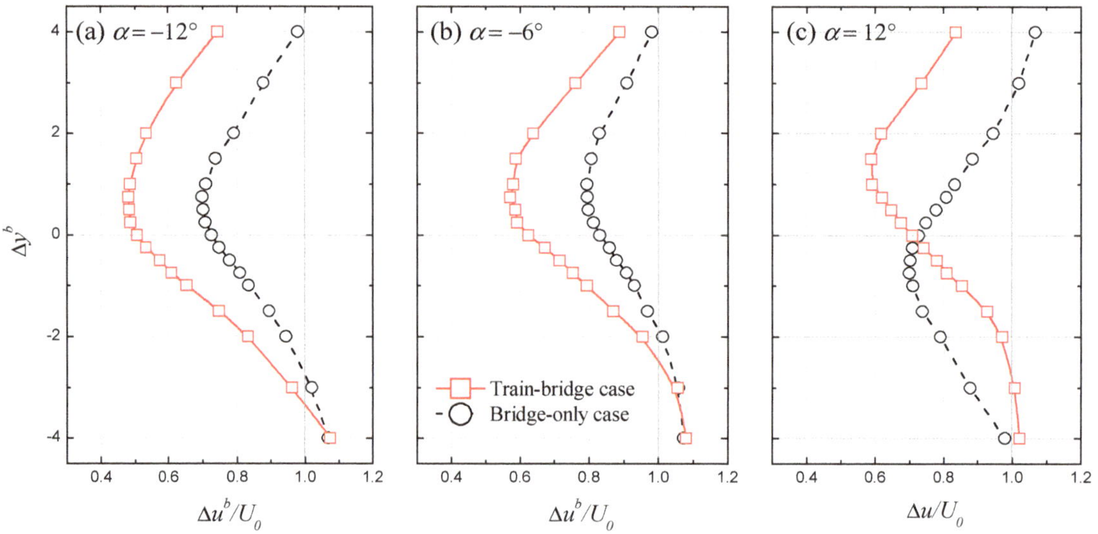

Figure 7. Wake profiles of the bridge-deck model at angles of attack of −12°, −6°, and 12°.

The above comparisons of the instantaneous flow structures and near wake profiles between the train-bridge and bridge-only cases suggest that the intensity of the flow separated from the train model is much stronger than that of the bridge-deck model, namely, the flow separation intensifying effect.

Compared to the study of [34], the space downstream from the train model is expanded, which results in enough room for the development of the roof separated flow of the train model. Thus, the flow separation intensifying effect of the present study is much stronger as is evidenced by the wider and deeper near wake.

4. Aerodynamic Characteristics

This section presents the aerodynamic force coefficients of the present models. Compared to several previous studies, the present experiment is verified. The aerodynamic properties of the downstream case reported by [34] are also illustrated here to reveal the influences of the occupied track variation on the aerodynamic features of the train-bridge system.

4.1. Aerodynamic Force Coefficients

The aerodynamic force coefficients are calculated via the pressure field test results. $\overline{C}_D^t$, $\overline{C}_L^t$, $\overline{C}_M^t$, $\widetilde{C}_D^t$, $\widetilde{C}_L^t$, and $\widetilde{C}_M^t$ are the mean and fluctuating drag, lift, and moment coefficients of the train model, whereas $\overline{C}_D^b$, $\overline{C}_L^b$, $\overline{C}_M^b$, $\widetilde{C}_D^b$, $\widetilde{C}_L^b$, and $\widetilde{C}_M^b$ are the mean and fluctuating drag, lift, and moment coefficients of the bridge-deck model, respectively. The detailed definitions about the above force coefficients are available in [34].

Figures 8 and 9 present the aerodynamic force coefficients of the train and bridge-deck models, respectively. For verification, the experimental results of the train-only case, bridge-only case, [37–39] are also illustrated here. As demonstrated in Figure 8a,b,e and Figure 9b,c, the results of the train-only and bridge-only cases show good agreements with those from previous studies when taking into consideration differences of the experimental models. Because all the present cases were carried out under similar conditions, the above comparisons verify the present investigation.

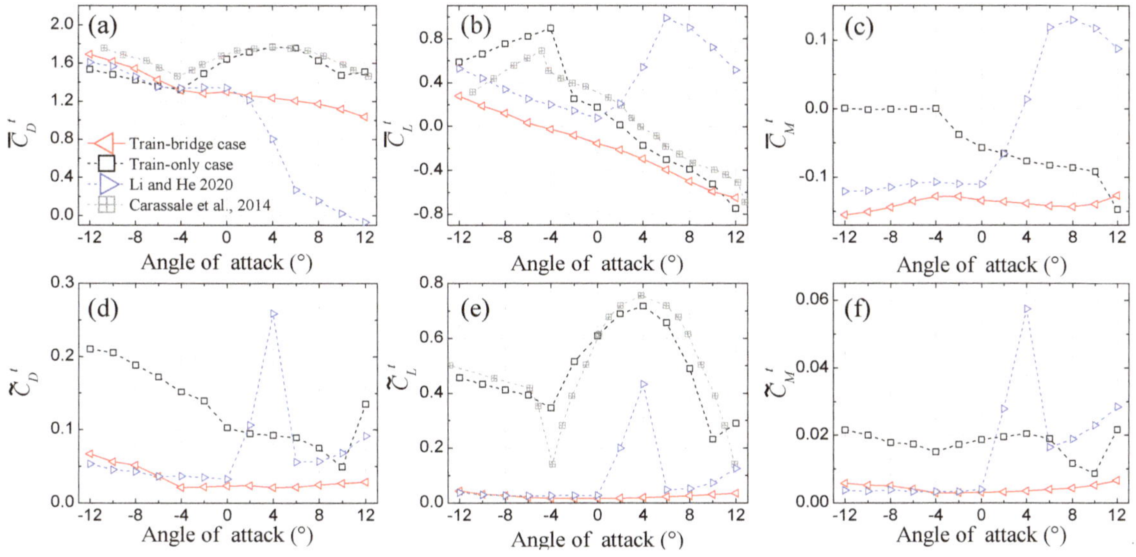

Figure 8. Aerodynamic force coefficients of the train model: (**a**) Mean drag coefficients; (**b**) Mean lift coefficients; (**c**) Mean moment coefficients; (**d**) Fluctuating drag coefficients; (**e**) Fluctuating lift coefficients; and (**f**) Fluctuating moment coefficients.

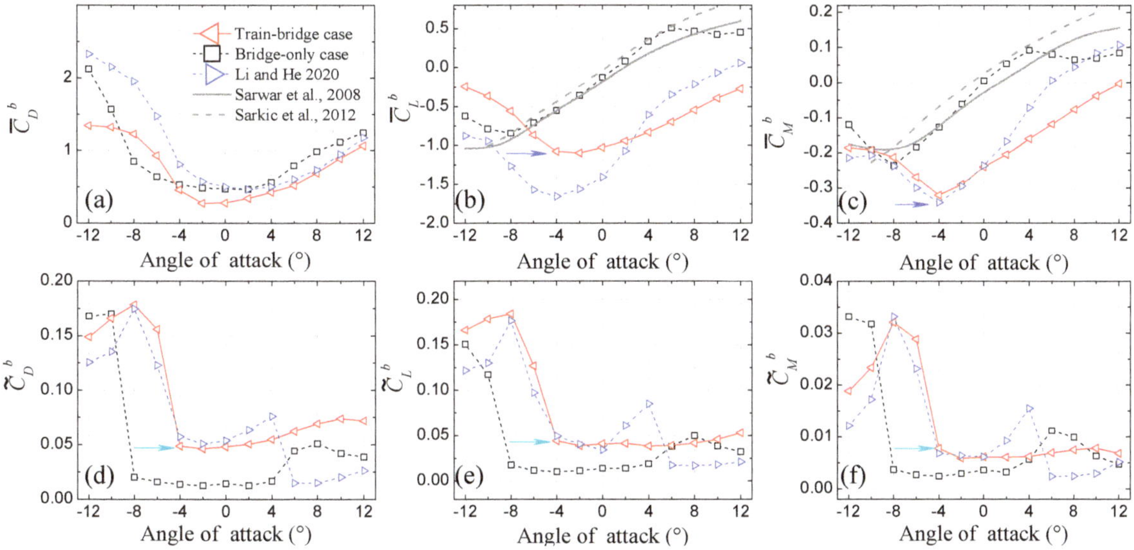

Figure 9. Aerodynamic force coefficients of the bridge-deck model: (**a**) Mean drag coefficients; (**b**) Mean lift coefficients; (**c**) Mean moment coefficients; (**d**) Fluctuating drag coefficients; (**e**) Fluctuating lift coefficients; and (**f**) Fluctuating moment coefficients.

As shown in Figure 8a, $\overline{C}_D^t$ in the train-only and train-bridge cases is in good consistency with each other in $\alpha = [-12°, -6°]$, while $\overline{C}_D^t$ in the train-bridge case is much smaller than that in the train-only case in $\alpha = [-4°, 12°]$, suggesting a strong bridge-deck shielding effect. The above two typical behaviors of $\overline{C}_D^t$ in different ranges of α are in fair agreement with the flow visualization and flow profile test results as shown in Section 3.1.2. As demonstrated in Figure 8d–f, $\widetilde{C}_D^t$, $\widetilde{C}_L^t$, and $\widetilde{C}_M^t$ in the train-bridge case are much smaller than those in the train-only case, which shall be attributed to the underbody flow restraining effect as concluded in Section 3.1.1.

In comparison with the study of [34] the effects caused by the occupied track variation on the aerodynamic force coefficients of the train model can be concluded by the following two aspects as shown in Figure 8:

(a) In $\alpha = [0°, 12°]$, no abrupt changes, i.e., drag crisis (Figure 8a), jumping and recovering of lift, moment, and three fluctuating force coefficients (Figure 8b–f), can be observed in the present study. As concluded in Section 3, these differences shall be chiefly caused by the disappearance of the quasi-Reynolds number effect.

1. In $\alpha = [-12°, 0°]$, the $\overline{C}_D^t$, $\overline{C}_L^t$, and $\overline{C}_M^t$ in the two cases with train model mounted on the upstream and downstream tracks share almost the similar behaviors, as described in Figure 8a–c. However, the $\overline{C}_L^t$ in the present study is reduced by 0.23 in the range of $\alpha = [-12°, 0°]$.

As demonstrated in Figure 9b–f, the stall angle of the bridge-deck model in the train-bridge case is about $\alpha = -4°$, which is about 4° larger than that of the bridge-only case as highlighted by two blue arrows in Figure 9b,c. The variation of the stall angle shall be attributed to the flow transition promoting effect as summarized in Section 3.2.1. Moreover, the $\widetilde{C}_D^b$, $\widetilde{C}_L^b$, and $\widetilde{C}_M^b$ are enhanced with the appearance of the train model as shown in Figure 9d–f, which shall be ascribed to the flow separation intensifying effect as concluded in Section 3.2.2.

Similarly, the effects caused by the occupied track variation on the aerodynamic force coefficients of the bridge-deck model can also be summarized into two aspects as follows:

(a) In $\alpha = [0°, 12°]$, no abrupt changes, i.e., jumping and recovering of $\overline{C}_L^b$, $\overline{C}_M^b$, $\widetilde{C}_D^b$, $\widetilde{C}_L^b$, and $\widetilde{C}_M^b$ can be observed in the present study relative to those of [34]. Obviously, this difference shall also be caused by the disappearance of the quasi-Reynolds number effect as concluded in Section 3.1.2.

(b) In $\alpha = [-12°, 0°]$, $\overline{C}_M^b$, $\widetilde{C}_D^b$, $\widetilde{C}_L^b$, and $\widetilde{C}_M^b$ in the two cases with train model mounted on the upstream and downstream tracks display similar behaviors, while the values of $|\overline{C}_D^b|$ and $|\overline{C}_L^b|$ in the present study are much smaller than those of [34].

4.2. Strouhal Numbers

The Strouhal numbers (St^t, St^b, and St^w for the train, bridge-deck, and whole train-bridge models, respectively) are calculated via the power spectral densities (PSDs) of the model lifts based on d, D, and $d + D$, respectively.

Figure 10a,b depicts the St^t and St^b in the range of $\alpha = [-12°, 12°]$. The results of an infinite prism tested by [40] and the train model in the train-only case are also demonstrated here for comparison. Generally, the train-only results highly mimic those of [40], again suggesting a validation of the present investigation.

The Strouhal numbers of the present models in the train-bridge case displays similar behavior but with different values as described in Figure 10a,b, respectively. That is expected because the train, bridge deck, and the whole train-bridge share a similar vortex shedding frequency.

As shown in Figure 10, the behaviors of the Strouhal numbers in the train-bridge case in all the test ranges of $\alpha = [-12°, 12°]$ can be classified into three aspects. In $\alpha = [-12°, -10°]$, the Strouhal numbers of the present models are about 0.04. In $\alpha = [-8°, -6°]$, no Strouhal numbers can be identified both on the train and bridge-deck models. If a rectangular bluff body is governed by the impinging leading-edge vortices flow pattern (as mentioned in Section 3.2.1) at large Reynolds number, the periodic vortex shedding from the model is almost imperceptible [31,41]. For the present train-bridge system, the lower half bridge-deck is governed by the ILEV flow pattern in this range of α as summarized in Section 3.2.1, and the inflow Reynolds number based on d is 1.25×10^5 belonging to the large Reynolds number [41]. Thus, the underlying mechanism of this Strouhal number disappearance shall be this ILEV flow pattern. In $\alpha = [-4°, 12°]$, both the Strouhal numbers of the train and bridge-deck models decreases gradually.

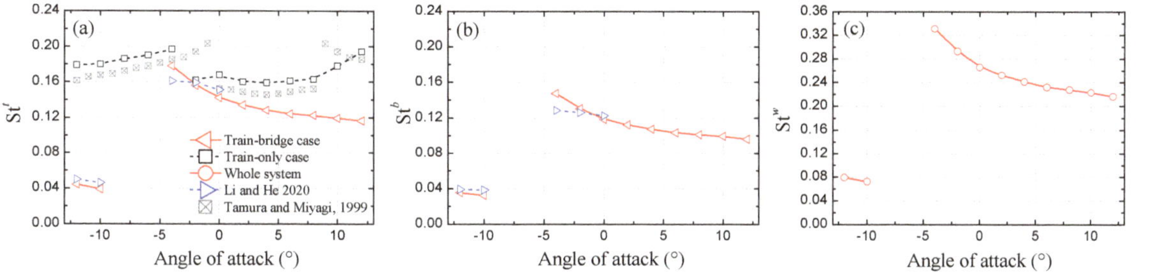

Figure 10. Strouhal numbers of the models: (**a**) the train model; (**b**) the bridge-deck model; and (**c**) the whole system.

Upon comparing with the study of [34] with the train model mounted on the downstream track, the present study with the train model mounted on the upstream track can still identify the Strouhal numbers in $\alpha = [0°, 12°]$. As pointed out by [42], the periodic vortex shedding of a circular cylinder tends to cease with an increase in Reynolds number in critical, transcritical, and supercritical regimes. As discussed in [34], the quasi-Reynolds number effect highly mimics the classical Reynolds number effect. Moreover, the above quasi-Reynolds number effect governs the main flow features of the train-bridge system with the train model mounted on the downstream track in $\alpha = [0°, 12°]$. Thus, due to the disappearance of the quasi-Reynolds number effect in the present study, the detectable Strouhal numbers in $\alpha = [0°, 12°]$ here are reasonable.

5. Concluding Remarks

The aerodynamic features of a train and flat closed-box bridge system with a train model mounted on the upstream track was investigated via wind tunnel tests. Compared to the train-only and bridge-only cases, the aero interactions of the train and bridge system were examined. Compared to the same train and bridge system but with the train model mounted on the downstream track, the effects of the occupied track (with the train model mounted on) variation of the train model on the aerodynamic properties of the train and bridge system are revealed. Core findings are summarized as follows.

(1) The aero interactions of the train and flat closed-box bridge system can be roughly categorized as the underbody flow restraining effect (UFR), bridge deck shielding effect (BDS), quasi-Reynolds number effect (Q-Rey), flow transition promoting effect (FTP), and flow separation intensifying effect (FSI). Among these, the FTP was insensitive to the occupied track variation of the train model, whereas the UFR, BDS, FSI, and Q-Rey were dependent on the occupied track variation, especially for Q-Rey, which disappeared when the train model was mounted on the upstream track. The Q-Rey disappearance was mainly attributed to the speed-up leading-edge shear layer of the bridge-deck not reattaching on the rounded shoulder of the train model mounted on the upstream track.

(2) The aerodynamic features of the train and flat closed-box bridge deck models are very sensitive to aero interactions and occupied track variations. The aero interaction resulted in obvious reductions in the aerodynamic forces of the train at the wind angle of attack α of $[-4°, 12°]$ and in the static stall angle of the bridge, and led to sensible increases in the absolute values of the aerodynamic forces of the bridge in α of $[-4°, 12°]$. The occupied track variation caused the disappearance of the drag crisis and other saltatory aerodynamic behaviors of the train-bridge system in $\alpha = [0°, 12°]$.

Author Contributions: Conceptualization, X.H.; methodology, H.L.; experiments, H.W. and H.L.; software, H.W.; validation, H.L. and X.H.; writing, H.W. and H.L.; supervision, X.H.; funding acquisition, X.H. and H.L. All authors have read and agreed to the published version of the manuscript.

Funding: This research was funded by Natural Science Foundation of Hunan province under Grant No. 2021JJ40744, and by National Natural Science Foundations of China under Grant No. 51925808 and U1934209.

Institutional Review Board Statement: Not applicable.

Informed Consent Statement: Not applicable.

Data Availability Statement: Data are contained within this article.

Acknowledgments: The authors thank the reviewers for their great help on the article during its review progress.

Conflicts of Interest: The author(s) declared no potential conflict of interest with respect to the research, authorship, and/or publication of this article.

References

1. He, X.; Wu, T.; Zou, Y.; Chen, Y.F.; Guo, H.; Yu, Z. Recent developments of high-speed railway bridges in China. *Struct. Infrastruct. Eng.* **2017**, *13*, 1584–1595. [CrossRef]
2. Baker, C. The flow around high speed trains. *J. Wind Eng. Ind. Aerodyn.* **2010**, *98*, 277–298. [CrossRef]
3. Baker, C.; Johnson, T.; Flynn, D.; Hemida, H.; Quinn, A.; Soper, D.; Sterling, M. *Train Aerodynamics: Fundamentals and Applications*; Butterworth-Heinemann: Cambridge, MA, USA, 2019.
4. Li, Y.; Qiang, S.; Liao, H.; Xu, Y.L. Dynamics of wind–rail vehicle–bridge systems. *J. Wind Eng. Ind. Aerodyn.* **2005**, *93*, 483–507. [CrossRef]
5. Baker, C. A review of train aerodynamics Part 1–Fundamentals. *Aeronaut. J.* **2014**, *118*, 201–228. [CrossRef]
6. Olmos, J.M.; Astiz, M.A. Improvement of the lateral dynamic response of a high pier viaduct under turbulent wind during the high-speed train travel. *Eng. Struct.* **2018**, *165*, 368–385. [CrossRef]
7. Olmos, J.M.; Astiz, M.A. Non-linear vehicle-bridge-wind interaction model for running safety assessment of high-speed trains over a high-pier viaduct. *J. Sound Vib.* **2018**, *419*, 63–89. [CrossRef]
8. Xia, H.; Zhang, N.; Guo, W.W. Analysis of resonance mechanism and conditions of train-bridge system. *J. Sound Vib.* **2006**, *297*, 810–822. [CrossRef]
9. Xu, Y.L.; Ding, Q.S. Interaction of railway vehicles with track in cross-winds. *JFS* **2006**, *22*, 295–314. [CrossRef]
10. Xu, Y.L.; Zhang, N.; Xia, H. Vibration of coupled train and cable-stayed bridge systems in cross winds. *Eng. Struct.* **2004**, *26*, 1389–1406. [CrossRef]
11. CEN. Railway Applications-Aerodynamics-Part 6: Requirements and Test Procedures for Cross Wind Assessment. 2010, 14067-6. Available online: https://standards.iteh.ai/catalog/standards/cen/a9acedac-3968-4b8c-a6cc-5a035c91e9b3/en-14067-6-2018 (accessed on 24 December 2021).
12. Cheli, F.; Ripamonti, F.; Rocchi, D.; Tomasini, G. Aerodynamic behaviour investigation of the new EMUV250 train to cross wind. *J. Wind Eng. Ind. Aerodyn.* **2010**, *98*, 189–201. [CrossRef]
13. Gao, D.; Chen, W.; Eloy, C.; Li, H. Multi-mode responses, rivulet dynamics, flow structures and mechanism of rain-wind induced vibrations of a flexible cable. *J. Fluids Struct.* **2018**, *82*, 154–172. [CrossRef]
14. Gao, D.; Chen, W.-L.; Zhang, R.-T.; Huang, Y.-W.; Li, H. Multi-modal vortex- and rain–wind-induced vibrations of an inclined flexible cable. *Mech. Syst. Signal Processing* **2019**, *118*, 245–258. [CrossRef]
15. Gao, D.; Deng, Z.; Yang, W.; Chen, W. Review of the excitation mechanism and aerodynamic flow control of vortex-induced vibration of the main girder for long-span bridges: A vortex-dynamics approach. *J. Fluids Struct.* **2021**, *105*, 103348. [CrossRef]
16. He, X.H.; Li, H. Review of aerodynamics of high-speed train-bridge system in crosswinds. *J. Cent. South Univ.* **2020**, *27*, 1054–1073. [CrossRef]
17. Li, H.; He, X.H.; Wang, H.F.; Kareem, A. Aerodynamics of a scale model of a high-speed train on a streamlined deck in cross winds. *J. Fluids Struct.* **2019**, *91*, 102717. [CrossRef]
18. Li, X.-Z.; Wang, M.; Xiao, J.; Zou, Q.-Y.; Liu, D.-J. Experimental study on aerodynamic characteristics of high-speed train on a truss bridge: A moving model test. *J. Wind Eng. Ind. Aerodyn.* **2018**, *179*, 26–38. [CrossRef]
19. Baker, C. The determination of topographical exposure factors for railway embankments. *J. Wind Eng. Ind. Aerodyn.* **1985**, *21*, 89–99. [CrossRef]
20. Barcala, M.A.; Meseguer, J. Visualization study of the influence of parapets on the flow around a train vehicle under cross winds. *WIT Trans. Built Environ.* **2008**, *103*, 797–806. [CrossRef]
21. Barcala, M.A.; Meseguer, J. An experimental study of the influence of parapets on the aerodynamic loads under cross wind on a two-dimensional model of a railway vehicle on a bridge. *Proc. Inst. Mech. Eng. Part F J. Rail Rapid Transit* **2007**, *221*, 487–494. [CrossRef]

22. Bocciolone, M.; Cheli, F.; Corradi, R.; Muggiasca, S.; Tomasini, G. Crosswind action on rail vehicles: Wind tunnel experimental analyses. *J. Wind Eng. Ind. Aerodyn.* **2008**, *96*, 584–610. [CrossRef]
23. Guo, W.; Xia, H.; Karoumi, R.; Zhang, T.; Li, X. Aerodynamic effect of wind barriers and running safety of trains on high-speed railway bridges under cross winds. *Wind Struct.* **2015**, *20*, 213–236. [CrossRef]
24. He, X.H.; Li, H.; Hu, L.; Wang, H.F.; Kareem, A. Crosswind aerodynamic characteristics of a stationary interior railway carriage through a long-span truss-girder bridge. *Eng. Struct.* **2020**, *210*. [CrossRef]
25. Ogueta-Gutiérrez, M.; Franchini, S.; Alonso, G. Effects of bird protection barriers on the aerodynamic and aeroelastic behaviour of high speed train bridges. *Eng. Struct.* **2014**, *81*, 22–34. [CrossRef]
26. Suzuki, M.; Tanemoto, K.; Maeda, T. Aerodynamic characteristics of train/vehicles under cross winds. *J. Wind Eng. Ind. Aerodyn.* **2003**, *91*, 209–218. [CrossRef]
27. Wang, M.; Li, X.-Z.; Xiao, J.; Zou, Q.-Y.; Sha, H.-Q. An experimental analysis of the aerodynamic characteristics of a high-speed train on a bridge under crosswinds. *J. Wind Eng. Ind. Aerodyn.* **2018**, *177*, 92–100. [CrossRef]
28. Xiang, H.; Li, Y.; Chen, S.; Li, C. A wind tunnel test method on aerodynamic characteristics of moving vehicles under crosswinds. *J. Wind Eng. Ind. Aerodyn.* **2017**, *163*, 15–23. [CrossRef]
29. Xiang, H.; Li, Y.; Liao, H.; Li, C. An adaptive surrogate model based on support vector regression and its application to the optimization of railway wind barriers. *Struct. Multidiscip. Optim.* **2016**, *55*, 701–713. [CrossRef]
30. Gao, D.-L.; Chen, W.-L.; Li, H.; Hu, H. Flow around a circular cylinder with slit. *Exp. Therm. Fluid Sci.* **2017**, *82*, 287–301. [CrossRef]
31. Naudascher, E.; Wang, Y. Flow-Induced Vibrations of Prismatic Bodies and Grids of Prisms. *J. Fluids Struct.* **1993**, *7*, 341–373. [CrossRef]
32. Zhou, Y.; Alam, M.M. Wake of two interacting circular cylinders: A review. *Int. J. Heat Fluid Flow* **2016**, *62*, 510–537. [CrossRef]
33. Ma, C.; Duan, Q.; Li, Q.; Chen, K.; Liao, H. Buffeting Forces on Static Trains on a Truss Girder in Turbulent Crosswinds. *J. Bridge Eng.* **2018**, *23*, 04018086. [CrossRef]
34. Li, H.; He, X.H. Lateral aerodynamic interference between an interior train and a flat box bridge-deck. *Exp. Therm. Fluid Sci.* **2020**, *117*, 110115. [CrossRef]
35. Deniz, S.; Staubli, T. Oscillating Rectangular and Octagonal Profiles: Interaction of Leading- and Trailing-Edge Vortex Formation. *J. Fluids Struct.* **1997**, *11*, 3–31. [CrossRef]
36. Païdoussis, M.P.; Price, S.J.; De Langre, E. *Fluid-Structure Interactions: Cross-Flow-Induced Instabilities*; Cambridge University Press: Near York, NY, USA, 2010.
37. Carassale, L.; Freda, A.; Marre-Brunenghi, M. Experimental investigation on the aerodynamic behavior of square cylinders with rounded corners. *J. Fluids Struct.* **2014**, *44*, 195–204. [CrossRef]
38. Sarwar, M.W.; Ishihara, T.; Shimada, K.; Yamasaki, Y.; Ikeda, T. Prediction of aerodynamic characteristics of a box girder bridge section using the LES turbulence model. *J. Wind Eng. Ind. Aerodyn.* **2008**, *96*, 1895–1911. [CrossRef]
39. Šarkić, A.; Fisch, R.; Höffer, R.; Bletzinger, K.-U. Bridge flutter derivatives based on computed, validated pressure fields. *J. Wind Eng. Ind. Aerodyn.* **2012**, *104-106*, 141–151. [CrossRef]
40. Tamura, T.; Miyagi, T. The effect of turbulence on aerodynamic forces on a square cylinder with various corner shapes. *J. Wind Eng. Ind. Aerodyn.* **1999**, *83*, 135–145. [CrossRef]
41. Nakamura, Y.; Ohya, Y.; Tsuruta, H. Experiments on vortex shedding from flat plates with square leading and trailing edges. *J. Fluid Mech.* **1991**, *222*, 437–447. [CrossRef]
42. Norberg, C.; Sunden, B. Turbulence and reynolds number effects on the flow and fluid forces on a single cylinder in cross flow. *J. Fluids Struct.* **1987**, *1*, 337–357. [CrossRef]

applied sciences

MDPI

Article

Fully Convolutional Neural Network Prediction Method for Aerostatic Performance of Bluff Bodies Based on Consistent Shape Description

Ke Li [1,2], Hai Li [2], Shaopeng Li [1,2,*] and Zengshun Chen [1,2]

[1] Key Laboratory of New Technology for Construction of Cities in Mountain Area, Chongqing University, Ministry of Education, Chongqing 400045, China; keli-bridge@cqu.edu.cn (K.L.); zengshunchen@cqu.edu.cn (Z.C.)

[2] School of Civil Engineering, Chongqing University, Chongqing 400045, China; lhtm_study@cqu.edu.cn

* Correspondence: lisp0314@cqu.edu.cn

Abstract: The shape of a bluff body section is of high importance to its aerostatic performance. Obtaining the aerostatic performance of a specific shape based on wind tunnel tests and CFD simulations takes a lot of time, which affects evaluation efficiency. This paper proposes a novel fully convolutional neural network model that enables rapid prediction from shape to aerostatic performance. Its main innovations are: (1) The proposal of a new shape description method in which the shape is described by the combination of the wall distance field and the space coordinate field, which can efficiently express the influencing factors of the shape on the aerostatic performance. (2) A step-by-step strategy in which the pressure field is used as the model output and then the calculation of the aerostatic coefficient is proposed. Compared with the simple direct prediction of the aerostatic coefficient, the logical connection between input and output can be enhanced and the prediction accuracy can be improved. It is found that the model proposed in this paper has good prediction accuracy, and its average relative error is 9.42% compared with the CFD calculation results. Compared with the direct use of the shape as the model input, the accuracy is improved by 13.25%; compared with the direct use of the drag coefficient as the model output, the accuracy is improved by 10%. Compared with traditional CFD calculations and wind tunnel experiments, this method can be used as a fast auxiliary screening method for the optimization of the aerodynamic shapes of bluff body sections.

Keywords: deep learning; prediction; aerostatic performance; shape; convolutional neural networks

Citation: Li, K.; Li, H.; Li, S.; Chen, Z. Fully Convolutional Neural Network Prediction Method for Aerostatic Performance of Bluff Bodies Based on Consistent Shape Description. *Appl. Sci.* **2022**, *12*, 3147. https://doi.org/10.3390/app12063147

Academic Editors: Wenli Chen, Zifeng Yang, Gang Hu, Haiquan Jing and Junlei Wang

Received: 21 February 2022
Accepted: 17 March 2022
Published: 19 March 2022

1. Introduction

Slender structures, such as bridges, high-rise buildings, and others, are prevalent in the field of civil engineering. These types of structures have low rigidity, and their aerostatic performance is directly related to the safety of the overall structure under the action of strong wind. Taking the main beam of a bridge as an example, excessive resistance and lifting moments will cause the structure to produce excessive lateral displacement and torsion and, in severe cases, even wind-induced aerostatic instability, which will affect the safety of the structure.

Slender structures have typical two-dimensional characteristics, and their aerostatic performance can be determined by the shape of their cross-section. The aerostatic performance of the bluff body section is usually expressed by the three-component force coefficient, and the aerodynamic shape of the bluff body section is particularly important among the influencing factors. The section form and aspect ratio are different, and the corresponding three-component force coefficient and variation law are also different. Kazutosh et al. [1] conducted experimental studies and numerical simulations on bridge girder sections with three different shapes, considering the influence of the degree of central slotting on the three-component force coefficient. Simiu et al. [2] explained that the drag coefficient of a rectangular section shows a trend of first increasing and then decreasing when the aspect ratio of the rectangular section is between 0 and 2. When the aspect ratio changes from 2 to 4, the drag coefficient has a certain discreteness. When the ratio is greater than 4, the drag coefficient remains basically unchanged. Shimada et al. [3] studied the variation of the drag and lift coefficients with the aspect ratio of a rectangular section as the parameter and found that the drag coefficient reached the maximum value when the aspect ratio was 0.6.

To sum up, there is an extremely complex nonlinear relationship between the cross-sectional shape of a bluff body and its aerostatic performance. This relationship does not have a certain regularity, making it difficult to describe in the form of a specific mathematical formula. It cannot be quickly passed. The aerodynamic shape obtains the aerostatic three-component force coefficient. Therefore, it is not possible to quickly obtain the aerostatic three-component force coefficient using the aerodynamic shape.

At present, there are two common methods for carrying out aerodynamic studies. The first method involves putting the segment model into a wind tunnel for testing [4–7], and the other method involves computational fluid dynamics (CFD) [8–10]. Although these two methods are widely used at present, they also have their own shortcomings. Investing in wind tunnel test equipment is expensive, and significant manpower and material resources are required for the tests. CFD simulation calculation consumes a lot of computing resources, which greatly restricts the efficiency of wind resistance design. Due to its excellent data fitting ability and efficient large data application ability, deep learning can discover input–output mapping relationships that humans cannot perceive in a large amount of data and can improve the aerostatic performance efficiency of bluff body sections.

With the continuous development of neural network technology, artificial intelligence prediction technology has been gradually applied in the field of wind engineering in recent years [11]. Lute et al. [12] applied a support vector machine (SVM) to predict the flutter derivative of a cable-stayed bridge and to estimate the flutter critical wind speed. June et al. [13] built an artificial neural network model to identify the flutter derivatives of various main beam cross-section forms, and the prediction results were excellent. Liao et al. [14] accurately predicted the flutter critical wind speed of different streamlined box girders by comparing four machine learning methods. Hu et al. [15,16] accurately predicted the wind pressure and pressure coefficient on a cylindrical surface by comparing multiple machine learning methods. Guo et al. [17] utilized convolutional neural networks (CNN) to predict velocity fields for several geometries, with an accuracy rate of about 98%. Miyanawala et al. [18] proposed the use of a Euler distance field to express the aerodynamic shape to predict the lift and drag coefficients and innovatively generated the section shape in the form of pictures, which greatly reduced section shape expression time. Jin et al. [19] used convolutional neural networks (CNNs) to establish a model of the relationship between the pressure field and the velocity field of the cylindrical structure and predicted the wake velocity field based on the surface pressure distribution on the bluff body. Lee and You [20] introduced the flow field continuity condition into flow field prediction and analyzed the ability of multi-scale convolutional neural networks and adversarial neural networks to maintain mass conservation and momentum conservation in turbulence prediction.

In summary, scholars have begun to explore the successful application of deep learning to related fields of wind engineering, which also provides the basis for this study. When designing the neural network architecture, this paper focuses on the input design and output design of the network. In terms of the input, since simply using the shape as the input of the neural network as a parameter cannot express any complex cross-sections, it is not universal. Therefore, this paper draws on Miyanawala's idea of using the Euler distance to describe the aerodynamic shape and proposes to describe the shape as a combination of the wall distance field and the space coordinate field, which can distinguish the flow field characteristics at different positions in space and efficiently express the influencing factors related to aerostatic performance. In terms of the output, the logic of directly predicting the three-component force coefficient from the aerodynamic shape is too jumpy to be conducive to the training and testing of the neural network. Therefore, this paper proposes to divide the aerostatic performance prediction into two steps. First, the pressure field is predicted, and then, the aerostatic three-component force coefficient is obtained by traditional integral calculation. A fully convolutional network, with some improvements, is selected as the neural network. The optimization of the fully convolutional neural network suitable for classification problems is extended to the pixel regression problem, retaining the advantages of the efficient spatial feature extraction of the fully convolutional network, the high-dimensional mapping relationship from shape to pressure field, and the aerostatic performance prediction of bluff body cross-sectional shapes. In addition, considering that CFD numerical simulation can provide a large amount of flow field information as the input and output of a deep learning model, this paper uses CFD simulation to obtain the required data samples for model training and testing. Ribeiro et al. [21] built a DeepCFD model to predict U_x, U_y, and p by inputting the signed distance function (SDF) and flow region channel, and the prediction effect was excellent. This has some similarities with the research content of this paper. However, this paper pays more attention to the aerostatic performance of the bluff body section. Therefore, a weighted value is added to the wall distance field in the design input to ensure the prediction accuracy of the wall pressure. The space coordinate field is proposed to highlight the characteristics of the flow field. When designing the model, a weighted value is also used for the L2 loss function, and the loss value closer to the aerodynamic shape adopts a weighted value closer to 1.

Based on the above overall concept, the second and third chapters of this paper introduce the ideas and implementation details of input and output design. The fourth chapter introduces the structure of the fully convolutional neural network. The fifth chapter shows how the adopted CFD setting details and example design are used for the training and testing of the neural network model; the sixth chapter introduces the model hyperparameter optimization results; the seventh chapter analyzes the accuracy of the results and the comparison with other models.

2. Input Design

The input of the neural network model needs to be able to express the aerodynamic shape information of the bluff body section, and the information should have a consistent expression structure. For the expression of the aerodynamic shape of the main beam, the traditional method of using coordinate points will associate the structure of the input data with the type of shape, causing a data heterogeneity problem. With this problem in mind, this paper proposes an aerodynamic shape description method in the form of images, which can be effectively combined with CNN, achieve consistent representation of input data, and improve information transfer efficiency. Based on this method, the aerodynamic shape is described by two types of fields, namely, the wall distance field and the space coordinate field.

2.1. Wall Distance Field

In order to avoid the problem of data heterogeneity, this paper draws on Chen's [22] idea to express the shape in the form of images, that is, assigning different values to the internal and external spatial points of the shape. In Chen's original design, the inside of the shape is expressed by 1, and the outside of the shape is expressed by 0. Although this design can express any shape, the efficiency of conveying shape information is not high, and it cannot provide the distance information of spatial points and walls. Therefore, this paper proposes the wall distance field as a description of the aerodynamic shape, which can provide the relative positional relationship between the spatial point and the wall for the limited receptive field of the CNN and directly introduce the shape pair into the input of the neural network, influencing the factors of aerostatic performance. In the wall distance field I_{Dist}, which is one of the model inputs (I), the assignment of the i spatial point is shown in Equation (1):

$$I_{Dist,i} = \beta \times e^{-1 \times \frac{min(R_{i,\Gamma})}{B}} \quad (1)$$

In the equation, $R_{i,\Gamma}$ represents the set of distances between the i spatial point and the set of aerodynamic shape boundary points Γ, and min $(R_{i,\Gamma})$ refers to the final minimum value at all distances. For the consideration of dimensionless shapes, B is introduced to represent the characteristic length of the shape. Because the space near the wall has a greater impact on the pressure field, we use a negative exponential form to increase the weight near the boundary layer while limiting the input to 1 to reduce the hidden danger of gradient explosion. β expresses the coefficient inside and outside the shape. When the i spatial point is located on the wall or inside the shape, it is set to 0, and otherwise, it is set to 1. The wall distance field I_{Dist} of an example ellipse with an aspect ratio of 4 is shown in Figure 1.

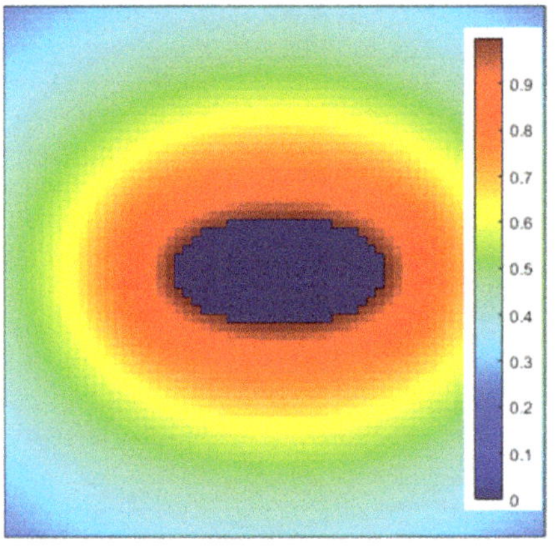

Figure 1. Wall distance field of an ellipse with AR = 4.

2.2. Space Coordinate Field

Simply using the wall distance field will result in the loss of some shape information. To illustrate this problem, the shape expressed in Figure 2 can be taken as an example. The red frame line is the receptive field of the convolutional neural network in a hidden layer. When it moves to the two positions shown in the figure, it will receive the same wall distance field information. However, these two positions are upstream and downstream of the flow field and have different effects on the pressure field and aerostatic performance. In order to solve this problem, we also need to highlight the flow field characteristics of the aerodynamic profile of the bluff body section.

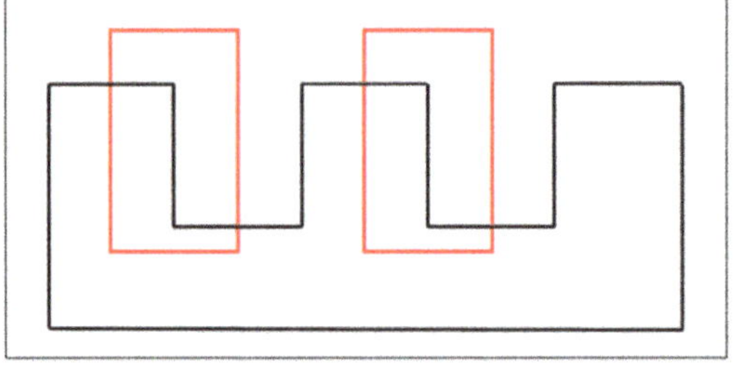

Figure 2. Relationship between receptive field and shape.

To this end, we propose to use the coordinate field describing the upstream–downstream relationship and the coordinate field describing the crosswind position as complementary information to the wall distance field as the other two inputs to the neural network.

For the coordinate field I_X, describing the upstream-downstream relationship, the assignment of the i spatial point is shown in Equation (2):

$$I_{X,i} = \beta \times \frac{1}{1 + e^{\frac{-X_i}{B}}} \tag{2}$$

where X_i represents the downwind x coordinate of the i spatial point. The origin of the x-coordinate is at the center of the section. The settings of the exponential part can achieve both the size difference between the upstream and downstream data and the normalization of the input.

Similarly, for the coordinate field I_Y, describing the position of the crosswind direction, the assignment of the i spatial point is shown in Equation (3):

$$I_{Y,i} = \beta \times e^{\frac{-|Y_i|}{B}} \tag{3}$$

where Y_i represents the y coordinate of the crosswind direction of the i spatial point. The origin of the y-coordinate is located at the midpoint of the windward end: for circles and hexagons, it is the only windward end point, and for rectangles, it is the midpoint of the windward surface. The settings of the exponential part can achieve both the difference between the distance of the crosswind and the distance as well as the normalization of the input.

Figures 3 and 4 show the values of I_X and I_Y in the spatial coordinate field for an example ellipse with an aspect ratio of 4.

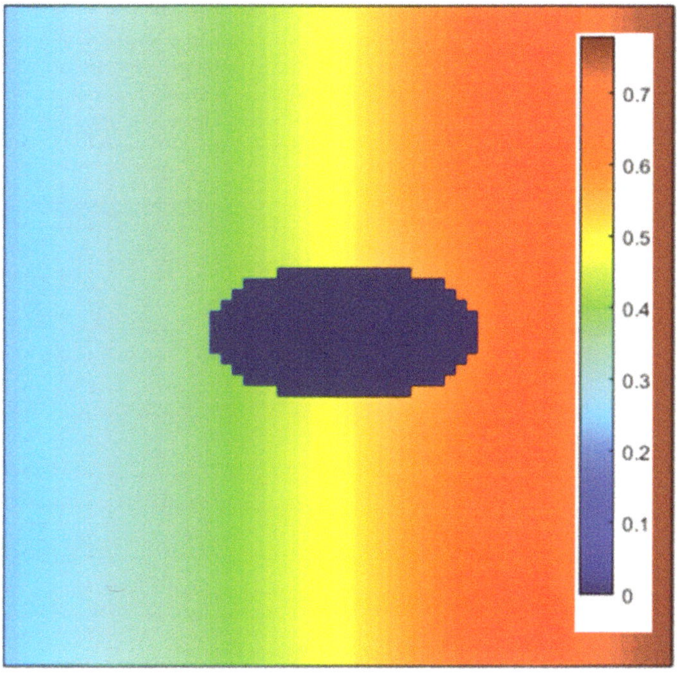

Figure 3. I_X field of an ellipse with AR = 4.

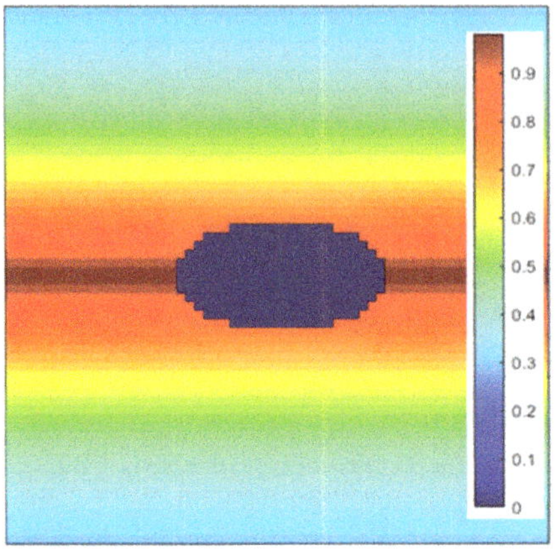

Figure 4. I_Y field of an ellipse with AR = 4.

3. Output Design

In terms of output design, the simplest and most direct approach is to use the aerostatic three-component force coefficient as the output of the prediction model. However, the mapping logic is very complicated, and it directly skips multiple data processing links such as local sampling of the pressure field and surface pressure integration. Although deep learning networks have the ability to implement the mappings, this is bound to increase the burden on the network, requiring more hidden layers and more complex structures. On the other hand, the pressure field can not only provide information on the aerostatic three-component force coefficient but also on the surrounding flow field structure, providing support for other scientific research scenarios such as flow control and heat source analysis. Therefore, this paper adopts a two-step strategy to obtain the aerostatic performance of bluff body sections. First, the neural network is used to output the pressure field $O_{pressure}$ of the steady flow field, and then, the surface pressure is integrated to obtain the aerostatic three-component force coefficient. Figure 5 shows the output of the neural network for an example rectangular section with an aspect ratio of 4.

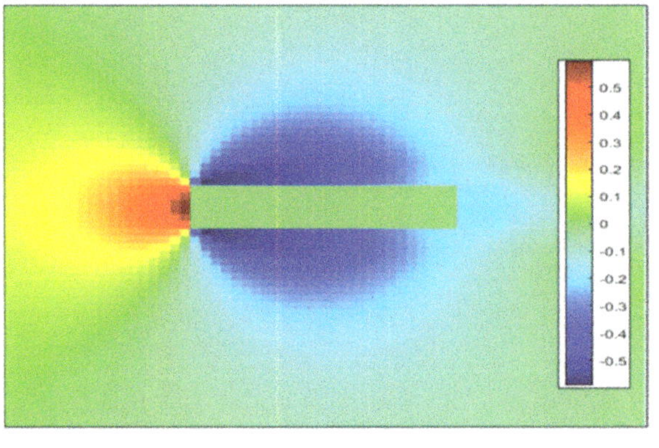

Figure 5. Pressure field $O_{pressure}$ of a rectangular section with AR = 4.

4. Neural Network Framework Design

Convolutional neural networks have been proposed since 2012 and have achieved many application results in the fields of image classification and image detection [23,24].

The multi-layer convolutional structure of CNNs can automatically learn features. Shallower convolutional layers can learn some local region features, while deeper convolutional layers have larger receptive fields [25–27]. These properties allow CNNs to learn more abstract features, and, therefore, it is widely used in image processing problems [28]. The fully convolutional network developed on the basis of CNN is very suitable for solving the problem of mapping from image to image [29]. FCN replaces the last fully connected layer of CNN with a convolutional layer and uses deconvolution to the last convolutional layer. The output result is unsampled, and the size of the output result is gradually restored to the same size as the input image so that a prediction value can be generated for each pixel point. The spatial information of the original input image is also preserved.

FCN is good at pixel-level classification problems, and this paper predicts the pressure field or continuous pressure value at each pixel position, which is a regression problem. Although pixel-level regression problems have higher requirements from neural network models than classification problems, they generally require training using generative neural networks. Considering that the pressure field involved in this paper has a gentle gradient and continuous smoothness, the fully convolutional network can be extended from solving pixel-level classification problems to solving pixel-level regression problems. The UNet network proposed by Ronneberger [30] in 2015 can complete end-to-end training with very simple architecture, and the network inference speed is very fast, so a modified UNet network (Figure 6) is adopted to solve the pixel-level regression problem in this paper. Figure 6 shows the image width, height, and number of channels after convolution, pooling, and feature fusion. D represents the number of channels, S represents the number of convolutions, and N represents the number of poolings. These parameters will be studied in Chapter 6. Meanwhile, the deep learning network structure will be determined. In order to ensure that the pressure data in the aerodynamic shape is 0, a mask layer consisting of 0 and 1, with a bluff body section shape, is passed when the prediction result is output. Next, the innovative design of the UNet structure and loss value is explained.

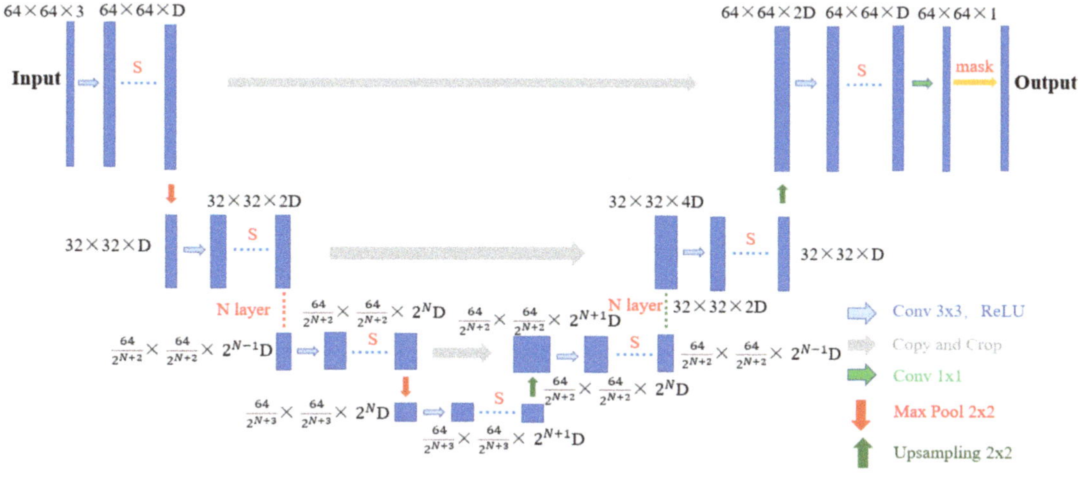

Figure 6. Modified UNet network structure.

First, the deep learning model needs to match the input and output data sizes. The input involved in this paper is the wall distance field (I_{Dist}) and the spatial coordinate field (I_X, I_Y), and the number of channels is 3. Considering the information content and training time contained in the image, the image pixels are set to 64 × 64. The output is a pressure field with the same size as the input image, which is a single channel picture. Since the regression problem of continuous variables needs to be predicted, it is necessary to modify the output layer of the UNet and remove the Softmax layer.

After modifying the input and output layers, the hidden layer details of the network structure are also required. The network architecture used in this paper is based on the transformation of the UNet and also consists of three parts: downsampling, upsampling, and skipping structure. The downsampling part includes the two basic components of the convolution layer and pooling layer, and the feature extraction of the input is realized by repeating the convolution-pooling process many times. The convolutional layer uses a 3 × 3 small convolution kernel, and the padding size is 1. Using max pooling makes the image twice as small after each pooling. After multiple downsamplings, the shape information of the bluff body section contained in the wall distance field and the spatial coordinate field is highly abstracted into multiple information channels, and the upsampling part is responsible for restoring the pressure field image based on this information. In order to retain more detailed information in the downsampling process, the new feature map in the upsampling process of each level and the feature map at the corresponding downsampling position are fused on the channel, which is the skip structure. In order to extract more detailed features before the final output, a 1 × 1 convolution is used to open up all information channels to achieve the regression prediction of each pixel. The convolution-pooling process is performed N times in total, and there are S convolutions each time. Both N and S will be determined in the model optimization in Chapter 6.

In the model training process, due to the different measurement standards of the classification model and the regression model, it is necessary to change the loss function used to deal with the classification problem in the original UNet network to an L2 loss function to deal with the regression error. At the same time, considering that the prediction results in this paper pay more attention to the accuracy of the flow field data around the aerodynamic shape, the L2 loss function is multiplied by the weight value. The loss value closer to the aerodynamic shape is used, with a weight value closer to 1, and for the loss value farther away from the aerodynamic shape, the weighted value is closer to 0. The weighted value of the loss function $loss_{weight}$ is generated according to Equation (4):

$$loss_{weight} = I_{Dist,i} \quad (4)$$

5. Test Design

5.1. Computational Case Design

This chapter mainly shows how to obtain the data for training and verifying the model. A certain number of aerodynamic shapes are designed as an example. The shape information and the corresponding pressure field calculation results are used as the input and output of the neural network. According to different basic shapes, the calculation examples are divided into four types: regular hexagon, circle, rectangle, and regular rhombus. Each basic shape is subdivided into different shapes according to the aspect ratio and corner position, for a total of 105 shapes (Table 1). In order to facilitate the use of a dimensionless form to determine the shape, the horizontal length of all shapes is fixed as 1 and the aerodynamic shape changes with the vertical height.

Table 1. List of the shape conditions.

Basic shape	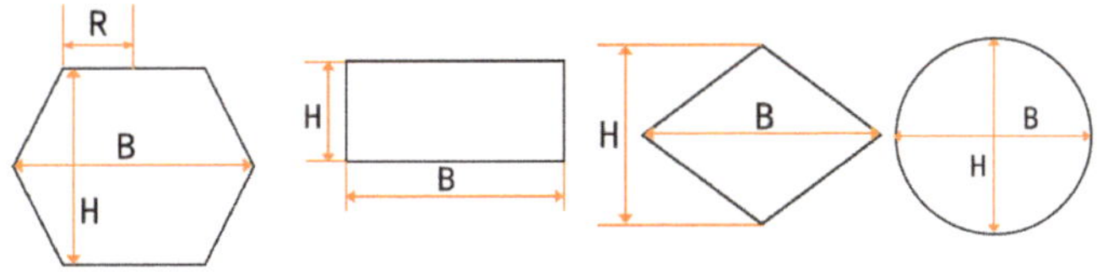			
Distance R from upper corner to X = 0	[0.1, 0.2, 0.3, 0.4]	—	—	—
B		1		
AR = B/H		[0.5, 0.56, 0.625, 0.71, 0.83, 1, 2, 3, 4, 5, 6, 7, 8, 9, 10]		
Quantity	60	15	15	15

5.2. Data Preparation

In this paper, CFD numerical simulation is used to obtain the steady-state pressure flow field of the bluff body section. The computational domain was rectangular with a blockage ratio of 0.9%. It is shown in Figure 7, where B is the length of the bluff body section. In order to keep the mesh from changing too fast, three computational regions were defined [31,32]. The rigid zone kept the mesh shape constant, outside of which the deforming zone underwent smoothing and remeshing. The outermost fixed zone was to satisfy the blockage ratio requirement. The details of the mesh are illustrated in Figure 8.

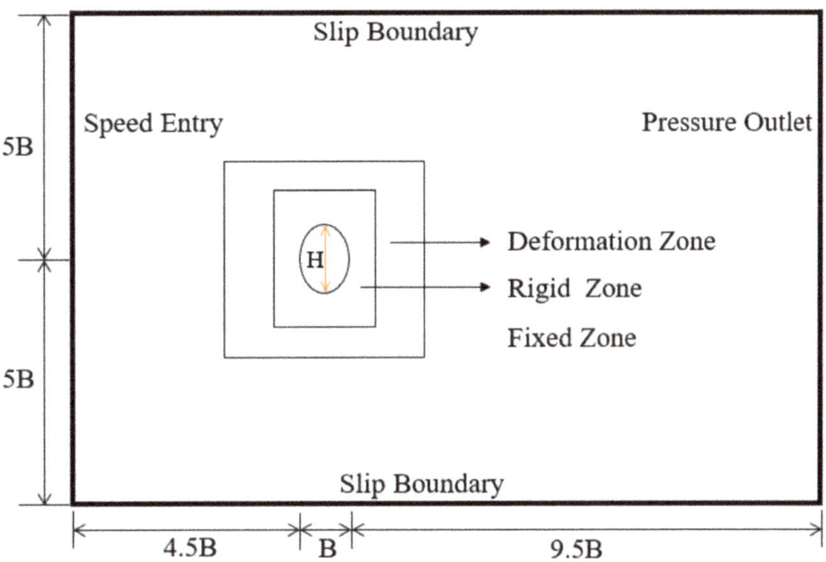

Figure 7. Sketch of the two-dimensional (2D) computational domain.

Figure 8. Detailed demonstration of the computational grid within the rigid zone.

In order to reduce the calculation time while ensuring the calculation accuracy, the turbulence model uses the Reynolds-averaged Navier–Stokes equations (RANS). A blending k–ω SST turbulence model was used to reduce sensitivity to the inlet boundary condition. The governing momentum equation of incompressible flow is given as

$$\frac{\partial u_i}{\partial t} + u_j \frac{\partial u_i}{\partial x_j} = \frac{-\partial p}{\rho \partial x_i} + \nu \frac{\partial^2 u_i}{\partial x_j \partial x_j} - \frac{\partial \overline{u_i' u_j'}}{\partial x_j} \tag{5}$$

where $i, j = 1, 2$ is the two-dimensional analysis, p denotes the pressure, and $\overline{u_i' u_j'}$ is the Reynolds stress component, in which the superscript refers to the fluctuating parts of the velocity p. With the k–ω SST turbulence model, two additional equations are introduced to obtain the Reynolds stress component, where k is the turbulent kinetic energy and ω is its rate of dissipation (Equations (6) and (7)). The details of the involved parameters are explained by Menter [33]:

$$\frac{\partial(\rho k)}{\partial t} + \frac{\partial(\rho u_j k)}{\partial x_j} = P - \beta^* \rho \omega k + \frac{\partial}{\partial x_j}\left[(\mu + \sigma_k \mu_t)\frac{\partial k}{\partial x_j}\right] \tag{6}$$

$$\begin{aligned} &\tfrac{\partial(\rho \omega)}{\partial t} + \tfrac{\partial(\rho u_j \omega)}{\partial x_j} \\ &\quad = \tfrac{\gamma}{\nu_t} P - \beta \rho \omega^2 + \tfrac{\partial}{\partial x_j}\left[(\mu + \sigma_\omega \mu_t)\tfrac{\partial \omega}{\partial x_j}\right] \\ &\qquad + 2(1 - F_1)\tfrac{\rho \sigma_{\omega 2}}{\omega}\tfrac{\partial k}{\partial x_j}\tfrac{\partial \omega}{\partial x_j} \end{aligned} \tag{7}$$

The simulations were conducted using Ansys Fluent. The semi-implicit method pressure-linked equations consistent (SIMPLEC) algorithm was used to solve the velocity–pressure coupling problem. To obtain the appropriate numerical accuracy and stability, a second-order scheme was used for the pressure equation, and the second-order upwind scheme was used for the momentum equation, turbulent kinetic energy equation, and specific dissipation rate equation. The inlet boundary was set as a velocity inlet, the incoming turbulence intensity was set as 4%, and the viscosity ratio was set as two. The outlet boundary was set as a pressure outlet. Slip conditions were applied to the top and bottom boundaries for all variables. No-slip conditions were applied to the surfaces of the bluff body section, which means that the flow state at the object's surface was equal to the motion of the bluff body section. Considering that the section of the bluff body with a small aspect ratio in this paper is difficult to achieve the pressure field under the steady flow field when the Reynolds number is large, the Re was maintained as 4000 (using the reference length B) in all cases. The calculation results are shown in Figure 9.

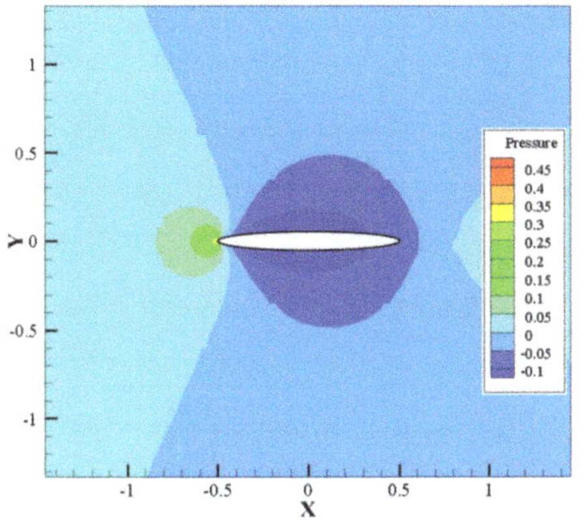

Figure 9. CFD calculation of pressure cloud map.

The pressure field data is acquired once the flow field has stabilized after several iterations. Each cycle saves 10 instantaneous pressure field data, reads 5 cycles, and averages 50 instantaneous pressure field data to obtain the pressure field of the steady-state flow field. The pressure field after data processing is resampled to obtain a 64 × 64 single-channel image. The X and Y coordinates of the sampling range are [−1.26, 1.26], and the sampling size interval is 0.04. It is worth mentioning that this article mainly focuses on the construction method, so there is no need to strictly verify the accuracy of the CFD calculation results. The model training and testing are based on different data: 80% are randomly selected as the training data set, 10% are used as the validation data set, and the remaining 10% are used as the test data set. The dropout rate for the trained model is 0.2 to prevent overfitting.

6. Model Performance Evaluation and Optimization

The hyperparameters of a deep learning model directly affect the performance of the model. Taking model depth as an example, an insufficient number of hidden layers will make the model less complex and unable to capture the nonlinear mapping between input and output. However, too many hidden layers will cause unstable gradient calculations. In order to quantitatively evaluate the performance of the model, we optimize the number of channels (D), the number of convolutions (S), and the number of pooling times (N) in Figure 6 based on the mean absolute error MAE (Equation (8)) of the extracted drag coefficient based on the predicted pressure field.

$$MAE = \frac{1}{m}\sum_{i=1}^{m}|t_i - y_i| \tag{8}$$

Pytorch performs very well in scientific research, which is mainly reflected in the very Pythonic style of Pytorch. Pytorch reduces the difficulty of getting started and can be built layer by layer when building a deep learning model, which is convenient for real-time modification. Therefore, this paper uses Pytorch to build the deep learning model.

6.1. Optimization of the Number of Hidden Layers

The number of hidden layers has a greater impact on the model performance than the size of the hidden layers. Therefore, this paper determines the suitable model depth for the prediction of the aerodynamic performance of the bluff body section [34,35]. The convolutional layer, pooling layer, and ReLU layer are considered to be downsampling components, and the number of these components (that is, the number of pooling times (N) in Figure 6) is optimized. Considering that the size of the data image is 64 × 64, after 6 iterations of max pooling, the input information is converted into multiple channels of 1 × 1, and, therefore, the maximum number of downsampling components is set to 6. When naming the model, UNet_2-3 represents two downsampling components, and each downsampling component layer contains three convolution operations. Using the same training and validation sets, the comparison of different models is shown in Table 2.

Table 2. Model performance comparison—number of downsampling components.

Model Settings	UNet_2-2	UNet_3-2	UNet_4-2	UNet_5-2	UNet_6-2
MAE	0.2313	0.2038	0.1942	0.2190	0.1593
Parameter quantity	1.016×10^6	4.262×10^6	1.724×10^7	6.915×10^7	2.768×10^8

From the results in Table 2, it can be seen that as the number of model layers increases, MAE first decreases, then increases, and finally decreases to its lowest value. The model obtains the smallest drag coefficient MAE when using six downsampling components. However, the data in Table 2 are based on two convolution operations. Therefore, it is necessary to continue to observe the effect of the number of convolution operations (Table 3).

Table 3. Model performance comparison—number of convolutions.

Model Settings	UNet_2-3	UNet_3-3	UNet_4-3	UNet_5-3	UNet_6-3
MAE	0.1801	0.1175	0.1696	0.1456	0.1234
Parameter quantity	1.533×10^6	6.401×10^6	2.587×10^7	1.037×10^8	4.152×10^8

Combining the observations in Tables 2 and 3, it can be seen that when three convolutions are used, the MAE is significantly lower than when two convolutions are used. At the same time, the UNet_3-3 model has the lowest MAE and fewer parameters than the other models, and its computational cost is lower. Considering that increasing the number of convolution operations can significantly improve the model performance, in order to determine the optimal number of convolution operations for the UNet_3 model, UNet_3-4 is added for comparison. The calculation shows that the prediction result of UNet_3-4 is $MAE = 0.1675$ and the parameter quantity is 8.540×10^6. It can be seen that when the number of downsampling components is 3, the MAE does not decrease but increases after the number of convolution operations is increased to 4. In summary, considering the performance and the amount of computation, this paper chooses 3 as the number of downsampling components and uses the UNet network model with 3 convolutions.

6.2. Hidden Layer Size Optimization

After determining the depth of the hidden layer, it is necessary to further optimize its size, that is, the number of channels (D) in Figure 6. The number of hidden layer channels of the first downsampling component in the selected UNet network for comparison is set to 16, 32, and 128. The number of hidden layer channels in the subsequent downsampling component is twice the model width of the previous layer. The same training set and validation set are used, and the comparison of different models is shown in Table 4. When naming the model, UNet_3-3-16 represents that the number of hidden layer channels of the first downsampling component is 16.

Table 4. Model performance comparison—number of convolutions.

Model Settings	UNet_3-3-16	UNet_3-3-32	UNet_3-3-64	UNet_3-3-128
MAE	0.1531	0.0942	0.1175	0.1700
Parameter quantity	4.010×10^5	1.601×10^6	6.401×10^6	2.559×10^7

It can be seen from the results in Table 5 that as the number of hidden layer channels increases, the MAE first decreases and then increases. When the initial hidden layer channel number D is 32, the MAE is the smallest and the number of model parameters is moderate. Therefore, UNet_3-3-32 is selected for research in subsequent chapters.

Table 5. Average relative error of predicted drag coefficient after changing input.

Shape	R = 0.1 Hexagon	R = 0.2 Hexagon	R = 0.3 Hexagon	R = 0.4 Hexagon	Circle	Rectangle	Diamond
Average relative error (%)	7.690	22.20	45.33	26.72	17.26	21.37	18.13

7. Pressure Field Prediction Effect

The pressure field prediction accuracy directly affects the subsequent extraction of the drag coefficient, so it is necessary to check the prediction results of the flow field. After the model training is completed, the wall distance field and spatial coordinate field of the test data set are input into the model to obtain the predicted value of the pressure field. It is then converted into a visualized image and compared with the CFD calculation results, as shown in Figure 10. Due to space limitations, representative shapes of pressure fields are selected for presentation.

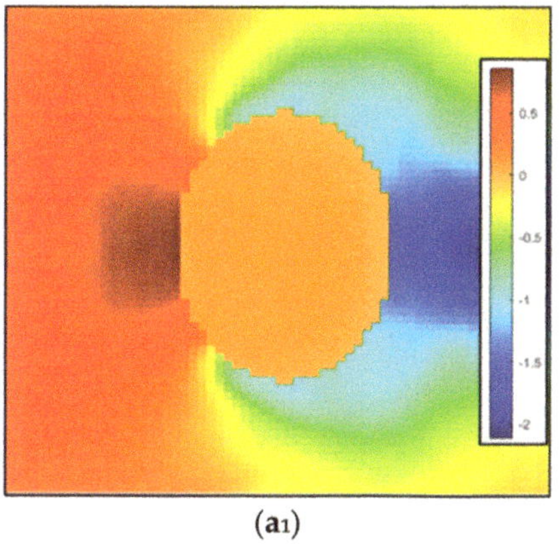

(a1)

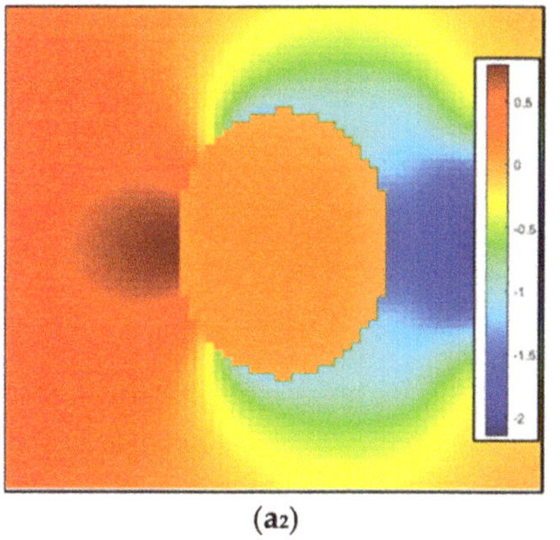

(a2)

Figure 10. *Cont.*

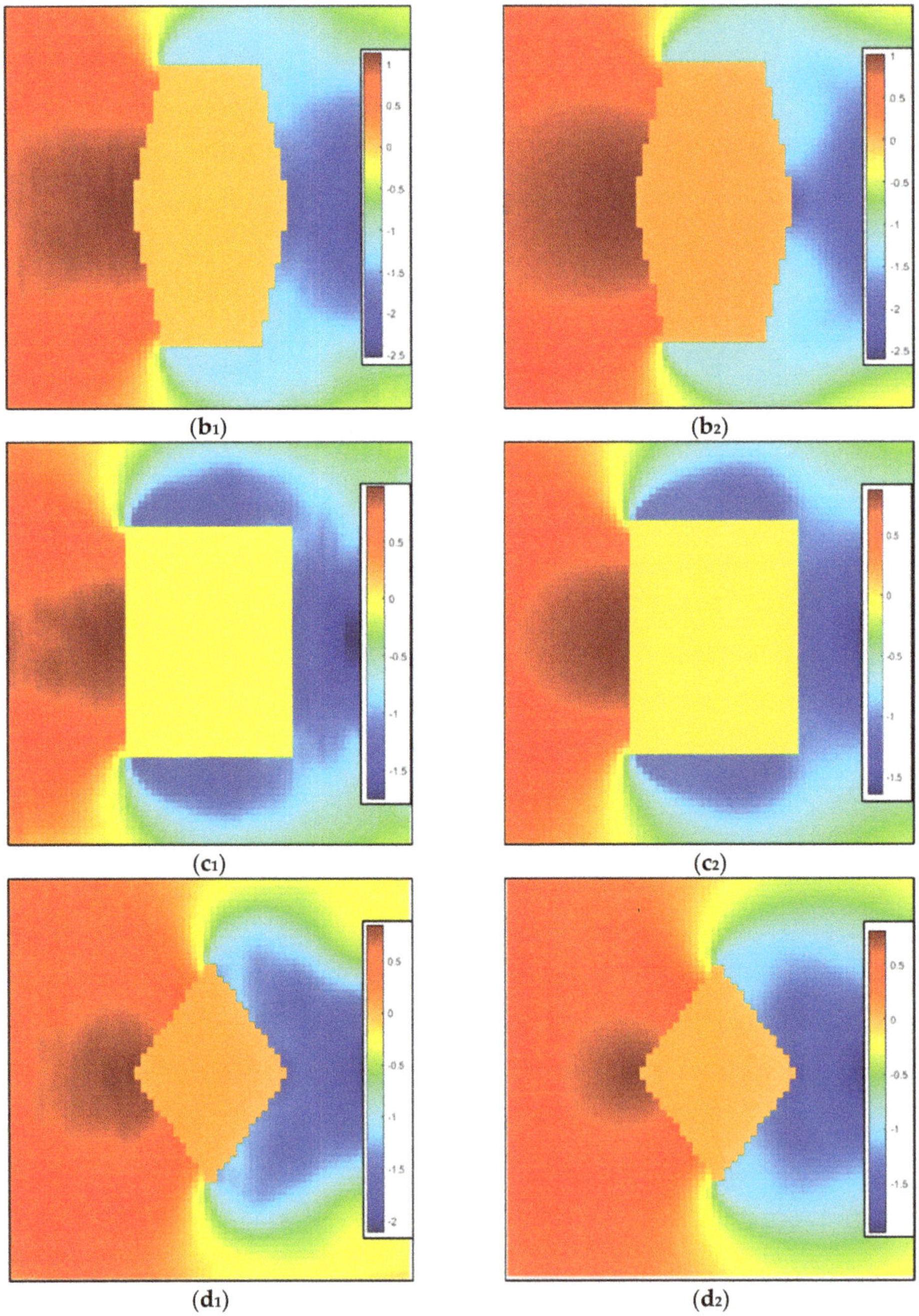

Figure 10. Comparison of pressure field prediction effects of four typical shapes. The first column is the predicted value. (**a$_1$**) The ellipse with AR = 0.71. (**b$_1$**) The hexagon with AR = 0.56 R = 0.3. (**c$_1$**) The rectangle with AR = 0.71. (**d$_1$**) The diamond with AR = 0.71. The second column is the true value. (**a$_2$**) The ellipse with AR = 0.71. (**b$_2$**) The hexagon with AR = 0.56 R = 0.3. (**c$_2$**) The rectangle with AR = 0.71. (**d$_2$**) The diamond with AR = 0.71.

The analysis and prediction results show that the larger part of the prediction error for the ellipse, hexagon, and diamond generally exists in the negative pressure area that deviates far from the shape, and the predicted pressure is larger than the pressure field calculated by CFD. The larger part of the error of the rectangular section exists in the positive pressure area that deviates far from the shape, and the predicted pressure is smaller than that calculated by CFD. The authors believe that this phenomenon is mainly caused by two factors. First, when designing the loss function, the aerostatic performance is considered to be more closely related to the pressure field near the wall. Therefore, in order to predict the wall pressure value more accurately, the loss function gives more weight to the prediction accuracy close to the wall and has more relaxed requirements for the prediction value of the grid points farther away from the wall of the bluff body section. Secondly, the improved UNet model is still too simple, and it is necessary to use a more complex generative neural network in subsequent research. However, the improved Unet network can still achieve rapid prediction of the wall pressure and third-component force coefficient to a certain extent, as will be discussed in subsequent chapters.

8. Wall Pressure Prediction Results

Wall pressure is of great significance in the evaluation of aerostatic performance and can also be used to extract third-component force coefficients. Figure 11 shows the comparison of the wall pressure prediction results and the CFD calculation results. Due to space limitations, only four typical shapes and two for each shape are shown. Since the shapes used in this paper are all symmetrical up and down, only the wall pressure in the upper part is shown.

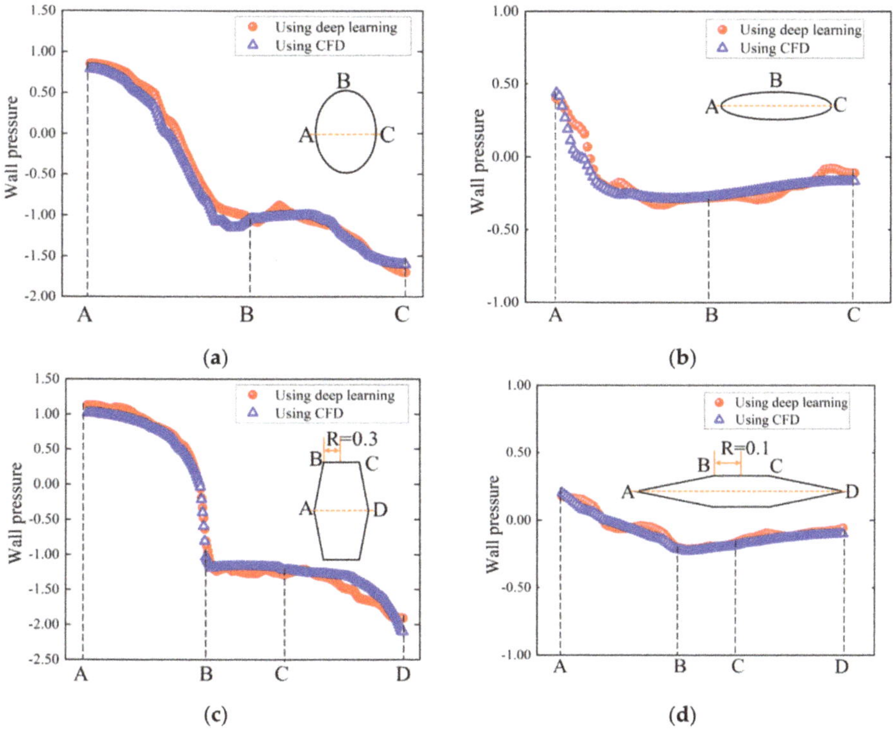

Figure 11. *Cont.*

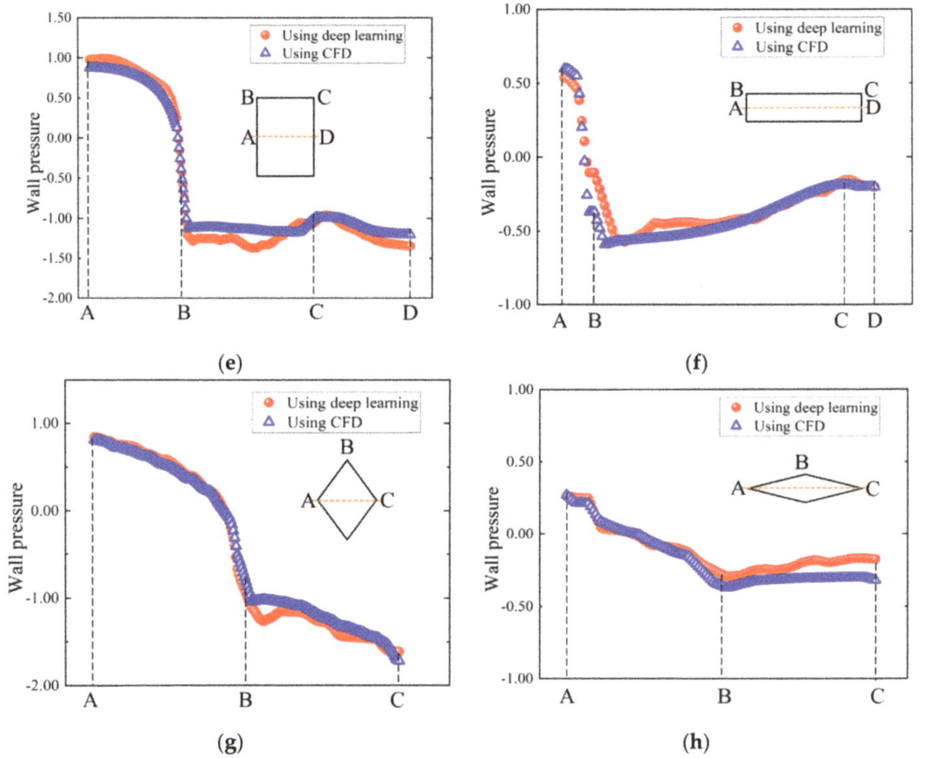

Figure 11. Comparison of the prediction effects of the wall pressure field of the four shapes. (**a**) The ellipse with AR = 0.71. (**b**) The ellipse with AR = 4. (**c**) The hexagon with AR = 0.56 R = 0.3. (**d**) The hexagon with AR = 7 R = 0.1. (**e**) The rectangle with AR = 0.71. (**f**) The rectangle with AR = 4. (**g**) The diamond with AR = 0.71. (**h**) The diamond with AR = 4.

In general, the analysis of the wall pressure prediction results shows that the basic trend of the wall pressure distribution can be simulated based on the model proposed in this paper, but there are certain errors. The prediction results of the positive pressure area are generally better than those of the negative pressure area. Among them, the area with the largest error is mainly concentrated in the negative pressure area generated by flow separation and the negative pressure area near the wake. The absolute error value does not exceed 0.2 at the most, and the range is small. Meanwhile, the prediction errors of the rectangle and rhombus are generally larger than those of the ellipse and hexagon. The author believes that this is mainly caused by the uncomplicated UNet model. The original UNet model is mainly used to deal with classification problems. Although this paper has modified it, the modified UNet model still has a lot of room for improvement when dealing with regression problems.

9. Drag Coefficient Prediction

The drag coefficient under the shape can be obtained through the wall pressure of each aerodynamic shape, and the aerostatic wind load in the wind resistance design can be obtained from the drag coefficient. Therefore, the prediction of the drag coefficient is very important, and the input and output of the deep learning model also directly affect the prediction accuracy of the drag coefficient. In this paper, the three distance fields (I_{Dist}, I_X and I_Y) are used as the input of the model, and the output is the pressure field. By comparing the accuracy of the drag coefficient, it is shown that the innovative input and output proposed in this paper can effectively improve the prediction accuracy.

9.1. Compared to Using Shape as Input

Donglin Chen sets the internal aerodynamic shape to 1 and the external to 0 to express the aerodynamic shape, then uses it as the model input for training. Since the sampling area in this paper is large to ensure the normal training of the model, the internal aerodynamic shape is set to 0 and the external is set to 1 to express the aerodynamic shape as the model input. In this paper, three distance fields (I_{Dist}, I_X, and I_Y) are innovatively proposed to express the aerodynamic shape and flow field information. This paper uses the UNet model determined in Chapter 6 and exactly the same parameters for training and compares the accuracy of the drag coefficient to illustrate that the three distance fields (I_{Dist}, I_X, and I_Y) proposed in this paper can effectively improve the prediction accuracy of the model. Taking an ellipse with an aspect ratio of 4 as an example, the 0–1 distance field is shown in Figure 12. The average relative error of the drag coefficient of each basic shape predicted by the model is shown in Table 5.

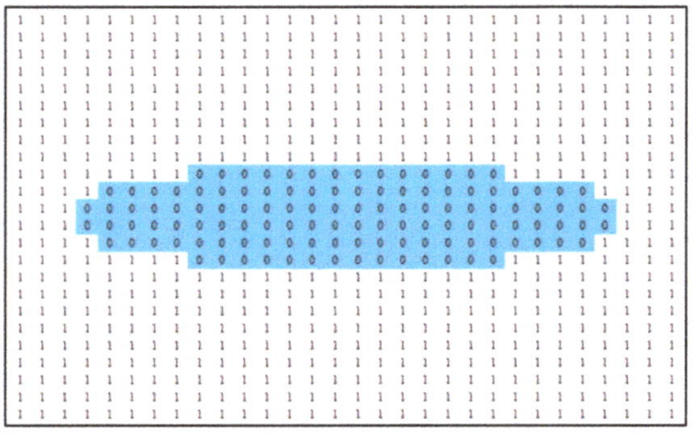

Figure 12. Partial 0–1 distance field.

It can be seen from Table 6 that the average relative error of the R = 0.3 hexagon reaches 45.33%, and the average relative errors of the R = 0.4 hexagon and rectangle also exceed 20%, making the total average relative error reach 22.67%. In contrast, the average relative error of the three distance fields (I_{Dist}, I_X, and I_Y) acting as the model input is 9.42%, which is 13.25% higher than the average relative error of the drag coefficient when the 0–1 distance field is used as the model input. It can be seen that the wall distance field and spatial coordinate field can make the model learn and predict more efficiently.

Table 6. Average relative error of predicted drag coefficient after changing output.

Shape	R = 0.1 Hexagon	R = 0.2 Hexagon	R = 0.3 Hexagon	R = 0.4 Hexagon	Circle	Rectangle	Diamond
Average relative error (%)	3.450	7.740	12.14	5.010	9.110	13.08	15.42

The average relative error when using the 0–1 distance field as the model input is much higher than the average relative error of the wall distance field and the spatial coordinate field proposed in this paper. The main reason for this is that the 0–1 distance field is too simple to express the aerodynamic shape and cannot provide effective distance information and flow field information, while the drag coefficient is closely related to the distance information and flow field information. It is, therefore, difficult for deep learning models to establish a relationship between the aerodynamic shapes and the drag coefficient.

9.2. Compared to Using the Drag Coefficient as Output

There are two ways to predict the drag coefficient of the bluff body section through the deep learning method. One is to directly predict the drag coefficient as the output; the other is to predict the pressure flow field and then pass it through the data to get the drag coefficient.

The direct method still uses the three distance fields of I_{Dist}, I_X, and I_Y as the input of the model and the processed drag coefficient as the output. The CNN model consists of the convolutional layer, pooling layer, ReLU layer, and fully connected layer. The CNN model, with a model depth of 3 layers and an initial model width of 128, was selected by comparing the size of the MAE and the number of parameters, as shown in Figure 13.

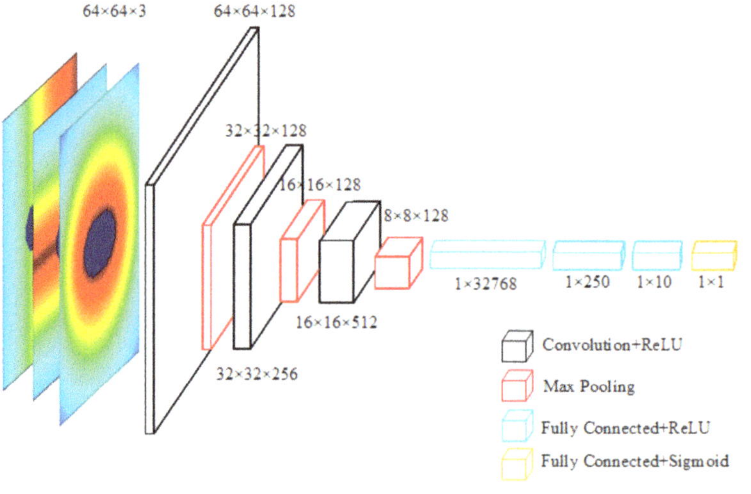

Figure 13. Determined CNN model.

By directly predicting the drag coefficient through the CNN model, the overall average relative error of the test set reached 19.64%, the average relative error of the diamond reached 38.81%, and the average relative errors of the R = 0.2 regular hexagon and the R = 0.3 regular hexagon were less than 10%.

The author believes there are two main reasons for the large error in the direct prediction of the drag coefficient through the CNN model. The first reason is that there are very obvious differences between the basic shapes, which makes it difficult to accurately associate the shape input with the drag coefficient in the deep learning model. Therefore, we can only fit relatively satisfactory results under different shapes, which leads to large errors in some shapes. The second reason is that it is too simplistic to directly use the drag coefficient data as the output and directly use the image as the output. When predicting a number such as a drag coefficient, it is difficult for a deep learning model to establish a certain connection between the input and output. The randomness is also large, which makes the model prediction less stable.

In reference to the first reason, better deep learning model architecture can be selected by changing the structure of the model. However, the CNN model is the optimal solution for the direct prediction of the drag coefficient in this paper. Therefore, the output needs to be redesigned for the second reason. The drag coefficient can be obtained indirectly by predicting the pressure field of the steady-state flow field. The input of the model is still the three distance fields of I_{Dist}, I_X, and I_Y, and the output is the stable pressure flow field. Due to the different input and output models, the UNet network is selected and the optimal model is determined, as shown in Figure 6.

Considering that the drag coefficients of different shapes are quite different, the average relative error is used to comprehensively evaluate the prediction accuracy of different shapes, as shown in Table 6. At the same time, the comparison between the predicted drag coefficient and the CFD-calculated drag coefficient for each shape is shown in Figure 14.

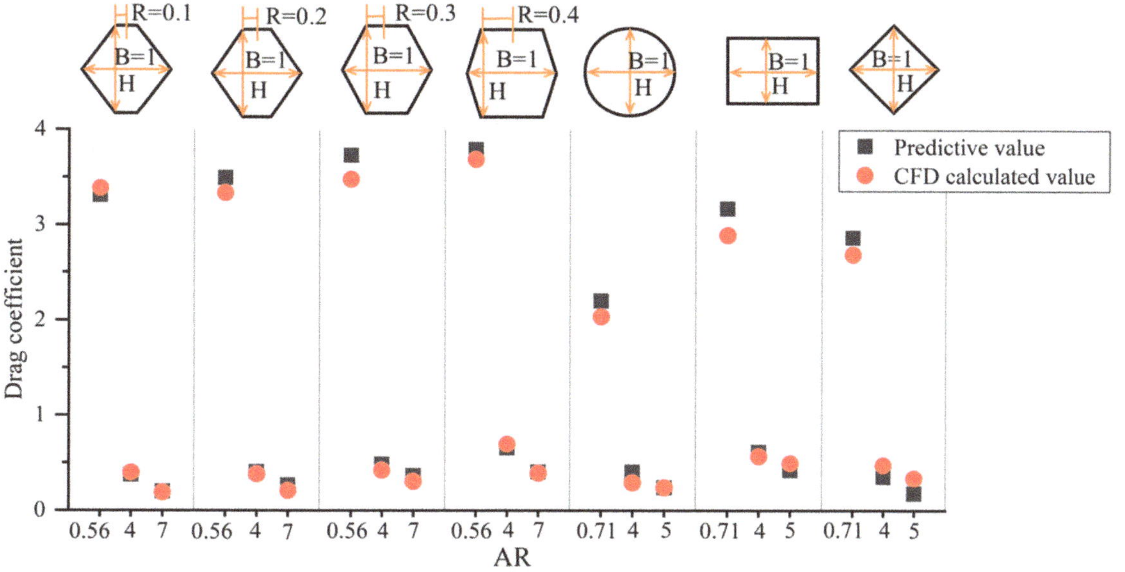

Figure 14. Drag coefficient prediction results.

From the prediction results given in Table 6, it can be seen that the average relative errors of the drag coefficients of the R = 0.1 hexagon, R = 0.2 hexagon, and R = 0.4 hexagon and circle are less than 10%. A direct relationship is maintained between the accuracy of the wall pressure prediction and the prediction accuracy of the drag coefficient. The relative errors of the drag coefficients predicted by the R = 0.3 hexagon, rectangle, and diamond exceed 10% but do not exceed 20%. The average relative error is the largest for the diamond, reaching 15.42%. The average relative error for all shapes is 9.42%. It can be seen that the UNet model adopted by the indirect method has better performance than the CNN model that directly predicts the drag coefficient, and the average relative error of the drag coefficient is increased by 10.22%. It can be seen that the indirect method of predicting the drag coefficient strengthens the connection between the input and the output. The prediction of the flow field can also provide applications for flow field control and other scenarios, and the practical performance of the model is greatly improved.

10. Conclusions

This paper proposes a novel, fully convolutional neural network model that enables rapid prediction from shape to aerostatic performance.

(1) The overall average relative error of the drag coefficient is 9.42%.
(2) The shape is described by the combination of the wall distance field and the space coordinate field. This improves the prediction accuracy by 13.25% compared to when the shape is directly used as the model input.
(3) A step-by-step strategy in which the pressure field is used as the model output is proposed. The model prediction accuracy is improved by 10.22% when using the pressure field as the model input compared to directly predicting the drag coefficient.

Author Contributions: Conceptualization, K.L.; methodology, K.L.; software, H.L.; validation, H.L.; formal analysis, H.L.; investigation, K.L.; resources, K.L.; data curation, Z.C.; writing—original draft preparation, H.L.; writing—review and editing, K.L.; visualization, H.L.; supervision, S.L.; project administration, S.L.; funding acquisition, K.L. and S.L. All authors have read and agreed to the published version of the manuscript.

Funding: This research was funded by the National Science Foundation for Young Scientists of China grant number 51808075; the National Science Foundation of China grant number 51978108, the 111 project of the Ministry of Education and the Bureau of Foreign Experts of China grant number B18062; the Natural Science Foundation of Chongqing China grant number cstc2020jcyj-msxmX0773 and cstc2020jcyj-msxmX0937; the Fundamental Research Funds for the Central Universities grant number 2020CDJ-LHZZ-018, 2021CDJQY-025 and 2020CDJ-LHZZ-016; the Chongqing full-time postdoctoral exit and stay in the Chongqing Project grant number 2020LY07.

Institutional Review Board Statement: Not applicable.

Informed Consent Statement: Not applicable.

Data Availability Statement: Not applicable.

Acknowledgments: This paper is supported by the National Science Foundation of China (51808075, 51978108), 111 Project (B18062), the Natural Science Foundation of Chongqing, China (cstc2020jcyj-msxmX0773, cstc2020jcyj-msxmX0937), the Fundamental Research Funds for the Central Universities (2020CDJ-LHZZ-018, 2021CDJQY-025 and 2020CDJ-LHZZ-016), and the Chongqing full-time post-doctoral exit and stay in the Chongqing Project (2020LY07).

Conflicts of Interest: The authors declare no conflict of interest.

References

1. Matsuda, K.; Tokushige, M.; Iwasaki, T. Reynolds number effects on the steady and unsteady aerodynamic forces acting on the bridge deck sections of long-span suspension bridge. *IHI Eng. Rev.* **2007**, *40*, 12.
2. Simiu, E.; Scanlan, R.H. *Wind Effects on Structures: Fundamentals and Applications to Design*; John Wiley: New York, NY, USA, 1996.
3. Shimada, K.; Ishihara, T. Prediction of aeroelastic vibration of rectangular cylinders by K-e model. *J. Aerosp. Eng.* **1999**, *12*, 122–135. [CrossRef]
4. Liu, S.Y.; Zhao, L.; Fang, G.S.; Hu, C.; Ge, Y.J. Investigation on aerodynamic force nonlinear evolution for a central-slotted box girder under torsional vortex-induced vibration. *J. Fluids Struct.* **2021**, *106*, 103380. [CrossRef]
5. Chen, W.-L.; Li, H.; Hu, H. An experimental study on the unsteady vortices and turbulent flow structures around twin-box-girder bridge deck models with different gap ratios. *J. Wind Eng. Ind. Aerodyn.* **2014**, *132*, 27–36. [CrossRef]
6. Chen, W.-L.; Zhang, Q.-Q.; Li, H.; Hui, L. An experimental investigation on vortex induced vibration of a flexible inclined cable under a shear flow. *J. Fluids Struct.* **2015**, *54*, 297–311. [CrossRef]
7. Laima, S.; Li, H.; Chen, W.-L.; Li, F. Investigation and control of vortex-induced vibration of twin box girders. *J. Fluids Struct.* **2013**, *39*, 205–221. [CrossRef]
8. Mannini, C.; Sbragi, G.; Schewe, G. Analysis of self-excited forces for a box-girder bridge deck through unsteady RANS simulations. *J. Fluids Struct.* **2016**, *63*, 57–76. [CrossRef]
9. Xu, F.Y.; Wu, T.; Ying, X.Y.; Kareem, A. Higher-order self-excited drag forces on bridge decks. *J. Eng. Mech.* **2016**, *142*, 06015007. [CrossRef]
10. Li, H.; Chen, W.-L.; Xu, F.; Li, F.-C.; Ou, J.-P. A numerical and experimental hybrid approach for the investigation of aerodynamic forces on stay cables suffering from rain-wind induced vibration. *J. Fluids Struct.* **2010**, *26*, 1195–1215. [CrossRef]
11. Hongye, G.; Biao, Y.; Hui, H.; Rui, X.; Chang, L.; Yu, L.; Qianhui, P. State-of-the-art review of bridge informatization and intelligent bridge in 2019. *J. Civ. Environ. Eng.* **2020**, *4*, 14–27.
12. Lute, V.; Upadhyay, A.; Singh, K. Support vector machine based aerodynamic analysis of cable stayed bridges. *Adv. Eng. Softw.* **2009**, *40*, 830–835. [CrossRef]
13. Jung, S.; Ghaboussi, J.; Kwon, S.-D. Estimation of aeroelastic parameters of bridge decks using neural networks. *J. Eng. Mech.* **2004**, *130*, 1356–1364. [CrossRef]
14. Liao, H. Machine learning strategy for predicting flutter performance of streamlined box girders. *J. Wind Eng. Ind. Aerodyn.* **2021**, *209*, 104493. [CrossRef]
15. Hu, G.; Kwok, K. Predicting wind pressures around circular cylinders using machine learning techniques. *J. Wind Eng. Ind. Aerodyn.* **2020**, *198*, 104099. [CrossRef]
16. Hu, G. Deep learning-based investigation of wind pressures on tall building under interference effects. *J. Wind Eng. Ind. Aerodyn.* **2020**, *201*, 104138. [CrossRef]

17. Guo, X.; Li, W.; Iorio, F. Convolutional Neural Networks for Steady Flow Approximation. In Proceedings of the 22nd ACM SIGKDD International Conference on Knowledge Discovery and Data Mining, San Francisco, CA, USA, 13–17 August 2016; pp. 481–490.
18. Miyanawala, T.P.; Jaiman, R.K. A Novel Deep Learning Method for the Predictions of Current Forces on Bluff Bodies. In Proceedings of the ASME 2018 37th International Conference on Ocean, Offshore and Arctic Engineering, Madrid, Spain, 17–22 June 2018.
19. Jin, X.; Cheng, P.; Chen, W.-L.; Wen-Li, C. Prediction model of velocity field around circular cylinder over various Reynolds numbers by fusion convolutional neural networks based on pressure on the cylinder. *Phys. Fluids* **2018**, *30*, 47–105. [CrossRef]
20. Lee, S.; You, D. Data-driven prediction of unsteady flow over a circlular cylinder using deep learning. *J. Fluid Mech.* **2019**, *879*, 218–254. [CrossRef]
21. Ribeiro, M.D.; Rehman, A.; Ahmed, S.; Dengel, A. DeepCFD: Efficient Steady-State Laminar Flow Approximation with Deep Convolutional Neural Networks. *arXiv* **2020**, arXiv:2004.08826.
22. Chen, D.; Gao, X.; Xu, C.; Chen, S.; Fang, J.; Wang, Z. FlowGAN: A Conditional Generative Adversarial Network for Flow Prediction in Various Conditions. In Proceedings of the 32nd International Conference on Tools with Artificial Intelligence (ICTAI 2020), Baltimore, MD, USA, 9–11 November 2020.
23. Sermanet, P.; Eigen, D.; Zhang, X.; Mathieu, M.; Fergus, R.; LeCun, Y. OverFeat: Integrated Recognition, Localization and Detection using Convolutional Networks. *arXiv* **2013**, arXiv:1312.6229.
24. He, K. Spatial Pyramid Pooling in Deep Convolutional Networks for Visual Recognition. *IEEE Trans. Pattern Anal. Mach. Intell.* **2015**, *37*, 1904–1916. [CrossRef]
25. Szegedy, C.; Liu, W.; Jia, Y.; Sermanet, P.; Reed, S.; Anguelov, D.; Erhan, D.; Vanhoucke, V.; Rabinovich, A. Going Deeper with Convolutions. In Proceedings of the IEEE Conference on Computer Vision and Pattern Recognition, Columbus, OH, USA, 23–28 June 2014.
26. Dan, C.; Alessandro, G.; Luca, G.; Jürgen, S. Deep Neural Networks Segment Neuronal Membranes in Electron Microscopy Images. *Adv. Neural Inf. Process. Syst.* **2012**, *25*, 2852–2860.
27. Zeiler, M.D.; Fergus, R. Visualizing and Understanding Convolutional Networks. In *European Conference on Computer Vision, Proceedings of the Computer Vision–ECCV 2014, Zurich, Switzerland, 6–12 September 2014*; Springer: Cham, Switzerland, 2014; pp. 818–833.
28. Girshick, R.; Donahue, J.; Darrell, T.; Malik, J. Region-Based Convolutional Networks for Accurate Object Detection and Segmentation. *IEEE Trans. Pattern Anal. Mach. Intell.* **2015**, *38*, 142–158. [CrossRef]
29. Shelhamer, E.; Long, J. Fully Convolutional Networks for Semantic Segmentation. *IEEE Trans. Pattern Anal. Mach. Intell.* **2017**, *39*, 640–651. [CrossRef] [PubMed]
30. Ronneberger, O.; Fischer, P.; Brox, T. U-Net: Convolutional networks for biomedical image segmentation. In *Medical Image Computing and Computer-Assisted Intervention 2015, Proceedings of the Medical Image Computing and Computer-Assisted Intervention–MICCAI 2015, 18th International Conference, Munich, Germany, 5–9 October 2015*; Navab, N., Hornegger, J., Wells, W.M., Frangi, A.F., Eds.; Springer International Publishing: Cham, Switzerland, 2015; pp. 234–241.
31. Miranda, S.D.; Patruno, L.; Ubertini, F.; Vairo, G. On the identification of flutter derivatives of bridge decks via RANS turbulence models: Benchmarking on rectangular prisms. *Eng. Struct.* **2014**, *76*, 359–370. [CrossRef]
32. Patruno, L. Accuracy of numerically evaluated flutter derivatives of bridge deck sections using RANS: Effects on the flutter onset velocity. *Eng. Struct.* **2015**, *89*, 49–65. [CrossRef]
33. Menter, F.R. Zonal two equation k-omega turbulence models for aero-dynamic ows. In Proceedings of the 24th Fluid Dynamics Conference, Orlando, FL, USA, 6–9 July 1993.
34. Eldan, R.; Shamir, O. The Power of Depth for Feedforward Neural Networks. In Proceedings of the Conference on Learning Theory, New York, NY, USA, 23–26 June 2016; pp. 907–940.
35. Lu, Z.; Pu, H.; Wang, F.; Hu, Z.; Wang, L. The Expressive Power of Neural Networks: A View from the Width. In Proceedings of the Advances in Neural Information Processing Systems 30 (NIPS 2017), Long Beach, CA, USA, 4–9 December 2017.

MDPI
St. Alban-Anlage 66
4052 Basel
Switzerland
Tel. +41 61 683 77 34
Fax +41 61 302 89 18
www.mdpi.com

Applied Sciences Editorial Office
E-mail: applsci@mdpi.com
www.mdpi.com/journal/applsci

www.ingramcontent.com/pod-product-compliance
Ingram Content Group UK Ltd.
Pitfield, Milton Keynes, MK11 3LW, UK
UKHW060123010826
14089UKWH00006B/112

9 783036 546391